35/35

McFarlin Library
WITHDRAWN

Prevention, Fishing and Casing Repair

Prevention, Fishing and Casing Repair

James A. "Jim" Short

PennWell Books

Copyright © 1995 by
PennWell Publishing Company
1421 South Sheridan Road/P.O. Box 1260
Tulsa, Oklahoma 74101

Short, J.A.
    Prevention, Fishing and Casing Repair/James A. Short.
        p.   cm.
    Includes bibliographical references and index.
    ISBN 0-87814-439-0
    1. Oil well drilling — Accidents — Prevention. 2. Oil wells — Maintenance and repair 3. Oil wells — Equipment and supplies — Fishing.
TN871.2.S5373   1995   94–48577
622' ,338'0288 — dc20   CIP

All rights reserved. No part of this book may be reproduced, stored in a retrieval system, or transcribed in any form or by any means, electronic or mechanical, including photocopying and recording, without the prior written permission of the publisher.

Printed in the United States of America

1 2 3 4 5 6 — 95 96 97 98 99

DEDICATION

This book should include a coauthor: my wife, Catherine "Miss Kitty" Leona (Campbell) Short, a major contributor in almost too many ways to mention. Therefore, we dedicate this book to our grandchildren with hugs "OOO" and kisses "XXX."

        Danny Lee Butler – "Danny Boy"
        Catherine Christina Mann – "Christie"
        Courtney Elizabeth Short – "Princess"
        Erica Ashley Pratt – "Erka"
        Rachel Mayra Pratt – "Ra Ra"
        Benjamin Joseph Mann – "Bengie Butch"

Bless each of you. We love you dearly for what you are, what you mean to us, and what you have given of yourselves. Spread your wings, take the future in your hands, and mold it to a bright and glorious dream.

In some musing moment of reminiscence, please remember

        Maw Ma and Paw Pa
        Who Love You

        James "Jim" Arthur Short

        Catherine "Miss Kitty" Leona (Campbell) Short

CONTENTS

List of Examples —————————————————————————— xix
Preface to the First Edition ————————————————— xxiii
Disclaimer ————————————————————————————————— xxv
Introduction ————————————————————————————— xxvii

Chapter 1 — CAUSE AND PREVENTION

PREVENTION 1
  Human Error ——————————————————————————————— 2
  Spud Conference ————————————————————————— 4
  Cost and Severity ——————————————————————— 5
  Fishing Cost Survey —————————————————————— 6

OPERATIONS 7
  Questionable Operations ———————————————————— 8
  Slip- and Tong-Area Failures ————————————— 8
  Connections —————————————————————————————— 9
  Bits and Cones ——————————————————————————— 11
  Tripping Tubulars ——————————————————————— 14
  Shutdown —————————————————————————————————— 15
  Pulling Wet Drillpipe ————————————————————— 16
  Drilling Out Cement —————————————————————— 16
  Open-Hole Testing ——————————————————————— 17
  Other Precautions ——————————————————————— 19
    Undergauge Holes ——————————————————————— 19
    Strapping ——————————————————————————————— 19
    Pipe Wiper —————————————————————————————— 21
    Drill-Collar Clamps ——————————————————— 22
    Summary Precautions ——————————————————— 23

MECHANICAL FAILURES 24
  Rig Selection ———————————————————————————— 24
  Unmatched Rigs ——————————————————————————— 25
  Pump Failure ————————————————————————————— 27
  Hoisting Equipment ——————————————————————— 29
  Blowout Preventer ——————————————————————— 31
    Testing ————————————————————————————————— 31
    BOP Drills —————————————————————————————— 32
    Other Precautions ——————————————————————— 32
  Downhole Failures ——————————————————————— 33

vii

MUD 33
    Changes -- 36
    Aerated Mud -- 36
    Losses -- 38
    Hole Cleaning -- 39
    Emusifier-Wetting Agents ------------------------------ 41
    Barite -- 42
        Bridging --- 42
        Barite Sag --- 43
        Barite vs. Iron Ores ------------------------------- 43

LOST CIRCULATION 44
    General Procedures and Actions ------------------ 46
        Point Of Loss --------------------------------------- 47
        Dry Drilling -- 48
    Remedial Actions --------------------------------------- 48
        Low-Rate Lost Circulation ------------------------ 49
        Moderate-Rate Lost Circulation ------------------ 49
        High-Rate Lost Circulation ----------------------- 51
    Cementing Lost Circulation ------------------------- 52

FORMATIONS 57
    Influence on Hole Problems ----------------------- 57
    Problems --- 58
        Surface Formations -------------------------------- 58
        Shallow Gas -- 58
        Swelling Clays ------------------------------------- 59
        Fluid-Sensitive Shales ---------------------------- 59
        Geopressured Shales ------------------------------ 60
        Fractured Formations ----------------------------- 61
        High-Permeability Reefs ------------------------- 63
        Other Formations ---------------------------------- 64
    Combination Formation Problems --------------- 64
    Depleted Reservoirs ----------------------------------- 67

DEVIATION, DOGLEG, AND CROOKED HOLE 71
    Cause and Correction -------------------------------- 72
    Recommended Deviation and Dogleg ------------ 73

REFERENCES AND SUGGESTED READING 76

Chapter 2 — GENERAL RULES AND PROCEDURES

PERSONNEL 79
BE PREPARED BEFORE FISHING OCCURS 80

viii

WHEN A FISHING JOB OCCURS 81
    Take Immediate Action ------------------------------- 81
    Learn All Details --- 85
    Develop a Plan of Action -------------------------------- 87
TIME SPENT FISHING 89
WHEN AND HOW TO QUIT 90
POST-ANALYSIS 96
GUIDES TO FISHING 97
    Work on the Fish from the Top Down ------------------- 97
    Select the Strongest Catch Tool -------------------------- 98
    Pull the Fish out of the Hole without Delay ---------- 98
    Do Not Rotate When Pulling a Fish --------------------- 98
    Expedite Fishing Operations ------------------------------ 98
    Eliminate the Cause of Fishing if Necessary --------- 98
    Don't Arbitrarily Assume the Fishing Job Is Easy -- 99
    Priority of Operations -------------------------------------- 99
    Operations Recorders -------------------------------------- 99
    Stuck Condition of Fish ----------------------------------- 99
    Keep Top of Fish Clean --------------------------------- 100
    Anticipate Mechanical Repairs ------------------------- 100
    Summary Guides to Fishing --------------------- 100
    Take Reasonable Risk – Improvise -------------- 101
PORE PRESSURES AND TRANSITION ZONES 102
    Pore Pressure --- 102
    Pore-Pressure Plots ------------------------------------ 103
    Transition Zones --------------------------------------- 106
KICKS 109
    Detection -- 109
    Control --- 110
REAMING 112
    Operations --- 112
    Assembly --- 112
    String Reamer --- 114
DRAG AND TORQUE 118
WALL STICKING 121
KEYSEAT 126
    Detection --- 128
    Prevention -- 132
    Removal --- 132
    Wireline Keyseat ------------------------------------- 133
REFERENCES AND SUGGESTED READING 133

Chapter 3 — TOOLS AND ASSEMBLIES

FISHING TOOL SELECTION 137
TUBULAR FISHING TOOLS 139
- Overshot — 139
  - Single Bowl — 139
  - Double Bowl — 144
  - Triple Bowl — 144
  - Trap Door — 145
  - Side-door — 145
  - Continuous — 145
  - Casing Bowl — 145
- Basket — 145
  - Regular — 145
  - Reverse Circulating — 149
  - Poor Boy — 149
  - Boot — 150
- Spear — 153
  - Pipe — 153
  - Washpipe — 154
  - Packer Retriever — 156
- Screw-in Sub — 158
- Taper Tap — 158
- Die Collar — 159
- Mills — 160
  - Junk Mills — 164
  - Skirted Mills — 165
  - Cement Mills — 165
  - Pilot Mills — 165
  - Reamer Mills — 166
  - Tapered Mills — 166
  - Pineapple Mills — 166
  - Throated Mills — 166
- Mill Shoes — 166
- Pipe Cutters — 168
- Jars — 170
  - Hydraulic Fishing — 172
  - Hydraulic Drilling — 173
  - Mechanical Fishing — 173
  - Mechanical Drilling — 173
  - Accelerator Jars — 173
- Bumper Sub — 176
- Combination Hydraulic Jar and Bumper Sub — 176

x

Surface Bumper Sub --- 177
Washpipe --- 177
Safety Joint --- 177
Knuckle Joint --- 180
Bent Sub --- 181
Bent Joint --- 181
Wall Hook --- 182
Circulating Sub --- 183
Ported "Side-door" Back-off Sub --- 184
Shock Sub --- 184
Shear Rams --- 184
Tools Run Inside Casing --- 184
    Pulling Tool --- 184
    Reversing Tool --- 184
    Roller --- 185
    Whipstock --- 186

TUBULAR OR WIRELINE FISHING TOOLS   186

Casing Patch --- 186
Casing and Tubing Swage --- 186
Chemical Cutter --- 188
Grab --- 189
Jet Cutter --- 190
Junk Catcher --- 190
Junk Shot --- 191
Magnet --- 192
Socket --- 193
Wireline Spear --- 193

WIRELINE FISHING TOOLS   195

Bailers --- 197
    Bucket Bailer --- 198
    Dart-Bottom Bailer --- 198
    Dump Bailer --- 198
    Sand Pump --- 198
    Chipping --- 199
Blind Box --- 199
Friction Socket --- 200
Cutters and Perforators --- 201
    Wireline Cutter --- 201
    Wireline Tubular Cutter --- 201
    Mechanical Wireline Pipe Perforator --- 204
Impression Block --- 205
Hydraulic Jars --- 206
Mechanical Bumper Jars --- 206

Sinker Bars ------- 206
Casing and Tubing Swage ------- 206
Explosives ------- 208
Rope Socket ------- 208
Overshot ------- 208
ASSEMBLIES AND EQUIPMENT 208
Surface Equipment ------- 209
Drill and Work Strings ------- 209
Drillpipe ------- 210
Aluminum Drillpipe ------- 211
Tubing ------- 211
Coil Tubing ------- 212
Assembly Equipment and Tools ------- 212
Heavyweight Pipe ------- 212
Drill Collars ------- 213
Keyseat Wiper ------- 215
Reamer ------- 216
Stabilizer ------- 216
Double-Drilling Bumper Jars ------- 217
Assemblies ------- 219
Drilling ------- 220
Overpull ------- 221
Handling and Inspection ------- 222
Fishing ------- 224
Milling ------- 225
Wash Over ------- 225
Reaming ------- 225
Clean Out ------- 226
Wireline Assemblies ------- 227
REFERENCES AND SUGGESTED READING 228

Chapter 4 — FISHING PROCEDURES

MILLING, DRILLING AND WALLING OFF 229
Selecting the Mill ------- 230
Milling Operations ------- 230
Small Fish ------- 234
Large Fish ------- 236
Dressing Off the Top of the Fish ------- 237
LEAKS 237
WORKING STUCK TOOLS 242
Free Pointing and Free Pipe ------- 243
Pipe-Stretch Method ------- 244

xii

  Electrical Free-Point Pipe Logs ------------------------- 245
 Begin Working Immediately ---------------------- 247
 How Hard to Work ---------------------------------- 247
 How Long to Work ---------------------------------- 253
 Procedures and Techniques ---------------------- 256
  Working Results Indicate Working Method -------- 256
  Working Force Levels ------------------------------- 257
  Initial Working Direction ------------------------------ 257
  Working to Establish Movement and Circulation - 258
  Work Up -- 263
  Work Down --- 265
  Final Working Procedure ------------------------------ 266
 Working Wirelines ---------------------------------- 268

PARTING THE FISH 269
 Where to Part --------------------------------------- 270
 Backing Off -- 272
  Regular Back-off ------------------------------------- 272
  Outside Back-off ------------------------------------- 278
  Blind Back-off -- 278
 Cutting -- 281
  Chemical Cutter ------------------------------------- 281
  Jet Cutter -- 283
  Mechanical Cutter ----------------------------------- 283
  Explosives -- 284
  Shear Rams -- 284

CATCHING THE FISH 284
 Cleaning Inside the Fish -------------------------- 287
  Damaged Fish Top ----------------------------------- 287
  Cleaning --- 287
 Screw In -- 289
 Swallowing -- 290
 Outside Catch -------------------------------------- 290
 Inside Catch -- 296
 Washing Over -------------------------------------- 297
  Operations --- 298
  Cased vs. Open Holes ------------------------------- 299
  Wash Beside a Fish --------------------------------- 301

PLUG BACK AND SIDETRACK 302
 Plugging Back -------------------------------------- 303
 Dressing Off -- 304
 Sidetracking -- 305

MILLING CASING AND TUBULARS 306

REFERENCES AND SUGGESTED READING 309

xiii

Chapter 5 — OPEN-HOLE FISHING

BIT CONES AND BITS 313
- Cones — 313
  - In Soft Formations — 317
  - In Hard Formations — 318
- Bits — 321
  - Stuck Bits — 322
  - Parted Bits and Large Junk — 322

DRILL-COLLAR ASSEMBLIES 324
- Without Drilling Jars and Bumper Sub — 324
- With Drilling Jars and Bumper Sub — 325
- Sticking the Drill-Collar Assembly — 326
  - Working, Back Off, and Clean Out — 326
  - Recovery in Multiple Sections — 327
  - Fishing and Washing Over — 327
- Pumping Out Pipe — 328
- Inside Catch — 330
- Fishing Neck — 331

STUCK DRILLPIPE 333
- Perforating for Circulation — 333
- Parted or Twisted Off — 335
- Out-of-Gauge Hole — 338
- Accidental Sidetracking — 338

KEYSEATS 338
- Where and How the Assembly Is Stuck — 339
- Pipe Free to Move Downward — 340
  - Working and Rolling Out — 341
  - Wearing Out the Keyseat — 341
  - Reaming Out the Keyseat — 343
  - Set Pipe on Bottom — 343
- Assembly Firmly Stuck — 345
  - Keyseated at the Bit — 346
  - Stuck at the Top of the Bottomhole Assembly — 347
  - Back Off and Recover — 347

WALL STICKING 350
- Working — 351
- Releasing Fluids — 351
  - Usage — 353
  - For Oil Mud — 358
- Reducing Hydrostatic Head — 359
  - Lightweight Fluid — 359

  Nitrogen --- 360
  Compression Packer ------------------------------------ 360
FISH ON A FISH 361
FORMATION PROBLEMS 362
MISCELLANEOUS 368
  Dissolving Agents ------------------------------------- 371
  Provisional Plugs ------------------------------------- 372
  Fishing Under Pressure ----------------------------- 373
  Bridging --- 373
REFERENCES AND SUGGESTED READING 375

Chapter 6 — CASED HOLE, WIRELINE, AND SURFACE FAILURES

TEST TOOLS, PACKERS, AND PLUGS 377
  Malfunctions and Prevention --------------------- 378
  Wireline Packers and Plugs ----------------------- 381
  Test Tools -- 383
  Pressure Below Packers and Plugs ------------- 381
  Fishing with a Taper Tap -------------------------- 383
SMALL-DIAMETER TUBULARS 390
  Fishing Operations ----------------------------------- 391
  Tubingless Completions ---------------------------- 391
  Concentric and Parallel Configurations ------- 393
  Very Small Tubulars --------------------------------- 394
  Coiled Tubing -- 395
SANDED-UP TUBULARS 399
TUBULAR FAILURES AT THE SURFACE 403
  Causes of Failures ----------------------------------- 403
  Wellhead Security ----------------------------------- 404
  Tubular Bent, Not Parted -------------------------- 405
    Handling Bent Tubulars ------------------------- 407
    Moderate Bend --------------------------------------- 407
    Severe Bend --- 408
  Tubular Bent, Not Caught in Slips -------------- 409
  Types of Catches ------------------------------------- 414
  Transfer Pipe String Weight ---------------------- 416
  Special Situations ----------------------------------- 417
WIRELINE OPERATIONS 418
  Fishing with Tubulars vs. Wirelines ----------- 418

Types of Wirelines ---------------------------------- 418
Operating Precautions ---------------------------- 419
    Running Close-Tolerance Tools ----------------------- 420
    Jumping the Sheave ------------------------------------ 421
    Worn Wirelines -- 421
    Flagging Wirelines ------------------------------------ 421
    Swabbing Precautions --------------------------------- 422
    Crippled Rope Socket --------------------------------- 424
    Wireline Keyseats -------------------------------------- 426
Operating Problems --------------------------------- 426
WIRELINE FISHING 428
Fishing Assemblies --------------------------------- 429
Fishing for Wirelines ------------------------------- 429
    Wireline Inside Pipe ----------------------------------- 430
    Grab and Spear --------------------------------------- 430
Stripping Over Stuck Wireline ------------------ 435
Parting Wirelines ------------------------------------ 440
Fishing for Wireline Tools ----------------------- 441
REFERENCES AND SUGGESTED READING 444

Chapter 7 — CASING AND LINER REPAIR

CASING PROGRAM 445
Program Considerations ---------------------------- 446
    Casing Sizes, Weights, and Grades -------------------- 446
    Operation Failures ------------------------------------- 447
    Change in Scope --------------------------------------- 447
    Casing Inspections ------------------------------------- 449
Running Casing -------------------------------------- 451
    Is the Hole Ready to Run Casing? --------------------- 451
    Precautions --- 455
End of the Tubing ----------------------------------- 456
FAILURE AND REPAIR 456
Types of Casing Failures --------------------------- 459
    Leaks --- 459
    Split or Burst -- 460
    Parted --- 462
    Collapsed -- 462
Factors Affecting Casing Repairs --------------- 464
Summary of Casing Repair Methods ----------- 465

CEMENTING CASING FAILURES 472
   Cement Applications in Casing Repair --------- 473
      General Review --- 473
      Squeezing Relation to Fracturing ---------------------- 474
   Squeeze Cementing -- 477
      Conventional --- 478
      Staging -- 480
      Cement Backed by Metal ---------------------------------- 481
      Multiple Consecutive Squeezes ----------------------- 483

DRIVE PIPE AND CONDUCTOR 489
   Casing, Rotary, Mast, and Crown Alignment 490
   Parted Conductor Casing ----------------------------- 495

SURFACE CASING 497
      Loose at the Surface -- 497
      Leaking Around the Surface ---------------------------- 498
      Wear at or Near Bradenhead --------------------------- 499
      Overall Wear --- 501
      Buckling --- 507

INTERMEDIATE CASING 507
   Parted Casing -- 508
      While Running -- 508
      In an Uncemented Section ------------------------------ 510
      In a Cemented Section ------------------------------------ 512
      Alignment Tool --- 513
   Loose Float and Shoe Joints ---------------------- 518
   Releasing Casing from a Bradenhead or Hanger Spool -- 523
      Normal Tensional Pull ---------------------------------- 523
      Casing Jacks --- 524
      Release by Cutting --------------------------------------- 524

COMPLETION – PRODUCTION CASING 525
   Split or Burst Casing -------------------------------- 526
   Collapsed Casing ------------------------------------- 531
   Liners --- 536
      Failed Intermediate Casing above a Liner ----------- 536
      Squeezing the Liner Top -------------------------------- 538

OLD WELLS 542

REFERENCES AND SUGGESTED READING 548

INDEX 549

xvii

LIST OF EXAMPLES

Chapter 1 — CAUSE AND PREVENTION

Cat Line over the Crown --------------------------------- 3
Spud Conference and Small Drill Tubing --------- 4
Bit Run Excessively Long ----------------------------- 13
Gas-Leaking Tool Joints While Testing ----------- 18
Drill-Collar Clamp on First Joints of Casing --- 22
Pump Failure -- 27
Split Crown Sheave ------------------------------------- 30
Emulsifier-Wetting Agent Usage -------------------- 42
Curing High-Rate Lost Circulation ----------------- 54
Summaries of Cementing Lost Circulation ----- 55
Spud In on Hillside ------------------------------------- 58
Shallow Gas --- 59
Pencil Cave Shale --------------------------------------- 60
Fractured Shales --- 62
Combination Lost Circulation and Gas Kicks - 65
Depleted Reservoir -------------------------------------- 68
Geopressured Below Depleted ---------------------- 69
Crooked vs. Straight Hole ----------------------------- 71

Chapter 2 — GENERAL RULES AND PROCEDURES

Recovering a Hole Opener Shank ------------------ 82
Examples of Squeezing before Continuing
   Fishing --- 90
Deep Completion over a Fish ----------------------- 91
Catching a Fish with an Unmatched Pin ------ 101
Deceptive Gas Shows in Oil Mud --------------- 106
Reaming in an "S" Type Directional Hole --- 113
Removing a Keyseat with a String Reamer -- 116
Reducing Drag and Torque with Walnut
   Hulls -- 121
Two- and Three-Arm Calipers Detect
   Keyseats --- 130

xix

Chapter 3 — TOOLS AND ASSEMBLIES

    Drill Collar Pin Failure ---------------------------- 223

Chapter 4 — FISHING PROCEDURES

    Milling Up a Bit and Junk ------------------------ 232
    Drillstring Leaks ------------------------------------ 240
    Working Stuck Tubing ---------------------------- 254
    Working Heavy Stuck Drill Tools --------------- 255
    Stuck Packer and Final Working Procedure - 267
    Cutting Tubing with a Chemical Cutter ------- 281
    Cut Drillpipe with a Chemical Cutter --------- 282
    Hitting an Obstruction on a Trip ---------------- 295
    Washing Beside a Fish ---------------------------- 301

Chapter 5 — OPEN-HOLE FISHING

    Loosing and Recovering Bit Cones ------------- 315
    Recovering Bit Cones in a Deep Well --------- 319
    Recovering a Bit and Sub -------------------------- 321
    Recovering Dropped Stand of Drill Collars -- 325
    Pumping Out Stuck Drill Collars ---------------- 329
    Fishing with a Fishing Spear --------------------- 331
    Catching a Fish in an Out-of-Gauge Hole ---- 337
    Releasing a Keyseated Bit ------------------------- 346
    Keyseated at the Top of the Drill Collars ----- 349
    Wall Sticking Combined with Keyseating ---- 355
    Wall-Sticking Examples --------------------------- 357
    Fishing with Formation Problems -------------- 363
    Miscellaneous Fishing Situations --------------- 368
    Dissolving Carbonates to Release a Fish ------ 371
    Fishing under Pressure ---------------------------- 373
    Effect and Severity of Bridging ----------------- 374

Chapter 6 — CASED HOLE, WIRELINE, AND SURFACE FAILURES

    Recovering a Plug by Removing Casing ------- 378
    Milling a Packer While Using a Packer
        Retriever -- 383

Milling Out a Stuck Squeeze Packer ---------- 384
Broken Packer Stinger and Seal Assembly -- 387
Fishing with a Taper Tap ------------------------ 388
Collapsed Casing in a Slim-Hole
    Completion ------------------------------------- 392
Recovering Macaroni Tubing Inside Larger
    Tubing --- 393
Recovering Tubing Sanded Up Iknside
    Casing --- 401
Picking Up Drillpipe Bent above the
    Rotary --- 408
Picking Up Bent Pipe Incorrectly --------------- 412
Welding a Lift Nipple on Drillpipe Parted
    at the Surface ---------------------------------- 416
Drilling a Preset Packer ---------------------------- 420
Wireline Operating Problems -------------------- 426
Stripping Over a Stuck Wireline ---------------- 437
Stripping Over a Keyseated Logging Tool ---- 437
Fishing for Wireline Tools ------------------------ 443

Chapter 7 — CASING AND LINER REPAIR

Inspection Failures ---------------------------------- 450
Multiple Squeeze Split Casing ------------------ 486
Misaligned Conductor ----------------------------- 492
Conductor Casing Failure ------------------------ 496
Repairing a Leak at the Bradenhead ---------- 500
Recovering Worn, Parted Surface Casing ----- 502
Casing Leak at 9,000 ft ---------------------------- 505
Casing Leak at 13,000 ft -------------------------- 506
Parted, Uncemented Casing --------------------- 510
Repair Parted Casing with an Alignment
    Tool --- 513
Cementing Loose Shoe Joints -------------------- 522
Opening Loose Shoe Joints with a String
    Reamer --- 522
Repairing Split Tubing and Burst Casing ----- 527
Packing Off Milled Collapsed Casing ---------- 533
Running a Liner Through Collapsed
    Production Casing ---------------------------- 535
Cementing a Dropped Liner --------------------- 537
Reentering an Old Well --------------------------- 545

xxi

PREFACE TO THE FIRST EDITION

Fishing and Casing Repair covers fishing and downhole remedial operations encountered in drilling-casing-completion operations. The reader should be familiar with standard operating practices and procedures. Common terminology is used without special explanations except where there may be some conflict due to local usage.

While researching the subject, I noted that most fishing- tool equipment suppliers have good information about their product, but otherwise there is very little published information on fishing and casing repair. Therefore, the main sources of references are personal experience and contact with others in the industry.

Failures are very costly — in some cases approaching 25% of the drilling budget. Competent, experienced personnel can prevent or efficiently complete most fishing and downhole remedial operations, especially in regular operations and special activities such as fishing where there is a higher risk of incurring another fishing job.

The fishing job starts when a fish is left in the hole or stuck pipe cannot be released. A plan of action with contingency plans is developed; general rules on fishing are used as guidelines. Action includes procedures such as soaking, milling, washing over, parting, catching, and working the fish. Casing failures and remedial operations are treated similarly.

In this book, specific examples of operations are presented. Background conditions causing the problems are discussed, including possible solutions, why specific alternatives were chosen, and their results. Both common problems and those that occur less frequently are included.

Operational practices and procedures that have been used and their results are presented. These do not apply to all areas and different conditions but may work in many cases. All procedures in drilling, casing, completion, fishing, and remedial repair must be evaluated individually. The operator must determine the proper course of action.

This text is intended to disseminate information to help prevent fishing and to improve remedial operations. If it succeeds, the industry will have benefited.

xxiii

Addendum

The preface for the book *Fishing and Casing Repair* is equally applicable to this sequel, *Prevention, Fishing and Casing Repair* with other significant additions. The word "Prevention" in the new title represents an underlying theme that is repeated throughout the new text whenever applicable. The importance of prevention has been emphasized by additional years of experience and contacts with others. The same is true of many related activities, such as working stuck pipe. Work it longer and harder before taking other action. Remember that fishing does not start until there is actually a fish in the hole. More people need to be involved in and have a clearer understanding of the overall fishing process.

J.A. "Jim" Short
January 22, 1995

DISCLAIMER

This text contains descriptions, statements, procedures, methodology, interpretations, reports, and other written matter and information, hereinafter collectively called "contents," that have been carefully considered and prepared as a matter of general information. The contents are believed to represent situations and conditions reliably that have or could occur but are not represented or guaranteed as to their accuracy or application to other conditions or situations, and the author has no knowledge or control of their interpretation. Therefore, the contents and all interpretations and recommendations made in connection therewith are presented solely as a guide for the user's consideration, investigation, and verification. No warranties of any kind, whether expressed or implied, are made in connection therewith. The user is specifically cautioned, reminded, and advised that any use or interpretation of the contents and resulting use or application thereof are made at the sole risk of the user. In consideration of these premises, any user of the contents agrees to indemnify and save harmless the author from all claims and actions for loss, damage, death, or injury to persons or property.

INTRODUCTION

Prevention, Fishing and Casing Repair is a sequel to *Fishing and Casing Repair*. It follows the format of the original book, but otherwise has been revised, expanded, and updated with new material. In summary, this revised edition is a more detailed, in-depth study of fishing

As noted in the original text, "Almost all, if not all fishing can be prevented." Additional years of experience and contacts with experienced personnel have strengthened this conviction. Prevention is strongly emphasized. The first chapter is devoted primarily to prevention and the theme is followed throughout the text. Prevention reduces fishing and also identifies weak points requiring emphasis and attention. The responsibility for preventing fishing and successfully conducting a fishing job is shared by all connected with the drilling operation, from the rig crew to the highest level of authority.

Some preventive measures described in the text may appear obvious or elementary. Nevertheless, they all are necessary to avoid fishing. Likewise, some of the text appears to be instructive and in some cases it is. The astute reader will recognize that these "preventive measures" focus on areas that are prone to accidents and failures and are major causes of fishing. These range from simple operations such as reaming and cleaning the hole to higher-risk operations such as detecting and controlling a kick. Also, it is all too common for fishing to lead to fishing, with one or more additional fish in the hole. Preventing the first fish would also prevent the extra fish.

The fishing tool section has been expanded to cover over 90 versions of 55 basic tools. Descriptions and illustrations of the tools and their operation are included for those who are less experienced.

Additional examples have been included for a total of over 85 examples and various summaries that cover a broad range of sticking and fishing situations. They are a quick reference, showing what necessitated the fishing job, how it was conducted, and what others have done. They also clarify many sticking situations and emphasize the importance of prevention and being prepared.

Common problem areas are covered in more detail. These include cleaning the hole, excess drag and torque, and formation problems during fishing such as combination high pressure and lost circulation zones. A new section covers drilling depleted formations, including the use of pore pressure and fracture gradient predictive methods.

xxvii

Coverage of detecting transition zones, setting casing at the correct point in a transition zone, and controlling kicks is expanded. Wall sticking is covered in more detail, including additional methods of releasing pipe stuck in this common sticking situation. Special problems such as barite sag and fishing in high-angle directional and horizontal holes are discussed. Cementing is covered in more detail because of its wide use in lost circulation, plugging back, and casing repairs. Special attention is given to temporary plugging, plugging back, and sidetracking.

There is an increasing awareness of the importance of preventing fishing and conducting fishing operations correctly. One indication of this is that there were very few, if any, references on the subject when the first book was written as compared to the relatively large number of references included in this book. I believe *Fishing and Casing Repair* contributed to this and optimistically hope that *Prevention, Fishing and Casing Repair* will make an additional contribution.

Chapter 1

CAUSE AND PREVENTION

Fishing is any operation or procedure to release, remove, or recover a fish. The fish is any obstruction or equipment, usually tubular, in the wellbore that restricts or adversely affects current or future drilling, casing, completion, or production operations. Fishing occurs in straight, deviated, and horizontal holes, both open and cased. Fish are removed with a wide variety of fishing tools and related services, such as logging and cementing. Casing and liners fail for various reasons. They are repaired with various tools and by different procedures, including cementing.

Severity and cost of fishing ranges from removing bit cones in a few hours of rig time to a blowout, loss of equipment, danger to personnel, and drilling kill wells requiring hundreds of rig days. Preventive procedures can eliminate these.

Drilling a straight, vertical hole is one of the best preventive actions and minimizes hole problems. The importance of a straight hole increases with depth. Sometimes this kind of drilling is not possible because of the well plan, formation conditions, or other reasons. Use the best design and procedures for deviated, high angle, and horizontal holes.

Fishing and casing failures cause drilling inefficiency and increased costs. It may be necessary to perform a fishing job, but fishing does not contribute directly to drilling, casing, or completion. Prevention reduces cost and increases operating efficiency.

PREVENTION

Prevention is the process of recognizing the causes of fishing and taking preventive measures. Prevention covers a broad range, basically anything relating to the well from spud to abandonment.

Almost any incorrect practice may cause fishing problems. Prevention calls for preventive maintenance so exercise care and foresight before problems arise.

Most, if not all, fishing is preventable. There are cases where it may appear that the cause of fishing could not be anticipated, but these are very few. A prime example is the well fire debacle in Kuwait. It can be said that these were not preventable. However, what if the wells had been equipped with storm chokes, even modified for the high flow rates normal to the area? Various shipwrecks with disastrous oil spills, refinery fires, and pipeline blowouts do not improve the industry's image. Although beyond the scope of this book, all are preventable.

The subject of cause and prevention of fishing is so important that it warrants a substantial part of all personnel time, including all levels involved in drilling and completion operations. The severity and cost of fishing is probably not recognized by the industry, since much of it is "small jobs" such as recovering bit cones. However, fishing is not a normal operation. Personnel are not familiar with it, and as a result, FISHING OFTEN LEADS TO FISHING.

A concentrated, industrywide prevention program has not been started for several reasons. Fishing is not a subject of general interest except to a few people in the industry. Personnel injuries are relatively uncommon. Blowouts or other newsworthy actions are seldom associated with fishing. On the contrary, fishing is fairly routine. The only headlines it makes are on the red ink side of the ledger column. The cost of fishing jobs is either unrecognized, overlooked, or accepted as part of the cost of doing business.

A properly prepared drilling and casing program is an important preventive measure. Develop a program that prevents or alleviates fishing problems. This includes giving full consideration to operations, providing adequate hole and casing sizes, and safety factors. Give special consideration to potential formation problems, especially lost circulation. Provide for excess drag and torque in crooked, highly deviated, extended reach, and horizontal holes. Be aware of the special problems and requirements of high angle and horizontal wells.

One of the first preventative steps is recognizing the actual cause of fishing. For example, if a hole is lost during fishing, the loss is attributed to the fishing job. Although this may be the direct cause, the indirect and real cause may be due to a number of different factors.

Human Error

Many fishing jobs reportedly are caused by human error. This is not recognized here as a valid cause for fishing. The actual cause is probably due either to lack of experience or poor judgment. Correct the problem with improved supervision and good training under experienced supervisors. Higher motivation to improve job performance can be a major contribution. Hurrying and lack of time are not valid reasons for the many problems that result in fishing jobs.

As an example, a wear bushing was positioned incorrectly while nippling up the blowout preventers. This caused a major fishing job involving over 10 days' rig time for a 20,000-ft depth capacity rig. It occurred with experienced crews and supervisors. Failure to measure "strap" drillpipe on the last trip before running casing or a liner may cause errors in depths and other problems. Another example of human error is included in the following example.

Cat Line over the Crown

A step-out well had 9⅝-in. casing set at 1,500 ft and was drilling an 8¾-in. hole at 4,000 ft with a conventional drilling assembly. The hole was being drilled underbalanced to maximize the penetration rate. The area had a history of shallow, moderately high-pressure, low-volume gas sands, so as a precautionary measure a rotating head was used as a diverter. While drilling at 4,000 ft, the pump pressure declined, indicating a leak. The pipe was started out of the hole to locate the leak. After pulling ten stands, and with the eleventh stand in the elevators, pulling operations were shut down to remove the rotating head.

The cat line was used to pull the split master bushing so that the rotating head could be picked up above the rotary. Either the pickup tool was incorrectly hooked into the master bushing, or the cat line was incorrectly tied to the bushing pickup tool. One end of the cat line came free, and the other end was removed immediately from the cat head. However, the cat line was overbalanced and continued to run over the crown. The driller and crew left the floor just as the traveling block and pipe dropped. The elevators lodged in the rotary against the ring that retained the master bushing, and the drilling line unspooled.

After the crew returned to the floor, they installed a safety valve in the top of the drillpipe and closed the pipe rams. The hole had been making a small amount of gas and was shut in to prevent the gas from displacing the mud and exposing the preventers to a higher pressure than necessary. This would also make it easier to kill the well in case the pipe parted high in the hole.

The rig was repaired in 36 hours. The safety valve was opened and did not have any pressure. A drillpipe single was screwed into the drillpipe and, after picking up some weight, the pipe rams were opened. The assembly was free, but it had lost weight, indicating a fish in the hole. When the assembly was pulled, three stands of drillpipe and the drill-collar assembly were missing. The drillpipe had parted in the slip area, indicating a new tension failure.

An overshot was run to catch the fish. At this time, the fish had been on bottom without circulating for about 48 hours. It was caught and worked free in about one hour. When the fish was recovered, the bit cones were found sheared and the bit shanks split. A junk basket was run to recover the cones but did not make any recovery, so the cones were milled.

The overall fishing job was conducted properly with satisfactory results. Although human error was the cause, this type of mistake is extremely difficult to prevent.

Common examples of human error include dropping a sledge hammer, crowbar, tong jaw, stilson wrench, or other small tools into the open hole. These cause fishing jobs that are relatively easy if conducted correctly. However, it is not uncommon for these to lead to an extended fishing job and in some cases even loss of the hole. The original cause was dropping the tool into the hole, but the real cause could probably be attributed to improper fishing operations. Again, as hindsight, keeping the hole covered would have prevented the problems.

Spud Conference

A good method of solving a problem is to discuss it with other interested people. This may lead to an obvious solution. Another method is to hold a "spud conference" to discuss potential problems. Include anyone familiar with the situation, such as operating personnel and service company representatives. Limit the number of people attending the meeting to create an effective group, and confine the discussion to the subject.

Spud conferences are common on most critical wells and those where special problems are expected. In one case a spud conference was held prior to deepening a well and proved very beneficial, as illustrated in the following example.

Spud Conference and Small Drill Tubing

A spud meeting was held on a deep, critical well. The purpose of the meeting was to advise the personnel of the existence of hydrogen sulfide (H_2S), recommend safety procedures, and hold a preliminary training session. This was conducted in the normal manner.

The well was a deep, high-pressure oil well with a high gas/oil ratio. It had 17-in. conductor at 50 ft, 9⅝-in. surface casing at 2500 ft, 7-in. intermediate casing at 11,000 ft, and a 5-in. flush joint (FJ) liner at 15,000 ft. A high productivity zone was found at 16,500 ft in subsequent field development. Sidetracking in the 7-in. casing was not a reasonable option because of the producing zone's high productivity. The alternatives were to drill a twin well, a very costly procedure, or to deepen the existing well.

A program was designed to deepen the well. It included shutting off the existing productive zones by puddle cementing to minimize formation damage during deepening. The drilling program included deepening with a tapered drillstring of 3½-in. internal flush (IF) drillpipe inside of the 7-in. casing, 2⅜-in. drill tubing inside the 5-in. liner, and drilling a 4-in. hole with oil mud. The well was to be completed with a small flush joint liner. Other completion equipment included a wireline packer with a mill-out extension and plug receptacle nipple seated in the bottom of the 5-in. liner. The production string included 2⅞-in. tubing fitted with special couplings and a compression packer with about a 1,500-ft extension. It was spaced to latch the extension into the wireline packer with the compression packer seated in the 7-in. above the existing producing zones in the 7-in. and 5-in. liner. They would be produced through the sliding sleeve and commingled with the deeper production.

Studies indicated five wells had tried this or a similar deepening procedure. Four were lost due to failures in the 2⅜-in. drillstring before penetrating the deeper producing zone. One was partially successful, deepened with water-base mud and completed through stuck 2⅜-in. drill tubing.

Another purpose of the meeting was to discuss the well plan. The contractor's operating personnel said that the main cause of failures was overstressing the drill tubing while making connections, even with hydraulic drill tongs. A driller suggested very careful drilling practices. A derrick man suggested using a jerk line on the handle of small tubing tongs fitted with a torque gauge and tightening the connections carefully with the cat head to give maximum control. This would take more time but was indicated as the safest procedure. The extra time was not too significant since only two trips were expected with long running diamond bits.

The well was deepened successfully in an economical manner as a high productivity well.

Cost and Severity

While anyone involved in drilling operations knows of several fishing jobs, the industry as a whole probably is not fully aware of the severity of the problem and total cost of fishing operations. The severity of these losses emphasizes the need of prevention. A few examples of fishing and associated losses are illustrated in the following incidents:

- A development well was drilling an 8¾-in. hole at 11,000 ft. The drillpipe pin on top of the drill collars failed, leav-

ing the drill collars in the hole. The first overshot slipped off with a 50,000 lb. overpull. The drill collars were worked loose and recovered with a second overshot run.
- A bit and bit sub were left in the hole and could not be recovered with an overshot. A joint of washpipe was run to wash over the fish. The bumper jars failed, leaving the lower half of the bumper jars and a joint of washpipe in the hole. The washpipe was recovered with an overshot, and the bit and bit sub were recovered with a second overshot. The bit cones were removed by milling.
- A tool joint on the bottom of the top drill collar failed, leaving the drill collar assembly in the hole. Washpipe was run after an unsuccessful overshot run. The washpipe stuck, and the hole ultimately was sidetracked.
- A tool joint washed out on a reamer located 69 ft above the bit. The bit locked up, probably due to lack of circulation. The drilling assembly was pulled. It had parted at the reamer, leaving 60 ft of fish in the hole. While going in the hole with a fishing assembly, the drillpipe parted in the slip area 2½-ft below the tool joint and dropped the 14,000-ft fishing assembly. The fish was caught with an overshot, worked free, and recovered.
- A second fish was left in the hole when a tool joint in the top of the bumper sub in the fishing string failed due to a cracked drill-collar pin. The damaged fishing string was laid down, and a new fishing string picked up. An overshot was run, and the skirt broke off, leaving a third fish in the hole. All fish were recovered after extensive fishing.
- Washover pipe was run to wash over a fish but became wall stuck after washing over a short distance. It was soaked free and released. The hole was sidetracked after extensive fishing operations.

Many fishing jobs occur in drilling and casing operations as described throughout the text.

Fishing Cost Survey

An informal survey was conducted to learn how many rigs were involved in fishing operations and appropriate monetary losses. The survey included about 12% of the rigs operating in the continental United States and covered all types of drilling operations at various depths. This was in an active drilling area during a boom period.

Over 16% of the rigs were conducting some type of fishing operation during the two-month survey period. Total daily rig operating costs during fishing are about 75% higher than during normal drilling operations. Therefore, almost 25% of the total drilling expenditure was wasted during the survey period. The loss is considerably higher when considering sidetracking, lost holes, and redrilling.

Actual drilling costs were approximately 13% higher than the original cost estimate during the same period. Therefore, after allowing 10% for contingencies, the wells may have cost as much as 23% more than the original estimate. Such overexpenditure frequently is blamed on increased costs. But most of the 13%, or 23%, increase is attributed to fishing and related operations. The similarity of 23% here and the previous 25% are significant.

OPERATIONS

Many problems occur during drilling and other operations. Frequency of failure increases with increasing operational difficulty. Successful and efficient operations have the least number of problems. When one does occur, a detailed review of the problem and associated factors leads to the best solution. Always consider alternate solutions. The best prevention is to conduct operations in a prudent, orderly manner.

Competent, experienced personnel design and implement correct procedures. Successful operators conduct all operations in a prudent manner with full consideration of potential fishing. Don't overlook important items while trying to expedite the job. Consider possible consequences before undertaking a course of action. Know how to fish out any equipment before running it into the hole. Consider other ways of doing the job if the tool cannot be fished out. If there is no alternative available except possibly sidetracking, use the tool as a calculated risk.

Handle equipment as smoothly as possible. Be aware of the problems caused by operations, such as running full-gauge bits into under-gauge hole and unintentionally backing off or sidetracking.

Always use the minimum mud weight with allowance for a margin of safety. Hole problems frequently are associated with higher mud weights. These include wall sticking, cuttings removal, and control of mud properties. Problems often are more severe in soft to medium formations. Higher mud weights also generally decrease penetration rate. The importance of minimum weight mud often is overlooked.

Questionable Operations

Questionable operations may be conducted as a calculated risk. Evaluate the risk vs. the gain, and select the best course of action based upon good judgment. Some common examples of this are:

- Laying down drillpipe before running casing.
- Running casing after logging without a cleanout trip.
- Dropping about 12,000 ft of 3½-in. drillpipe a short distance on the last bit run before logging and pulling the crooked pipe out of the hole. After logging, a cleanout trip was made with the same pipe before running casing. This saved the time and transportation expense of picking up a new string of pipe for one cleanout trip.
- Conducting logging operations for five days on a 13,000-ft well without a cleanout trip because logging tool action did not indicate it was needed.
- Running a bit for an extensive period may be justified in special cases. A common example is drilling near total depth with a worn bit. If the depth is not critical, the operator may risk losing bit cones by running the bit longer, rather than making the trip to change bits. Depending on conditions, this may be justified for several reasons. If the depth is not critical and bit cones are lost, then stop drilling and leave the cones in the hole. Normally, removing the cones is not difficult if it becomes necessary to drill deeper. There is a higher risk of problems during tripping than during drilling. Also, the bit may drill to total depth. Therefore, the operator could save the cost of a new bit, the time spent changing the bit, and reduce the amount of higher risk operations required.

Slip- and Tong-Area Failures

The slip area, located 2 to 4 ft below the tool joint, often is subject to a high failure rate. The area may be slip cut, caused by setting the pipe down too hard, not setting the slips evenly, rotating the pipe with the slips, or not using back-up tongs on trips when the pipe weight is still too light. All these cause points of weakness.

The drillpipe may be bent slightly in or a short distance above the slip area. Common causes include (1) setting the tool joint too high in the rotary during connections, (2) making the tool joint excessively tight, requiring excess pipe tong torque to break the joint, or (3) breaking out the tool joint in the rotary with one pipe tong.

Avoid these practices and check the area frequently. Bent tool joints can be detected easily when rotating the drillpipe on trips. In some cases similar problems occur in the tong area. Some of these failures are illustrated in the following.

- A long, heavy string of C-135 drillpipe was dropped on a trip in the hole using double elevators. It parted in the slip area while the elevators were being set down. The failure and resulting fishing job were caused by setting the elevators down too hard and/or using faulty pipe damaged in the slip area.
- While drilling at 13,000 ft, 4½-in. drillpipe parted 2 ft above the tool joint in the tong area, leaving 30 stands of drillpipe and the drill collar in the hole.
- Most of the drillstring had been pulled from a 12,000-ft hole when the pipe parted about 1 in. below the top of the slips while breaking out a stand of drillpipe.

Eliminate slip- and tong-area failures by using good operating practices and handling the pipe correctly. Take the pipe out of service when it is damaged.

Connections

Making connections is a common operation, but bit plugging and sticking are not. The drilling assembly usually sticks at one of two points during a connection — while lifting the assembly off the slips after making the connections, or while lowering the bit to resume drilling. Severity may increase due to caving, which may occur in recently drilled formations that have not stabilized. A less common problem is breaking the kelly.

Circulating and reaming down 5–10 times before making the connection reduces the risk of sticking. However, this extends normal connection time of 3–5 minutes to 10 minutes or more, depending on depth. Often it is not necessary to make even one reaming pass before the connection, although at least one pass is recommended. However, the best approach is to have a competent drilling supervisor and crew who recognize when it is necessary to ream before a connection.

Bit plugging and sometimes sticking are most common when forcing the bit into caving while setting the kelly drive bushing into the rotary. The bit may be pushed into cuttings or caving material with extra weight in order to insert the kelly drive bushing into the rotary.

This may plug the bit and cause sticking, especially if the pump is not running. Reaming before making a connection helps ensure the hole is full gauge and smooth. Move bit cuttings off bottom and a sufficient distance uphole to help prevent settling and sticking the bottom hole assembly while at rest. If cuttings or caving material on bottom prevent lowering the kelly to the point where the kelly drive bushings cannot be inserted into the rotary, then remove the joint of drillpipe and clean out again before rerunning the single.

Figure 1-1 Long vs. short kelly

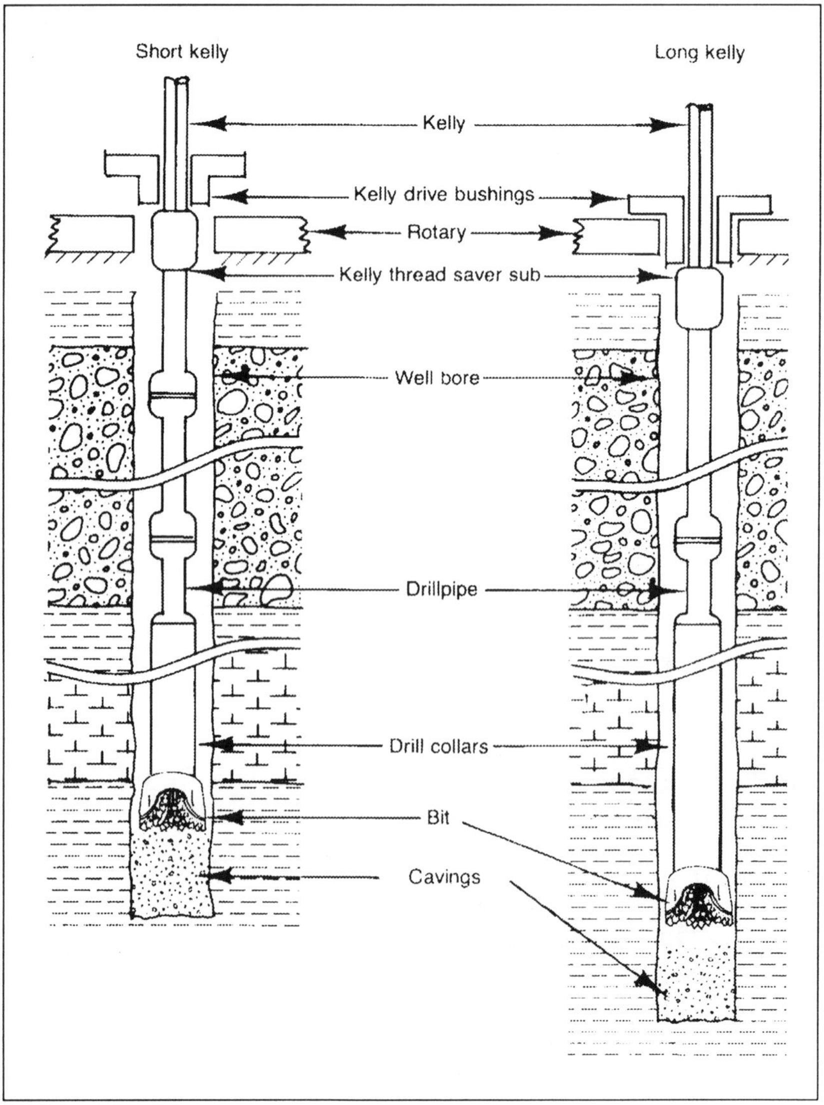

Depending upon the rig equipment, some procedural and equipment changes may help prevent sticking during connections. A longer kelly allows drilling the hole deeper before making the connection (Fig. 1–1). Both holes have the same amount of fill or caving material. The short kelly causes the bit to sit on the cavings before the kelly drive bushings enter the rotary. The long kelly allows drilling a relatively deeper hole, so the kelly drive bushings are in place in the rotary while the bit is still some distance above the cuttings. This reduces the risk of plugging or sticking when inserting the kelly bushings into the rotary.

The operator frequently shuts the pump down as the bit is pulled up off bottom. This dissipates pressure inside the drilling assembly for a reasonably dry connection. It also increases the chance of cuttings accumulating at the bottom of the hole. A better procedure is to shut the pump off after the kelly is midway out of the hole and then use the mud box if necessary. Very little pumping is required because of the small annular space in the drill collar section.

Kellys may be bent in the upper two or three feet of the square. Causes include excess weight on the kelly, and rough handling when picking up, setting back, or making connections. A bent kelly and a heavy swivel may vibrate or swing as the bent section enters the drive bushing. This can cause the kelly to break while drilling, depending on the severity of the bend and the rotary speed. The driller usually prevents this by stopping drilling with the bent section of the kelly out of the bushing. This reduces the distance between the bit and the bottom of the hole on connections, thus increasing the risk of plugging and sticking. Prevent this by using a long, straight, heavy-duty kelly.

Start the pump immediately after making the connection and before lifting the drillstring. This helps remove the cuttings or cavings and distributes them uphole (Fig. 1–2). Then lift the drillstring the minimum amount and pull the slips.

Bits and Cones

Operate bits within design specifications; do not exceed manufacturer recommendations. Bit companies publish recommended operating conditions based on formation types. However, predicting the type of formation may be difficult. For this and other reasons, such as running too long, lost bit cones are one of the more common downhole equipment failures.

Roller bit cones are one of the most common causes of fishing. They loosen due to wear, which is generally indicated by uneven, erratic, low-level, fluctuating torque. This is harder to detect in deeper holes and those with high drag and torque. Severity increases as the cones loosen, with a substantial increase when they lock or drop off.

Figure 1-2   Cuttings on top of the drill collars or bit
Note: If the assembly is pulled up into the packed cuttings, it can stick. The correct procedure is to start the pump slowly and then move the assembly slowly downward.

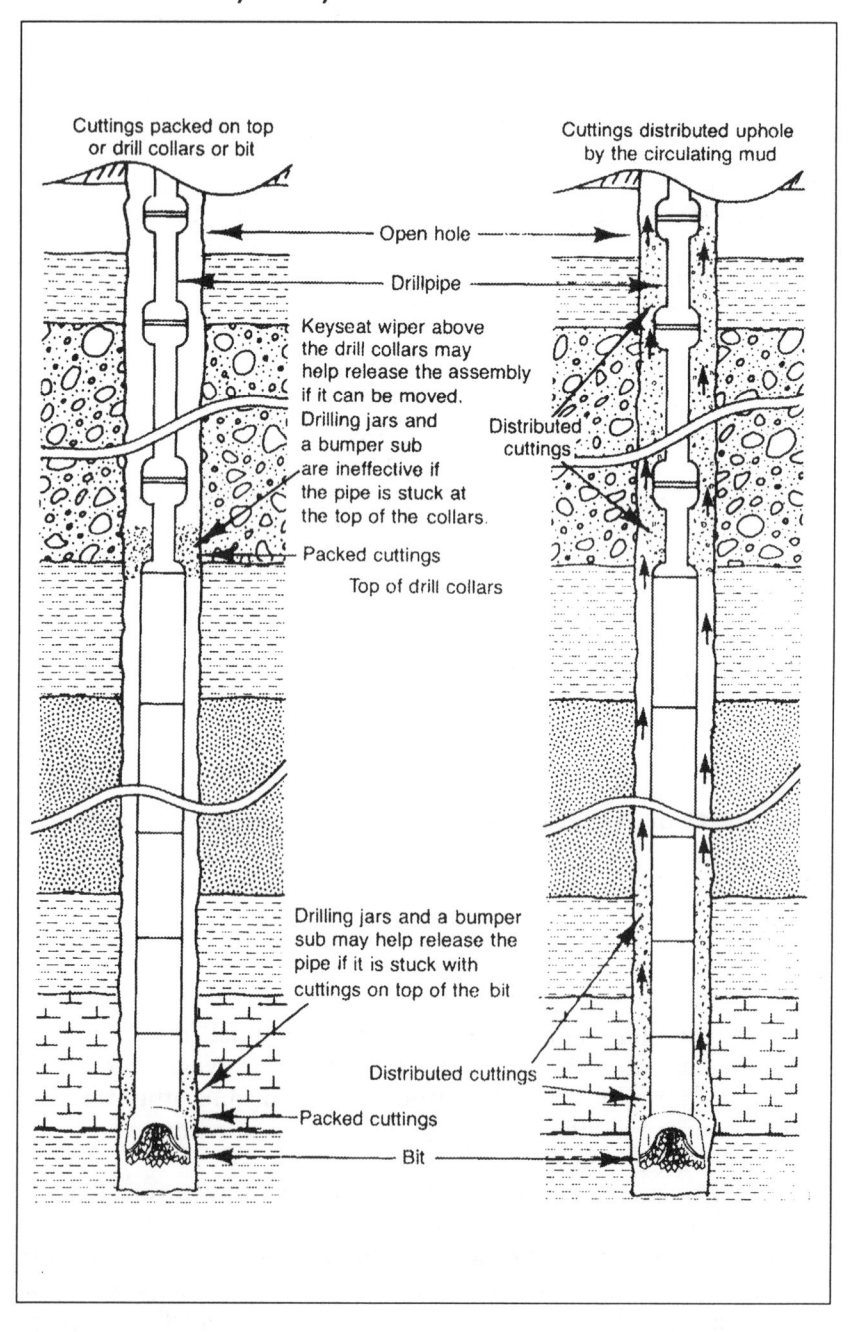

12   Chapter 1

Fluctuating torque may also be caused by increased bit weight and drilling in fractured formations or those containing chert or pyrites. Experienced operators detect loose or locked bit cones by the described actions and pull the bit before losing the cones and causing a fishing job.

Sometimes, excessive time is required to begin drilling because a new roller or drag bit must first form a cutting pattern on the bottom of the hole. The drilling action in this instance often is similar to drilling in fractured formations or formations containing chert or pyrites. If this is the case, select the smoothest running combination of bit weight and rotary speed for roller bits and continue drilling. Diamond or drag-type bits normally drill fractured formations satisfactorily, although the running action may be rough. Otherwise, pull them if drilling chert or pyrites or if there is suspected junk in the hole.

A bit entering a harder formation may slow down or stop drilling. It may be balled up, so spud and ream to remove the ball. If the drilling rate reduces slowly when a tooth-type roller bit is in the hole, this may indicate the teeth are becoming worn, so normally you should pull the bit. Insert-type roller bits often are less affected by formation changes. A drilling rate reduction associated with increased torque indicates worn or locked rollers; again, you should pull the bit. Sometimes the rollers torque briefly and lock while the bit rotates relatively smoothly, although the drilling rate is negligible and torque increases. The rollers are locked and sliding on the bottom of the hole, so pull the bit before the journals wear and break, dropping off the cones.

A partially worn diamond or drag-type bit may slow down, with a relatively constant or slightly increasing torque, when it drills into a harder formation or a formation for which the bit is not designed. Torquing may also indicate either fractures or formations containing chert or pyrites, which destroy the diamonds. One piece of chert may wear out an entire row of diamonds, causing a ring out. The diamonds may embed completely when entering a softer formation, resulting in a slight pressure increase.

One of the main causes of fishing for bit cones is running the bit excessively, as illustrated in the following example.

Bit Run Excessively Long

A hole was being drilled at 18,500 ft with an 8½-in. hard formation roller bit. The bit, which had a normal life of about 75 hours, had been running about 50 hours with a relatively smooth, moderate level of torque. The bit began to torque heavily for about 10 minutes, then smoothed out to a torque level slightly higher than the torque during prior drilling. After about 45 minutes, torque level again increased and began fluctuating. Prior to initial torquing, the bit had been drilling at about 4 ft/hr. During the final hour, it drilled about 1 ft. The bit was run at a moderately high level of fluctuating torque for another 3 hours, during which it drilled 1 ft. When the bit was pulled, the rollers had been lost in the hole and the bit shanks were worn almost to the jets.

The bit action was indicated clearly on the depth recorder. During the first period of the torque, the bit cones locked. The bit was rotated with the cones in the locked position for about 45 minutes, at which time the cones probably released from the body of the bit and created a heavy torque for about 15 minutes. During the remainder of the time, the bit body was rotating on the loose cones in the bottom of the hole. The bit obviously was overrun and should have been pulled after the first strong indication of torque, even if it had been in the hole only for about 50 hours. The operating personnel were negligent in not recognizing this.

Bit selection is beyond the scope of this text, but the following contains general comments:

- Polycrystalline bits make exceptional runs under favorable conditions. These bits should definitely be considered for long shale sections, especially when drilling with oil mud or mud with a high oil content.
- When running roller bits, allow for the higher speed of the mud motor in evaluating overall drilling time.
- Generally, bits used in directional drilling have slightly smaller jets because of the pressure drop used in the mud motor and the reduced amount available for the bit. This leads to bit plugging and requires special actions such as running a screen in the drillpipe.
- Do not run a used bit except for minor operations such as limited cleanout. Ensure the bit is in good condition.

Tripping Tubulars

Many failures occur during tripping, not all directly associated with tripping. Examples include dropping the blocks, running into the crown or rotary, setting slips improperly, handling drill collars, and making up the drill and other tools. Take all precautions, such as running a pipe wiper while tripping in the hole.

Check the condition of drillpipe rubbers during the trip; replace and/or add more as needed. Change drillpipe breaks and order of stands periodically on trips to distribute wear and prevent excess tightening of tool joints. Note the condition of all tool joints as they are disconnected. Replace pipe that has overly tight tool joints. Mud in a fresh broken tool joint indicates a leak. The best practice is to lay the two joints down and have them inspected and/or refaced. Pin failures can cause severe fishing jobs. For example, the first 3½-in. drillpipe pin above the drill collars parted while milling on two 6½-in. bit cones at 15,000 ft, causing a moderately severe fishing job.

Periodically change 5 or 10 joints of pipe on top of drill collars to an uphole position. Avoid running the same section of drillpipe in deviated hole sections for extended periods — 200 to 500 drilling hours for normal conditions; shorter intervals for rough, high torque drilling, reaming, or drilling on junk. Remove and replace bent pipe.

DO NOT make special "hole wiper trips" during long bit runs except as required by special conditions.

Shutdown

Occasionally, the rig will shut down while drilling or tripping due to mechanical problems. Repair these as soon as possible. The main consideration is how to prevent the drillpipe from sticking during the shutdown. One example is a leaking hydraulic line where the rig could still be operated for a few minutes.

When a shutdown must occur and there is ample warning, the best procedure is to pull the drillpipe up into the last string of casing. Check the annulus periodically to ensure it remains full. It is not necessary to circulate constantly, but if the drillpipe is very deep, it is a good practice to circulate slowly at about 20–40% of the normal circulation rate.

If a shutdown occurs during drilling, pick up the kelly as high as possible. Lower the drillpipe 6 in. to 1 ft at 10–30 minute intervals, depending upon depth, the distance available for lowering, and the amount of time the rig is expected to be shut down. Ensure that the amount lowered each time is sufficient to move the entire drilling assembly as shown by the weight indicator.

Do not lower the tool joint below the kelly into the rotary. This allows the drillpipe rams to be closed if necessary and leaves a connection at a convenient point if one is needed for additional operations. A high pump rate can be used for a short period of time, but pumping at a high rate for an extended period of time with limited drillpipe movement can cause undesirable well-bore erosion.

During tripping, install the kelly and pick up as high as possible. Then lower periodically as described. If the kelly cannot be installed, install the drillpipe inside preventer and lower the drillpipe slowly as described.

There have been cases where the drillpipe has been lowered in this manner for over 24 hours and was still free when operations were resumed. In some special cases, a Caterpillar type tractor has been attached to the dead line to move the drillpipe when the pipe load was not excessive.

Pulling Wet Drillpipe

Wet pipe on trips has many disadvantages. Normally it can be prevented by slugging the pipe correctly. Mud may be lost, even with a good collection pan and drain system. Mud also may cause the rig floor and other working surfaces to be slippery, increasing the risk of an accident. Mud often contains chemicals that are harmful to humans, so limit exposure for safety reasons. The drillpipe can be pulled wet by using the mud bucket, but this takes longer, which increases operating time, and still involves some risk.

Since it is more difficult to measure accurately the amount of fill up required when pulling a wet string, the risk of a flow from the well bore and possible development of a kick situation is increased. Loosening excessive amounts of mud may increase the difficulty of filling the hole completely and correctly. This increases the risk of sloughing and possible kicks due to reduced hydrostatic head.

Drilling Out Cement

Drilling out cement is a common operation. Normally, you should use a medium-long tooth bit or mill on a limber drill collar assembly. However, running the bit and workstring into green (soft) or contaminated cement and sticking while tripping into the hole is preventable. The action probably also plugs the bit, another disadvantage since circulation is often a major aid in releasing stuck pipe. It normally is not economical to wash over or mill over 200–400 ft of cemented pipe, often much less in a cased hole. The problems are similar to washing over sanded-up tubing.

The pipe can be inadvertently run into green cement for one or a combination of reasons. The top of a cement plug often is contaminated with mud so it will not harden regardless of the time spent waiting on the cement. Hole diameters may be smaller than estimated or calculated, so the top of the cement is at a correspondingly higher level. Overdisplacement, or using excess cement and cement channeling through the mud, causes a higher cement top than originally projected. Squeeze jobs can flow back if not held by back pressure, or if the cement was still soft due to the use of excess retarder. There have been cases where the operator overlooked the amount of pipe run into the hole.

Observe standard precautions while tripping into the hole to drill cement. Know the length of pipe in the hole and the depth to the top of the cement. Begin checking for drag about 500–1000 ft above the calculated cement top. Watch the weight indicator closely when lowering the bit, but do not rely upon it completely: It fluctuates rapidly during tripping and may be too insensitive. Normally, it is not sufficiently sensitive to indicate sticking in this situation before the pipe is stuck.

When nearing the top of the cement, lower a stand about 30–50 ft and then pick it up a short distance to ensure that it is free. Repeat this procedure until near the top of the cement. Circulate and ream the last 200–400 ft above the predicted cement top. Usually there is extra time available at this point while waiting on the cement to harden. HURRYING AND LACK OF TIME are not valid reasons to risk sticking the drillpipe.

Clean out completion casing with a casing scraper above the bit or mill. This removes cement and debris adhering to the casing wall and prevents sticking packers and/or presetting plugs. Some prefer running the bit and following with a second trip to run the scraper because of the perceived high risk of sticking with the bit and scraper combination. The recommended procedure is to run the scraper with the bit or mill except in extenuating circumstances. Use all precautions to ensure that the mill is not run too deeply into cement coating the walls of the casing. Ream periodically and circulate to clean the bit and scraper.

It is common after primary cementing to drill cement, float collar and float shoe, and about 10 ft of new hole with a bit and drill collars. The purpose is to reduce casing wear and minimize the possibility of backing off the bottom casing joints. Stabilizers are not used for similar reasons. This procedure often necessitates an extra trip to run the correct bit and assembly for drilling below the casing.

The recommended procedure for most situations, excluding running diamond bits, is to run the assembly and bit that is the most efficient for drilling below the casing. Drill the float and guide shoe and then drill ahead with this assembly. Observe precautions such as making up the joints with epoxy cement and cementing the shoe joints correctly, as described in the section "Loose Float and Shoe Joints" in Chapter 7. Otherwise, drill out with reduced rotary speed and handle the drilling assembly smoothly until the top of the drill collars are below the bottom of the casing. This procedure has been used successfully in a variety of situations, including running stiff assemblies and drilling out with insert and Polycrystalline Diamond (NOT regular diamond) bits. It saves time by eliminating the extra trip.

Open-Hole Testing

There is a higher risk of sticking and a resulting fishing job during open-hole testing. There are areas where the test tool can be left seated and the formation tested for a week without sticking the tool.

In other areas the test tool will stick if it is left seated for more than 10 minutes.

Test tools most commonly stick at the packer elements. Select a hard formation as a packer seat. Soft formations tend to wash or erode around the packer elements, especially when testing with high pressure differentials.

Limit or minimize the pressure shock when opening the test tool. If it is used, run a minimum water cushion but allow sufficient hydrostatic head to prevent collapsing the pipe. Allow sufficient clearance between the rotary and surface manifold to lower the assembly in case the tool moves downward in the final seating process. Use an equalizer valve when running dual packer elements. Consider using a nitrogen blanket when testing high-pressure zones, but provide for the hydrostatic head if there is minimal fluid inflow to prevent collapsing the pipe. Use a minimum water cushion or add water before releasing the pressure.

Testing dry gas, especially at high pressure for an extended period through drillpipe, may involve some risk. There is a tendency for gas to escape through drillpipe tool joints. This can cause gas-cut mud and, if not handled properly, may increase the difficulty of controlling the well. Testing dry gas should normally be limited to a very short time interval, as noted in the following example.

Gas-Leaking Tool Joints While Testing

A high-pressure gas well drilled with 15 lb/gal mud had 7 in. casing set at 13,000 ft. A thin sand stringer at 12,700 ft was to be tested before completing the main prospective interval at 12,500 ft. The sand stringer had questionable productivity and was possibly water bearing, so it was to be tested using a retrievable test tool with a full column of water and 15 lb/gal mud in the annulus.

The zone was perforated, the test tool was run on 3½-in. drillpipe, and seated, and the pipe was filled with a full cushion of saltwater. The pipe and surface equipment were tested to 10,000 psi and the annulus pressured to 2,500 psi. The test tool was opened and had an immediate pressure of about 4,500 psi, which increased as the water flowed out of the hole through a choke. The well was allowed to flow dry gas through double chokes for 10 minutes and was then shut in for a 10-minute pressure buildup. Surface pressure at the end of the buildup was about 8,500 psi. The tool was closed, and the drillpipe was loaded with mud and circulated extensively through the circulating valve. Because the annulus was loaded with gas, it was necessary to close the rams and circulate the well through the choke manifold. It could not be reverse-circulated because of the excess annular pressure that had to be released through the choke manifold to prevent bursting the casing.

In hindsight, the thin sand stringer should have been tested with tubing. Because of an approaching deadline completion date, it was tested with the drillpipe as the fastest method.

Other Precautions

There are many other details, sometimes apparently small ones, that can prevent problems which in turn often lead to larger problems. For example, check for metal cuttings in the mud with a ditch magnet. Clean the magnet periodically. Examine the metal fragments for indications of casing and other metal wear. Check the drillpipe rubbers and replace them as necessary. Otherwise, continued wear can lead to a hole in the casing or a backed-off casing shoe. There are many other similar examples.

Undergauge Holes

Undergauge holes may present risks of problems such as pinched bits and sticking. The main cause of undergauge holes is drilling with a worn, out-of-gauge bit. Another cause is the failure to run stabilizers or reamers, especially near bit reamers when drilling hard formations. Muds that form thick filter cakes and swelling formations also cause undergauge holes. Treat the mud and ream as necessary.

Corrections for undergauge holes are indicated by the cause. Always run a near-bit stabilizer with fixed blade or diamond bits. In harder formations, run bits with more gauge protection. Ensure that all reamers and stabilizers are full gauge. Use roller reamers with hard formation cutters for hard, abrasive formations.

Ream the last 60–90 ft before reaching bottom with a new bit. This removes cuttings and caving material that may cause sticking. It also helps prevent running a new, full-gauge bit into a tight, undergauge hole. Although this prevents premature roller bit failure by pinching the cones, it may also cause lost cones, requiring milling and fishing. An immediate fishing job may be required if the bit is wedged in the tapered hole section (Fig. 1–3).

Strapping

It often is ADVISABLE to strap the pilot bit or bullet nose below a hole opener, using light or heavy strapping depending on the application (Fig. 1–4).

In one case a bullet nose on an 8½ x 12¼-in. hole opener was not strapped and backed off. It was recovered with a short-skirted overshot dressed with a basket grapple using a mill shoe heavily coated with tungsten carbide.

Cause and Prevention

Figure 1-3   Bit stuck in tapered hole

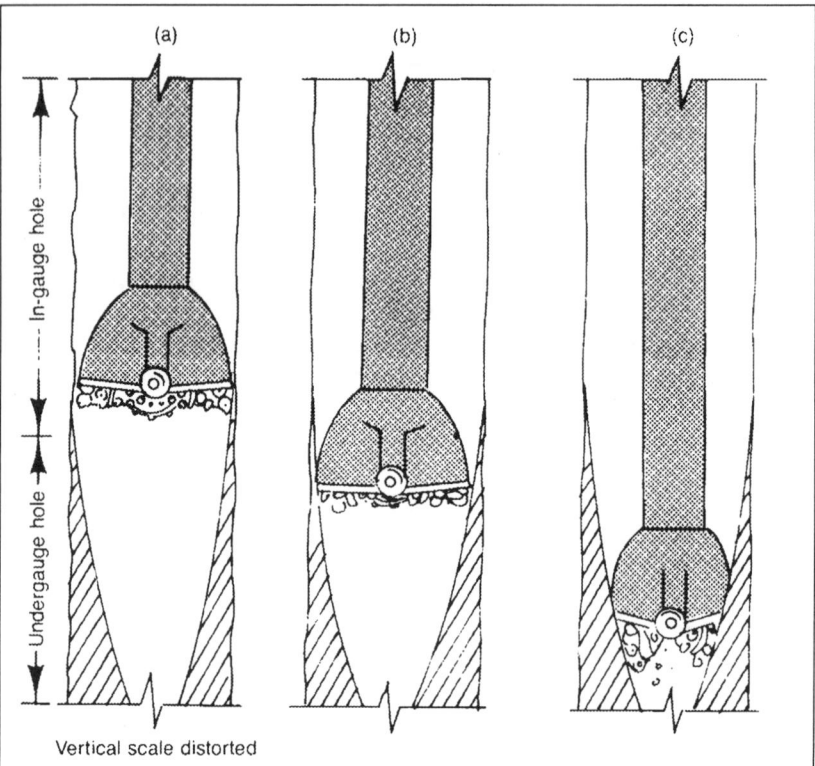

Vertical scale distorted

(a) New bit approaching the undergauge hole section; ream old hole from this point to total depth.

(b) New bit jammed or wedged into tapered old hole; could require backing off and a fishing job. If drilling jars are run, the pipe probably can be released; however, the cones probably are damaged.

(c) New bit strongly jammed or wedged into tapered old hole. The cones (sometimes with part of the bit shank) have been sheared, requiring a fishing-milling job.

In another case an 8¾-in. hole was sidetracked in a hard, abrasive formation using a maximum kickoff angle with a small, shallow penetration that was opened to 8¾ in. Prior attempts to sidetrack the hole conventionally were unsuccessful, but this method was effective in the hard formation. The bullet nose on the hole opener was heavily strapped to prevent it from backing off. The strapping wore thin and had to be replaced due to the high angle and the abrasive formation.

Figure 1–4 Strapping a tool joint

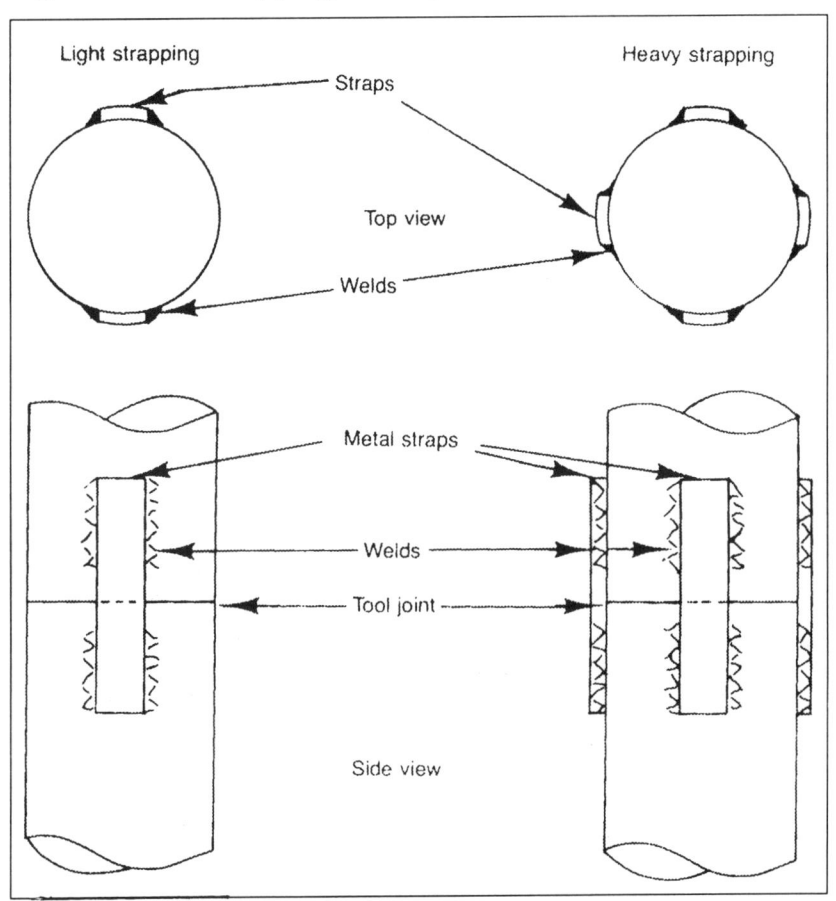

Pipe Wiper

Small items such as nuts, bolts, tong dies, and hand tools are sometimes lost in the hole. This usually occurs while pulling the drill collars through the rotary, changing the bit or other tools on the drill-collar assembly, or running the assembly in the hole. These items are seldom lost in the hole while pulling the assembly because of the pipe wiper. The pipe wiper is removed on the last stand of drillpipe above the drill collars to keep the wiper from being split or damaged when pulling out the larger diameter drill collars.

As an important preventive measure, a pipe wiper should be used on the assembly at all times, especially when handling drill collars in the rotary and tripping in the hole. (It may be omitted during drill-

Cause and Prevention

ing if a rotating head or similar piece of equipment is used.) When the good pipe wiper is removed from the assembly, install a worn wiper. The drill collars and other large diameter tools can be stripped through it. Anything dropped through the rotary will hit the wiper and fall to the side, keeping it out of the hole. Use the same procedure when running the assembly in the hole, especially when running diamond or other fixed blade bits. It only takes a very small piece of metal, such as a tong die, to ruin a diamond bit.

Always close the blind rams when the drillpipe or drill collars are out of the hole. In one case a small wrench was dropped in the hole on top of double pipe rams. One ram section had to be opened to recover the wrench. In another case a bit was hanging below the bradenhead under a stand of drill collars. A slip segment dropped in the hole and lodged in the bradenhead. The pipe was not picked up, since the slip segment might have bypassed the bit and fallen downhole. A joint of drillpipe was screwed into the top of the collars and lowered below the slip segment to see if it could drag the segment uphole, but it was unsuccessful. The blowout preventers had to be disconnected and picked up to recover the slip segment. Using a pipe wiper would have prevented the problem.

Drill-Collar Clamps

Another preventive action is to use drill-collar clamps on all downhole tools that do not have a relatively large diameter difference between the area where the slips are set and the section above the slips. Place drill-collar clamps on all drill collars, even bottleneck types. In special situations, bottleneck drill collars can vibrate or jar and fall through regular slips if the clamp is not used.

In one case, jars were lowered through the rotary and the bottlenecked drill collars were set in regular slips without the clamp. The jars opened, causing a vibration, and the drill collars vibrated and dropped through the slips, which created a fishing job.

When running casing, a drill-collar-type clamp should be used for the first several joints. This is especially important when using spring-loaded elevators and slips. Lock these down and/or use drill-collar clamps on light loads as illustrated in the following example.

Drill-Collar Clamp on First Joints of Casing

Casing (10¾-in., SFJ) was being run in a deep hole. The float shoe and float collars had been made up on the shoe joint. The float-collar end of the joint was picked up and latched in the slip-type, spring-loaded elevators and started in the hole. When the bottom of the joint was about 20 ft below the rotary, the joint began to float. This released the weight on the spring-loaded elevators, and the elevator slips jumped open, allowing the joint to drop. The float collar had a thread protector with a rim that extended abut ⅛ in. over the edge of the semiflush tool joint. This rim caught in the top of the rotary slips, closing them flush on the top of the joint.

A short bar was wedged into the top of the thread protector. The cat line, or lifting line, was tied to this, and the joint was picked up about 2 ft by dragging it through the closed rotary slips. A drill-collar clamp was then placed on the joint. The drill-collar clamp should have been placed on the joint before placing the joint in the elevators.

Most operators use the drill-collar clamp for 20–30 joints of casing or until the casing weight is sufficient to ensure it will not let spring-loaded slip-type elevators open. Automatic fill-up equipment often does not function correctly until several joints have been run and there is an adequate differential pressure. These light and possibly alternating casing loads increase the risk of accidently dropping the casing, especially with spring-loaded running equipment, so the safety clamp should be used as a safety factor.

A few operators use the drill-collar clamp while running the entire string of casing. This probably is unnecessary and takes extra time. The best procedure is to use the drill-collar clamp until the pipe has gained sufficient weight so it will keep the slips from springing open and releasing. After the casing load becomes relatively high (over 50,000 lb), the drill-collar clamps probably would not hold the casing if it were dropped. As prevention, though, the clamps should be used on the first few joints of casing run.

Summary Precautions

Other precautionary items are summarized in the following list and are included throughout the text.

- Tie the blind-ram operating lever with a breakable wire to prevent accidental use.
- Obtain an analysis, or fingerprint, of the ions, cations, elements, etc. of saltwater, usually used for workover.
- Have a full-opening inside diameter on all drill tools where possible.
- Cut the correct bell or taper on the casing landed in a casing spool or head.
- Always place drillpipe tongs on the tool joint.
- Run diamond and PDC bits in a clean hole.
- Never rerun a dropped roller bit.
- Use work-string pipe instead of tubing when possible.
- Cripple the rope socket in most cases.
- Run gauge rings before running full-gauge wireline tools.

- Use no-go restrictions on swab bars.
- Do not set the casing above the bit cones if there are plans to drill deeper.
- Keep a full-opening inside preventer (in the open position) available on the rig floor at all times.
- Do not allow a backlash and unintentionally back off by rapidly releasing the high right-hand torque when operating under high drag.
- Do not arbitrarily use check valves, kelly fluid saver, and similar tools, even those with knock-out or pump-out features, because of the risk of malfunction.
- Ensure that the inside diameters of crossover subs and other tools do not have internal restrictions, such as a reduced inside diameter, which may prevent the running of string shots and similar tools.

MECHANICAL FAILURES

Mechanical failures frequently cause fishing jobs. Almost any type of normal drilling problem, such as a hazardous formation, open-hole testing, or the reaming out of keyseats, can become a fishing operation if equipment fails. Therefore, mechanical problems create, both directly and indirectly, a need for fishing. The solution is to select the correct equipment, operate it correctly, and keep it in good operating condition.

Rig Selection

Rig selection offers another opportunity for preventive measures. Select a rig that has the proper specifications for drilling the well. Normally these include the type of rig, power, hoisting equipment, pumps and the depth capacity of the mast and substructure, floor space for pipe set-back, mud-handling capacity, blow-out preventers and choke manifold, and similar items.

Rig selection is important for preventing problems because of the possibility of lack of power, mechanical failures, and generally unreliable equipment. The best rigs for deeper and more critical holes are usually equipped with a silicon-controlled rectifier (SCR) system. Alternating-current (AC) generators, usually driven by diesel engines, are very efficient. The bus bar provides variable power distribution to larger and efficient high-torque, direct-current (DC) motors. AC current is distributed to small motors (up to 10 Hp), and most rigs have a smaller AC stand-by light generator.

Diesel electric rigs are less flexible. Diesel engines drive DC generators that are less efficient. Power distribution is limited to individual generator units. Smaller AC generators supply power for lights and small motors.

Mechanical rigs are somewhat less flexible, but can be highly efficient, especially in shallower, less demanding drilling. Diesel engine power passes through torque converters into a compound drive group. In most cases this power is distributed to the drawworks, the pumps, and sometimes the rotary.

Check all available records before selecting a specific rig. Downtime is very important. The operator usually pays for a specific amount of downtime and the rest is at the expense of the contractor. However, this does not help in the case of ongoing expenses such as mud costs and ancillary services, including tool rental and third party work. Other cases such as lost circulation, caving, and stuck pipe almost invariably deteriorate with time. Review downtime on tour sheets from prior operations. Note that many repairs made during rig operation do not cause downtime and may not be noted on the tour report. Checking the record of equipment repairs should indicate if the rig has had excessive repairs or if some component has failed repeatedly. The cut and slip and bit records, mud reports, and similar records should also be checked.

A less obvious cause of fishing jobs is using undersized, inadequate equipment. One example is drilling to 9,000 ft with a 6,000-ft nominal depth capacity rig. Innovative drilling personnel can extend the depth to which a rig can drill (not the nominal depth capacity) by increasing the number of lines on the traveling block (Fig. 1–5), running smaller drillpipe, or using smaller pump liners to increase pump pressure. However, this either overstresses the equipment or overburdens the rig capacity, and is not recommended except in special cases.

A similar problem occurs when the depth objective is extended appreciably after starting drilling operations. This also can overstress the equipment and increase the risk of a fishing job. Additional risk may be incurred with reduced hole sizes. This increases both the risk of a fishing job being needed and the risk that it cannot be completed successfully.

Unmatched Rigs

Unmatched rigs have one or more components that are either under- or oversized for the nominal depth capacity. A common example is oversized pumps, which can cause problems because

Figure 1-5 Pull decreases with more lines

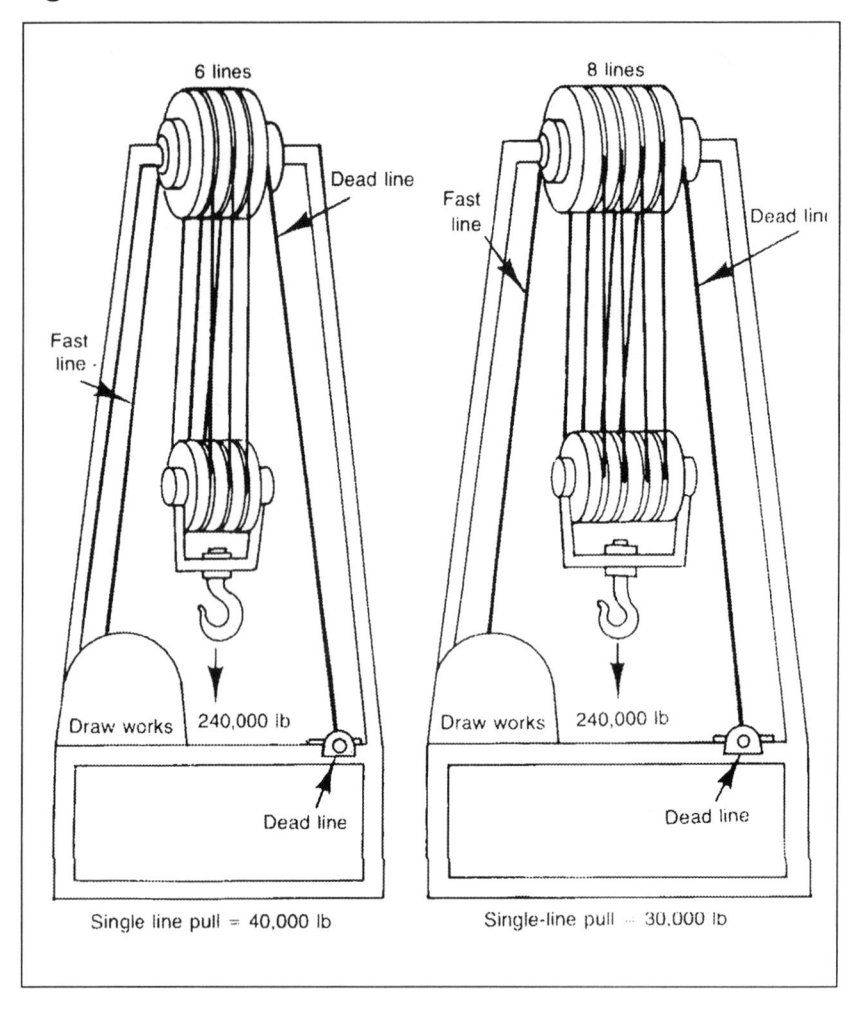

normally they are run within specified limits based upon pressure, volume, and strokes per minute. Sometimes the output volume may be too high, causing excess pressure on the downhole equipment, borehole erosion, especially around the bottom-hole assembly, and difficulties using mud motors and turbines. These problems can occur even when running the minimum-size pump liner or plunger. In many cases the pumps cannot be throttled down because of the minimum SPM requirements based on the pump-operating specifications. Throttling valves or adjustable- or fixed-vent chokes are seldom satisfactory. The problem can be prevented by knowing the volume and pressure requirements of the hole during drilling and selecting a rig with the correct pumping equipment.

Pump Failure

The most common mechanical problem is a pump failure. This creates many problems if it occurs in a situation such as cleaning out heaving shales or circulating out a kick. Normally, the rig is equipped with a standby pump for use when the main pump fails. However, the standby pump may also fail. A pump failure that almost caused a fishing job is illustrated in the following example.

Pump Failure

A wildcat well was drilling a 9½-in. hole at 19,000 ft with 5,000 ft of open hole below the last casing string. The very hard, abrasive sand shale sequence was drilled with an insert bit in a dispersed polymer mud containing 4% emulsified oil. The packed hole drilling assembly included 7¾-in. drill collars, integral blade stabilizers, a shock sub, drilling jars, a keyseat wiper above the drill-collar assembly, heavyweight pipe, and a combination 5 and 5½-in. drillstring.

A trip had just been made to change bits, and the mud viscosities and gels were slightly high (funnel viscosity 63 sec, PV 63, YP 13, gels 5/7). The calculated pore pressure was 17.5 lb/gal, and the 16.8 lb/gal mud had an equivalent circulating density of 17.3 lb/gal. The slightly underbalanced mud system was being used to improve penetration rates, which averaged about 5 ft/hr.

The hole had been logged at 18,000 ft and was shut down for 48 hours, from the time drilling stopped until it resumed. Hole deviation had begun increasing slowly from 2° at the casing shoe to 7° at 17,500 ft. At that time, a packed pendulum assembly was run and the hole deviation was reduced evenly to 2°. The average maximum absolute dogleg was less than 1.5°/100 ft. There was some minor keyseat formation in the interval from 16,000 to 18,000 ft, indicated by the pipe drag of up to 30,000 lb and the difference between the two-arm and three-arm calipers' hole diameters.

The hole diameter increased gradually from bit size at TD to about 13 in. diameter below the casing shoe. The normal increase in hole diameter was caused by mud erosion. The hole was being circulated at 340 gpm and 2,800 psi. The lifting quality of the mud was satisfactory to slightly low for cuttings removal in out-of-gauge hole sections, as indicated in Table 1–1.

Cause and Prevention

Table 1-1 Upward cuttings velocity

Hole diameter, in.	13	16	19	22
Annular mud velocity, ft/min	67	36	25	18
Slip velocity, ft/min	8	7	6	5
True Cuttings velocity, ft/min	59	29	19	13
Time, min, to move cuttings upward 1000 ft.	17	35	53	77

In several places, though, the diameter was greater than the maximum caliper reading of 16 in.

The on-line pump failed while drilling, and the second pump was placed on line while the first was repaired. After drilling about 1 hour, the second pump failed. Since the first pump was expected to be repaired shortly, the driller decided to pick the pipe up and move it slowly (normal drag = 25,000 lb) instead of pulling the entire drilling assembly up into the casing. When circulation stopped, the 42-ft kelly was 30 ft down. The pipe was moved by picking up about 60 ft off bottom (kelly and one single) and lowering it about 5 ft every 5 minutes. After about 1½ hr, the pipe drag began to increase. With the kelly about 5 ft into the rotary, the pipe could not be moved upward with 50,000 lb of overpull. The pipe was left hanging about 25 ft off bottom with normal weight and without trying to pull the pipe free. Pump repairs were completed in another hour.

The pump was started slowly. The pipe was worked to establish circulation, which took about 20 minutes. By stopping the pipe above bottom, it was possible to move it downward, which allowed some free movement and helped establish circulation. If the drillpipe had been pulled harder into the cuttings, it could have been stuck tighter, possibly necessitating a fishing job. After working the pipe for another 30 minutes, the drilling assembly was freed. The hole was circulated for another 30 minutes, while working the pipe slowly, and drilling operations were resumed.

Another point regarding the example is whether the pipe should have been pulled up into the casing. There was no apparent reason to pull the pipe into the casing when the first pump failed since a second pump was available. But when the second pump failed, the best course of action was not as easy to determine.

The apparent safest procedure would have been to pull the pipe up into the casing. This probably would have prevented a stuck pipe and the need for a fishing job. However, the well was being drilled in a near-balanced condition, so there could have been a kick while tripping — one of the most dangerous times for a kick to occur. Since the pumps were inoperative, the pipe could not be slugged and would have to be pulled wet, a slow, slightly dangerous procedure. The delay would increase the risk of a kick. There would be no way to fill the hole while pulling up into the casing except with the mud-mixing pump. Mud lost from pulling a wet string and inability to add mud could lower the mud column, reduce the hydrostatic head, and increase the risk of a kick. This probably was a major factor in the operator's decision to leave the pipe near bottom. The risk of wall sticking was minimal because of the near-balanced mud system. The obvious problem was sticking due to cuttings and slough material.

There are several preventive measures that could have been used had the operator decided to pull the drilling assembly up into the casing. Pulling slowly would have reduced the risk of swabbing. This would also have reduced the mud loss when the stands were disconnected. The annulus could have been filled with water, which, though a less preferred fluid, would have provided some hydrostatic head. If the well had kicked, it would have given a more positive indication compared to air and would have helped prevnt fluid or gas entry into the borehole.

The primary problem was inadequate pumps. Preventive measures should have included improved pump maintenance to ensure the pumps were in good operating condition. If pumps require excessive repair, they should be replaced. However, this may be difficult because of contractual obligations and other considerations.

In summary, the operator selected a course of action in this situation that involved some risk, but it was successful and saved time.

Hoisting Equipment

One of the most serious mechanical problems is a hoisting-equipment failure. The most severe problems are a broken drilling line, a failed crown block or traveling block, drawworks drum, or brake. There is a danger of dropping the drilling assembly, which could result in a fishing job and a high risk of injury to personnel. There is NO EXCUSE for dropping the blocks.

Ton-mile records and good cutoff procedures have almost eliminated broken drilling lines. However, a prudent operator will check the drilling contractor's records to ensure that good line slipping and cutoff practices are used. These records should also be checked any time an operation calls for the drilling assembly to be downhole for an extended period. If the drilling line is approaching a cutoff point, the line should be cut before starting the downhole operation.

Sometimes, it is necessary to stay in the hole longer than expected. This can cause severe line wear that could require either: (1) shutting down to cut the line at a time when a fishing job almost certainly would occur (such as for drilling and cleaning out a caving formation); or (2) continuing operations (such as jarring) with an increasing risk of a drilling-line failure. Frequently checking the condition of the drilling line is an important preventive action.

Jarring and heavy pulling, especially with high hook loads, accelerates drilling-line wear. Worn sheaves can further accelerate wear. Working the fish for a long time without moving the downhole assembly appreciably can cause concentrated line wear at one or more points. The line often can be slipped to change the point of wear with minimum downtime.

The major components of the hoisting system, such as the brakes, sheaves, and bearings, normally are checked periodically and changed when necessary. However, during long periods of operation such as deep drilling, the equipment should be checked more frequently.

There have been instances in which the drilling assembly was dropped due to brake failure. In one case during a trip to change bits, the drilling assembly was run in the hole to the bottom of the casing and set in the slips, the rams were closed, and preparations were made to cut the drilling line. Either the line clamps slipped or the traveling block was not hung correctly. The blocks dropped, causing an extended, expensive fishing job. Therefore, it is important to take the proper precautions when cutting or slipping the drilling line.

Split Crown Sheave

An on-structure wildcat well had 7-in. intermediate casing set at 12,000 ft. A $6\frac{1}{8}$-in. hole was being drilled at 14,000 ft with a diamond bit and $3\frac{1}{2}$-in. drillpipe. There were several kicks while drilling the interval from 13,000 to 14,000 ft. The 14.5-lb/gal mud slightly overbalanced the pore pressure.

A trip was to be made to change bits. The kelly was set back, an extra single was pulled out and laid down, and the first stand was being pulled out of the hole. When the elevators were about 30 ft above the rotary, a crown sheave split, cutting the drilling line and dropping the traveling blocks and drilling assembly. Eight lines were strung in the derrick, and as the lines unstrung through the blocks the assembly dropped faster. When the elevators hit the rotary, the bottom of the elevator ears broke. But the elevators held, and the drilling assembly remained intact. The traveling block fell to the floor on one side of the rotary. The unraveling drilling line did some damage to the mast bracing and drawworks cover, but there were no injuries to personnel. The drillpipe rams were closed and locked, and a safety valve screwed into the top of the drillpipe.

The damaged equipment was repaired and replaced in about 48 hours. The crown sheave had cracked near the base of the wireline groove, and a piece of the crown-sheave flange broke off. This apparently let the drilling line slip out of the crown sheave and cut or break as it dropped over the main crown-shaft housing.

After the repairs were completed, the kelly was connected to the drilling assembly and circulation was established without difficulty. Shortly thereafter, the drilling assembly was picked up and found to be free, and was pulled and inspected.

Blowout Preventer

The blowout preventer (BOP) system, which includes the choke manifold, accumulator, fill-up line, control stations, and accessory equipment, is one of the most important equipment items on the rig. Operators should ensure that preventers have an adequate pressure rating. A general guide is that the working pressure rating should be equivalent to the maximum pressure expected in the well. The most common preventer stack has three pipe rams and a bag-type preventer. Some wells may require shear rams, a rotating head, or a diverter. A rotating head is commonly used when drilling long sections underbalanced.

A variety of standard precautionary procedures should be observed. Pipe rams should *always* be in the lowest preventer. Some operators use an extra set of blind rams in the lowest position, especially on a four-ram stack. If an extra set of pipe rams is kept on location, the upper blind ram can be replaced immediately when required. Control stations must be conveniently positioned so they can be safely operated under all possible conditions. The blind ram control should frequently be fit with a breakable restraining device to prevent accidental closing. Before running casing, one pipe ram should be changed as a casing ram.

An inside blowout preventer should be on the rig floor at all times, open and ready for use, including available crossover subs (such as for running casing). Place the opening and closing tool in a convenient, known place. Many operators hang it on the side of the BOP control panel.

Good well head security is important when running casing. This procedure normally includes changing pipe rams to fit the casing for annular control. The preventers normally are tested after changing rams, especially on critical wells. Have an inside preventer with a changeover sub to fit the casing available on the rig floor.

Testing

A careful operator will maintain a complete testing and maintenance program, and will record the tests on the tour report. Tests should include: closing and opening pipe rams once per day; actuating the choke manifold controls and closing and opening all rams once a week or on the next trip; and testing the complete BOP system, including the accumulator, any time there is a question about preventer integrity. Normally testing should be done after changing rams, at 20–30 day intervals, or in compliance with regulatory agency requirements. BOPs and the choke manifold should always be tested after heavy usage. Use the correct test plug to prevent sticking the plug in the seating receptacle. Also check BOP nuts and bolts for tightness at weekly intervals.

BOP Drills

BOP training and drills should be conducted as required. Practice tests should be performed during drilling, while making a connection, and while tripping.

Conduct a BOP drill at 2- to 3-day intervals for each crew while drilling until time-to-close is satisfactory. Time-to-close is the time period from the start of the drill (signaled with a horn or whistle) until the well is shut in. Test intervals can then be extended to 3–7 days per crew. Satisfactory times are in the range of 1 minute while drilling at 5,000 ft to about 2–3 minutes while drilling at 15,000 ft. For tests during tripping, add about 2 minutes for time-to-close.

Other Precautions

Standard precautions, such as full-opening and double valves, should be taken in most cases. On critical wells, install an extra kill line to the edge of the location. Do not use the kill line for a fill-up line. The choke manifold flow line should have double valves, including a manual valve next to the preventer stack. To eliminate bends, use piping that has full-opening lines and that is 3 in. or larger. Take adequate precautions to prevent freezing during cold weather.

The choke manifold should be set as far from the rig as is convenient. Bends should be minimized, with plugged "T's" used when they occur. An adequate-capacity, well-designed mud gas separator with a return line to the mud tank is essential. Install a pilot light at the end of the blowdown or blooie line. Have the end of the line a safe distance from the wellhead, usually downwind from prevailing winds.

Most rigs have a safety device such as a siren to warn of low pit levels and low flow-line flow rate. It is common to shut the siren off during connections. In at least one case, this was unintentionally left off after the connection, and a kick occurred shortly thereafter. Fortunately, the kick was detected by other indicators. Failure to turn the siren back on could have been catastrophic. One highly recommended solution is to install a blinking yellow or orange light that turns on automatically when the siren is shut off.

Downhole Failures

Downhole failures include lost bit cones, parted pipe, mechanical failures in downhole tools, washed out tool joints, cracked pins, and split boxes. All rig equipment should be maintained carefully, inspected at correct intervals, and operated within design limits.

The fact that many severe fishing jobs are caused by downhole conditions and the way the downhole assembly is operated points up the importance of using good procedures and exercising extra care when drilling in problem formations.

A twistoff is a relatively common failure that invariably causes fishing. The main cause of a twistoff is torque in excess of the strength of one or more components of the downhole tubulars. In many cases it can be prevented by drillstring maintenance and careful handling procedures. Worn pipe and fatigued or damaged tool joints are major causes of twistoffs. Prevention measures include inspection and replacement as necessary. Fine-threaded connections, such as those on an overshot bowl, or test tool and some casing and tubing threads are less strong and therefore susceptible to twistoff problems associated with over-torquing.

The risk of twisting off is higher when running combination work strings. Common pipe combinations are 5 in. x 4½ in., 3½ in. x 2⅞ in., and 2⅞ in. x 2⅜-in. Higher-risk combinations include 4½ in. x 2⅞ in. x 3½ in. x 2⅜ in., and 2⅞ in. x 1½ in. The larger, heavier pipe at the top of the string has a relatively higher momentum during rotation as compared to the smaller pipe in the lower part of the assembly. If the bit or tool on the bottom of the assembly hangs up, this higher momentum can twist off the pipe body or tool joint of the smaller, weaker pipe. Stopping the pipe rotation very quickly at the surface can cause a similar failure, especially when using large drill collars relative to the size of the pipe.

Under normal conditions, combination assemblies generally can be operated satisfactorily. The frequency of failure increases as the difference between the two pipe sizes increases. Problems can be prevented by minimizing the use of combination strings, especially with large size differences. When these combinations must be used, they should be operated carefully with minimum torque.

MUD

One of the most important preventive procedures is to select and maintain a good mud system with good physical properties. Do not overlook this, even while combating either lost circulation, sloughing, kicks, fishing, or any combination of these. Many of the following precautions are self explanatory, but they warrant repeating here because of their common occurrence.

Cause and Prevention

Table 1-2 Mud property changes signal contamination

Contamination	Funnel viscosity	Plastic viscosity	Yield point	Gel strengths	pH	Fluid loss	P_f/M_f	P_m	Cl−	Ca++	Solids	CEC (MBT)
High Solids Contamination	Slight increase	Increase	Slight increase	Slight increase	Same	HTHP Could increase	Same	Same	Same	Same	Could increase	Slight increase
Contaminant	Funnel viscosity	Plastic viscosity	Yield point	Gel strengths	pH	Fluid loss	P_f/M_f	P_m	Cl−	Ca++	Solids	CEC (MBT)
Cement contamination	Increase	Slight increase	Increase	Increase	Increase	Could	Increase w change	Increase in Pl	Normally	Could same	Same increase	Same
Contaminant	Funnel viscosity	Plastic viscosity	Yield point	Gel strengths	pH	Fluid loss	P_f/M_f	P_m	Cl−	Ca++	Solids	CEC (MBT)
Gypsum and/or anhydrite	Increase	Slight increase	Increase	Increase	Decrease	Increase	Decrease	Decrease	Same	Increase	Same	Same
Contaminant	Funnel viscosity	Plastic viscosity	Yield point	Gel strengths	pH	Fluid loss	P_f/M_f	P_m	Cl−	Ca++	Solids	CEC (MBT)
Rock Salt	Increase	Slight increase	Increase	Increase	Decrease	Increase	Decrease	Decrease	Increase	Slight increase	Slight increase	Same

Contaminant	Funnel viscosity	Plastic viscosity	Yield point	Gel strengths	pH	Fluid loss	P_f/M_f	P_m	Cl^-	Ca^{++}	Solids	CEC (MBT)
High Temperature Gelation	Increase at flow line	Slight increase	Increase	Increase	Decrease	Increase	Decrease	Same	Same	Same	Same	Same
Contaminant	**Funnel viscosity**	**Plastic viscosity**	**Yield point**	**Gel strengths**	**pH**	**Fluid loss**	P_f/M_f	P_m	Cl^-	Ca^{++}	**Solids**	**CEC (MBT)**
Carbonate	Increase	Same	Increase	Increase especially 10 min.	Could change	Could change	Increase	Slight increase	Same	Could change	Same	Same
Contaminant	**Funnel viscosity**	**Plastic viscosity**	**Yield point**	**Gel strengths**	**pH**	**Fluid loss**	P_f/M_f	P_m	Cl^-	Ca^{++}	**Solids**	**CEC (MBT)**
Bicarbonate	Increase	Same	Increase	Increase especially in 10 min.	Could change	Could change	Increase in M_f	Same	Same	Could change	Same	Same

Source: Roseland, courtesy OGJ
Legend: shaded areas show the most reliable indicator of a particular contaminant.

Cause and Prevention

Mud-handling procedures include all of the operations and equipment used to work with the drilling fluid as it moves through the surface and downhole circulating system. Use of good operational procedures with correctly sized, well-maintained equipment will help maximize the penetration rate, prevent fishing, and enhance fishing operations.

Using a low mud weight and the minimum amount of low gravity solids will increase penetration rates, reduce treating costs, prevent sticking, and increase the chances of releasing stuck pipe or completing a fishing job. Twin, double-deck shakers with small-sized screen openings are almost standard equipment on larger rigs drilling deeper holes.

The interface between mud and other fluids can cause chemical contamination, dilution, and settling of weight material. All three of these problems can increase the cost of mud treatment and the risk of sticking the assembly. If barite settles out at the interface, it can accumulate and stick the pipe. Interface problems can be solved by using the correct volume and type of spacers.

Changes

Changes in mud properties are good indicators of conditions such as contamination or increasing salinity from saltwater flows. These conditions are listed in Table 1–2.

The recovery of drill cuttings usually depends on the drilling rate, a figure that tends to be relatively constant. Recovery may be reduced by a lack of cleaning or by accumulation in washed-out hole sections. An increase in cuttings recovery may be due to caving, viscous sweeps, or drilling into a higher pressure zone. The shape and volume of cutting may indicate a possible kick situation, as described in the section "Kicks" in Chapter 2.

Aerated Mud

Aerated mud has a frothy appearance somewhat similar to gas-cut mud. If the mud appears to be gassy and yet no gas is detected by the mud-gas analyzer, then the mud probably contains air. An operator should be careful to differentiate between air and gas in the mud, because gas is an important indicator of a kick, while aerated mud seldom is a problem. Severe cases may affect pump pressure and operation. A common cause of aerated mud is a leaking suction or mud-gun nozzle whipping or mixing air into the mud. Mud may also be temporarily aerated by pumping it into empty drillpipe or casing. Air trapped in mud can be removed with sub-surface mud mixers and de-gassers.

Pump pressure normally fluctuates in a range of about 2% or in the order of 50–100 psi. To maintain a constant pump pressure, check to see that the pump throttle setting has not changed and check the pump output to ensure it is operating correctly.

Decreasing pressure commonly indicates a leak. A surface leak usually is easy to detect. However, if this is not the cause, stop pumping and immediately pull the pipe to prevent parting, such as a washed-out tool joint, and leaving a fish in the hole. Other causes of pressure decreases are reduced mud weight, gassy or aerated mud (in rare cases due to an air leak), and a partially plugged or starved pump suction. A kick may also cause pump pressures to decrease slightly.

Increasing pump pressures often are associated with a partially plugged or balled bit or a worn diamond bit (usually due to a ring-out). Other causes include increasing mud viscosity (which decreases pump SPM), excessive formation caving or sloughing, cuttings accumulation due to improper hole cleaning, and deteriorating mud properties, such as an increasing equivalent circulating density.

Since mud volume seldom fluctuates, any changes should be investigated. Mud volume increases may be caused by a kick, or by fluid, liquid, or gas flow into the mud system. Chlorides and mud weight should be checked to detect a flow of saltwater or, less commonly, of fresh water. Water may enter the mud system at the surface through a leaking valve or other means. The most common cause of reduced mud volume is lost circulation, followed by surface losses. A considerable amount of makeup mud is required to fill the new hole for fast shallow drilling in large diameter holes.

The most common weight fluctuations are caused by an out-of-balance system. These fluctuations can be corrected by increasing mixing-action or water-back heavy streaks and adding weight material to light streaks. Weight may fluctuate due to entrapped air or gas. Barite sag causes fluctuating mud weight, especially in directional and horizontal wells, as described in the section "Barite Sag" in Chapter 1. Weight fluctuations may cause mud to flow back, which increases the difficulty of operations and can cause a plugged bit.

Increased solids and deteriorating mud properties may cause mud weight increases. Pills or trip slugs due to frequent tripping also may cause weight increases. These may be controlled by decreasing the size of the slugs and increasing water additions. To ensure its accuracy, verify the mud balance.

Dilution causes decreases in mud weight, most commonly from water added at the surface or water entry into the well bore. This can cause weight material to settle out of the mud.

Cause and Prevention

Displacing mud is seldom a problem in cased holes except where there is channeling and contamination. However, problems in the open hole also include release of wall cake, mud losses, and contamination, all of which may cause sticking. Channeling can be very severe. In one case, 12-lb/gal water-base mud was displaced with 13-lb/gal oil mud inside of $9^5/_8$-in. casing set at 8,500 ft. Oil mud returns were first observed when the water-base mud lacked 150 bbls for full displacement, indicating channeling in a 3,000-ft annular section. Displacement can cause problems, such as three days' lost work time due to severe bit plugging because the mud was not screened correctly. Therefore, conduct operations in a prudent manner.

Always consider salvaging mud behind casing and liners. However, remember that salvaging is secondary, and the primary objective is to cement the casing or liner correctly. Water is the most economical displacement fluid for salvaging. Nevertheless, it may cause an excessive reduction of the hydrostatic head, which could lead to a blowout, cement contamination, etc. Other methods of salvaging include using a diverting tool in the casing string above the calculated top of the cement, and displacing with a lower quality mud or using scavenger cement.

Losses

Among the natural ways that mud is lost are as a coating on the cuttings, as filtrate, and through equipment leaks and minimal evaporation. Miscellaneous mud losses include pulling wet pipe, connection losses, not using a drain pan, accidental losses over shaker screens, and flow surges due to fast reaming-down on connections. Miscellaneous losses normally are a very small percentage of the total volume. They are more common at shallower depths, usually because of lower quality mud and higher make-up volume associated with a larger-sized hole. A special case of mud loss is covered in the section "Lost Circulation."

Losses of mud cannot be readily calculated, but they can be estimated from make-up volumes during normal drilling. Although it is not a true loss, make-up volume is included here. A rough rule of thumb for make-up volume is 5 bbls/100 ft drilled or 15 bbls/day for a shallow to medium-depth $8^3/_4$-in. hole in medium-drillable, normal-pressured formations.

Because the volume of mud in the circulating system is constantly changing, the magnitude of mud losses are not always readily apparent at the time they occur. A summary mud volume balance can be calculated for a section or for the entire well, as listed in Table 1–3.

Table 1-3 Drilling well mud volume balance

Liquid mud delivered to rig	1,450 bbls
Diesel additions	550
Water additions	170
Liquid equivalent of solid additions	50
Total volume of additions	2,220 bbls
Initial displacement losses	200 bbls
Casing shoe squeeze losses	290
Normal losses (15 bbls/day for 22 days)	330
Tripping losses	140
Calculated salvage loss	160
Salvage displacement loss	130
Salvaged (recovered) mud	970
Total losses and salvage	2,220 bbls

Normal losses include the make-up for a new hole. Calculated salvage loss is mud left in the hole for operational reasons during plugging. This and the normal losses and salvaged mud totaled 1,460 bbls, meaning that 65% of the total volume was utilized efficiently. The remaining losses of 760 bbls, or 34%, were theoretically wasted — ie., they represent mud that could have been saved.

Hole Cleaning

A clean hole is important to efficient operations. This normally is not a problem in cased holes, but may be a problem in open holes. A clean hole helps prevent sticking and facilitates most fishing. Drill cuttings and cavings should be flushed out of the hole conventionally with the circulating mud. Factors that strongly affect hole-cleaning efficiency include the type, physical condition, and chemical condition of the mud; the circulation rate; and the hole condition. Hole condition includes deviation and diameter, especially of washed-out sections (Fig 1–6).

Cause and Prevention

Figure 1-6 Viscous sweep mud action in a washed-out section

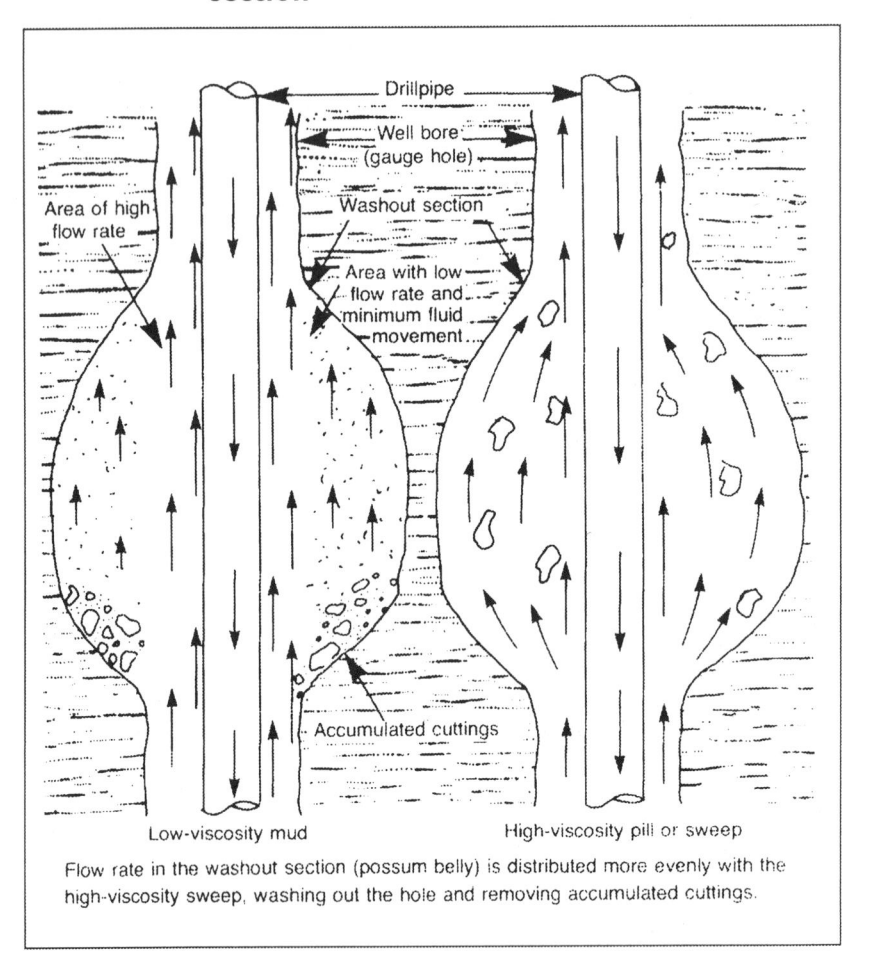

Flow rate in the washout section (possum belly) is distributed more evenly with the high-viscosity sweep, washing out the hole and removing accumulated cuttings.

Deviated holes are harder to clean compared to vertical holes. Mud solids tend to remain on the low side of the inclined wellbore; they roll and drag on the low side of the hole as they move upward. There is less fluid movement on the side of the wellbore compared to the center, so it takes longer to circulate and move the cuttings to the surface. For example, it takes about three times the circulating volume to clean a hole inclined at 45° compared to a vertical hole and even larger volumes at higher drift angles.

Figure 1-7   Drill cuttings slump

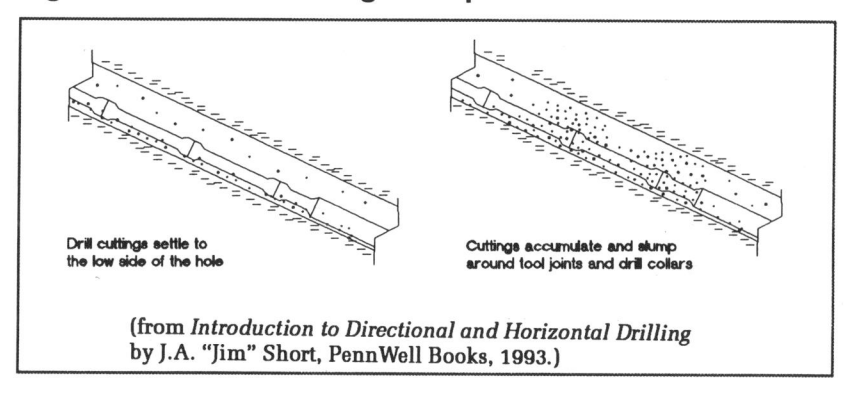

Drill cuttings settle to the low side of the hole

Cuttings accumulate and slump around tool joints and drill collars

(from *Introduction to Directional and Horizontal Drilling* by J.A. "Jim" Short, PennWell Books, 1993.)

As the cuttings move up the hole, the rolling, dragging action reduces their size. The rotating and reciprocating action of the drill tools grind and crush mud solids and drill cuttings against the wall of the wellbore, further reducing their size. This reduction increases the problem of solids separation and causes mud properties to deteriorate. Increased drilling time contributes to the problem. The action is more pronounced in deviated holes.

Increased circulating time causes hole erosion. Mud moves through enlarged hole sections or washouts at a slower rate, thereby increasing cuttings accumulation. Accumulated particles slump or slide downward due to gravity and can cause sticking or lost circulation (Fig. 1-7). The problem is more severe in high-angle and horizontal holes. Remove the cuttings with extended circulation, viscous sweeps, or in some cases, increased gel strength or higher density. Equivalent circulating density (ECD) aids in evaluating the circulating conditions in the wellbore. The importance of good circulation and large-capacity, efficient solids-removal equipment cannot be overemphasized.

Emulsifier-Wetting Agents

Emulsifier-wetting agents are important additives in oil mud systems. They should be used with careful judgment: excess usage increases mud costs, while restricted usage can lead to settling of solids and barite and may cause bridging and, possibly, stuck pipe. The general tendency is to add excess wetting agents because of the severe consequences of under-treatment. The problem is difficult to prove by definitive testing or data, as illustrated in the following example.

Cause and Prevention

Emulsifier-Wetting Agent Usage

Three wells were drilled in the same area with similar drilling, mud, and casing programs and were drilled to about the same depth in the same amount of time. Approximate emulsifier-wetting agent additions were as follows:

Table 1-4 Emulsifier-wetting agent additions

Well Number	A	B	C
gal/ton of barite	4.8	4.6	4.9
gal/ton of cuttings	5.4	8.2	11.5

Well A kicked and was killed with additions of barite and emulsifier wetting agents. The well was brought under control and drilled ahead without problems. Well B took a kick, and weight material was used to kill the well. Drilling was resumed without problems. Emulsifier-wetting agents were not added with the first part of the barite, but the mud had been pretreated a few days prior to the kick. Well C also kicked and was controlled with barite additions without the use of emulsifier-wetting agents. The pipe was found stuck after the kick, and a fishing job had to be performed. The mud reverted, requiring extra treatment and make-up. Water-wet barite was recovered during cleanout.

Barite

Barite is the standard mud-weighting material used in the drilling industry, usually supplied within American Petroleum Institute specifications. In rare cases it may contain excess impurities such as sand, calcite, or limestone. Sand is abrasive, and other impurities may contaminate some mud systems. Probably the main disadvantage of barite is using it in very high-weight mud systems, where it tends to cause a high concentration of solids. It is also susceptible to attrition, causing small-sized mud solids over long periods of time.

Other weight materials such as salt (NaCl), calcium chloride (CaCl), potassium chloride (KCl) and, to a lesser extent, magnesium chloride (MgCl) are used in special situations, such as with completion fluids. Dolomite, limestone, and iron ores have been used to a limited extent as weighting material.

Bridging

Barite may settle out of the mud and form a plug or bridge. The plug can be almost as hard and tough as cement. The action is similar to bridging with sand, cavings, or drill cuttings. Common causes are low mud viscosity and gel strength.

There also is a tendency for barite to settle out of mud at a mud/water interface, causing bridging. This can stick the pipe and packer, possibly resulting in a need for fishing. Water commonly is spearheaded ahead of cement, so sometimes the cement and mud should be separated with a spacer. A good spacer in many cases is gel, with water containing 20 lb of bentonite/bbl in place of plain water.

Barite Sag

Barite sag occurs when dispersed barite particles in the mud system settle and collect on the low side of the wellbore. It is not significant in vertical and low-angle directional holes, but it is more common in high-angle and horizontal holes where it may cause serious problems, especially with heavy-weight mud. Severity increases with increasing angle and higher mud weights.

Under certain conditions the semifluid barite slumps or slides down the wellbore and accumulates in a manner similar to that of cuttings. This increases or decreases the mud weight in localized areas. The barite concentrations redistribute in the mud when circulation resumes. The net result is areas of high and low mud density. The most common problem is lost circulation, but the slumping barite may cause sticking.

Barite sag often is not recognized. It can be detected by mud-weight variations while circulating after shutdowns when the mud column has not been moved for some time. Common shutdown situations include tripping, after running casing, while recording measurements, and similar operations. Variations in mud weight greater than 4 lb/gal have been noted.

Measures to prevent barite sag include maintaining good mud properties so the barite particles stay in suspension; minimizing noncirculating time; stopping periodically while tripping into the hole and circulating for short periods; stagging into the hole; and circulating at shorter intervals and increasing circulation time for more severe cases. Observe similar precautions when running casing.

Barite vs. Iron Ores

Iron ores are the only other materials that have been seriously considered as an alternative to barite. Hematite and illmenite are most commonly used. Some vendors supply a blend of iron ore and barite, usually containing from 30 to 75% iron ore. The primary reason for using iron ores is higher density and hardness as listed in Table 1–5.

Table 1-5 Weight material specifications[1]

Material	Principal Component	Specific Gravity	Hardness (Moh's scale)
Hematite	Fe2O3	4.9–6.5	5.5–6.5
Magnetite	Fe3O4	5.0–5.2	5.6–6.5
Iron Oxide[2]	Fe2O3	4.7	
Illmenite	FeO.TiO	4.5–5.1	5.6
Barite	BaSO4	4.2–4.5	2.5–3.5

1. Source: Gray, George R., and Darley, H.G.H., Composition and Properties of Oil Well Drilling Fluids, Fourth Edition, Gulf Publishing Co., Houston, Texas, Table II-1, p. 529.
2. Manufactured

Iron ores are more abrasive than barite. Excluding all other factors, an abrasive fluid drills faster than a nonabrasive fluid. However, the increase in speed is questionable from a practical viewpoint. Also, excessive abrasiveness definitely causes excessive wear on the pumps and circulating equipment. The main objection to iron ores is their abrasiveness, which has caused as much as a 20% increase in pump repair costs. Other reported problems include excess settling in mud tanks, a need for higher ultimate-yield mud to keep the ores (hematite) in suspension, excess shaker screen plugging, mud loss due to runover, difficulty in cleaning tool joints properly, and possibly associated tool joint wear that is excessive.

Claims that iron oxides increase penetration rates have not been substantiated in the field or in published literature. In most cases the reduced volume of solids in the mud is given as the main reason that penetration rates are increased. However, it is uncertain whether the reduction in solids overcomes the other objections.

LOST CIRCULATION

Lost circulation, a major problem in many drilling operations, is the loss of fluid from the borehole. The severity of lost circulation ranges from extra operating time to a blowout, or lost hole. It also can occur in combination with other conditions, causing additional problems. Lost circulation is divided into two broad classifications: the more common *natural* lost circulation and the less common(?) *induced* lost circulation.

Induced lost circulation is caused by operations. It is commonly caused by excessive mud weight and pressure surges fracturing or breaking down the formation. After initiating, a fracture propagates at a lower pressure, as described in the section "Cementing Casing Failures" in Chapter 7. Pressure may inadvertently be applied with the mud pump. Poor mud is probably a more common cause than expected, which emphasizes the importance of good rheology.

Other causes of induced lost circulation include starting the mud pump against closed pipe rams after a BOP safety drill, circulating out a kick incorrectly, and spudding the ball off a balled-up bit. Cuttings caving into the hole or bridging around the drilling assembly can cause a similar problem. Surging and swabbing, due to slacking off or picking up the drillpipe too fast, may cause alternating fluid movement in and out of the formation, thus reducing its stability, especially in fluid-sensitive shales. This action can also loosen lost circulation material (LCM) lodged in mud flow path(s), causing a reoccurrence of lost circulation. Equivalent circulating densities (ECD) and surge pressures usually reach a maximum effective value at or below the drill collars. The net result is that the entire openhole section is subject to induced lost circulation. These problems emphasize the importance of having experienced, alert personnel who observe proper precautions.

Natural lost circulation occurs in formations that take fluid naturally under normal drilling conditions. The most common cause is that the hydrostatic head exceeds the pore pressure. Operational procedures can cause or increase the severity of this problem. For example, drilling soft formations at a high rate of penetration increases the volume of cuttings suspended in the mud, which in turn increases the apparent weight of the mud and also the hydrostatic head. A similar problem occurs when cleaning out cuttings rapidly.

Natural lost circulation occurs in almost all formations and at all depths, especially with heavyweight mud. It is most common at shallower depths in sand, gravel, and other unconsolidated formations. The loss rate often is relatively constant, sometimes decreasing with time and with filter-cake buildup. Except for natural fracturing, these losses are often fairly easy to restore.

Lost circulation may depend upon the depth to the water table and the surface relief. Lost returns in the formations above the water table are relatively common. In some areas it is common to obtain mud returns to the surface around the rig, as described in the section "Leaking around the Surface" in Chapter 7. Mud returns may also occur at lower elevations. In one extreme case, when a well was drilled near the edge of a canyon, mud flow was observed in the talus slope at the base of the cliff, about 800 ft below the rig elevation.

Natural fractures and voids can be one of the most difficult losses to cure. They are more common in the older, harder, more consolidated rock, usually found at deeper depths and near faults and areas subjected to tectonic forces. Voids may occur as vugs in limestones and some dolomites. Fractures and voids frequently can be detected by torque and bit action.

Fracturing often is characterized by high rates of mud loss — sometimes reported as "the bottom fell out." The rate of loss may increase with continued drilling, sometimes after establishing circulation, indicating an exposure to additional fractures. Closed fracture systems may cause a quick mud loss of a few to several hundred barrels, followed by full returns.

General Procedures and Actions

Lost circulation is a main cause of fishing. While some lost circulation can be prevented, in many cases it can only be controlled. Nevertheless, good preventive actions and complete understanding of the causes of lost circulation can prevent or reduce most problems. Some general precautions are included in Table 1–6.

Table 1–6   Lost circulation considerations

- Allow for lost circulation in the drilling and casing program when information indicates it may occur.
- Drill the lost circulation zone, and below, with a minimum-weight mud with optimum flow properties.
- Drill lost circulation sections with gas, air, foam, aerated mud, or, if hole conditions permit, dry drilling. Be aware of the cuttings-lifting capacity of the different fluids and their inability to control pressure.
- Drill lost circulation zones and lower sections with a maximum-clearance drilling assembly to minimize the pressure on the formation.
- Lost circulation may result in the need for a fishing job, so drill with assemblies using the minimum number of tools such as reamers, stabilizers, and drillpipe rubbers to facilitate possible fishing.
- Run a drilling jar and bumper sub.
- Running an extra string of casing is expensive, but in some cases it is the most economical way to operate. At the least, an extra string of casing should be provided for if severe lost circulation is expected.
- Provide for extra storage to ensure a sufficient mud supply. Have an extra solids-settling pit when lost circulation material must be carried in the mud system for an extended period. Maintain an ample supply of water, mud-mixing material, and lost circulation material. Consider setting up a mud-mixing plant to reduce the time required by rig equipment. Do not overlook the need for sufficient supervisory personnel, including reliefs.
- Prevent plugging by removing the jets from the bit or run a core head type bit with a large opening. Open-ended drillpipe may be used, but the core head is better in case cleaning out is necessary.

- Wellbores frequently are very sensitive to pressure changes after establishing circulation and may start losing mud again. Minimize pressure surges and increase mud weight slowly if necessary.
- Lost circulation is more common at the bit, but induced lost circulation may occur at any weak point in the borehole.
- In general, treat lost circulation, and most other formation problems, from the top down. Work with the assembly lower in the hole when kicks and water flows occur with lost circulation. The location of the assembly with respect to kicks and the safety of the hole take precedence over other problems. Otherwise, work on lost circulation with the drilling assembly either higher in the hole or with the bit near the point of lost circulation, depending upon well and loss conditions.
- Allow the hole to cure by resting without circulation if applicable. Do not overlook this important action.
- Severe lost circulation may require cementing. Select a mud system that resists contamination and can be treated satisfactorily.
- Report lost circulation material (LCM) additions in lb/bbl of mud in the system.
- Report LCM concentration in the mud in percent by volume. Measure by catching 1,000 cc of mud and straining out the LCM, damp dry it, and return it to the cup. Read the percentage from the calibration on the cup. For example, 50 cc of material in the cup is 5% by volume of LCM in the mud. A rough conversion is 1% by volume equals 2.5 lb/bbl based on LCM density of 45 lb/cu ft.

Point Of Loss

Lost circulation frequently occurs at the bit. Otherwise, the point where the loss occurs in the borehole may not be known. This information may be needed in order to seal the point efficiently, such as spotting LCM pills. The point of loss may be located with regular and differential temperature surveys, which are most commonly run inside of drillpipe. This is a safety procedure both to prevent sticking the small tools and to control the well better if high-pressure zones are exposed in the wellbore.

Spinner surveys are less common and may require large volumes of mud. In some cases, the survey instruments can be run through and out the drillpipe, with the end of the pipe near the estimated point of loss. Some tools, such as spinner surveys, may be affected by the amount of LCM in the mud system.

Cause and Prevention

Dry Drilling

Dry or blind drilling is the procedure of drilling ahead without mud returns at the surface. There is a moderate to high risk depending upon the situation. The drilling assembly should remain free to prevent sticking. Cases in which 50–100 ft of new hole have been dry drilled are relatively common, and dry drilling of over 1,000 ft has been reported.

Fluid should be pumped down the drillpipe at 20–50% of the normal rate while drilling ahead slowly. The fluid passes through the bit, picks up the cuttings, and carries them into the lost circulation zone. The cuttings either pass into the lost circulation zone or stack up in the borehole, depending upon relative sizes and flow rates. In the latter case, there is a high risk of sticking the drilling assembly. This can be prevented by drilling ahead about 5 to 20 ft, picking up to ensure the assembly is free, then lowering the bit, checking for fill-up, and resuming drilling or redrilling the cuttings. Redrilling the cuttings reduces their size so that they enter the lost circulation zone easier. The cuttings often seal the lost circulation zone efficiently.

It is not uncommon to regain circulation while dry drilling. To ensure that the drilling assembly is free, continue alternately drilling and picking up. Observe other guides to sticking such as pipe weight, drag, and rotary torque. Normally dry drilling should be done with water. It may be necessary to build mud volume in case of partial circulation.

Dry drilling can be highly efficient for high-rate loss situations. It may also allow continued drilling through the entire lost circulation zone so the complete interval can be treated. In severe cases it may eliminate the need of setting an extra string of casing.

Remedial Actions

When lost circulation first occurs, the problem should be analyzed in order to determine a course of action The first action may be readily apparent; if so, initiate it immediately. In most cases the pump rate should be reduced to reduce the loss. Prudent operators drill with a minimum-weight mud but reduce it if conditions permit. There is a high risk of caving and sticking during lost circulation so, as the first step, prevent sticking by pulling the bit up at least 50 ft or preferably above any formations that caused problems in the past. This allows the drilling assembly to be worked both up and down if sticking occurs and it also improves the chance of releasing pipe stuck on bottom. Then work the pipe down the hole, reaming and cleaning out as needed, after the hole stabilizes. If conditions permit, pull the bit inside the last string of casing.

Reducing either actual or apparent mud weight can have a strong effect on lost circulation. Consider a well that is losing 11 lb/gal mud at a constant rate in a zone at 10,000 ft with a pore pressure gradient of 0.46 psi/ft and a differential pressure of 1,120 psi. Lowering the mud weight 9% to 10 lb/gal decreases the differential pressure 46% to 600 psi and reduces the rate of fluid loss by a similar amount. Therefore, the circulation loss can be reduced substantially with a small weight reduction, even with a relatively low mud weight.

In some cases, increasing viscosity a small amount may help reduce lost circulation. Flow rate is approximately inversely proportional to viscosity. A fluid with a viscosity of 2 cp flows at about one-half the rate of a mud with a viscosity of 1 cp under the same conditions. Higher viscosity mud may help regain circulation in a few cases if it does not cause an increase of apparent mud weight.

Low-Rate Lost Circulation

Low-rate lost circulation is defined as a mud loss of 10–15 bbl/hr or less. It often occurs while drilling and often declines with time. Low rate lost circulation generally is easy to cure, and often can be corrected while conducting operations.

To determine if the loss occurs without drilling, first pick up the bit 10–20 ft and shut down the pump. Reduced loss or no loss while circulating indicates the new hole should be sealed off. Assuming no hole problems, drilling and circulating should be continued while observing the rate of loss. Continue drilling if the loss is sustainable. Frequently it is better to build more mud in inexpensive mud systems than to use LCM. Drill cuttings and mud solids are moderately efficient sealing agents under many conditions.

If the loss is too high or is increasing, add small volumes of smaller particle size LCM, in the range of 2–3 lb/bbl and of fine or medium size — frequently half and half, or mica, paper, or fibrous material. This treatment normally cures minor cases of lost circulation. For higher losses, LCM additions should be increased. A total concentration of 6–8 lb/bbl with a wider range of particle sizes, including coarse material, normally seals off most low-rate lost circulation. LCM may affect some measuring instruments, such as Measurement While Drilling (MWD) equipment in directional and horizontal holes, but most tools will tolerate a limited amount of LCM.

Moderate-Rate Lost Circulation

Moderate-rate lost circulation is defined as a loss in excess of about 15 bbl/hr but not a complete loss of circulation. There is a moderate risk of sticking the drilling assembly due to caving formations, so operators should pick it up a safe distance and check the loss as described earlier. If necessary the hole should be filled.

First, LCM should be added, as described for low-rate lost circulation. Use higher concentrations of up to about 10 lb/bbl, with a higher percentage of larger particle sizes for higher rates of loss. There is a high risk of plugging the jet bit nozzles at LCM concentrations above about 15 lb/bbl. It may be necessary to make a trip and change the bit for one with larger ports, or to remove the jets or run a corehead-type bit.

Sometimes adding LCM should be combined with healing, especially if the mud contains fresh bentonite. The hole may heal or stop losing mud if circulation is stopped or reduced to a very low rate. This allows mud in the lost circulation zone to be stationary so it gains additional gel strength. This healing process increases the resistance to flow and helps reduce the loss into the formation. Normal healing times range from 3–8 hours, depending upon the mud properties. Healing the hole takes time and requires patience. Rotate or move the pipe periodically to prevent sticking — possibly picking it up into the last string of casing.

If the foregoing is unsuccessful, LCM pills should be spotted. The best procedure is to spot the slurry through a core head or open drillpipe. If the pipe cannot be pulled because of hole conditions, then the jets should be blown out of the bit with a string shot. They are not difficult to drill out later with a short-tooth insert bit. Set the bottom of the pipe opposite the point of loss, after first cleaning it out if necessary. Use a low circulation rate to prevent additional mud loss. Pick the pipe up periodically to ensure that cuttings are not accumulating and bridging around it. If conditions permit, consider drilling a small amount of hole to ensure that the entire lost circulation zone has been penetrated. For very severe lost circulation where there is danger of the hole caving, spot the LCM pill with the pipe at a safe distance above the lost circulation zone. Generally, the LCM pill will fall down the hole into the lost circulation zone.

An LCM pill containing about 40 lb/bbl of LCM material should be mixed and pumped in 50–100 bbl of mud. The average pill contains about 20% fine, 40% medium, and 40% large-particle-size LCM. Larger sizes act as a bridge and smaller particles act as a sealer. The pill should weigh a few pounds less than the mud. Add bentonite just before pumping the pill. Spot the pill rapidly so the bentonite will seep into the lost circulation zone before it completes hydrating. This can be very helpful when used with healing.

Commonly the pill should be spotted using the balanced columns technique, displacing about half of the pill into the formation or annulus and an equal height column into the drillpipe. If there is any question on displacement volumes, the pill should be underdisplaced. Then pull the pipe up a safe distance, shut down, and wait for the hole to heal. Pumping small volumes of clean mud into the casing annulus places a low pressure on the LCM pill and possibly moves it a short distance into the formation. Water may be used, but it is better to add about 20 lb/bbl of bentonite to the water.

Wait about two more hours and if the hole is not full, fill it slowly. Then begin pumping slowly down the drillpipe. If the hole circulates, then add LCM to the mud as necessary and continue circulating until there is full circulation. Resume operations, generally cleaning out to the bottom first. Take all precautions to prevent the lost circulation from reoccurring.

If the hole does not circulate but stands full, wait for additional healing. If it still does not circulate, another LCM pill should be spotted. The second LCM pill is mixed with higher concentrations of LCM to as much as 75 lb/bbl for very severe lost circulation and a larger or smaller volume pill depending upon the rate of loss and how successful the first pill was. Rig mud pumps usually begin missing at concentrations of 50–60 lb/bbl, due to plugging screens and sticking valves. Pump and displace the pill in a manner similar to that for the first pill and wait for the hole to heal. Procedures can be modified as indicated by results of the first pill.

It frequently takes two or more pills to regain circulation. Once it is regained, the hole should be allowed to heal for a few more hours to help ensure that lost circulation does not occur again after operations are resumed. The lost circulation material should usually be left in the mud system for at least 12–24 hours after regaining circulation. This allows time to determine whether the lost circulation has healed and may also help prevent losses in lower zones after starting drilling. One recommended procedure is to leave the LCM in the mud for about 8–12 hours, then screen out half of it by alternately flowing over and bypassing the shaker at even time intervals for one circulation. Screen out the remainder several hours later.

If circulation has not been reestablished after spotting about three LCM pills, then the procedures described for high-rate lost circulation should be considered.

High-Rate Lost Circulation

High-rate lost circulation is defined as a complete loss of returns. The hole may stand full or the mud level may drop out of sight. Lost circulation rates at well over 500 bbl/hr have been recorded.

Under these conditions there is a moderately high risk of an aboveground or underground blowout from even moderate pressure formations as fluid levels in the hole drop. There is also a high risk of sticking, so all precautions should be taken to prevent it. The best method of doing this is to pull the pipe into the casing or, if possible, at least above any zones that may cave.

If the fluid level drops, it is important to know how far it is downhole. Under some conditions, the weight indicator may give the depth to the top of the fluid. For example, consider drilling at 10,000 ft with 900 ft of 6¾ x 2¼-in. 109 lb/ft drill collars and 9,100 ft of 4½-in. 16.60 lb/ft drillpipe. The weight of the drilling assembly in air is 249,000 lb, and 208,000 lb in 11 lb/gal mud, with a buoyancy factor of 0.8350. If reservoir pore pressure were 0.434 psi/ft, and complete loss of returns balanced this pressure, the fluid level would drop to 2,413 ft below the surface. At this depth, the drilling assembly would give a hook load of 214,000 lb, compared to 208,000 lb in a hole full of mud. The difference in the two drilling assembly weights is the lack of buoyancy acting on 2,413 ft of pipe. Unfortunately most weight indicators are not sufficiently sensitive to be reliable in this case.

The most common method is to load the annulus with water. Calculate the length of the water column based upon the volume of water pumped and assuming that none of the water is lost. The top of the mud helps determine how severe the lost circulation is and what procedures are necessary to regain circulation. Having the hole full of water, or mud, also serves as a safety precaution because it will show if the formations are starting to flow and will aid in controlling a possible kick situation should one develop.

Try regaining circulation by dry drilling if possible. Another method is to use LCM pills with larger-particle-size LCM. Field experience indicates that large fractures may absorb extremely large volumes of fluid and be very difficult to seal with LCM. Ultimately, large fractures may be sealed by filling them either with a solid material or with a material having an extremely high gel strength so it will not continue to flow through the fracture. A note of caution: There may be some claims about the effectiveness of certain LCM materials that are difficult to support.

Materials that can be used for sealing lost circulation include cement, polymers, gels, and diesel oil containing cement or bentonite. They all serve the same purpose, and placement techniques generally are similar. Polymer and gels usually are mixed as a thin slurry that thickens after displacement into the lost circulation zone. Diesel oil slurries of cement or bentonite (called gunk squeeze) keep the additives inert until the slurry is in the formation. When the cement or bentonite contacts water in the formation, the dry cement forms a slurry that hardens, and the bentonite hydrates to form a highly viscous mud.

The use of these materials involves some risk of sticking the drilling assembly. In all cases, it is important to ensure that the material does not circulate around the drill collars and drillpipe. Always pull the pipe out of the mixture before it begins to harden or set up.

Cementing Lost Circulation

Cementing is a positive approach to curing lost circulation. Normally, it is a last resort procedure. Cementing is widely applicable;

lost circulation zones have been successfully cemented at depths of 18,000 ft with mud weights in excess of 15 lb/gal and bottomhole temperatures around 300° F.

Some operators hesitate to cement lost circulation, probably because of unfamiliarity and because it is a time-consuming, higher-risk operation. The risk can be minimized by cementing in a prudent manner with the correct assembly. Cement contaminates some muds, but this can be handled by pretreatment, dilution, replacement, or discarding.

The tendency of some to underestimate the effectiveness of cementing to seal off lost circulation is unfortunate, since cementing frequently is the most convenient and efficient solution. The next reasonable alternative usually is to run a liner or casing. This is seldom a good solution because of the risk, higher cost, and the need for changing the well plan (such as reducing the hole size), which may restrict deeper drilling. Also, casing may not run to the bottom, so lost circulation still occurs. However, casing may eliminate uphole problems so that deeper zones can be cemented.

Cement with a good slurry of Class A cement. Various other cements, such as thixiotropic cements, have been used in special situations. Generally, use the procedures for mixing and displacing the slurry described in the section "Squeeze Cementing" in Chapter 7. Always consider using accelerators, but do not use retarders unless absolutely necessary. Design the slurry for a minimum thickening time, and add about 10 lb/bbl of LCM. Use normal to minimum slurry weight and possibly lighter weight bentonite slurry for longer cement columns.

Normally cementing should be done through open-ended pipe, a core-head-type bit, or a washover shoe with the end of the pipe at or below the lost circulation zone. Mix the slurry and pump it into the pipe. About half of the slurry should be displaced out the bit with mud, or in severe cases, gel water. The pipe should be picked up a safe distance, allowing the rest of the slurry to fall out of the pipe. For very severe lost circulation and if there is a risk of sticking, set the end of the pipe a safe distance above the lost circulation zone. Displace the slurry with a sufficient volume to ensure clearing the pipe.

If the hole stands full or begins circulating while displacing the cement plug, then pick up the drillpipe and wait a short time before adding LCM and beginning to clean out. Sometimes close the pipe rams and apply a small amount of pressure — usually an equivalent mud weight increase of about ¼-lb/gal. Gradually increase the pressure at 10–20 minute intervals until the pressure is the equivalent of the normal mud weight.

If the hole does not stand full, wait for the equivalent of 1½ thickening times and begin loading the annulus slowly. If it holds, add LCM to the mud system and start circulating slowly, increasing the

Cause and Prevention

rate slowly as the hole stabilizes. Then drill out the cement carefully. DO NOT RUN THE BIT INTO GREEN CEMENT. If the lost circulation is cured, drill ahead. Otherwise cement again, adjusting the procedures and slurry volumes as indicated by prior cementing results.

It is not uncommon to cement very severe lost circulation several times, especially wide fractures. All fractures and other large voids or conduits should be cemented at or below the fracture propagation pressure or else the fracture will continue to grow. Cement these by staging. In one case lost circulation was so severe that staging was actually started during displacement by adding small increments of displacing fluid down the pipe behind the slurry.

Placing slurry in large fractures and vugs and in the wellbore one time, then drilling out will theoretically stop lost circulation. However, from a practical viewpoint, this is seldom the case. Field experience indicates that complete sealing in one stage is highly unlikely. One procedure is multi-stage cementing. A modification is to spot the slurry, pick up and wait until it reaches an initial set, then apply moderate pressure, either by filling the hole or closing the pipe rams, which usually restores the lost circulation. The cementing procedure should be repeated until a satisfactory pressure is obtained. In one case, a 4-ft mine shaft was cemented successfully in this manner.

Cementing lost circulation often is effective, as described in the following example.

Curing High-Rate Lost Circulation

A 6,000-ft development well was drilled in an area where there was a high risk of lost circulation. Surface casing was set at 1,500 ft and an 8¾-in. hole was drilled below it with native mud weighing 9.5 lb/gal.

A fracture was drilled at 2,487 ft, which caused a partial loss of returns. It was sealed with a 100-bbl pill, containing about 15-lb/bbl of LCM. A small fracture at 2,508 ft caused partial loss returns, and complete returns were lost at 2,514 ft. The hole was dry drilled to 2,631 ft and a 250-bbl, 20-lb/bbl LCM pill was mixed and spotted. The drillpipe was picked up into the surface casing during a 4-hour wait for the hole to heal. The hole could not be filled, so pipe was run back to the bottom and the hole was cleaned out by dry drilling to 2,615 ft. A second 250-bbl LCM pill was ineffective.

A 150-sack cement plug was spotted at 2,615 ft, the pipe was pulled up to 1,960 ft, and a 100-sack plug spotted. Both used Class G cement with 4% gel and 2% CaCl. After a wait on the cement, the hole was cleaned out to total depth with full circulation.

Circulation was lost at 2,640 ft and regained with LCM. It was lost again at 2,710 ft, regained with LCM, and lost again at 2,730 ft. The hole was dry drilled to 2,800 ft. A 150-sack cement plug (no. 3) was set at 2,800 ft, the pipe was picked up, and a 100-sack cement plug (no. 4) was spotted at 2,150 ft.

After a wait on the cement, the hole was cleaned out with good returns until it was drilled to 2,740 ft, where all circulation was again lost. The hole was cleaned out to total depth and dry drilled to 2,856 ft. A 100-sack plug was set at 2,850 ft (no.5), the pipe was picked up, and a 75-sack plug (no. 6) was set to 2,470 ft. About 10 bbl of mud were pumped in the hole 20 minutes after spotting the last plug.

After a 6-hour wait on the cement, the hole was loaded with mud. The first cement bridge was found at 2,490 ft. Various bridges and stringers were cleaned out to 2,760 ft, at which point circulation was lost again. The hole was cleaned out to 2,805 ft by dry drilling. A 50-sack plug (no. 7) of thixiotropic cement with 3% CaCl, 10 lb of Gilsonite/sack, and ¼ lb of cellophane flakes/sack was spotted at 2,800 ft. The pipe was picked up to 2,520 ft, and a 125-sack cement plug (no. 8) was spotted. The hole was loaded from the top and held fluid.

After another wait on the cement, the hole was cleaned out to total depth with full circulation. It was circulated and conditioned while adding lost circulation material, then drilled ahead with full circulation. No additional lost circulation problems were encountered and the well was completed satisfactorily.

Overall, this was a good operation to plug off a severe, long lost circulation section. Lost circulation pills were tried until they were found unsuccessful. At that time, the method was changed from plugging to cementing, and later to cementing with double plugs. The cement plugs were squeezed by dumping water on the top of the hole, and later were tested by loading the hole from the top — a good procedure.

A number of wells were drilled in the area by various operators. Different procedures were tried, including using an extra string of casing. However, the most economical procedure was to use lost circulation pills, which were frequently successful, or otherwise followed by cementing.

Summaries of Cementing Lost Circulation

A summary of several examples of curing lost circulation is justified because of the severity of lost circulation.

- A well was designed for an extra string of casing. It lost circulation at 3,600 ft. About 100 ft were drilled in 10 days because of water and gas flows and caving. The 8¾-in. hole was opened to

Cause and Prevention

9⅞-in. in 10 days and dry drilled 3,950 ft in another 3 days, including waiting on water. Intermediate 8⅝-in. flush-joint casing was run, was stopped at 2,500 ft, and was washed down to 3,850 ft in 48 hours. It was cemented without returns, using lightweight cement followed by regular cement. After drilling out, there was about 10 ft of cement below the casing shoe, showing there were no lost circulation zones in the lower open hole. Then the hole was drilled and completed in a normal manner.

- A well had 7-in. casing set at 7,800 ft and was drilling a 6¼-in. hole at 9,000 ft with 10.5-lb/gal mud through a massive limestone section. The bit drilled into a cavern or fissure and dropped 9 ft before taking weight, and all circulation was lost. A minor weight loss indicates a relatively high fluid level.

Mud weight was reduced until about 25% of returns and a very small amount of drill cuttings recovery were obtained with about 9.3 lb/gal. After about 100 ft was drilled with partial returns, mud weight was reduced to 8.9 lb/gal without increasing returns. A zone about 100 ft below the first zone also was suspected of taking fluid. A higher mud weight was needed to control a deeper productive zone.

Two single-stage cement plugs of 200 and 250 sacks were set, followed by two gunk squeezes and three single-stage cement plugs in about 12 days without shutting off the lost circulation. Multiple-stage cementing was used to place one 250-sack plug, three 600-sack plugs, and three 300-sack plugs in three days. A final plug of 300 sacks sealed the lost circulation. The well was drilled to 14,000 ft with 12-lb/gal mud and completed as a gas well without further problems.

- A 6,000-ft development well had 9⅝-in. casing set at 340 ft. Lost circulation at 865 ft was cured with LCM. At 1,770 ft complete lost circulation occurred in a 10-ft fractured interval. Past experience indicated that LCM would not be effective. Eight cement plugs of 50–100 sacks each were set. While cleaning out at 1,720 ft, circulation was lost again. Two pills of LCM were not effective. Five more plugs of 100 sacks each were set, including some with thixiotropic cement. The well was dry drilled to 1,770 ft and four additional cement plugs were spotted, which cured most of the lost circulation. About 20-lb/bbl of LCM were added and the well was drilled to TD with only minor lost circulation.

Intermediate 7 in casing was not set because of the strong possibility of deeper lost circulation. The 50-sack plugs were probably too small. All plugs except the last were mixed with Class G cement, with 12% Calcium, 3% CaCl, and a 10-lb/sack of Gilsonite. Apparently, accelerated common cement would have been as effective.

FORMATIONS

One of the most important preventive measures is understanding formations and how they affect drilling, fishing, and other operations. The most common formations are shales, sand, sandstone, limestone, dolomite, salt, and various miscellaneous formations. They range from thin stringers to massive sections and often occur in combination, such as sandy shale, shaley lime, or limy sand. Problems caused by or related to one formation may occur in another.

Shales are especially important because they comprise over 50% of the formations drilled and probably cause a correspondingly higher percentage of hole problems and fishing jobs. Shale chemistry is complex and physical states vary, including massive, laminated, fractured, and unconsolidated with varying degrees of hardness.

Various formation conditions can affect the mud. High temperature formations usually occur at deeper depths due to the normal temperature gradient. A few areas have higher gradients so high temperatures occur at shallower depths. Some mud-treatment chemicals break down at higher temperatures, so those which are least affected should be chosen. High temperatures can affect the electrical components in electric logs and directional measuring instruments. Insulating tools are available for logging under high-temperature conditions.

Most formations contain oil, gas, saltwater, or a mixture of these. Some contain carbon dioxide (CO_2) or hydrogen sulfide (H_2S), which corrode tubulars and drilling tools. CO_2 may cause suffocation. H_2S causes both corrosion and embrittlement, and IS EXTREMELY POISONOUS, even in very low concentrations. Precautions for working with H_2S should include warning systems, masks, portable air packs, and good training.

Influence on Hole Direction

Formations may influence hole direction. Massive, flat, soft-drillability formations have little effect; gravity is the controlling factor. Formation influence on hole direction increases with increasing hardness or drillability — i.e., the number of fractures, striations, laminations and bedding planes, and alternate layering of hard and soft formations.

In relatively flat formations, a bit tends to drill in a slow right-hand spiral, clockwise looking downward. This turns the hole about 5–20 horizontal degrees per thousand vertical feet drilled in medium-drillability formations. In dipping formations, the bit tends to drill in the up-dip direction. The effect increases with increasing formation dip up to about 45° relative to the axes of the wellbore. There is some evidence that the effect reaches a maximum at about 20°. The bit tends to drill down-dip at relative angles over 45°. It drills in a curved direction if the hole direction is parallel to the strike of the formation. Less bit weight and higher rotary speed will reduce the formation effect. Otherwise, there is little if any evidence that the type of bit strongly changes general formation effects.

Problems

Formations directly or indirectly cause many problems, which lead to fishing jobs and further problems during fishing. To prevent sticking and/or adverse effects on fishing, drill and control them properly, using correct procedures. A straight hole eliminates most crooked-hole problems. It may be necessary to run casing. Take special precautions in deviated holes.

Surface Formations

Surface formation problems include lost circulation, caving, sloughing, and cleaning out gravel. These problems can be corrected by using heavier mud, higher viscosity, dry drilling, and/or setting large conductor. Drilling on boulders may cause a crooked hole. Drill the conductor hole with a rat-hole digger, or dig it with a backhoe and set a large-diameter conductor as described. This also allows additional drilling weight to drill deeper boulders more efficiently.

Very hard surface formations are difficult to drill because of the lack of bit weight. Drilling a pilot hole may be applicable. Sometimes, an operator can take advantage of the surface terrain, as illustrated in the following example.

Spud In on Hillside

Part of the Appalachian Basin has a layer of well-cemented sandstone exposed on the valley bottoms. Spudding into the sandstone frequently requires extra time drilling the first 50–100 ft, due to the lack of drilling weight. When possible, the location should be built on the hillside several hundred feet above the valley bottom. These softer formations drill relatively fast with a minimum bit weight. When the bit reaches the hard sandstone layer, there is enough drill-collar length and weight to drill the harder sandstone layer efficiently. The cost of the extra hole and the extra cost to build the location on the hillside is more than offset by the savings in rig time.

Shallow Gas

Gas under normal to moderately high pressure may occur at shallow depths. Therefore, there may be very little warning or reaction time when a kick occurs. The normal flow line increase or pit gain usually is followed immediately by a gas flow. Always use a diverter in these situations.

The danger is illustrated in the following example.

Shallow Gas

A development well was being drilled in an area known to have shallow gas. The drilling program included 9⅝-in. conductor at 60 ft, 7-in. surface casing in an 8¾-in. hole at 2,500 ft, and a 6¼-in. hole to 9,500 ft. Gas normally occurred in the interval from 900–1,600 ft. The worst problem encountered in about 25 wells drilled in the area was a small kick that flowed mud about 5 ft above the derrick floor before depleting.

The conductor was set and the flow line welded to the top of the conductor in the normal manner. The well blew out while drilling at 1,400 ft with 9-lb/gal native mud. All of the mud was blown from the hole and the gas column ignited. The heat weakened the legs of the mast and it collapsed. The gas flow subsided, the fire went out, and the hole filled with water from shallow water sands in an elapsed time of about 25 minutes. No injuries occurred, but the mast was a complete loss and the remainder of the rig required extensive repairs.

The problem could have been eliminated with a diverter, such as a rotating head. A conventional preventer could not have been used efficiently, except possibly with an open blooie line connected below the preventer. Otherwise, a closed preventer would probably have been blown out of the hole, since there was an insufficient length of conductor to anchor it.

Swelling Clays

Swelling clays are unconsolidated clay-type formations that frequently occur near shorelines such as the Texas-Louisiana Gulf Coast. Most swelling clays are drilled rapidly. The drill cuttings have rounded edges and are soft, often dispersing easily into the mud system causing extra treatment. Balled bits are common. Cuttings may overload the mud system and cause an apparent mud-weight increase. Maximize the circulation rate and reduce the drilling rate if necessary to keep the hole clean. Inert mud systems such as saltwater or oil mud have been very successful in removing the clays.

Fluid-Sensitive Shales

Fluid-sensitive, swelling, disintegrating, bentonitic, and mud-making formations with a high montmorillonite content are all similar classifications. Most shales are sensitive to water-base fluids to some extent, but fluid-sensitive shales can be highly reactive. Cuttings range from small and well rounded to harder shales with fracture and cleavage planes and more pronounced edges.

Fluid-sensitive shales cause deteriorating mud properties and increased treating costs. In severe cases use low fluid loss, inhibited, or isolating polymer-type mud systems. Drilling problems can including hole cleaning, lost circulation, sticking due to sloughing and caving, out-of-gauge holes, and bridging. Use standard preventive actions and careful operations.

Selecting the correct mud is critical. Severe problems with massive sections along the Texas-Louisiana Gulf Coast using common mud systems have been reduced or eliminated by using oil emulsion mud with super-saturated saltwater in the water phase.

Fluid-sensitive shale problems may be time related, as illustrated in the following example.

Pencil Cave Shale

The Pencil Cave shale overlying the Big Lime (Mississippi limestone) is widespread in the Appalachian Basin. It commonly causes problems about 24 hours after drilling with water-base fluid and continues caving and sloughing for several days thereafter.

In earlier drilling, the Pencil Cave shale was dump cemented. Later it was drilled with air with less difficulty, but it did cave due to natural fracturing. Logging tools set down on bridges in the open, air-filled hole. This was corrected by running a tapered rubber extension on the logging tool, which helped move the bottom of the tool off the ledge. Casing caused problems that were eliminated by running the casing in the air-filled hole. The hole was loaded with mud and cemented, minimizing the time the formation was exposed to drilling fluid.

The poker chip shale in Oklahoma acts similarly.

Geopressured Shales

Geopressured shales, or popping shales, are formations with a combination of pore pressure and physical characteristics that cause fragments to spall or pop off into the wellbore. It is most common in shales, but limited field evidence indicates that it may occur in very fine-grained, well-cemented sands and other formations with very low permeability.

With geopressured shales, penetration rates frequently increase and the mud may be gas cut. Drill cuttings vary in size, but they have a distinctive long, tabular shape with relatively sharp edges and occasional concoidal fracture faces. Often there is a pronounced increase in cuttings volume. *Note of caution:* These cutting are similar to those recovered while drilling pressure transition zones, so always be aware of the potential of a deeper, high-pressure zone.

The spalling action normally is attributed to a pore pressure that is greater than the combined structural strength holding the fragment in place and the differential pressure between the hydrostatic mud pressure and the pore pressure acting on the equivalent of the cross-sectional area of the fragment. Shale has very low permeability, so pressures cannot equalize through the fragment and are contained by the surface of the particle and the contact surfaces between the particles. The particle is released and moves into the wellbore with an assumed popping action — hence the term "popping shales."

Excluding natural fracturing, this explanation leaves some question, since many formations support higher-pressure differentials into the wellbore without spalling. Air-drilled holes through normally pressured shales at 10,000 ft have a pressure differential into the wellbore of over 4,000 psi without appreciable spalling. A gas-drilled hole through a geopressured (0.55 psi/ft) sand shale sequence at 9,000 ft had a differential pressure into the wellbore of 4,500 psi without any spalling.

Preventive procedures include recognizing the problem early, and taking precautions to prevent sticking while continuing drilling. Spalling generally decreases with time. For more severe cases, an operator should clean out, pick up and wait for the shales to spall, then clean out again. Repeat the procedure until the spalling slows. Increase the mud weight in small increments of 0.2–0.3 lb/gal if continued spalling prevents drilling, then circulate and clean out to determine if the weight increase is effective. Theoretically, the only way to cure the problem without casing is to increase the mud weight until it approaches the pore pressure; however, it seldom is necessary to increase the mud weight this high. The final alternative is to run casing.

Fractured Formations

Fractured formations are naturally fractured. Calcite and other evaporates commonly occur in the fractures and may be found free in the cuttings or adhering to the fracture face. Fracture widths range from microfractures to very large fractures that can cause stuck bits. Fracture orientation ranges from well ordered to random.

Fractured formation drill cuttings sizes range from very small (similar to drill cuttings) to large blocks or tabular pieces, with sharp, angular edges and corners. Some are so large that they must be broken up by drilling before they can be circulated out of the hole.

Problems associated with fractured formations include caving and potential sticking, out-of-gauge holes, ledges and bridges, and high fluctuating torque with the risk of twisting off or sidetracking. Treat these problems similarly to those with geopressured shales.

Other, less common procedures include increasing the mud weight, dumping cement, and running casing as a last resort. Penetrating severe fractured formations is illustrated in the following example.

Fractured Shales

A 21,000-ft wildcat well was drilling an 8½-in. hole at 18,500 ft with a packed-hole drilling assembly and a premium-grade, sealed-bearing button bit that had an average life of 40 hours at 2.5 ft/hr. Hole deviation was 4° at the last casing shoe at 14,000 ft. Formation dip was 20° and expected to be 30° at total depth. The formation was a naturally fractured, extremely hard and abrasive sandstone, containing chert and pyrites. Hole sections from 15,000 to 20,000 ft and from 16,500 to 17,000 ft had severe caving problems and required lengthy cleanout before stabilizing. It was common to clean out 200–300 ft after a trip.

The hole was sidetracked while cleaning out a bridge at 16,600 ft. Repeated efforts to enter the old hole were unsuccessful. This hole was drilled to 16,750 ft, where a survey showed a 5° deviation, further verifying the new hole. The old hole had an 8° deviation at this depth. While running back in the hole, a bridge was encountered at 16,600 ft. While cleaning this out, the bit entered the original hole, going to 17,800 ft with some reaming. Entering the old hole after accidentally sidetracking is an infrequent problem caused by incorrect operations.

It was decided that in subsequent trips the bit would most likely reenter the new hole. The new hole had a lower angle of deviation, probably by sidetracking the original hole on the low side; therefore the original hole needed cementing, even if in this case the cement actually was softer than the formation. This procedure required abandoning over 1,000 ft of original hole, but it represented the least risk of running the 7-in. intermediate drilling liner into the wrong hole.

The hole was cemented after cleaning out caving and bridges several times. A packed-hole assembly with the bottom two reamers as stabilizers was run and the cement was cleaned out with a high rotary speed and low bit weight. This further increased the stiffness of the assembly, helping to ensure that it entered the new hole. After cleaning out, the assembly was picked up, and it encountered caving and a minor bridge running back to bottom. The sidetrack section was reamed to straighten the new hole as much as possible, reduce the dogleg, and ensure entry and continued operations in the new hole. The hole was cemented again and cleaned out with the same procedure used earlier, including reaming with a stiffer assembly. Running this assembly represented some risks, especially in view of the caving formations. However, at this point it was necessary to ensure that one hole was open to run the pipe through and that the hole would stay open for running casing or a liner.

The hole was drilled ahead, maintaining the deviation at 4–6°. Fracture zones were drilled, including some bridging. At 20,000 ft, a 7-in. intermediate liner was run without difficulty.

High-Permeability Reefs

Reefs with high permeability generally also feature vugular porosity or other large voids. Pinnacle reefs, such as those in Michigan, are similar to high-permeability reefs and characteristically have thick sections. Some of over 500 ft have been reported, often with a small areal extent.

Reefs often cause either lost circulation, high pressure flows, or a combination of the two. Difficulty increases with increasing permeability and higher pressures. Drilling into the reef usually results in a kick or lost circulation. Lost circulation may be followed by kicking after reducing the hydrostatic head. Reefs filled with oil generally cause less problems than those filled with gas. In one case, while drilling a long pinnacle reef section, over ten kicks occurred and it was necessary to close the well in and circulate it dead. Another pinnacle reef with a gas cap caused so many problems that serious consideration was given to plugging back and sidetracking to try and enter the oil column.

The method of drilling a reef depends upon the specific conditions. One solution seldom applies to all cases. Generally, the casing should be set near the top of the reef, especially if overlying shales are fluid sensitive. Otherwise the lost-circulation-and-kick sequence of problems decreases the stability of the shales, causing additional problems and higher risk of sticking. Lost circulation material seldom is effective except in lower-permeability reefs.

Drilling very high permeability reefs with a floating mud column has had varying degrees of success. They should be dry drilled, with precautions taken to prevent sticking by cutting accumulations. Control the annulus by maintaining a lightweight mud cap, adding mud periodically. Run a back pressure valve for pressure control while making connections and tripping. Use a retrievable-type valve to allow wire-line work and place it near the bit so the drillpipe can be pulled out of the hole to a point where the annular mud column controls the pressure. Mud should be added to the annulus while tripping. Log through a lubricator while continuing mud additions in the annulus as needed. The blow-out preventer should be fitted with blind rams and rams to fit the drillpipe and drill collars.

Other Formations
Various other formations or situations can also cause problems.

- Salt formations dissolve in freshwater muds, creating the problems associated with cavities and washed-out sections. Salt contaminates freshwater muds, so salt formations should be drilled with a saturated salt or inert mud. Some salt sections flow or move with time and may cause tight hole while drilling, along with associated problems. This movement may collapse casing; therefore, an operator should design for stronger casing through salt formations.
- Earthquake zones may cause ruptured or collapsed casing, a condition that is difficult to prepare for. Precautions may include stronger or larger casing, with smaller casing to absorb part of the movement. Safety shutoff valves might also prove useful. See what other operators do in this area.
- Coal beds, especially those over a few feet thick, usually are highly fractured and slough into the wellbore. Lost circulation in coal beds is relatively common. Treat them as fractured formations.
- Fracture zones, or zones that are highly fractured, differ from fractured formations in that they are generally thinner, have larger fractures, and frequently exhibit high pressures, often with lost circulation. It is not uncommon for the bit to slip into a fracture and become stuck, so use standard cleanout procedures, repeating them until the formation stabilizes. Take precautions to prevent sticking the drillstring.
- Anhydride or gypsum formations frequently cause mud contamination, which is similar to cement contamination but with a reduced pH. For thin sections, pretreat the mud with barium carbonate to precipitate out the calcium, phosphates, or soda ash. It usually is best to drill thick sections with a gypsum lignosulfonate or inert mud.
- Saltwater flows deteriorate the mud by increasing the chlorides, causing sloughing and swelling of fluid-sensitive shales. Signs for detecting them include increased chlorides, increased pit volume, continuing flow at the flow line after the pump has been shut off, and deteriorating mud properties. The mud should be treated to restore good properties. Increase the mud weight to shut off the flow, but be aware of potential lost circulation.

Combination Formation Problems
Combination formation problems are those with two or more problem formations. The most common include any combination of

fluid-sensitive and geopressured shales, fractured formations, high-pressure zones, and lost circulation. Sometimes they are considered as a single problem, since they may be difficult to distinguish from single formation problems. Recognizing the differences is important for treating purposes. Combination formation problems often are more complicated and involve a higher degree of risk than single formation problems.

Correcting a problem in one formation may cause a problem in another. The most common occurrence is related to lost circulation. Raising mud weight to control a high-pressure zone may cause lost circulation in another zone. Lost circulation may cause a kick from a higher-pressure zone or may decrease wellbore stability, causing caving and sloughing.

Treat combination formation problems based upon specific conditions. The most severe individual formation problem should generally be treated first, with additional treatment for the secondary formation problem. Use the procedures described for the individual formation. Always consider how the treatment of one formation will affect the other.

The combination of lost circulation and gas kicks is one of the more severe formation problems. This problem, with the associated pore pressures, is explained in Fig. 1–8. The problems with this combination is illustrated in the following example.

Combination Lost Circulation and Gas Kicks

A 17,500-ft wildcat well had 13 3/8-in. surface casing at 1,450 ft, 9 5/8-in. intermediate casing at 8,000 ft, and was drilling an 8 3/4-in. hole at 14,700 ft. Sands with pore pressures of 18-lb/gal were expected in the gross interval from 15,000 ft to TD. Pore pressures had increased from about 12.5-to 13.8-lb/gal over the previous 500 ft, indicating a long pressure-transition zone. Circulation loss was minor, except for a 400 bbl loss, probably in the shallow formations. Caving shale problems were increasing.

Intermediate 7-in. casing was run even though it was recognized that lost circulation might occur below it while drilling the deeper 18-lb/gal zones. However, the upper hole was already causing problems, and setting casing would protect the upper hole.

The 7-in. casing ran freely, but with limited mud returns while running it into the lower hole, and about 500 bbl of mud were lost. The hole could not be circulated and was cemented without returns. Later, cement bond logs indicated good cement in the lower section. The casing was cleaned out and tested, and 15 ft of new hole was drilled and tested to a leak-off of 16-lb/gal.

Several gas shows were encountered while drilling and all returns were lost at 15,000 ft. Three LCM pills were unsuccessful and about 1,000 bbl of

Figure 1–8   Combination high-pressure and lost-circulation zones

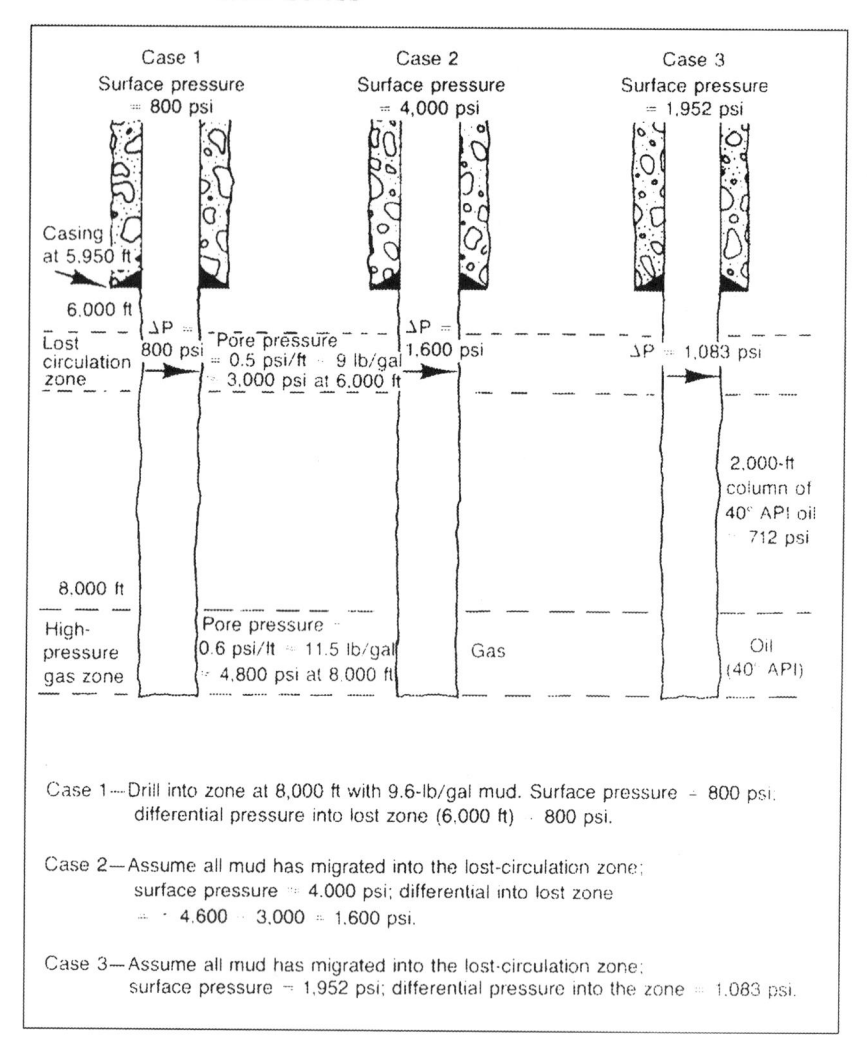

```
        Case 1                    Case 2                    Case 3
    Surface pressure         Surface pressure          Surface pressure
       = 800 psi                = 4,000 psi               = 1,952 psi
```

Casing at 5,950 ft

6,000 ft
Lost circulation zone

ΔP = 800 psi — Pore pressure = 0.5 psi/ft = 9 lb/gal = 3,000 psi at 6,000 ft

ΔP = 1,600 psi

ΔP = 1,083 psi

2,000-ft column of 40° API oil = 712 psi

8,000 ft
High-pressure gas zone

Pore pressure = 0.6 psi/ft = 11.5 lb/gal = 4,800 psi at 8,000 ft

Gas

Oil (40° API)

Case 1—Drill into zone at 8,000 ft with 9.6-lb/gal mud. Surface pressure = 800 psi; differential pressure into lost zone (6,000 ft) = 800 psi.

Case 2—Assume all mud has migrated into the lost-circulation zone; surface pressure = 4,000 psi; differential into lost zone = 4,600 − 3,000 = 1,600 psi.

Case 3—Assume all mud has migrated into the lost-circulation zone; surface pressure = 1,952 psi; differential pressure into the zone = 1,083 psi.

mud were lost. The pipe was pulled into the casing and the mud weight was reduced to regain circulation. The well kicked when the mud weight was reduced to 13.6-lb/gal. It was shut in with 1,100 psi on the drillpipe and 2,450 psi on the casings. It was circulated on a choke. The mud weight was increased to 14.8-lb/gal mud with about 30% returns when all returns were again lost.

The well was squeezed with a diesel oil bentonite (gunk) and circulation was established. The drillpipe was pulled and rerun open-ended. Circulation was lost while washing out within 4 ft of the bottom. The formation was squeezed with 250 sacks of cement, full returns were established, and drilling operations were resumed.

During the next several months, the well was drilled to about 16,500 ft. Less than 20% of the work time was spent drilling. The major part of the time was spent mixing and spotting lost circulation pills, squeezing the formation with cement (and once with a diesel oil bentonite squeeze), circulating out gas on the choke, and mixing mud. About 15.1-lb/gal mud would hold the pressures, but mud losses ranged from 100–800 bbl/day.

The well was deepened about 120 ft in a month, with increasing lost circulation and kicks, until operations were effectively at a standstill. After a considerable delay, which included several cement jobs, circulation was established at total depth. A 5-in. Fj liner was run and cemented with limited returns. The top of the liner tested satisfactorily, and the well subsequently was completed as a commercial producer.

This combination of lost-circulation and gas kicks is probably one of the most difficult, risky, and expensive drilling operations that can be attempted. Before setting the 7-in. casing, the last 3,000 ft of hole was drilled at an average of about 90 feet per day. The lower hole below casing was drilled at an overall average rate of 20 feet per day. The lower formations were slightly harder, but the extra time was caused primarily by combating lost circulation, circulating out gas kicks, and mixing mud.

That operations were surprisingly free of stuck pipe and similar problems is attributed to setting the 7-in. casing and employing experienced, qualified operating personnel. The obvious problem here was the underdesigned casing program. As a note of interest, subsequent wells were drilled in the area by extending the casing depths, although they did allow the option of an extra string of casing.

Depleted Reservoirs

The term "depleted reservoirs," as used here, refers to those reservoirs where the pressure is less than the original, often substantially so. Depleted reservoirs are becoming an increasingly common and important consideration for deeper drilling operations in older fields. Associated problems include sidetracking and redrilling junked production wells, and drilling infield wells, disposal wells, and wells for secondary and tertiary recovery.

There is very little information in the literature about detecting depleted formations from pore-pressure plots. Operators intuitively expect conventional pore-pressure plots to predict the original pressure condition. Based on limited evidence, this expectation is valid. For example, a geopressured gas reservoir was produced, so pressures declined from 30% to 60% of original pressure. Wells drilled later had pore-pressure characteristics through the overlying transition and geopressured zones similar to the original wells.

Lost circulation is the main problem when drilling into and through depleted formations. It may be necessary to protect the depleted zone. Problems are minimal if the zone is cased before drilling deeper; however, drilling deeper without setting casing frequently causes various combination problems, generally including lost circulation.

Main considerations are the original and current pressure of the depleted reservoir, the pressure of the objective zone, and the nature of the formations. The required mud weights should be determined based on formation pressures. Fracture prediction helps determine if the depleted formation will sustain the mud weight. Otherwise, the problems should generally be handled in a conventional manner. It may be necessary to set casing through the depleted zone before drilling deeper, usually because of high pressure differentials between the depleted and objective zones. Casing off the depleted zone before drilling deeper solves most problems.

Theoretically, mud flows into the formation whenever the hydrostatic head causes a pressure differential in the formation, even by a small amount. Excluding cases of fracturing, this seldom occurs because of viscosity and the barrier formed by entrained solids and gels. This type of seepage lost circulation normally can be controlled and is not a major detriment to drilling depleted formations, even with a relatively high pressure differential into the formation.

Generally, fracturing is the controlling factor. Calculate fracture initiation and propagating pressure in the depleted zone and the maximum pressure expected while drilling deeper formations with a high pressure differential. These determine the criteria for mud weights and for the drilling and casing program.

An interesting method of handling lost circulation while protecting the formation is illustrated in the following example.

Depleted Reservoir

A shallow, high-volume gas well was drilled and completed in a normal-pressured, high-permeability sand at 2,200 ft. After an extended producing life, the well was effectively junked when a packer and some tubing were lost in the hole during a workover. Bottomhole pressure was 175 psi (1.53-lb/gal, 0.0795-psi/ft) and the formation had sufficient remaining recoverable reserves to justify drilling a replacement well, providing the formations were not damaged.

Replacement well plans included setting 5½-in. casing on top of the sand, drilling a 4½-in. hole through the sand with clear water containing potassium chloride (KCl) (8.8-lb/gal, 0.4576 psi/ft), hanging a slotted liner hung on a pack-off-type hanger, and completing the well through 2⅜-in. open ended tubing. Only a minimum amount of water would be lost to the formation — part of it would be recovered by swabbing and the remainder would dissipate into the formation without causing appreciable damage.

The well was drilled and completed according to the plan. There was intermittent, partial circulation while drilling through and below the sand, recovering some cuttings. After drilling to total depth and picking up, the crew checked the hole for cuttings and had negligible fill-up after one hour. The slotted liner was run and seated without difficulty. The well was swabbed in through tubing, and flowed intermittent heads of water, gas, and finally, dry gas. An estimated 20% of the water lost in the hole was recovered. The well was placed on production and produced satisfactorily without stimulation.

The depleted formation had a pore-pressure equivalent of 1.53-lb/gal and was drilled with 8.8-lb/gal mud, a considerable difference. However, the fracture pressure of the depleted formation was calculated at 11.39-lb/gal by Eaton's method. This indicated the well could be drilled with normal mud weight without lost circulation due to fracturing. Therefore, the only apparent risk was seepage-type lost circulation, which was shown earlier to be of minor concern.

Rigorous fracture-pressure formulas use a combination of the pore pressure and rock stress, usually the horizontal stress. In most cases of depleted reservoirs, the rock stress is not reduced to the point where it would cause a substantial reduction in the fracture pressure below the normal pressure gradient.

Predicting fracture pressures may also permit substantial savings in drilling costs, as illustrated in the following example.

Geopressured Below Depleted

The Gulf Coast Vicksburg often is geopressured and has a number of producing horizons. One area had two main sands at about 9,500 ft and 11,000 ft through a depth interval of increasing pore pressure. The gross sand sections were relatively consistent over the area, but actually consisted of a series of sand shale sequences. The normal completion in the area had 13^3/$_8$-in. surface casing in a 15-in. hole at 1,200 ft, 9^5/$_8$-in. intermediate casing in a 12^1/$_4$-in. hole in the upper part of the transition zone at about 6,500 ft, a 7^5/$_8$-in. flush joint drilling liner in an 8^3/$_4$-in. hole below the upper sand, and 5^1/$_2$-in. completion casing at a total depth of 12,000–14,000 ft. The field had been developed on a somewhat random basis over a period of years due to completions in different intervals under different operators and ownership.

A location was selected that was partially drained in the lower sand at 11,000 ft but was believed to have substantial recoverable reserves that would not be economically recovered by the offsetting wells. The upper sand at 9,500 ft had high pressure but relatively low permeability, so it was a questionable producer. However, the pressure had to be considered for purposes of drilling deeper with possible lost circulation in the lower sand. A conventional casing program was designed that included setting a drilling liner through the upper sand and completion casing at 12,000 ft (TD).

Omitting the drilling liner was evaluated as an option since it would reduce costs by about 25%, a substantial savings. The upper sand had a pore pressure of at least 15.5-lb/gal, and possibly 15.8-lb/gal. These figures defined the minimum mud weight for omitting the liner. Calculating the fracture gradient of the lower sand at 11,000 ft was more difficult because the stage of depletion was unknown, offsetting wells were not evenly spaced, and different permeabilities, production rates, and original and current bottom-hole-pressure data was questionable or not available. Fracture pressures were calculated with the best available data and compared to hydraulic-sand-fracturing data to give the estimated correct value listed in Table 1–7.

Table 1–7 Calculated fracture gradients

BHP, psi	Fracture gradient, lb/gal.
7,630	17.00 (Original)
6,000	16.34
4,000	15.53

The best estimated current bottomhole pressure was about 5,000 psi, which indicated that the well could be drilled without the liner. However, mud weights were critical, so equivalent circulating densities were calculated at different yield points and circulating rates at 10,000 ft, as listed in Table 1–8.

Table 1–8 Equivalent circulating density, at 10,000 ft

Circulation Rate, GPM	Yield Points 5	10	15
100	15.76	15.78	15.81
200	15.82	15.85	15.89
300	15.87	15.91	15.96
400	15.98	15.98	16.03

Because of the wide margin of error, it was decided to use the conventional program and make the final decision to set the liner after drilling to the liner-setting depth.

The well was spudded and drilled in the normal manner. While drilling the upper sand, a minimum mud weight was maintained primarily by monitoring the gas in the mud and carefully considering drill gas, which at times may be confused with gas that flows into the wellbore due to an underbalanced mud system.

The upper sand was drilled with a maximum mud weight of 15.6 lb/gal. About 50 bbls. of mud were lost almost immediately after drilling about 50 ft into the lower sand. Circulation was regained after a short time and the hole was drilled to total depth with 15.7-lb/gal mud. Casing was run and cemented and the well completed in the normal manner. The lost circulation in the lower sand was attributed to either a depleted, high-permeability sand stringer or loss into a fracture.

DEVIATION, DOGLEG, AND CROOKED HOLE

The severity of problems caused by deviation, dogleg, and crooked hole during drilling increases with increasing angle, higher changes of angle, and deeper depths. Problems may include keyseats, wall sticking, increased drag and torque, excess wear on the drilling assembly, accidental sidetracking, hole cleaning, hole stability, difficulty running and cementing casing, and casing wear while drilling deeper. Problems may also occur during the producing life of the well, including casing, tubing, and rod wear, difficulty with artificial lift, and future remedial work.

The main preventive measure is recognizing the problem. Measurements should be recorded at shallow depths and repeated at intervals short enough to recognize trends and know the hole condition. Preventive action should be taken as necessary before the hole condition becomes a problem. The importance of drilling a straight vertical hole in a careful, prudent manner is illustrated in the following example.

Crooked vs. Straight Hole

Two deep wildcat wells were drilled as vertical holes in the same general area. Both used similar drilling fluids, penetrated similar formations, and had similar drilling programs. Both programs included surface casing at 5,000 ft, 9⅝-in. intermediate casing in a 12¼-in. hole at 15,000 ft, and a projected total depth of 18,500 ft.

In the first well, intermediate casing stuck at about 12,000 ft (3,000 ft off bottom) and was subsequently cemented. The open interval caused problems in deeper drilling, requiring another string of casing to be set. Drilling operations were suspended before reaching the projected total depth because of reduced hole size and increasing formation problems. Overall well costs were higher than projected, and some prospective horizons were untested.

The second well was drilled to the projected total depth without difficulty and within projected costs.

The first well had several absolute doglegs with a maximum angle over 3°, compared to a maximum of 1.5° in the second hole. The first hole also had a higher average deviation. There may have been other contributing factors, but successfully running intermediate casing string and completing the second hole is attributed, to a great extent, to the straight hole.

Dogleg angles here may not be considered significant. However, smaller angles at shallower depths can be very important when drilling deep holes with long open-hole sections over long periods of drilling. This illustrates the severity of the drilling problems that can occur in highly deviated and crooked holes.

Cause and Correction

A common cause of excess dogleg and deviation is crooked hole formations. These are most commonly layered formations, frequently in which the layers are alternating hard and soft. Other crooked hole formations are areas of high or increasing dip and changes of dip, such as those that cross a fault.

Drilling practices contribute to or cause a dogleg. To some extent, all of the causes of doglegs can be attributed to poor drilling practices and related actions. These include failure to run the correct bottomhole assembly, rotating off bottom excessively, working the pipe without passing tool joints, and otherwise not operating the assembly correctly. Limber assemblies can build angle rapidly or are strongly affected by formation influence, especially while operating at a low RPM and high bit weight. Highly efficient pendulums can cause doglegs by dropping angle too fast. Deviation and sidetracking can cause doglegs by changing direction too sharply.

One of the first corrective actions is early recognition and the taking of remedial steps accordingly. The problem seldom goes away on its own. The obvious preventive measure is "drill a straight hole," but this is not always possible. Maintaining an excessively low deviation may cause increased cost. If there is a natural tendency for the hole to deviate and it does not adversely offset operations, allow it to deviate. Control the deviation only if it is excessive. Formation conditions may increase the cost of maintaining deviation and dogleg within acceptable limits. However, this additional cost often is justified in view of potential hole problems and the possibility of a lost hole, especially in deeper holes. The main point is to be aware of the risk, evaluate the problem, and take action accordingly.

Use the correct assembly and operate it correctly. Do not wait too long to take the positive preventive action. First, try reducing bit weight and increasing rotary speed. Otherwise, run a packed hole assembly, which tends to drill a straight hole in the same direction as the prior hole. This reduces crooked hole but may not help deviation. The stiffness of the assembly should be adjusted based upon formation and hole conditions. Deviation can be reduced or maintained by drilling with a regular packed-hole or forced pendulum. Use an angle-building assembly if hole angle drops excessively. Remove or reduce excessive crooked hole by reaming.

Running a deviation assembly can correct the hole direction if horizontal control is needed or if other procedures are ineffective or cause excessive drilling time. DO NOT overcorrect with an excessive change of angle that may cause a dogleg. Ream after deviating to remove sharp bends and smooth the curvature. Eliminate sidetracking or deviating whenever possible. Deviate on the low side of the hole instead of blind sidetracking to minimize the sidetracking angle. Set casing at or very near bottom to eliminate long sections of larger-diameter hole.

Highly deviated, extended-reach and horizontal holes create special problems. Evaluate the risk before drilling and then take all precautions. Otherwise, eliminate crooked holes whenever possible.

Permissible dogleg severity varies for different areas of operation. The acceptable deviation and dogleg limits should be defined in the drilling program.

Recommended Deviation and Dogleg

Allowable dogleg cannot be predicted accurately. However, the forces acting on a dogleg can be approximated, as illustrated in Table 1–9. Combining these data on torque and drag with experience and results in the area helps determine the maximum allowable dogleg.

Table 1-9   Lateral force exerted by drillpipe tool joints

Depth Below Keyseat, ft.	Pipe Wt Below Keyseat, 1,000 lb.	1	2	3	4	5	6
		|←		Angle Of Deviation, degrees			→|
		|←		Lateral Force, 1,000 lbs.			→|
2,000	75	0.41	0.82	1.23	1.63	2.04	2.45
4,000	87	0.47	0.95	1.42	1.90	2.37	2.84
6,000	113	0.62	1.23	1.85	2.46	3.08	3.69
8,000	140	0.76	1.53	2.29	3.05	3.81	4.57
10,000	167	0.91	1.82	2.73	3.64	4.55	5.46
12,000	194	1.06	2.12	3.17	4.23	5.28	6.34
14,000	220	1.20	2.40	3.60	4.80	5.99	7.19
16,000	274	1.49	2.99	4.48	5.97	7.46	8.95

The table is based on 16.6-lb/ft, range 2 drillpipe in 16-lb/gal mud and 800 ft of drill collars with a buoyant weight of 60,000 lb. The table assumes the hole above the dogleg is vertical. This is sufficiently accurate for vertical holes, but should not be used for highly deviated or crooked holes, since accuracy decreases with increasing deviation. The table further assumes static conditions with the pipe hanging free below the dogleg, and does not allow for drag or drillpipe rigidity. It also assumes all lateral forces are exerted by the tool joints as a single force. The table gives minimum values, since it is based upon changes of angle in the single plane. The forces could be calculated for a wide variety of conditions, but this is not justified because of the many unknown factors. The table is not precise; nevertheless it is included to show the order of magnitude of the forces exerted by the pipe against the wall of the hole.

Table 1-9 illustrates several points, such as the digging force of the tool joints during tripping. It is easy to visualize two or more deviated or crooked-hole sections and how the additional force to pull the drillpipe through a lower section is amplified in an upper section. This also accounts for the high tendency to develop keyseats by the wearing, dragging, and gorging action of the tool joints and the increased drag and torque.

Table 1-9 indicates the lateral tool joint force is in the range of 1,000–1,500 lb for angles of 2–3°. Field experience has indicated that few problems are encountered with dogleg angles in this range. The selection of a lateral force of 1,000–1,500 lb depends upon experience in the area and drilling and formation conditions.

Table 1-10 Recommended maximum deviation and dogleg angles

Total Depth, ft	Depth Interval, ft	Maximum Deviation Angle, deg/100 ft	Maximum Dogleg Angle deg/100 ft
5,000	0 – 100	1	4
	100 – 1,000	2	4
	1000 – 5,000	4	4
10,000	0 – 100	½	3
	100 – 1,000	1	3
	1,000 – 5,000	2	4
	5,000 – 10,000	5	4
15,000	0 – 100	¼	3
	100 – 1,000	½	3
	1,000 – 5,000	1	4
	5,000 – 10,000	3	3
	10,000 – 15,000	8	3
20,000	0 – 100	<¼	3
	100 – 1,000	¼	3
	1,000 – 5,000	½	4
	5,000 – 10,000	3	2
	10,000 – 15,000	5	3
	15,000 – 20,000	8	3

Table 1–10 is a general guide to maximum dogleg and deviation angles. Select the maximum dogleg angle based on experience and knowledge about the formations in an area and the hole being drilled.

The table gives full weight to deviation and dogleg in the surface hole. This important factor is too often overlooked. Since tension in the pipe is critical, a shallow dogleg in a deep hole is a frequent source of trouble. To illustrate, the table indicates the hole could be drilled about 10,000 ft below a 1° dogleg and 4,000 ft below a 2° dogleg based upon a lateral force limit of 1,000 lb. For the 1,500 lb lateral force limit, the hole could be drilled 16,000 ft below a 1° dogleg and 7,000 ft below a 2° dogleg.

It will be recognized that the limiting angles listed in the table are restrictive in some cases. There is no reason that higher limits cannot be used if the operator is aware of the risk and experience in the area justifies higher angles. Many holes are deviated at much higher angles than those listed.

Cause and Prevention

REFERENCES AND SUGGESTED READING

Adams, N., and Kuhlman, L.G., "How to Prevent or Minimize Shallow Gas Blowouts," *World Oil*: Part 1, May 1991: 51–58; Part 2 (June 1991): 66.

Blattel, Steve R., and Rupert, J. Paul, "The Effect Of Weight Material Type on Rate of Penetration Using Dispersed and Non-Dispersed Water-Base Muds," SPE 10961, Presented at the annual Fall Technical Conference of the Society Of Petroleum Engineers of AIME in New Orleans, LA, Sept. 26–29, 1982.

Blomberg, Nits R., and Nelberg, Berit, "Evaluation of Ilmenite as Weight Material in Drilling Fluids," SPE 11085, Presented at the annual Fall Technical Conference of the Society Of Petroleum Engineers of AIME in New Orleans, LA, Sept. 26–29, 1982.

Bradley, W.B., Jarman, K., Auflick, R.A., Plott, R.S., Wood, R.D., Schofield, T.R., and Cocking, D., "Task Force Reduces Stuck-pipe Costs," *Oil and Gas Journal,* May 27, 1991: 84–85.

Eaton, Ben A., "Fracture Gradient Prediction Techniques and Their Application in Drilling, Stimulation, and Secondary Recovery Operations," SPE No. 2163, American Institute of Mining, Metallurgical and Petroleum Engineers, Inc., Presented at the 43rd Annual Fall Meeting in Houston, TX, Sept. 29–Oct. 2, 1968.

Hanson, P.M., Trigg Jr., T.K., Rachal, G., and Zamora, M., "Investigation of Barite 'Sag' in Weighted Drilling Fluids in Highly Deviated Wells," SPE 20423, Society of Petroleum Engineers in New Orleans, LA, Sept. 23–26, 1990.

Hansford, J.E., and Lubinski, A., "Cumulative Fatigue Damage of Drillpipe in Dog Legs," *Transactions AIME*, 1966, 237: I–359.

Harvey, Floyd, "Horizontal Wells 4 — Fluid program built around hole cleaning, protecting formation," *Oil and Gas Journal,* Nov. 5, 1990: 37–41.

Hubbert, M. King, and Willis, D.G., "Mechanics of Hydraulic Fracturing," *Journal of Petroleum Technology,* 1957: 153–168.

Lubinski, A., "Maximum Permissible Dog-Legs in Rotary Boreholes," *Transactions AIME*, 1961, 222: I–175.

Mathews, W.R., and Kelly, Dr. John, "How to Predict Formation Pressure and Fracture Gradients," *Oil and Gas Journal*, Feb. 20, 1967.

Nederveld, L.S., and Vleaux, G.J., "A Barite Alternative," *Drilling-DCW*, April 1980: 80–84.

Saasen, Arild, Marken, Craig, Sterri, Njal, and Jacobsen, Jon, "Monitoring of Barite Sag Important in Deviated Drilling," *Oil and Gas Journal* Aug. 26, 1991: 43–50.

Short, J.A., "Jim," *Drilling and Casing Operations,* PennWell Publishing Company, Tulsa, OK, 1981.

Short, J.A. "Jim," *Introduction to Directional Drilling,* PennWell Publishing Company, Tulsa, OK, 1993.

Stein, N., "Resistivity and Density Logs Key to Fluid Pressure Estimates," *Oil and Gas Journal*, April 8, 1985: 81–86.

Tomren, P.H., Iyoho, A.W., Azar, J.J., "Experimental Study of Cuttings Transport in Directional Wells," SPE Drilling Engineering, Feb. 1986.

Walker, C.O., "Alternate Weighting Material" SPE 1116, Presented at the annual Fall Technical Conference of the Society Of Petroleum Engineers of AIME in New Orleans, LA, Sept. 26–29, 1982.

Wilson, G.J., "Dog-Leg Control in Directionally Drilled Wells," *Transactions AIME*, 1967, 240: I–107.

Zamora, Mario, and Hanson, Paul, "Rules of Thumb to Improve High-Angle Hole Cleaning," *Petroleum Engineer International*, Jan. 1991: 44–51; Feb. 1991: 22–27.

Chapter 2

GENERAL RULES AND PROCEDURES

Various general rules and procedures apply to most fishing situations. Operations, equipment and formations cause fishing and can restrict or prevent completion of fishing. All related factors are important but some warrant emphasis. Competent, experienced personnel are of prime importance. They prevent fishing and perform most fishing operations, including selecting and supervising a fishing specialist, if required. Always be prepared, since fishing often occurs when least expected. Know what actions to take, learn all details about the situation, and take immediate action. Do not overlook the importance of working stuck assemblies. Expedite operations and know when to quit or take an alternate action, such as sidetracking, redrilling, or abandonment. Perform a post-analysis to prevent a future occurrence or to learn how to handle it more efficiently. Observe the general guides to fishing and the common drilling operations that frequently cause fishing.

PERSONNEL

Competent personnel with experience in both drilling and fishing operations and with good communications skills are absolute necessities to prevent fishing and to conduct operations successfully when fishing occurs. A good understanding of fishing tools and their application aids in successfully completing fishing operations.

Fishing is not a normal drilling operation, but specialized, analogous to casing and cementing. Most personnel are less familiar with these procedures. Take extra precautions because unfamiliarity leads to mistakes which, in turn, may cause additional fishing or a second fish and the risk of losing the hole. The importance of experienced, conscientious personnel and good communications cannot be overemphasized. A well-qualified specialist in preventing fishing jobs would be of great value. There will always be occasions where high risk situations occur that may result in a fishing job. Moreover, most if not all of these can be prevented with proper planning and by conducting operations in a prudent, workmanlike manner.

Fishing as a non-routine operation places additional responsibility on supervisory personnel. An experienced team, including an on-site supervisor and an office manager, is vital to efficient drilling, and the preventing or completing of fishing jobs. Fishing frequently requires decisions that affect the status of the overall drilling project. These decisions often are needed on short notice. Accomplish this with a good channel of communication and established line of authority.

A fishing tool operator or a fishing specialist commonly is employed. Experienced operators have a list of available fishing tool personnel they have used in the past. An operator new to an area should investigate the available personnel, check references carefully, and compile a list of qualified fishing specialists.

Fishing specialists should have at least five years of widespread experience in fishing operations, including experience in the area where the job occurs. Also, they should be experienced in general drilling, completion, and rework operations. These are stringent requirements but justified by the importance and complexity of fishing.

BE PREPARED BEFORE FISHING OCCURS

Being prepared requires a good understanding of all conditions and general problems related to drilling in an area, including what caused prior fishing jobs and how they were handled. Other items are rig availability, specifications, capabilities, and condition. Equally important are the qualifications of the rig personnel, including supervision, experience, and proficiency. These requirements also apply to third party services such as logging, tool supply, cementing, and mud.

Study local drilling operations. Stay abreast of current activities through reports and discussions with other operators and service company personnel. A key part of being prepared is having qualified, experienced personnel.

Become familiar with area fishing jobs. Review daily drilling reports and mud and bit records. If the reports do not have sufficient detail, contact someone directly involved, such as the operator, well-site supervisor, engineers, or fishing tool operator.

Learn what caused the problem, how it was handled, and more important, how it could have been prevented. Note how the fish was stuck and recovered, and related conditions such as type of mud and properties. Other important information includes the fish and fishing string and details such as hole and drill collar sizes, jars, bumper subs, catches, safety joints, washpipe, and clearance. Also note formation conditions such as sensitivity to fluids, hole inclination and deviation, caving, tight hole, and wall sticking.

It often is a good practice to keep some fishing equipment on location during normal operations. Immediate availability may expedite fishing operations. It can increase the success of the fishing job in cases where caving, lost circulation, and similar problems occur. It is especially applicable in open-hole operations.

The amount and kind of equipment depends upon many factors. These include complexity of the drilling operation, frequency of fishing jobs, distance (in time) to a source of fishing tools and equipment, and its availability at the supply point.

There is a practical limit to the amount of equipment kept on the rig in this case. Bit cones are the most common fishing job, so consider a basket and one or more mills and junk subs. Other tools may include additional drill collars for the fishing assembly, bumper jars and bumper subs, and possibly overshots to fit the more common tubular diameters in the hole.

Following is a suggested equipment list.

- Mill — lost bit cones are a common cause of fishing
- Junk basket — bit cones and small junk
- Junk sub — run on mill or basket
- Overshot — OD's to work in current hole, change as hole size changes. Catch sizes include drillpipe body, tool joints, and top drill collar
- Back-off sub, bumper sub, jars

All equipment is sized to the equipment used in the well.

Sometimes fishing tool rental companies leave tools on location and charge only if the tool is used, with the understanding that the tool can be picked up for use in other areas if needed. Evaluate whether some fishing equipment should be maintained on the wellsite and what kind.

WHEN A FISHING JOB OCCURS

Experienced operators know what to do when fishing or sticking occurs. Generally take immediate action. Then learn all details and develop a plan of action, as well as an alternate plan.

Take Immediate Action

Take immediate action when a sticking situation or fishing job occurs. For example, when a downhole assembly first sticks, it often can be released with immediate, instead of delayed attention. The direction and type of work should be readily apparent. Otherwise, the risk of a fishing job grows as the time increases that the assembly remains stuck.

Taking immediate action is important, but always consider the safety of the hole. For example, if there is a well control problem, then consider it first. Ensure, as much as possible, that immediate action does not increase the risk. High angle directional and horizontal holes often create special problems.

One question may occur under certain circumstances, "Is there really a fishing job?" If an object is reported to have fallen in the hole, make sure it is not in the cellar or lodged in the BOPs. If the object is small, such as a bit cone, hammer, etc., often it can be walled off with a bit.

The operator should be sufficiently experienced to determine the immediate plan of action. Therefore, normally take action within 5 to 15 minutes or less after the problem occurs. The action required depends upon the specific conditions, but the course of action is often clear. Begin working stuck pipe or wireline tools shortly after sticking. Soon after work is begun, order the fishing tools and equipment needed to try and release or fish the stuck assembly, assuming the initial working procedure(s) is unsuccessful.

Limit taking a freepoint immediately. Usually take it after working the stuck assembly to tension levels needed for normal freepointing. This procedure is applicable to most vertical and low-angle directional wells. It may not be practical in high-angle directional and horizontal holes because of the increased risk of keyseating and wall sticking. However, consider it, even in these cases. Slug the pipe and start out of the hole if bit cones are lost, a leak develops, or the pipe string parts. It may be necessary to delay for some period of time such as pulling the drillpipe up into the last string of casing and waiting for the fishing tools. Try reconnecting a drillstring that accidentally backs off. Make five to ten attempts, if you are sure that the drill tool is backed off, before slugging and pulling out of the hole.

Taking action immediately is demonstrated in the following example.

Recovering a Hole Opener Shank

A wildcat well was drilling in an isolated area where 24–36 hours were required to obtain parts and tools. The program included $13\frac{3}{8}$-in.

Figure 2–1 Hole opener cutter and shank recovered in a junk basket

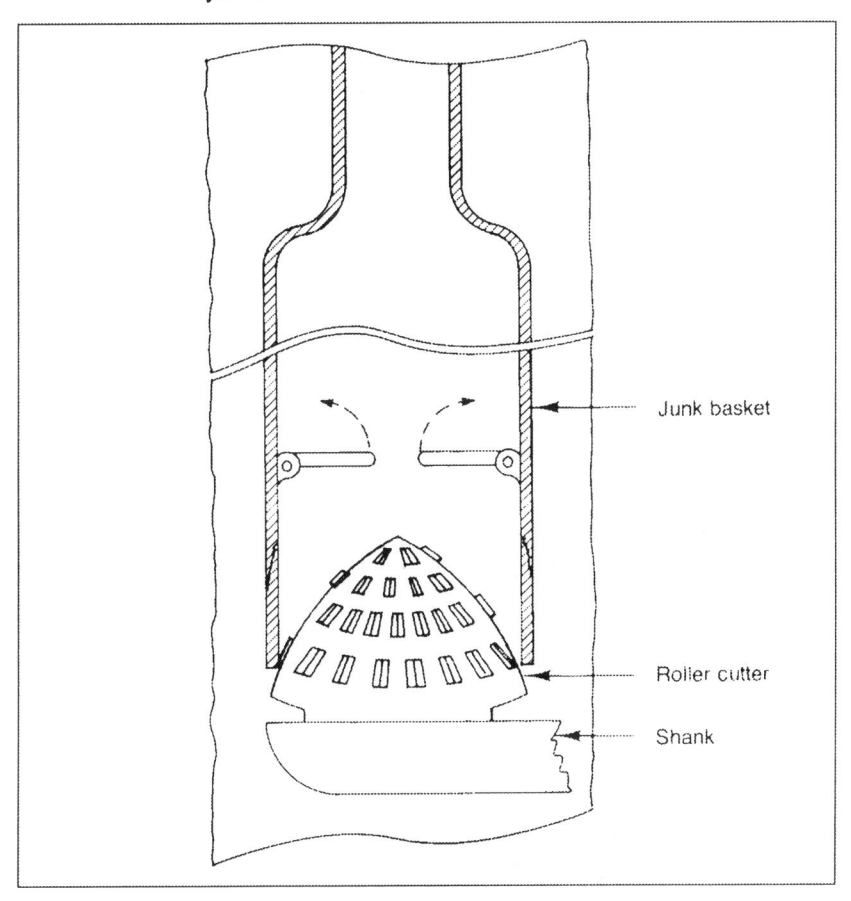

surface casing at 2,500 ft in a 17-in. hole, 9 5/9-in. intermediate casing in a 12¼-in. hole at 8,000 ft where a geopressured zone was expected, and completion in 5½-in. or 7-in. casing at 12,000 feet. There was no usable offset well data and only a limited amount of seismic information to predict pore pressure and formation tops. The large-size casing program allowed for an extra string of casing (7 inches) if needed.

The steeply dipping, highly fractured formations were very hard, and some were fluid sensitive. An 8¾-in. intermediate hole was drilled and opened to 12¼-in. Reasons for this included: (1) open-hole drillstem tests were expected and could be run more efficiently in the smaller hole; (2) the procedure allowed better control of hole deviation problems; and (3) the procedure allowed an overall higher penetration rate.

General Rules and Procedures

Figure 2–2   Sidetracking a hole while reaming

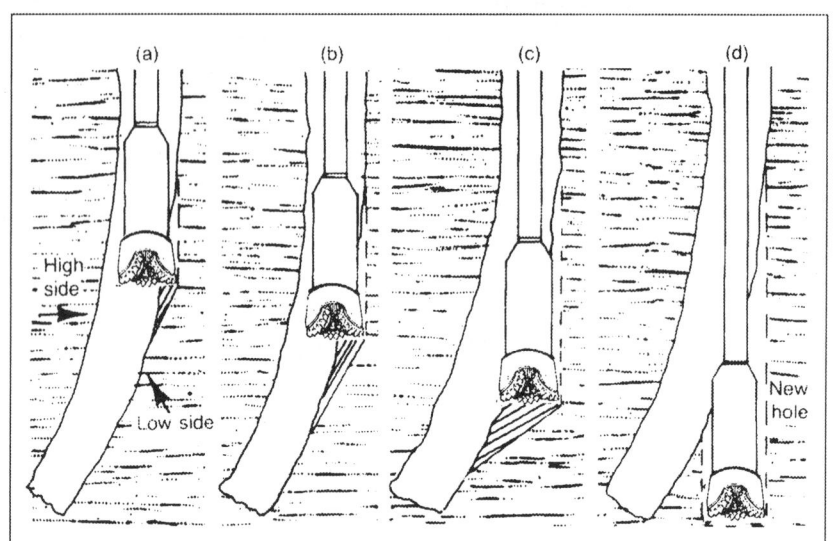

In (a), sidetrack started as a ledge; a light application of wt on the bit should crush the ledge, forcing the bit into the original hole. In (b) and (c), the ledge is increasing but probably can be eliminated. In (d), the hole is sidetracked; probably will be very difficult to reenter the original hole. The sequence from (a) through (c) can occur in a few minutes.

The 8¾-in. hole was being opened to 12¼-in. at 6,000 ft with a hole opener with roller cutters and an 8¾-in. pilot bit. The hard formations caused extremely high torque, and the hole opener failed. When it was pulled, one cutter had broken off at the body of the hole opener, leaving the roller cutter and shank in the hole.

Fishing tools were ordered and were expected to arrive within 18 to 24 hours. A bit was run to ensure that the cutter and shank had dropped to the bottom and were not hung on the ledge at the bottom of the 12¼-in. hole and to circulate the hole until the fishing tools arrived. A 7⅝-in. junk basket was available and was run to see if anything could be recovered, since there was a possibility that the cutter had broken from the shank. Otherwise, the entire piece of junk would be too large to enter the basket.

The junk basket was run. When it was set on bottom, the assemblies rotated freely, even with extra weight. This indicated the junk basket shoe was sitting down over the cutter and rotating. Additional weight was set on the junk basket to try jamming the cone into the junk basket. The pipe was pulled carefully, chaining out of the hole. The roller cutter was jammed into the bottom of the junk basket, exactly as envisioned, and the cutter and shank were recovered intact (Fig. 2–1).

This was a lucky(?) fishing job. Normally, the basket would not have caught the cone shank combination, or the fish would have been dragged out of the basket by the protruding edge of the shank as it rubbed on the wall while being pulled. Nevertheless, it was recovered, and considerable rig time was saved by taking immediate action.

Learn All Details

Learn all the details when a fishing job occurs, including what happened and what caused it. This information usually leads to the required remedial procedure. In the meantime continue with the immediate action. Be aware of complicating the fishing job, such as sidetracking while reaming (Fig. 2–2).

There are many different investigative procedures, most of which depend upon the situation and its related conditions. One of the first investigative steps is learning what is in the hole or what the fish is. Most of this information is known, but placing it on a schematic is helpful (Fig. 2–3). Make additional detailed sketches of special equipment. Update and revise these as required. It often is helpful to make a rough schematic of the downhole tools on a large scale electric log with the tools and pertinent comments located at their respective depth.

- List the drillpipe, drill tubing, or work string being used. Include the weight, grade, and condition. Note the tool joint type and OD. List other details such as hard banding, wear groove, etc. Repeat this for other pipe strings in combination assemblies.
- List detailed information about the drill-collar assembly, including overall length, type of connector, and ID. List the length of each collar and OD at each end and in the middle of the collar. Record the length and diameter of any recessed or enlarged sections.
- Make a detailed record of all other tools. Include type and size of tool joints, lengths, inside and outside diameters, length of individual pieces and overall length. Note all special tools such as keyseat wipers, drilling jars and bumper subs. Manufacturers' brochures often are very helpful and give detailed measurements, internal connections, and overall tool strength. Always verify part model or serial number.

The foregoing assumes recording accurate measurements before running the tools in the hole, which is a good standard procedure. Conventional measurement accuracy is either ⅛ in. or 0.01 ft for

Figure 2-3   Record of a fish

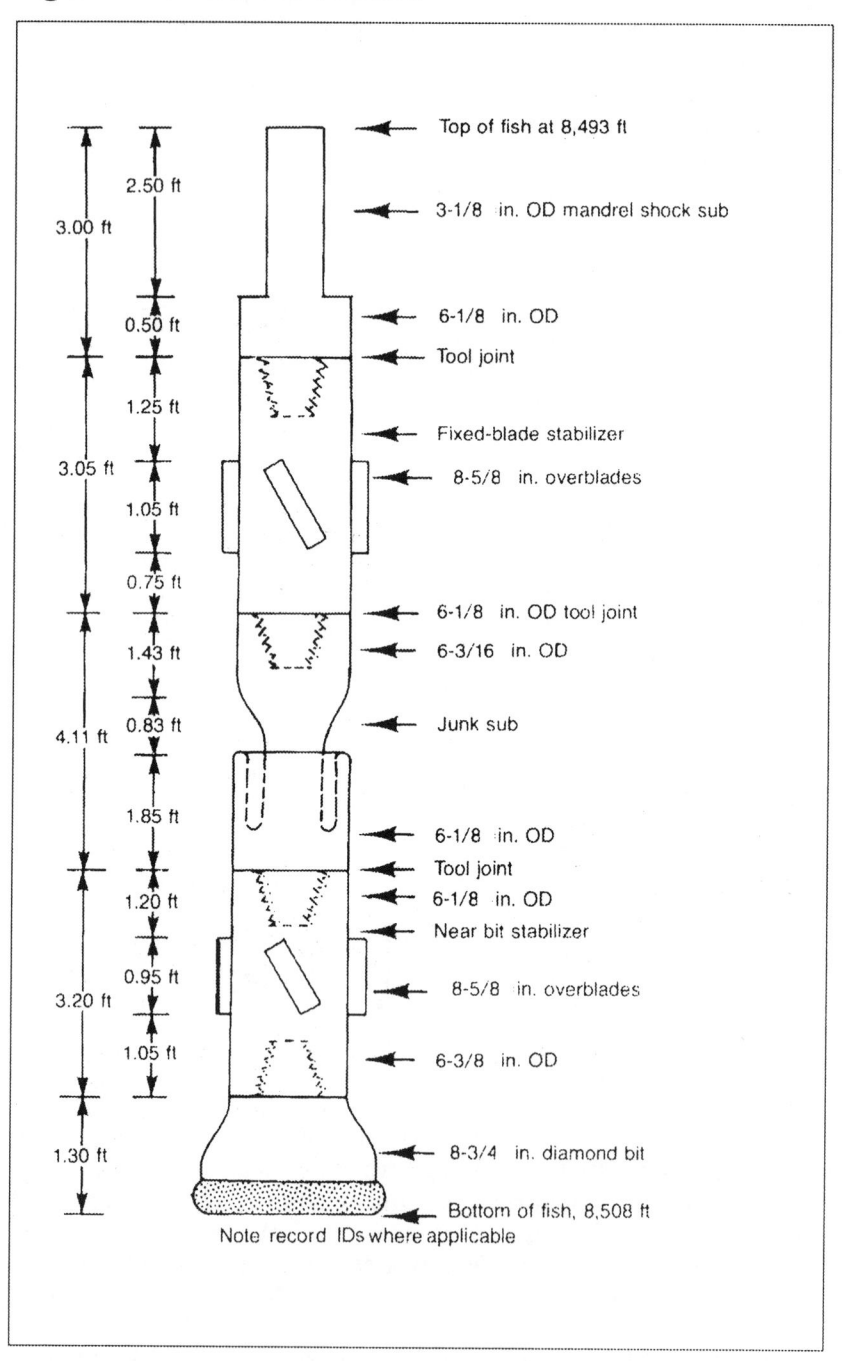

86　　　　　　　　　　　　　　　　　　　　　　　　　Chapter 2

length measurements and 1/32 in. for diameters. Measure diameters so there is no allowance for wear. Accurate measurements are very important because they help determine the type and size of the catch tool.

Then verify the overpull based upon the allowable strength of the weakest member in the complete assembly. Normally allow for buoyancy and calculate as provided for in the section "Overpull" in Chapter 3. This is a very important consideration for working a stuck fish. The overpull should be known, but verify it.

Next, learn all about the wellbore, formations, and related conditions. Normally the operating personnel will know these, but a review is warranted. Hole conditions include inclination, crooked hole, dogleg, keyseating, wall sticking, high torque and drag, etc. Formation conditions and characteristics include the type of formation, caving, fluid sensitivity, lost circulation, drillability, tendency to cause a crooked hole, increasing severity with time, and other items that adversely affect operations. Review the mud properties.

If the cause of the failure and the method of recovering the fish is not readily apparent, check all other sources of information including discussions with personnel. Records include drilling reports, well logs, mud and bit records, lithology logs, geolograph charts, etc. Geolograph charts (current and prior) are a good source of information on pipe weight, excess drag and torque, and keyseat development. They also help detect leaks in the drillstring. A mud report may show minor contamination that has recently become more severe.

This procedure is time well spent. A complete study of all information almost invariably leads to a plan of action and indicates the correct fishing procedure. It also provides information for the post-analysis.

Develop a Plan of Action

Develop a plan of action for the fishing job, including short, intermediate, long term and alternate plans as necessary. The plan will range from a partial plan to be supplemented later to a complete, detailed procedure. A fishing expert may be called in, depending upon the complexity of the fishing job and personnel experience.

Base the plan of action on information obtained while learning all the details. Commonly discuss it with others in meetings, conference calls, etc. Lower-level operating personnel frequently have good information and input. Prepare to submit the plan to supervisors if approval is required. Put it in writing. The plan is a continuing procedure that may be revised as necessary. Provide for flexibility, including all reasonable alternates, and modify as necessary. Implement it subject to the success of the original plan or changing conditions.

The immediate plan of action includes the work being conducted now and the next step. For example, assume that the lower part of the drilling assembly has been left in the hole and the remainder of the drill-collar assembly is being pulled. The immediate plan of action is to go back with a fishing assembly and either screw into the fish or catch it with an overshot. The tool used depends on the configuration of the top of the fish, ie., a failed drill collar connector or a backed-off connection. Normally, if the connection backed off, this may have been determined by sitting back down carefully on the fish and carefully rotating to the right. Therefore, the most probable immediate plan would be to go in with an overshot below a fishing assembly. An alternative might be a cleanout trip first if needed.

Always have an alternate plan, which would be the next probable action to be taken, assuming the present action is either unsuccessful or that continued operations require additional fishing. For example, a back-off is made near the bottom of the bottomhole assembly. This may occur in one of several types of connectors, depending upon the tools in the BHA. Plan ahead; do not wait until part of the fish is out of the hole. Have the necessary fishing tools to catch or screw into the different connectors. The proper tool usually will be only one of a few choices.

Several alternate plans may be required. If the first overshot did not recover the fish, one alternative would be to run a second overshot, to possibly catch the fish at a different place. If the second overshot is not successful, a subsequent alternative would be to run washpipe and wash over the fish. Always try to prevent waiting on fishing tools; have tools available for alternate plans. Normally the cost of these tools on a "stand by" basis is very small compared to the cost of the rig time while waiting for the tools to be delivered.

The long-range plan of action includes the courses of action following the immediate and alternative plans. If initial plans were successful, the long-range plan of action would be to continue drilling operations. If the alternatives were unsuccessful, the long-range plan would be to sidetrack the fish, redrill the hole, or complete the hole at the present depth. Also, the plan should include a reasonable estimate of how long current operations will be conducted.

The plan of action is a continuing procedure throughout the fishing operation. All plans must be flexible, allowing for various problems and changing conditions. Put the plan in writing as a guide and provide the information to others.

Always record complete dimensions of all equipment run in the hole. This is noted several times in the text and repeated here for emphasis. The initial equipment may be obvious. If so, arrange for delivery as early as possible, even before completing the plan of action if necessary. Provide for reserve equipment and subsequent equipment deliveries. Anticipate requirements and make arrangements accordingly. DO NOT WAIT ON FISHING EQUIPMENT except in extraordinary cases. DO NOT modify fishing plans because of equipment availability. This normally is not an acceptable practice, except as a last alternative, so use good judgment.

TIME SPENT FISHING

Carefully consider all operations during fishing before taking action. This is not a time to hurry. However, remember that recovering the fish quickly often reduces problems. Do not be over-cautious; this may be almost as undesirable as taking unnecessary risks. There is a higher risk of problems with longer operating time. Furthermore, fishing operations are hazardous, further increasing the risk of problems such as a fish on top of a fish. The old oil field truism, "Don't rush but hurry every chance you get," is especially applicable.

As an example, while fishing was being conducted through a long open-hole section, operations were hampered by alternating lost circulation and gas kicks and excessive time was lost while mixing mud. A mud-mixing plant was moved to the rig site. This provided adequate mud-mixing capacity while conducting operations, and the fish was recovered successfully without additional lost time.

Cementing off severe lost circulation is a commonly accepted procedure. Sometimes there is a tendency to continue combating lost circulation while conducting fishing operations. If there is a valid reason, then the approach is correct. Otherwise, cure the lost circulation by cementing if conditions are favorable. Some hesitate to cement above a fish because they believe that the cement will fall a long distance and effectively cement the fish. Normally cement will not fall more than 10 ft from the point of placement in a static fluid column, even if the column is clear water. Verify this by reviewing the bottom of the cement after squeezing perforations. Obviously cement will move downward if there is lost circulation and the fluid column is not static. If there is any question about the cement dropping down around the fish, spot a barite plug or a pill of high gel mud above the fish. Other situations that hinder fishing operations are described in the following example.

Examples of Squeezing before Continuing Fishing

A well had 7-in. completion casing set at 10,000 feet. Perforations at 9,000 feet were squeezed through 2⅞-in. tubing set in a squeeze packer. Cement channeled up behind the packer, cementing it and the lower part of the tubing in the hole. Fishing was started for the cemented tubing at 8,000 feet using a 12-hour workover rig. The well had several high-pressure zones. The well was shut in each night. The next morning it would be pressured up, requiring 2–6 hours to circulate and kill it before resuming fishing operations. Frequent high-pressure, low-volume gas kicks were common. The combination caused excessive delays.

It was obvious that there was a casing leak. It was verified by pressure testing above a packer. The packer was alternately moved and reseated until the leak was found in an area where there had been considerable milling during earlier fishing, evidently wearing a hole in the casing. Gas channeled up behind the casing from the lower, high-pressure zone and entered the wellbore through the leak. The hole was squeezed, successfully shutting off the casing leak. Fishing operations resumed, recovering the fish in a normal period of time.

In another case, a well had surface casing at 1,500 feet to shut off shallow gas sands and water flows. During deeper drilling, the drill collars stuck and subsequently were backed off, leaving the drill-collar assembly and about 1,000 feet of drillpipe in the hole. While conducting fishing operations, there were constant problems with lost circulation. The casing was tested and a leak was located at 400 feet. A retrievable bridge plug was set in the bottom of the casing, and the leak was squeezed off by bullheading cement down the casing. The casing was cleaned out, the bridge plug was recovered, fishing operations resumed, and the fish was recovered.

At times there is a tendency to repeat an unsuccessful operation. If the same procedure has been tried several times unsuccessfully, try another procedure. Always look for better, faster, safer ways to conduct operations.

WHEN AND HOW TO QUIT

An important part of fishing is knowing when to quit. Often there is a tendency to spend an excessive amount of time fishing. Part of this is the natural reluctance to leave a job unfinished. The decision of when to quit should be made early in the fishing job as part of the long-range plan. It may be based upon various actions. The most common are after a certain type of action or operation is unsuccessful, a fixed time limit, or a fixed expenditure limit. This plan does not necessarily have to be made as soon as fishing starts, but it should be developed early in the job. Like all plans, it must be flexible, yet followed rigorously unless new developments warrant a change. Recognize that some fishing jobs are unsuccessful and must be terminated before completion.

Consult with geologists, reservoir engineers, and others responsible for a well before starting an extensive fishing job. This may help determine whether fishing is warranted and provide guidance concerning the course of action. For example, the geological information found in the well may indicate that reserves above the fish may be sufficient to justify completion without fishing. The same information may also indicate that deeper reserves do not warrant fishing or justify drilling at a different surface location that has more potential.

Determine when to stop fishing early in the life of the job. Often there will be several more procedures or operations to try when the time to abandon a fishing job approaches. But unless the fishing job has been progressing satisfactorily and is near completion, these last-minute efforts are frequently unsuccessful. If there has been enough time to try all reasonable, practical methods, it is time to try another approach.

In some cases, fishing results may change the course of action, and the operator must select an alternate plan. Assume a drilling assembly stuck below prospective producing horizons. The plan could be to continue fishing for a period of time and, if results are unsuccessful, either complete through the stuck assembly or back off and run casing. If an additional fish were lost in the hole above the prospective producing horizon and plugged on the inside, this would eliminate the available alternative and require selecting a different, and often less favorable, long-range plan.

It is difficult to determine when a fishing job should be terminated, but the decision must be made and may lead to completion problems as illustrated in the following example.

Deep Completion over a Fish

A development well had 13 3/8-in. conductor casing at 800 ft, 10 3/4-in. surface casing at 6,000 ft, and 7 5/8-in. intermediate casing at 13,500 ft. The well was drilled to 16,000 ft. There were severe lost circulation and gas kick problems while drilling the last 1,500 ft of 9½-in. intermediate hole and the lower 6½-in. hole. A fish was left in the hole, bottom at 16,000 ft. An extensive fishing job recovered all of the fish down to about 15,000 ft. The fishing operations were hampered severely by the same problems that had restricted drilling operations.

There were a number of potential producing zones in the 6½-in. openhole section below 13,500 ft and others behind the bottom 1,000 ft of 7⅝-in. intermediate casing. The operator decided to run a 5-in. FJ production liner, set immediately above the fish, and complete first in the lower prospective zones through or around the fish. The lower formations were very hard and probably would stay open, permitting production from the zones opposite the fish.

The liner was run and cemented, the top was cleaned out and tested, and the float collar and shoe were drilled out with mills. There was no cement below the float shoe, and the hole was washed out down to the top of the fish at 15,000 ft with 15 lb/gal mud, which gave a small overbalance.

A wireline packer was run, but would not enter the liner, and was pulled. The liner was cleaned out again. A permanent packer with a pump-out plug below the mill-out extension was run and set about 50 ft above the bottom of the liner. The packer seated satisfactorily, but the setting tool would not release, so the running line was pulled out of the rope socket. The setting tool was recovered with an overshot.

A tapered 2⅞-in. by 2⅜-in. production tubing string was run with a sliding sleeve and stinger on bottom. The hole was displaced with 12 lb/gal packer fluid, and the tubing stung into the packer in the bottom of the liner. Either the packer, the sliding sleeve, or the seals on the pump-out plug leaked because both the casing and tubing had pressure. A blanking plug was run but stuck, and the wireline pulled out of the rope socket. Other completion problems were encountered, and finally the liner was cleaned out, including milling out the packer. A second completion assembly was run. It failed because of leaking and was pulled, including a wireline fish.

The well finally was completed by running completion tubing into the hole with a polished bore receptacle (PBR) seal assembly. The hole was circulated with completion fluid under pressure, the tubing was latched into the PBR on top of the 5-in. liner, a plug was set in the tubing, the blowout preventers nippled down, and a Christmas tree nippled up. The plug was removed and the well flowed to clean up. Drilling mud, formation debris, and some junk were recovered. The well almost died several times while flowing heavier mud, then cleaned up and produced clean gas. The whole operation required over two months to complete.

One main criterion for determining when to stop fishing or what other action to take often is controlled by economics. Calculate these in a regular manner. For example, sidetracking economics includes the cost to sidetrack and redrill, compared to the estimated cost to recover the fish. Sidetracking includes the cost to set a cement plug, dress off and sidetrack the hole and drill to the point where the hole was lost. Base the redrilling cost on the actual cost from the original well. Allow for problems if appropriate. However, economics may be less important in some situations. For example, if a wildcat well drilling a large favorable prospect is lost, then this may be taken as the risk loss of doing business and the prospect redrilled.

Prudent operations or obligatory considerations may override economics. It may even be mandatory to continue fishing. Sometimes the fish left in the hole allows communication between oil and gas zones so reservoir energy is lost, thus reducing recoverable reserves. The fish may provide an undesirable channel for fresh or saltwater into productive zones.

The fish may include a radioactive tool such as a neutron source. Regulatory agency rules normally require a diligent effort to recover the tool or isolate it with cement. Otherwise future operations must be conducted so they are a safe minimum distance from the radioactive tool. Normally regulatory agency approval is required prior to abandoning the hole.

Many other factors affect the decision to quit fishing. Some of these include length and depth of the open hole, the type of formation, hole sizes and depths. The overall effect of all factors must be considered when deciding whether to quit fishing. If it is determined that fishing will be unsuccessful, four alternatives are available: (1) plug back and sidetrack; (2) plug and redrill; (3) complete in, through, or above the stuck assembly; or (4) plug and abandon the hole.

Sidetracking often is an obvious alternative to continued fishing. Normally it is a relatively easy procedure, but it may be complicated by hole and formation conditions. Vertical, directional, and horizontal wells may be sidetracked in open or cased holes. There may be more problems sidetracking in high-angle directional and horizontal holes, especially in horizontal holes. Sidetracking in cased holes often is more difficult because of the need to mill part of the casing.

If sidetracking is a reasonable option, consider whether the casing was used to eliminate a formation problem, since the same formations may be exposed in the sidetracked hole. Sidetracking in or near pore pressure transition zones can be especially difficult, especially for casing set into the transition zone. This casing normally prevents lost circulation in the higher zones with lower pressure because of the heavier mud required to drill the higher pressured zone. Sidetracking here may repeat the problems of drilling transition zones. The option of setting another string of casing is costly and may not be reasonably possible because of reduced hole and casing sizes. Do not overlook these sidetracking problems. They are realistic and can be severe. Otherwise, problems and risks associated with sidetracking can range from hole problems due to doglegs to drilling back into the original casing.

Shallow oil wells less than 6,000–8,000 ft are seldom sidetracked. These wells will probably be placed on an artificial lift at a later date. The most common artificial lift is rod pumping and there are rod problems in sidetracked holes. Normally the most economical procedure is to plug the hole, move, and redrill. Gas wells may be sidetracked. Consider suspending fishing operations for holes in this depth range when the total cost of the fishing operations is approximately one-half the total cost of drilling to the point where fishing operations began.

There is a tendency to conduct more extensive fishing operations at shallow depths for holes projected to deeper depths (more than 15,000 ft). These normally have large casing strings at shallow depths, representing an appreciable investment so that additional fishing costs are justified. They may be sidetracked, but the operator should ensure that this will not affect deeper operations. There is less hesitancy to fish extensively or sidetrack these holes at deeper depths because of the large costs that have already been expended.

In special cases there may be an alternative to fishing: completing the well above, inside, or through the stuck assembly (Fig. 2–4). This depends on the location of the prospective producing horizons relative to the stuck assembly, the strength and size of the assembly, the relative severity of the fishing job expected, the type and volume of production expected, and other factors.

One objection to these types of completions is the cost of drillpipe and collar fish lost in the hole. Except for very easy fishing jobs, the cost of continued fishing operations will often offset the cost of the lost downhole assembly in a short time, especially after derating it to a used condition.

Completing through stuck drillpipe normally is not recommended. The completion and future operations are almost invariably hampered by the small hole because of the small diameter of the completion string and the necessary smaller diameter of the tools used to work inside this string. Other common problems include inadequate zone isolation, possible formation damage when cementing, and difficulty during stimulation. Installing pumping equipment for oil wells at a later date often is a problem.

Figure 2-4   Completing in or above a fish

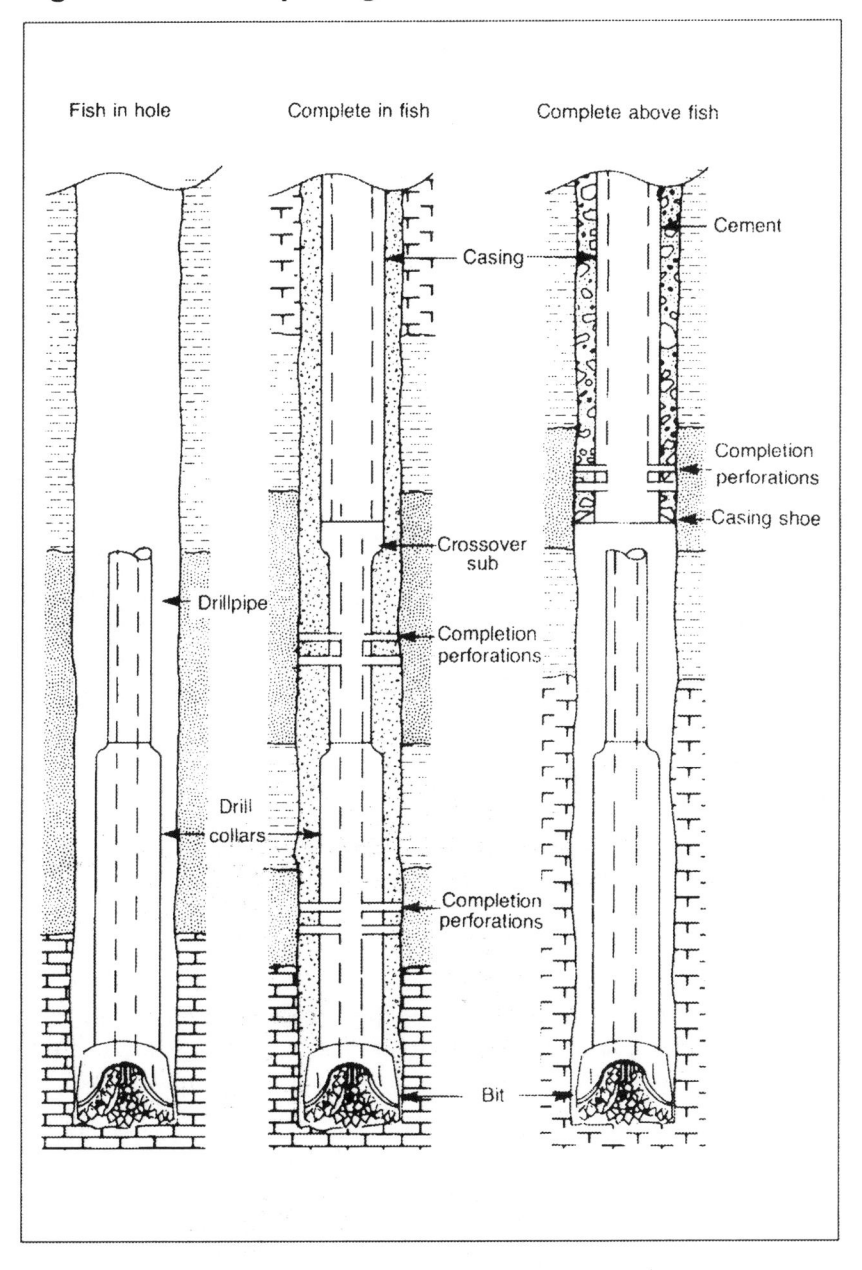

General Rules and Procedures 95

These completions normally are restricted to shallow oil and gas wells, medium-depth gas wells, and special circumstances. If the drilling assembly is stuck below the prospective producing horizon and cannot be recovered efficiently, then completion in the upper zone definitely should be considered, especially on a wildcat. Often it is possible to recover a sufficient amount of the downhole assembly so that casing can be run through the upper zone.

Single-zone gas wells at shallow to moderate depths are most commonly considered for completion through stuck drillpipe. The zone can be block squeezed and perforated if the fish cannot be circulated for cementing. Perforating guns are adequate with sufficient penetration, even for most drill-collar assemblies. Another alternative may be to part the drillpipe above the stuck point and run a tieback casing string.

POST-ANALYSIS

Post-analysis of a fishing job basically includes a report of what happened, how it was corrected, and what action will prevent it from happening in the future. The report is not intended to and should not be used to place blame. Any criticism should be constructive. Prepare the report a relatively short time after finishing the fishing job. The person directly in charge of the fishing job normally prepares the report. Also have it reviewed by the various personnel involved.

Much of the information for the report is found in the earlier investigations to learn all details, the fishing plan of action, and when to quit. Often there is a tendency to forget all about the fishing job when it is over. This is somewhat natural. But something should be learned from almost any fishing job.

Learning what happened may prevent a future occurrence. For example, a fishing job may be caused by a pin failure in the lower drill-collar assembly. Check the remaining drill collar tool joints. If one connector failed, the other connections were exposed to the same conditions and also subject to failure. The connectors could have failed from excess torque. This means restricting future operations to decrease torque or increasing the frequency of inspections to detect damaged connectors.

The Post Fishing Report is not "hindsight," but an attempt to learn the facts and make logical deductions. The following questions are examples of what should be considered.

- What circumstances led to the fishing job?
- What rig operation was being conducted? Where were the rig personnel, and what were they doing?
- Did the drillstring part, and if so where? Was it caused by an equipment failure or operational or formation problems?

- Was the drill- or work string stuck differentially, with drill cuttings or caving, an object dropped down the hole, or in a tapered hole or fracture, etc.?
- Have similar failures occurred or almost happened in the recent past?
- Was the mud weight being increased or decreased? Were the mud properties being changed, or did the mud contain dissolved gas?
- What are the formation characteristics (sands, washouts, doglegs, etc.)?
- Give a complete description of the fish with all measurements, where it was stuck or positioned, and other pertinent details.
- Was immediate action initiated? What was done and how effective was it? What other actions should have been taken?
- What equipment or materials were on location to aid in immediately freeing stuck pipe or initiating the fishing operations? Should additional equipment or materials be considered?
- Was a fishing job required? Were there obligatory reasons for fishing? Why was fishing terminated?
- Was a plan of action prepared and documented? Did it need revision?

GUIDES TO FISHING

Guides to fishing generally apply to all fishing or sticking situations. General guides are more detailed with a broad application. Summary guides generally are short, explicit, and to the point. These are similar to general preventive actions. Many guides are included in various places in the text. Some are summarized here as reminders, not necessarily listed in the order of importance.

Work on the Fish from the Top Down

Work on fishing and related situations from the top down and keep the downhole assembly above problem areas. For example, if a keyseat is developing in the open hole above the fish, consider it as part of the fishing job and remove it before continuing fishing. Run an assembly through a hazardous hole section to catch a fish only as a calculated risk.

One exception is when high-pressure zones are exposed in the wellbore, especially with lost circulation. It often is best in these situations to work with the assembly deeper in the hole so that weighted fluids control high-pressure zones more efficiently. This is especially important when a lost circulation zone occurs above the high-pressure zone or if there is a long interval between the two zones. The assembly may stick, but this is more acceptable than the risk of a blowout.

Select the Strongest Catch Tool

Select the fishing tool that makes the strongest connection, usually a screw-in connection. A second choice is an overshot connection. Normally use a die collar or taper tap as a last resort. They do not have a strong hold, are less strong, and are more likely to break. Future fishing operations can be severely restricted if the tools break, leaving the bottom part in the top of the fish.

Pull the Fish out of the Hole without Delay

Pull the fish carefully out of the hole immediately after catching it, subject to extenuating circumstances. The connection to the fish usually is weaker than other connections in the assembly. Most catch tools are stronger in tension than in compression. Also some catch tools such as the slip-type overshot are released by slacking off. Therefore maintain the catch in tension and pull it immediately to avoid the risk of losing the fish.

This is especially important in the case of a kick or other blowout situation. It often is difficult to control these situations with a fish on the bottom of the work string, so pull it as soon as possible. Pulling the fish immediately also applies to any assembly with questionable integrity. These include those with leaks, dropped work strings that may be bent, and similar situations.

Do Not Rotate When Pulling a Fish

DO NOT rotate the fishing string when pulling a fish. There is a risk of losing the fish by rotation, or shaking or vibrating it off. Disconnect the pipe as it is pulled with hydraulic drillpipe tongs or chain out.

Do not pull the fishing string high above the rotary, only high enough to set the slips. There is a risk of loosening the catch on the fish with a long slack-off.

Expedite Fishing Operations

Fishing is a high-risk operation, and the risk increases with time. This applies to actual fishing and regular operations. Don't rush to the point of endangering the operation, but speed up the work to finish the job in a prudent manner.

Eliminate the Cause of Fishing if Necessary

Often the cause of the fish may severely hamper fishing operations. For example, caving may cause a fishing job. Additional fish may be lost if the cause of the first fish is not eliminated. It is not uncommon in these conditions to lose a second or even third fish in the hole while fishing.

Don't Arbitrarily Assume the Fishing Job Is Easy

Many fishing jobs are not difficult and are completed with a minimum of difficulty. However, complications can develop quickly, often leading to extensive problems. It is not uncommon to have a second fish or "fish on top of a fish." Be aware of complicating factors.

Priority of Operations

Remember, the main objective is to recover the fish. Other operations may be required as a secondary objective. It may be necessary to complete the secondary objective, such as removing a keyseat by reaming, before returning to the primary objective of fishing.

Do not continuously repeat unsuccessful operations or procedures. A procedure may be indicated as the best way, but if it does not work after several attempts, then try something else.

Operations Recorders

There are various types of recorders used in drilling operations. Various parameters are recorded, such as drillstring movement and weight, rotary torque, pump pressure and strokes, pit volume, mud properties, and others, depending upon type and number of instruments.

These tools help both during drilling and in sticking and fishing situations, serving many useful purposes. They aid in checking for and verifying problems, and possibly in determining procedures for solving them. A close check of the mud pressure and volume and pump strokes helps detect leaks before they cause a fishing job. Mud gas detectors show the amount of gas in the mud, serving as an early warning of a kick. Pit volume and flow-line recorders help detect kick situations in time to control them with only minor difficulty and risk. Pit volume totalizers determine very small volumes of lost circulation and whether the hole is filling properly when pulling the pipe. The pipe depth and weight recorders help to determine whether keyseats are forming and the amount of pipe drag. Torque recorders indicate hole condition. Other parameters can be recorded depending upon need.

Stuck Condition of Fish

Don't worry unnecessarily about the stuck condition of the fish. It is already stuck and its condition normally will not deteriorate. Possible exceptions are differential pressure or wall sticking; moving or plastic or fluid sensitive, unstable formations; and excess circulation and work string movement immediately above the fish.

Keep Top of Fish Clean

Don't tag or otherwise touch the top of the fish with the work string too often; it isn't going anywhere. Tagging may damage the catch area, increasing the difficulty of catching the fish. Tagging the fish too hard or too often may cause it to fall to bottom if it is not on bottom already. One of the best examples of this is a fish caught in a keyseat.

Don't circulate excessively immediately on top of a fish in the open hole — only 5–10 minutes, then pull up 20–30 feet and continue circulating for regular hole cleaning. Otherwise, drill cuttings may settle around the top of the fish, sticking it tighter, and may possibly require washing over, a higher risk procedure. Circulating can wash out around the top of the fish, increasing the difficulty of catching it. Cuttings and formation debris may fall inside the top of the fish, creating a bridge. This may require cleaning out, a higher risk procedure. It restricts running tools such as a freepoint log or back-off shot inside the fish.

Anticipate Mechanical Repairs

Maintain the rig and mechanical equipment in good condition as a standard procedure. However, try to anticipate the mechanical condition of the equipment and make inspections and repairs as needed, especially before extended downhole operations. Check the blowout preventers, closing unit, and choke manifolds. Cut or slip the drilling line and make brake adjustments. Ensure that the pumps are in good condition. Check the surface circulating system, repack the swivel or top drive, and check all valves and surface lines.

Always consider the condition of the fishing string. Inspect it if its condition is questionable. Inspect all fishing tools, especially on deep, more critical fishing jobs.

Summary Guides to Fishing

Observe the following summary guides.

- Do not run anything in the hole that you are not prepared to fish out.
- Keep the fishing assembly in a clean hole.
- Do not underestimate the importance of working to release the fish.
- Always have good accurate measurements, including an accurate description of the fish with precise measurements.
- Minimize torque and drag to prevent obscuring surface indications of downhole fishing tool action.
- Always try to recover the fish in one piece.
- Do not wash over as the first fishing operation.

- Do not wait on fishing tools.
- Do not run unnecessary leak-off tests.
- Use close tolerance outside diameter skirts or "no-go" tools while fishing for wirelines.
- Use milling as a last resort.
- Maintain an accurate count of all tubulars on location and record in the drilling report.
- Do not shock the hole by surging and swabbing.
- Always maintain good, consistent, high-quality mud.
- Monitor all surface and downhole indications while fishing.

Take Reasonable Risk — Improvise

There may be a tendency to become overcautious while fishing. This is partially due to unfamiliarity. Caution is advised but in some cases may be almost as undesirable as taking unnecessary risks. Take reasonable risks based on good judgment.

For example there are times when drilling or milling an object, such as a plug, when the object will rotate under the bit so it cannot be cleaned out efficiently. Setting extra weight or spudding vigorously may break or stick the object so it can be drilled efficiently. In some fishing or sticking situations, it may be necessary to pull above the minimum yield too near the ultimate strength of the work string to release the fish. Another example is cementing or plugging lost circulation. If bit cones are known to be lost while drilling, then run on the bit shanks for 10–20 minutes. This breaks up the cones into rubble so they can more easily be caught with a basket. USE WITH CAUTION.

Situations may occur that should not happen, such as dropping the drilling assembly, running the wrong tool, etc. In these cases the main objective is to correct the problem. Sometimes the only way to do this is to improvise and look for better, faster, and safer ways to conduct operations.

Catching a Fish with an Unmatched Pin

A shallow 4,500 feet on-structure wildcat had 7-in. casing at 1,400 ft. Hard, abrasive formations were drilled in a 6¼-in. hole to TD. An attempt to log failed due to a fluid sensitive, caving shale near TD. A cleanout trip was made with the 3½-in. drillpipe, 10 stands of 5 X 2⅛-in. drill collars, and a hard-formation insert bit. The shale caved and stuck the drilling assembly while circulating with air on bottom.

The drillpipe was worked for about 6 hours without releasing it. Stretch data indicated sticking at or below the top of the drill collars. The hole

General Rules and Procedures

was loaded with lightweight mud, circulated, and the pipe worked another 6 hours without releasing it while waiting on fishing tools.

A free-point showed the bottom six drill collars 100% stuck and the seventh 25% stuck. The drill collars were backed off, leaving a fish of seven drill collars and a bit, with a 4½-in. regular box looking up. A fishing assembly was made up and run in the hole with a pin down to screw into the fish. After circulating, an attempt was made to screw into the fish. The pipe would torque and then jump free with an overpull of about 10,000 pounds. Checking surface tools indicated that a 4½-in. full-hole pin was run on the bottom of the fishing assembly by mistake.

Maximum torque was applied to the fish, including bumping back to jam the pin into the box. This could have damaged the box, but probably not severely because of the relatively low weights involved. Another consideration was that the location was isolated, and obtaining other tools would take 12 hours. The fishing assembly was torqued tightly with limited vertical bumping to help jam the pin into the box. The fish began to rotate slowly with the relatively high torque. The fish was released and recovered. Using the mismatched connection was successful.

The error in running the wrong pin was obvious. However, the INNOVATIVE procedure to recover the fish saved over 12 hours of rig time since waiting on the correct tools could have meant more time for the formation to cave. The procedure was a calculated risk, but successful. There was another, less obvious error. The full string of drill collars should not have been run on the cleanout trip. This is a common error in this situation but it is not recommended, based on the rule, "Don't run anything in the hole that is not needed."

PORE PRESSURES AND TRANSITION ZONES

Blowouts are one of the most severe problems encountered in drilling, the least serious result of which usually is fishing. Prevent blowouts with a good understanding of pore pressures, transition zones, and detecting and controlling kicks.

Pore Pressure

Pore pressure is the pressure of the fluid within the pore space of the formation. Normally it is equivalent to the hydrostatic pressure exerted by a column of water with a vertical height equal to the depth to the point of measurement. Normal pressure gradients depend upon the salinity of the water in the pore spaces. Salinity ranges from a freshwater gradient of 0.433 psi/ft or 8.33 lb/gal to more common saltwater, which has 20,000 ppm Cl, equivalent to 0.442 psi/ft or 8.5 lb/gal. Higher gradients of saltwater with 80,000 ppm Cl equivalent to 0.468 psi/ft or 9.0 lb/gal occur.

Abnormal pore pressures are those above or below the normal range. They commonly have a limited areal extent. The common order is increasing pressure with depth, but pressure reversals occur with a lower pressure below a zone of higher pressure.

The most common causes of abnormal pressures are geological conditions such as tectonic forces, depth of burial, rate of sedimentation, matrix strength, and varying degrees of repressured or depleted formations. They commonly occur in basins and areas containing sediments, such as shale beds over several hundred feet thick, that serve as a pressure seal. Sand sections within these geopressured shales frequently are over-pressured.

High-pressure formations have pore pressures in the range of 0.62 psi/ft or 12 lb/gal to 0.78 psi/ft or 15 lb/gal, but some exceed 1.10 psi/ft or 21.2 lb/gal. Subnormal or low pore pressure formations have pressures as low as 0.25 psi/ft or 4.8 lb/gal.

Abnormal pressure formations with high permeability cause problems ranging from blowouts to lost circulation. Otherwise they are not dangerous to drill as long as the formation has a very low permeability so fluids cannot move into or out of the wellbore at dangerously high rates. High-pressure formations containing gas are potentially more dangerous than those containing oil or water because of the greater compressibility and reduced density of gas compared to a liquid.

Pore-Pressure Plots

Pore-pressure plots help detect and predict pore pressure. They are a major aid in locating and evaluating transition zones and the magnitude of increasing or decreasing pore pressure. They help select the optimum mud weight to maximize penetration rates and maintain a margin of safety to prevent kicks. The plots are based primarily upon drilling and geological information as listed in Table 2–1.

Table 2–1 Source data for pore-pressure plots

Drilling and MWD Data	Well Logs and Other Measurements
Penetration Rate	Well Logs
Mud Temperature	Resistivity
Shale Density	Gamma
Gas in the Mud	Density
General Lithology	Sonic
Cuttings	Seismic
Shale Density	Shale Density
Chlorides	SP Log
Paleontology	Neutron Log
	Induction Electric Log
	Lateral Log

The data listed in Table 2–1 are used to develop the plots listed in Table 2–2.

Table 2–2 Pore-pressure plots and observations

Plots	Plots or Observations
Penetration Rate	Mud Rheological
D - Exponent (Modified)	Flow Line Temperature
Shale Density	Cuttings Lithology
Drill Cuttings Measurement	Mud Pit Volume
Neutron Logs	Mud Flowback
Sonic Logs	Bit Records
SP (Spontaneous Potential)	Drill Cuttings
Resistivity	Lithology
Mud Gas Content	Size and Shape
Background Gas	Volume
Drilled Gas	Torque and Drag
Connection Gas	Sloughing and Caving
Trip Gas	Tight Hole
Chlorides	

There are two general types of pore-pressure plots, each constructed in a similar manner. The construction and evaluation of these plots is well documented in the literature and summarized here. General pore-pressure plots are based on seismic and logging data, pore-pressure plots on wells in the area, and similar data. They serve as a guide for preparing the drilling and casing program and as a general reference to help interpret pore-pressure plots constructed while drilling. Often these are less accurate and definitive compared to drilling pore-pressure plots.

Drilling pore-pressure plots are prepared primarily from data and meassurements recorded during drilling. Measurement-while-drilling instruments can be very helpful because they record electric log-type data that is immediately available. Having immediate data is a major advantage over waiting until a section is drilled and logged. Data recorded after logging is normally too late to help with drilling the well.

Gather and plot data in a conventional manner. Plot at the correct depth by correcting for lag, usually on two- or three-cycle log paper. Common plotting scales include 1-in./100 ft for detecting trends and 1-in./10 ft for finer definition. Recognize that a base-line shift may be required at major geological boundaries.

Normally use the general area pore-pressure gradient as a base line. Plotted points give the pore pressure above or below this. Reference these from the base line with standard scales. However, there is always a question about the accuracy of these standard scales in each specific well because every well seems to be exceptional. Also, standard scales may not be available for wildcat wells. Calibrate and correct the standard scales by determining the pore pressure whenever possible.

Obtain a starting pore pressure with a leak-off test in the open hole immediately below the casing shoe. Conducting the test correctly will not hamper operations. Leak-off tests at greater depths may be unreliable because the point of leakage may be indeterminate due to longer openhole sections. An openhole drillstem test often provides pore pressure at a specific depth.

Various types of gas include gas shows, swabbed gas, connection gas, drill gas, trip gas, drill cuttings gas and background gas. Relate the depth of gas entry by lag time. Some of these are more accurate and others, such as swab gas, may give a false show. Connection and trip gas and some gas shows may indicate the pore pressure subject to the mud weight. Do not confuse connection gas with drill gas. Sand stringers in massive shale sections can be very good indicators of pore pressure, especially when drilling near balance. Use gas-cut mud data with caution because of the tendency of gas to migrate into the mud, especially oil mud. Check the effect of background gas due to the natural gas content in oil mud by measuring the amount of gas in weathered diesel.

Deceptive Gas Shows in Oil Mud
A well drilling with oil mud maintained an overbalance of 0.2 lb/gal using standard indicators such as trip and connection gas and several pore-pressure plots. A sand at 10,000 ft with a pore pressure of 13.8 lb/gal gave a small gas show while drilling with 14-lb/gal mud. This gas show continued to be identified as a false connection gas spike while drilling several thousand feet deeper with a mud weight of 16.3 lb/gal, an overbalance of 2.5 lb/gal at 10,000 ft, and verified by pore-pressure plots. The only reasonable explanation for the continued show of gas from the sand at 10,000 ft was some type of osmotic or solubility action, since the gas could not have flowed into the mud by normal pressure differential action.

Extrapolate pore-pressure plot trends to help predict pore pressures in the underlying formations. The reliability of the projection decreases with distance. There are no set rules on the distance that can be reliably projected. However it normally is at least 50 ft and can be several hundred or more.

Transition Zones

Transition zones are the depth interval in which the pressure gradient changes from the gradient in the upper zone to an abnormal pressure in the lower zone. They usually also represent the changing geological environment between the two pressure regimes. They are based on pressures but defined in terms of pressure gradients.

Prevent kicks and possible blowouts while drilling by detecting transition zones. Predict the extent of transition zones to select exact casing points, especially in situations requiring precise mud weight control. Wells have been either lost or abandoned before reaching total depth because of setting one or more casing strings too high, a costly, preventable mistake.

Pressure gradient changes across transition zones are relatively constant but variations occur. Several transition zones may occur in a wellbore but one is most common. Most zones have increasing gradients but decreasing gradients occur. Small gradient changes may not be significant, but larger changes can be very important.

Detect transition zones by their special properties, mainly using pore pressure plots (Fig. 2–5) and indicators described in the section on kick detection. Verify the transition zone by correlation with nearby wells if data are available. The nearest well may be a long distance away but pore-pressure plots tend to be similar over large areas. Detection in exploratory wells may be more difficult but normally should not be a major problem with the volume of potential information and the reliability of drilling indicators.

Figure 2–5   Pore-pressure plot

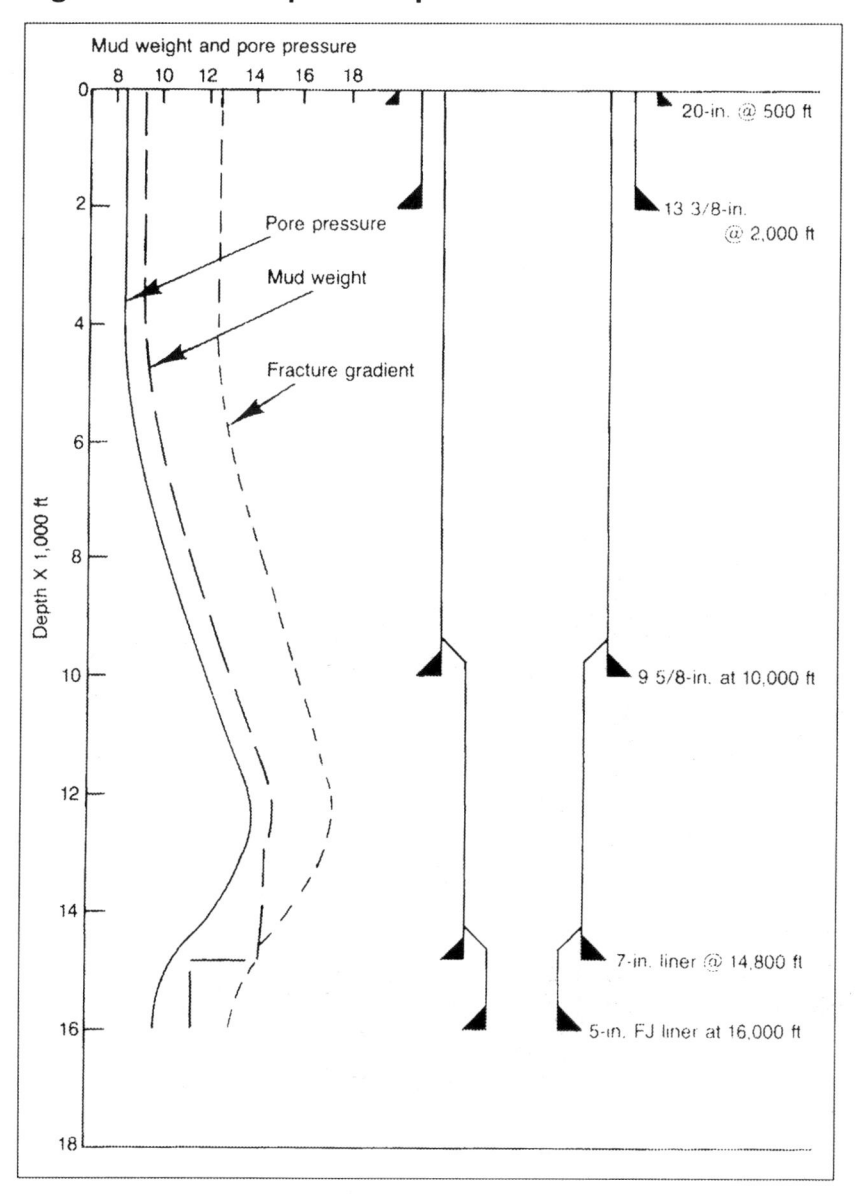

The comment has been made too often to the effect that a well was drilling ahead in a normal manner when "all of a sudden it drilled off into a high pressure zone and kicked!" The most probable happening in this case was failure to observe drilling indicators and pore-pressure plots. This should and can be prevented. Transition and high-pressure zones, like keyseats, warn the observant operator.

General Rules and Procedures 107

Table 2–3 illustrates pressure gradients and their length in transition zones. There should be sufficient time to recognize pore pressures and take action accordingly in most drilling situations. There could be a problem in a few cases of fast drilling and deeper wells with longer circulating and related lag time. Alternate actions in these cases might include reducing the drilling rate, circulating, and using a slightly higher mud weight as a safety factor.

Table 2–3   Transition zones — change of pore-pressure vs. depth

Top of Transition Zone Mud @ Depth, Wt., ft lb/gal	Bottom of Transition Zone Mud @ Depth, Wt., ft lb/gal	ft/psi	psi/100 ft	lb/gal/100 ft
Beaufort Sea				
9.0 @ 4,000 –to– 13.0 @ 5,800		450	0.44	0.22
10.0 @ 8,000 –to– 15.0 @ 9,200		140	0.42	0.42
Louisiana				
13.1 @ 7,600 –to– 17.0 @ 11,000		872	0.17	0.12
12.2 @ 8,700 –to– 15.3 @ 10,000		419	0.24	0.24
9.0 @ 12,200 –to– 14.0 @ 12,800		120	0.83	0.83
9.0 @ 12,300 –to– 15.0 @ 13,500		200	0.50	0.50
12.2 @ 13,000 –to– 17.0 @ 15,000		417	0.24	0.19
9.0 @ 11,200 –to– 11.0 @ 12,500		650	0.15	0.15
11.0 @ 12,500 –to– 15.0 @ 12,800		75	1.33	1.33
15.0 @ 12,800 –to– 17.5 @ 14,000		480	0.21	0.21
9.0 @ 11,300 –to– 16.3 @ 12,600		178	0.56	0.56
Tuscaloosa Trend (several states)				
9.0 @ 16,500 –to– 12.5 @ 18,000		428	0.23	0.23
12.5 @ 18,000 –to– 16.0 @ 18,700		200	0.50	0.50
16.0 @ 18,700 –to– 13.0 @ 20,000		–314	-0.30	–0.23
8.0 @ 13,000 –to– 12.0 @ 14,800		450	0.22	0.22
12.0 @ 14,800 –to– 15.0 @ 17,800		1,000	0.10	0.10
15.0 @ 17,800 –to– 9.0 @ 18,300		–83	–1.20	–1.20
9.5 @ 10,000 –to– 13.0 @ 11,800		514	0.19	0.19
13.0 @ 11,800 –to– 16.0 @ 12,700		300	0.33	0.33
16.0 @ 13,200 –to– 10.0 @ 14,000		133	0.75	–0.75
Texas				
9.0 @ 9,000 –to– 13.0 @ 10,200		300	0.33	0.33
9.0 @ 5,200 –to– 14.2 @ 8,700		673	0.15	0.15
Utah				
9.0 @ 14,700 –to– 14.0 @ 15,400		140	0.71	0.71
Unidentified				
13.5 @ 12,500 –to– 17.2 @ 13,500		270	0.27	0.37
9.0 @ 9,100 –to– 16.0 @ 12,800		529	0.19	0.19
12.0 @ 17,000 –to– 16.5 @ 18,200		267	0.38	0.38

The data in Table 2–3 illustrate general pore-pressure changes in transition zones. DO NOT USE for a specific well.

Blowout detection and prevention procedures are well established. As an important note of warning, maintain the blowout control equipment to keep it in good operating condition and adequately tested. The crews should be well trained with blowout drills.

KICKS

A kick occurs when formation fluids, usually gas, oil, saltwater (kick fluids) or a combination of these over-balance the mud column and begin flowing into the wellbore and toward the surface. They always create a high-risk condition. Most are preventable. However, they can be detected in sufficient time to control them, minimizing any serious problems. Otherwise they endanger personnel and can cause situations ranging from stuck pipe to a blowout and loss of the hole. In any drilling situation, and especially during kicks or blowouts, be aware of gas and oil, which are highly inflammable. Satisfactorily predicting transition zones and detecting and controlling kicks has been verified by extensive, reliable field experience.

Detection

Kick detection is the procedure of identifying a kick, the precursor of a blowout. It includes monitoring various indicators and recognizing when to take action. The most successful detection recognizes the kick very shortly after it occurs and in time to control it before a blowout develops. In the optimum case, it may not be necessary to shut the well in.

Detect kicks by monitoring various indicators. Prepare pore-pressure plots, update and review as necessary to detect transition zones and determine the correct, safe mud weight. Increased penetration rate, flowline increases and continued flow during connections have consistently been the best and earliest indicators of a kick, followed closely by flowline and pit volume increases, gas in the mud, and the shape and increased volume of cuttings and caving. Look for tabular, splintery-shaped cuttings with sharp edges, and sometimes conchoidal-shaped sides. Gas in the mud includes background, connection gas, trip gas, and cuttings gas. Gas analysis for the individual components gives a good indication of the type of kick fluids. Check micro-fossils or their environmental indicators if they are known to be associated with the transition zone. Increasing mud chlorides may indicate a pending saltwater flow.

Generally drill with a near-balanced mud system. Contrary to popular opinion, this is probably the safest and most economical procedure. Drilling generally is faster with fewer problems, and kick

General Rules and Procedures

indicators are more reliable and easier to detect. Improve indicator accuracy by maintaining constant parameters such as bit weight, rotary speed, and mud properties. Ensure that the rig crews have had adequate BOP and kill procedure training and are proficient at BOP drills.

Control

Controlling a kick is the procedure of circulating and removing kick fluids while maintaining full control of the well and restoring it to a safe operating condition. The following kill procedure is a brief summary.

Slow pump rates are used to help kill a kick. Normally two are recorded at about one-third and two-thirds of the normal pump rate while drilling. Record the pressure and pump strokes per minute (SPM) on the daily report. Kill procedures depend upon the situation when a kick occurs. A very minor kick may be circulated out without closing the pipe rams.

In most cases, stop drilling, shut down the mud pump, pick up the pipe until the bottom of the kelly is above the rotary, shut the well in by closing the UPPER drillpipe rams and record the shut-in pressures. ALWAYS close the UPPER pipe rams, leaving the lower pipe rams as a safety shut-off or as a replacement for the top pipe rams if they are damaged or become worn.

The shut-in drillpipe pressure determines the pressure and mud weight needed to kill the well. The initial pit volume increase indicates the volume of kick fluids entering the wellbore at the pressure conditions where the fluid enters the wellbore. Normally increase the mud weight and circulate the well with the correct mud weight to kill it, preferably in one circulation. Circulate through the drillpipe and out the choke manifold at one of the slow pump rates based upon the SPM. Control the flow rate through the choke manifold to obtain the same drillpipe pump-in pressure for the slow pump rate SPM being used. Two or more circulations may be required for high volume kicks or those in which there is a high gas content in the kick fluids.

Gas solubility in oil mud can increase the risk in some kick situations. Normal free gas expands due to the decreasing pressure as it moves up the hole. The increased gas volume is released through the BOP choke manifold as noted. However, additional gas may be solubilized in oil mud. Pressure reduces as the mud moves uphole and dissolved gas in the oil phase "flashes" out of solution. This gas increases both the amount of gas available for expansion and the expanded volume. Gas dissolved in the mud is released at an increasing rate as the pressure decreases. The released gas tends to lighten the fluid column which further reduces the hydrostatic pressure and releases more gas in a "chain reaction" that is self-accelerating. This requires handling a larger volume of gas at a faster rate when circulating out a kick with oil mud that contains gas in solution.

It may be necessary to flow through the choke manifold at a higher rate for a strong kick, usually because of a higher pressure differential or larger volumes of kick fluids entering the wellbore. This prevents a pressure buildup in the annulus and the risk of bursting the surface casing or causing lost circulation or an underground blowout. Either creates a severe situation. In this case it may be necessary to control the pump in volume based on judgment and experience, and aided by pressures and pit volumes. This situation increases the difficulty of circulating out the kick because of lighter wellbore fluids and reduced surface pressures may allow more kick fluids to enter the wellbore.

There is a risk of sticking while circulating out a kick, but this is preferable to a blowout. In some cases it may be possible to move the drillpipe if surface pressures are not excessive. Gas in oil mud may increase the problem. After circulating the well dead, circulate to ensure stabilized mud weight and wellbore conditions before resuming operations.

The period of time that a well can be shut in before starting to circulate out a kick often is the subject of considerable controversy. Many believe that kick fluids will migrate up-hole very rapidly if the well is left shut-in for any appreciable time. This is not supported by field evidence, even with a high gas content in the kick fluids. However, it is preferable to circulate the kick out as soon as possible because it represents a dangerous situation.

The length of time the well can be left shut in before circulating the kick out safely depends upon the type of kick fluids, mud weight, depth of the kick and related factors. However, experience has indicated that the well can be left shut-in for an extended period of time if surface pressures stabilize. Average kicks at about 10,000 ft have been left shut-in for over eight hours and then circulated out without difficulty. In one case, gaseous kick fluids may have migrated uphole about 2,000 ft in eight hours.

The volume of kick fluids entering the wellbore is a measure of kick severity. An average kick contains 10–15 bbls of kick fluids measured at reservoir conditions. Severity increases with increasing volume of kick fluids and gas content. This is an important reason for monitoring the pit volume closely.

General Rules and Procedures

REAMING

Reaming is the procedure of opening or smoothing the walls of the hole. It serves as a preventive action by reaming a tight or crooked hole. Reaming is a common fishing operation, such as opening the hole to the top of the fish. Reaming often is the only method of correcting a crooked hole, enlarging short bends and turns, smoothing the walls of the hole, and removing keyseats. Reaming does not directly reduce the change of angle. It tends to enlarge the hole to some extent, especially in smaller gauge sections and near the start and end of curved hole sections. The net result is an apparently larger hole with reduced curvature.

Operations

Reaming can be a high-risk operation. Reaming is probably the most common cause of unintentional sidetracking. Other risks while reaming include sticking the assembly and twisting off. The severity of reaming problems probably approaches that of keyseating and wall sticking. Ream only when absolutely necessary. ALWAYS ream with the correct type of assembly and operate it carefully.

Normally ream with the recommended assembly, usually a drill collar reaming, wiping assembly, or string reamers. Reaming with a regular assembly and bit is seldom effective except when reaming a newly drilled hole.

Assembly

The drill collar reaming assembly is the most common reaming tool. It is most commonly used for reaming near bottom, but is also used for reaming higher in the hole.

The assembly is similar to a packed-hole or stiff assembly. It has two stabilizers separated by two drill collars, with a reamer located between them. The reamer may be a vertical or spiral-blade type or three- or six-point roller reamer. Use full-gauge tools and large diameter drill collars to increase.

The two end stabilizers act as fulcrums or contact points. The drill collars act like a leaf spring, forcing the reamer(s) to cut into the inner wall. This helps straighten a crooked hole section or remove a keyseat. In severe cases it may be necessary to run a more limber assembly with smaller drill collars and possibly an extra collar between the stabilizers. A more aggressive tool may have two reamers separated by a drill collar, all between the two stabilizers. Usually drilling jars and a bumper sub are run on top of the drill-collar assembly.

There is a high risk of sidetracking when reaming off bottom with this assembly. Eliminate this with careful operations and by running a pilot bit or bull-nose type hole opener. These virtually eliminate the risk of sidetracking. DO NOT use an under-gauge or worn bit for severe reaming.

Normally ream with a slightly faster rotary. Do not ream too long at any one point. Several moderately fast, light reaming passes are much better than one slow pass. Try to smooth or open the hole over several runs, reaming completely through the crooked, deviated, or dogleg section. This reduces the risk of sidetracking and leaves a smoother wall.

Do not force the reaming assembly into a very tight hole with excessive weight because of the risk of sticking. If the reaming assembly is too aggressive and causes "very hard running," then pull it and run one that is less aggressive. After reaming, run the next assembly into the hole carefully. There may still be a risk of sticking.

It is very important to keep the pipe moving downward, picking up as necessary, while reaming with a reaming assembly or even a regular assembly. Otherwise there is a high risk of sidetracking, especially with a bit on bottom. The bit may form a ledge if the pipe begins to take weight and the reaming rate slows. This leads to sidetracking. Often a developing ledge can be eliminated by increasing the weight and reducing the rotary speed so the bit slips off the ledge and back into the original hole. Usually, the best practice is to ream with the heaviest weight possible, depending upon torque and collar weight. Use the combination of weight and rotary speed that gives the highest reaming rate. Reaming problems are illustrated in the following example.

Reaming in an "S" Type Directional Hole

A well was intentionally sidetracked at 7,000 ft and built angle at 2½°/100 ft to 30°. This angle was maintained to about 4,000 ft and then dropped with a pendulum in a conventional manner. Torque and drag began increasing when the top of the drill collars were below the second bend and continued increasing while drilling deeper. There also was some evidence of keyseat development in both bends. Replacing part of the drill-collar assembly with heavy drillpipe helped reduce the torque and drag problem.

The first remedial step was to ream the keyseat sections with the bit and stabilizers on the pendulum assembly when tripping to change bits. The tools were operated in a careful manner to prevent sidetracking or twisting off. This helped, but the keyseat continued to develop.

A keyseat wiping assembly was run. It had stabilizers at positions 30 and 90 ft, a spiral reamer at position 60 and a bullet-nose hole opener on bottom. The section was reamed but continued to have a heavy reaming action after the upper stabilizer passed below the upper curved section. The assembly should have moved freely downhole. This was continued until the top stabilizer was over 500 ft into the straight inclined hole section. The lower section had not caused problems and should not have required reaming.

The possibility of sidetracking was rejected because: (1) the bullet-nose hole opener could not sidetrack under any normal condition in a medium hard formation, (2) the hole was reamed at about 20 ft/hr with a moderate reamer weight, whereas the original hole was drilled at about 10 ft/hr with a heavier weight and slightly higher rotary speed, and (3) deviation surveys indicated that the hole had the correct course and angle.

The assembly was pulled and found to be in good condition except for worn rollers on the hole opener. Apparently the bottom stabilizer acted as a fulcrum and the lower drill collar served as a pendulum, and the hole opener was cutting and maintaining a ledge and drilling a small half-moon section on the low side of the hole. This had been considered during the latter part of the reaming run but light spudding and heavier weight would not break through. The assembly was rerun and the lower section reamed with heavier weight and spudding. A small amount of hole was reamed and the assembly was then run to bottom without difficulty.

String Reamer

String reamers are conventionally run in the drillstring. They ream while drilling or with the bit off-bottom to remove keyseats and, less commonly, to remove a crooked hole. One reamer is most common but two may be used, sometimes separated by a joint of pipe.

The string reamer is moderately efficient, but there is a high risk of twisting or backing off, or of tool joint failure, often due to metal fatigue. Run them only when the keyseat cannot be wiped out with a drill collar keyseat wiping assembly or even a regular keyseat wiper. String reamers are somewhat of a last resort tool because of the danger of failure while running.

A full-gauge, 360° contact, spiral-type reamer or a six-point reamer minimizes the fatiguing action on the reamer tool joints compared to a vertical blade reamer. In soft to medium-hard formations, a spiral blade stabilizer may provide sufficient cutting action. Use a three- or six-point roller reamer dressed with cutters designed for the formation hardness or for reaming in harder and or abrasive formations,

Figure 2-6   Removing a keyseat with a string reamer

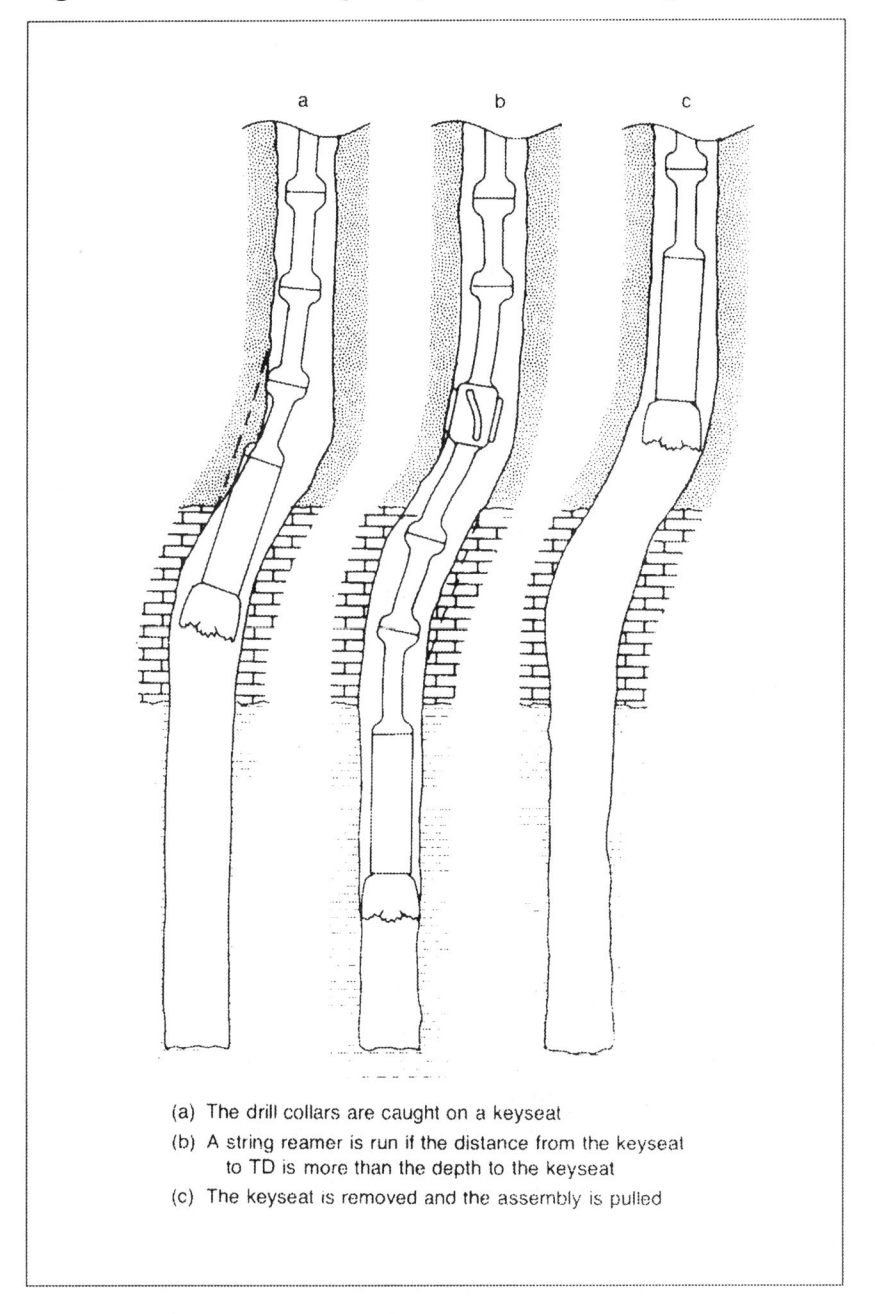

(a) The drill collars are caught on a keyseat
(b) A string reamer is run if the distance from the keyseat to TD is more than the depth to the keyseat
(c) The keyseat is removed and the assembly is pulled

General Rules and Procedures

For reaming while drilling, position the reamer in the drillstring so it is above the section to be reamed when the bit is on bottom. This reams the hole while the bit is drilling. Various reaming runs may be required. There is no risk of sidetracking in this case, but there is a risk of fatigue failure in the tool joints at the ends of the reamer or of hanging up the reamer and twisting off.

For reaming off-bottom, position the reamer in the drillstring, usually about 10 joints above the top of the drill collars. Then ream through the section carefully since the drilling assembly cannot be handled as efficiently with a string reamer. The risk of tool joint damage or hanging up and twisting off can be reduced with careful operations and minimum drill collar weight. Eliminate the risk of sidetracking with the procedures noted.

There is one special hole condition where keyseats can be removed effectively only with a string reamer. In this case the keyseat is located on the top side of a low-angle dogleg section of a highly deviated hole. The drill collar keyseat wiping assembly may not have sufficient stiffness to remove the keyseat. The natural sway of the drill-collar assembly may be more than the dogleg angle. Therefore remove the keyseat with a string reamer.

A string reamer may be used if the pipe is free to move downward but will not pull up past the keyseat, providing that the distance from the keyseat to total depth is greater than the depth to the keyseat (Fig. 2–6). If conditions are favorable, connect a string reamer, lower the pipe, and ream the keyseat section until it is open. If the string reamer wears out, pull it. Work the drilling assembly at this time to see if it can be pulled past the keyseat. If not, run another string reamer. Several reaming trips may be required.

Problems with a string reamer are illustrated in the following example.

Removing a Keyseat with a String Reamer

A 13,000-ft wildcat was drilled in an area with very limited drilling information. Formations were known to be hard and abrasive with expected crooked hole conditions. The original plan included 13 3/8-in. conductor in a 17 1/2-in. hole at 500 ft, 9 5/8-in. surface casing in a 12 1/4-in. hole at 4,000 ft, and completion with 5 1/2-in. casing in an 8 3/4-in. hole. This allowed for 7-in. intermediate casing string with completion in a 5-in. FJ liner, if necessary.

The conductor was set at 300 ft because of caving, unconsolidated surface formation, and shallow water flows. The surface hole deviated strongly to about 12° at 3,500 ft when a fishing job occurred. The 1,500-ft fish could not be recovered and was sidetracked at 1,600 ft. At 2,800 ft the 9 5/8-in. surface casing was run because of caving and deviation problems. It stuck at 2,500 ft, 300 ft off bottom, and was cemented. An unsuccessful fishing job occurred at 5,000 ft and the hole was sidetracked at 4,000 ft. Drilling continued to 12,000 ft, with numerous problems due to caving formations, lost circulation, high hole deviation, doglegs, and high drag and torque.

The hole was logged and data indicated the well would have to be drilled to 14,000 or 15,000 ft to test the objective horizons. Deviation increased to 15° at the total depth, and the hole was in bad condition. Four hole sections had an absolute dogleg over 4°/100 ft. Drill collar reaming assemblies and string reamers were used extensively for the crooked hole and to ream out keyseats. Pipe drag was excessive, and caving and lost circulation problems were becoming more severe.

A recommendation to set 7-in. casing was rejected, and the well was drilled ahead with increasing problems, including severe pipe drag. String reamers were run most of the time, and in one case two string reamers were run in different sections of the hole. Three days were spent working on the keyseat sections with a drill collar keyseat wiping assembly. A tapered drillstring (4½–4 in.) was run, and the number of drill collars reduced to the minimum to reduce overall pipe weight.

On a bit trip at 14,300 ft, the drill collars were worked up through several lower keyseats but could not be worked through one at 4,000 ft. After about 4 hours of work, it was decided to try to drill up through the keyseat with the wiper on top of the drill collars. This appeared to be cutting too deeply into the formation, and the pipe was trying to stick.

A string reamer was connected and the pipe was run in the hole. The keyseat section was reamed for about six hours, passing through the section at various times. While picking up, the reamer stuck in the keyseat where the top of the drill collars originally stuck. It was decided to try to drill upward and rotate the reamer through the section for a period of time before working the pipe down and backing off. The pipe was worked for about two hours when it parted about 6 in. above the top of the first tool joint above the reamer. A later examination of the break indicated a fatigue failure.

The fish did not drop as indicated by an impression block, which also showed the top of the fish forced into the wall of the hole. Various types of fishing tools were run but could not get over the top of the fish. Most of these tools had marks on the sides showing that they were passing the top and hitting the side of the fish. After about 10 days of fishing, the well was plugged and abandoned.

The direct cause of losing the hole was due to keyseats caused by a crooked hole. This could have been prevented by using better drilling techniques, including packed-hole assemblies. However, their limited use is somewhat justified by the excessive formation problems, and they caused or contributed to one of the fishing jobs and the sidetracking operations. Not running 7-in. casing was an obvious mistake.

General Rules and Procedures

DRAG AND TORQUE

Drag and torque is caused by friction as drill tools move, rub, and slide against the walls of the wellbore. Excess drag and torque causes problems in all drilling operations. These generally are more severe in deeper holes and in directional and horizontal drilling. Prevention starts with designing the well plan to minimize drag and torque. This includes casing programs, drillstring, tools, and operations. When possible, a straight hole is an important preventive measure.

Factors that increase drag and torque include a rough, irregular, uneven wellbore wall, crooked hole, changes of hole direction, doglegs, reduced clearance between the walls of the hole and downhole tool OD, tool weight, and poor mud properties. Severity increases at higher angles and with a high number of bends and turns. The dragging and plowing effect of drillstring projections such as tool joints and stabilizers contributes to the problem.

Drag and torque increases with heavier drillstring weight while drilling deeper. They appear to approach maximum sustainable values in high-angle extended-reach holes at measured depths in the range of 15,000 ft below a shallow kick-off point and subject to specific well conditions, including some cases where the deviated hole is cased. Deeper bends and turns cause amplified drag and torque in more shallow bends and turns. The greater drag and torque in lower bends and turns acts against upper bends and turns, causing an overall increased drag and torque.

Severe drag and torque may cause parting or twisting off. It contributes to or causes keyseats and increases the risk of differential pressure sticking. Excess drag and torque reduces effective overpull during fishing and may distort surface indications of downhole fishing tool action.

A torque field measurement is "rounds of torque," sometimes called "back torque." To illustrate, consider a 4½-in. drillstring in a vertical 8¾-in. hole 12,000 ft deep with less than 1½°/100 ft change of deviation, and a maximum drift of 2°. The drillstring normally has about 2 rounds of back torque, increasing to about 4 rounds while drilling with about 25,000 lbs of bit weight. Drag is about 10,000 lbs. Consider the same hole deviated at 6,000 ft at 2°/100 ft, the angle built to 45° and a straight, inclined hole to total depth. Free hanging torque is 4–5 rounds, rotating torque is 5–6 rounds, and drag is about 25,000 lbs under good conditions.

The best preventive action in operations is to recognize excess drag and torque before it causes problems. As a general guide, take action when drag exceeds about 15% of the hook weight of the work string, or when back torque exceeds about ¼ to ½ turns per 1,000 ft of open, straight, hole. These are approximate depending upon depth, length of open hole, deviation, dogleg and related factors.

Generally drag and torque can be reduced by correcting the factors causing them. Often this requires correction while drilling since some are difficult to correct later. Drill tool weight reductions can be effective, especially in the lower hole and for high directional and horizontal drilling. Split bottomhole assemblies with part of the drill collars in the vertical hole section are very effective in horizontal holes. Tapered drillstrings and replacement of some drill collars with heavyweight drillpipe may be helpful. Aluminum drillpipe reduces weight substantially but causes severe operating problems.

Good mud is essential. Increase water-base mud lubricity economically by adding 4–7% diesel oil and mixing well with emulsifying agents. This small amount of oil does not appreciably increase the lubricity of the drilling mud as measured by common tests. However, it has been field-proven to help reduce excess drag and torque and is strongly recommended. It also helps prevent sticking and aids in releasing stuck assemblies. Oil muds have good lubricating qualities and should always be considered for more demanding situations. Other additives used to increase mud lubricity include asphalt, alcohol-base lubricants, and graphite. However, field practice has indicated questionable results with these additives.

Granulated material such as glass, plastic beads, and ground walnut hulls have been effective in many cases for reducing excess drag and torque. Ground walnut hulls can be especially effective in this case. They also may prevent wall sticking and help reduce minor, leaking-type lost circulation.

The method(s) by which granulated materials reduce drag and torque is not well understood. One opinion is that granulated materials embed into the wall of the formation leaving numerous small points to hold the drillstring off the wall of the hole, thus reducing torque and drag and the risk of wall sticking. This appears to have the most merit, based on field experience.

Another opinion is that they act as miniature ball bearings, reducing the friction between the drillstring and the walls of the hole. Therefore they should be most effective in a cased hole. However, there are fewer problems in cased holes, and field experience does not support the use of granulated material in this case.

Granulated materials are applicable to most muds but must be retained in the system to be effective. This generally requires bypassing the shale shaker since particles are larger than efficiently-sized shaker screens, and sometimes other solids-separating equipment. This causes solids build-up and generally is not acceptable. High concentrations may cause bit plugging in a few cases or possibly interfere with the operation of fishing tools and some course- and angle-measuring instruments. Measurement-While-Drilling equipment cannot tolerate high concentrations of solids such as lost circulation material. Therefore, maintaining the material in the mud system and bypassing the solids-separating equipment is not a reasonable solution except for a short period of time.

Glass beads are the most expensive, especially in the volumes normally required. Expense can lead to limited use, so they are not as effective. Plastic beads are less expensive but subject to plastic distortion under heavy loads. They may soften and lose structural strength at elevated temperatures. Ground walnut hulls are the most economical, even with substantially larger additions, as compared to glass and plastic beads. They can be highly effective, but it is important to know how to use them correctly.

For moderate drag and torque, use periodic batch treatments of walnut hulls. Treat with 50% coarse grade and 50% medium grade in soft-to-medium drillability formations. Add 25% coarse grade, 50% medium grade, and 25% fine grade in medium-to-hard drillability formations. Always use at least 25% coarse grade. Common batch treatments use 0.5–2.0 lb/bbl and higher concentrations for more severe conditions. Distribute the additions over two or three circulations. Normally continue operations such as drilling, and bypass the shale shaker and other solids-separation equipment for a short period of time.

For moderately severe drag and torque, and after trying batch treatments, make continuous additions of walnut hulls while drilling or circulating, using the combination of particle sizes as described. A generalized treatment is 0.2–0.5 lb/bbl distributed over 24 hours. Leave the solids separation operating, even though the hulls are screened out and lost. An appreciable part of the hulls apparently embed in the formation, and the treatment has been very successful, substantially reducing drag and torque in many cases. For more severe drag and torque, it may be necessary to shut off the solids separation equipment for a period of time. Circulating through an extra solids- or sand-settling pit may help in this case.

Field evidence strongly supports the beneficial effects, as illustrated in the following example.

Reducing Drag and Torque with Walnut Hulls

An on-structure wildcat had 9⅝-in. casing set at 5,000 ft and was drilling an 8¾-in. hole at 12,000 ft in medium drillability formations with a standard 4½-in. drillstring with 7-in collars. Inclination ranged up to 5° with a maximum dogleg of less than 2°. Pickup drag increased to 25% over hook load and slack-off drag to 20% below hook load while making connections and working on bottom. Torque gradually increased to 7 rounds of back torque. The mud was treated with walnut hulls, first using a batch treatment followed by a continuous treatment. This reduced drag to 10% of hook load and torque to 2½ rounds of back torque.

NOTE: This is only one of many examples of the effectiveness of walnut hulls when used correctly.

Running casing is one of the final actions for correcting severe drag and torque. In some cases it will not cause a big reduction. However, it will save the hole drilled to this point. It eliminates problems of keyseating, differential pressure sticking, and the "plowing effect." Drillpipe rubbers are not used in open holes. They are of questionable value in reducing drag and torque in cased holes, but they do reduce casing wear.

WALL STICKING

Wall sticking, or differential pressure sticking, occurs when a section of the downhole assembly, usually the drill collars, lies next to and literally sticks to the wall of the wellbore. It can be one of the more difficult, and least understood, sticking problems. Consequently, there probably is a higher incidence of wall sticking than is generally recognized by the industry. One reason for this is the failure to recognize the type of sticking and to take preventive actions (Fig. 2–7).

Also, wall sticking is not consistent. As an example, conditions are frequently favorable for wall sticking when running casing, one of the main reasons for keeping the casing moving until it is cemented. However, there are instances where casing was run and left sitting for a relatively long time without sticking. In one case, 7-in. casing was run to 9,500 ft in an 8¾-in. hole, below 1,500 ft of 9⅝-in. surface

Figure 2–7 Wall sticking

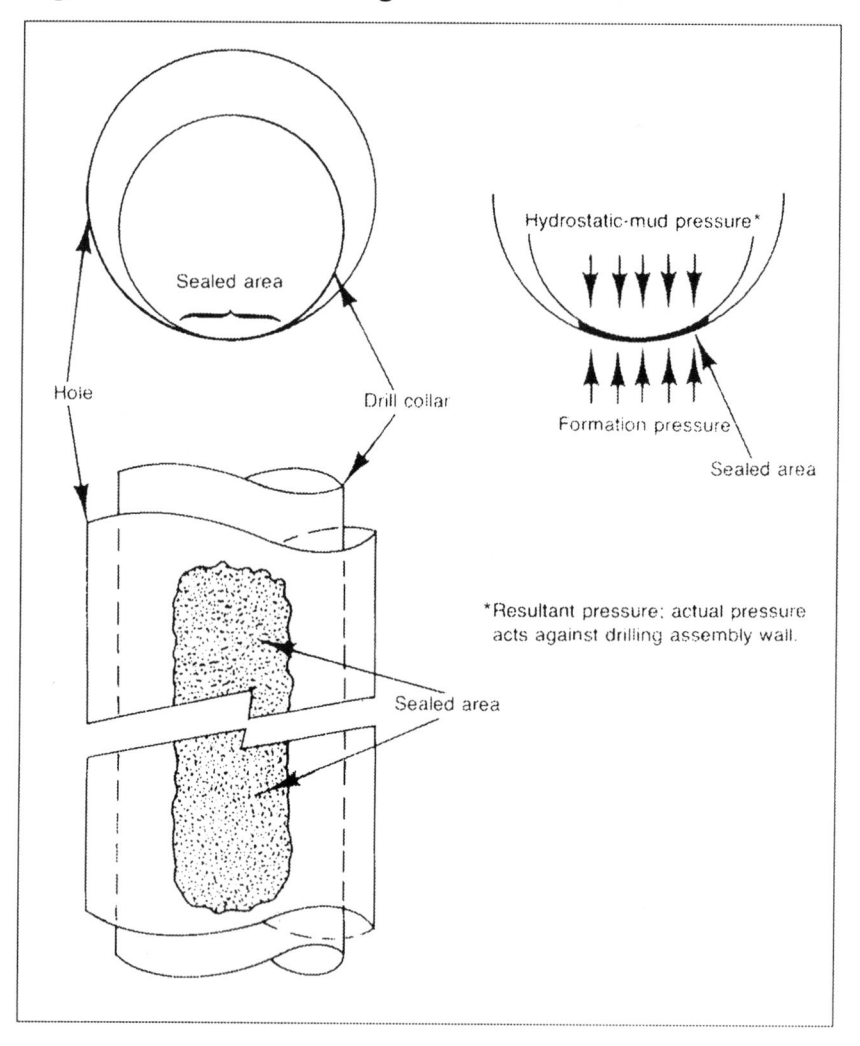

Labels in figure: Sealed area; Hole; Drill collar; Hydrostatic-mud pressure*; Formation pressure; Sealed area; Sealed area; *Resultant pressure; actual pressure acts against drilling assembly wall.

casing, left sitting on the slips for 1 hour and 40 minutes without circulating, and then picked up, rotated, and cemented without any problem. However, this is the exception. A similar situation often occurs in open-hole drillstem testing.

The primary indication of wall sticking is stuck pipe, often rapidly and completely, but with full circulation. General indications of wall sticking are similar to keyseating so it is important to distinguish between the two. Circulation is almost always unrestricted, similar to keyseating. Wall sticking often occurs after the drill tools remain motionless for a period of time, whereas keyseating generally occurs when the pipe is moving upward. Wall sticking is more common when the assembly is at or near bottom. Keyseating tends to occur on trips and when the pipe is an appreciable distance above total depth. Casing may become wall stuck, but it seldom, if ever, keyseats.

Wall sticking very commonly occurs after a drillstring keyseats. Most keyseating occurs at the top of the drill collars. The lower collars press against the side of the wellbore and remain there without movement, often for an extended period, so there is a high risk of wall sticking.

Wall sticking occurs when the side of the drill collars comes into contact with the wall of the open wellbore, when the pipe is at rest. There have been few, if any, documented cases of wall sticking in the drillpipe section. Generally, the hydrostatic head of the mud column exerts a differential pressure into the formation, a main cause of wall sticking. This differential first seals the perimeter of the contact area with wall cake, cuttings, and other mud solids. Then the differential pressure acts against the entire contact area to push the collars against the wall — hence the name of "differential pressure sticking."

The forces involved in wall sticking can be estimated. Assume a pressure differential of 0.1 lb/gal, 0.0052 psi/ft or 520 psi at 10,000 ft. Further assume a contact area 3-in. wide and 25 ft long. Then the contact area is (25 ft long x 12-in./ft x 6-in. wide) 900 square inches. The lateral force on pipe under ideal conditions is (900 in.^{2} x 520 psi/in.^{2}) 468,000 lbs. It is not uncommon to see overbalanced mud in excess of 1 lb/gal, which would represent an equivalent lateral force well into the millions of pounds. The vertical shearing and releasing forces would be less but still very high, depending upon the coefficient of friction. This illustrates one of the most important preventive measures: USE MINIMUM MUD WEIGHT.

The effect of a thick wall cake is demonstrated in Fig. 2–8. If a 4¾-in. OD drill collar is pressed through a ¹⁄₁₆-in. wall cake in a 6¼-in. hole, it could leave a minimum width of about 2 in. subject to wall sticking. Under the same conditions with a ⅛-in. wall cake, the width increases to over 3 in. This emphasizes the importance of a thin wall cake to help reduce the risk of wall sticking, so use good mud.

There is no convenient way to measure directly or predict these forces during operations except by field results, but they can be very large. Field instruments are available for measuring the sticking coefficient of mud filter cake. These can be helpful but are seldom used because of the many unknown forces, areas, and other related factors.

General Rules and Procedures

Figure 2–8 Wall sticking

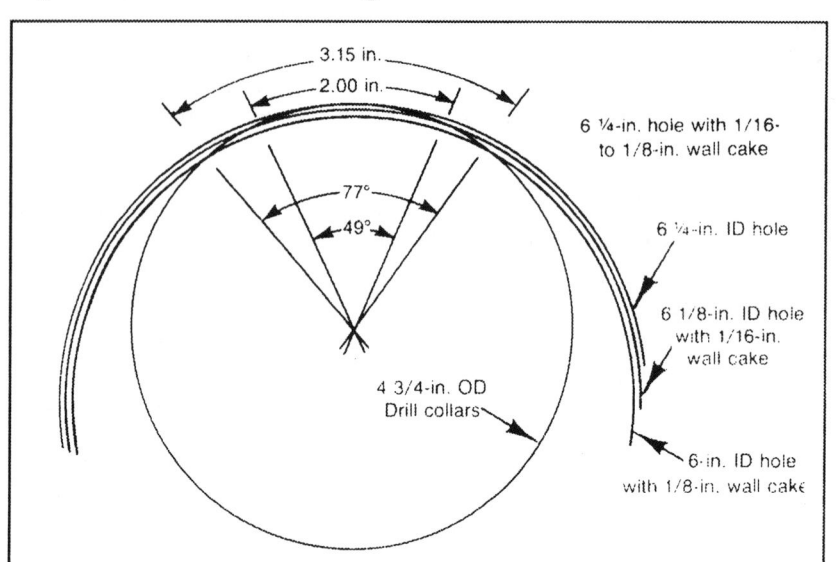

[Figure showing cross-section of drill collar against borehole wall with mud cake geometry. Upper diagram shows circular cross-section with labels: 3.15 in., 2.00 in., 77°, 49°, 6 1/4-in. hole with 1/16- to 1/8-in. wall cake, 6 1/4-in. ID hole, 6 1/8-in. ID hole with 1/16-in. wall cake, 6-in. ID hole with 1/8-in. wall cake, 4 3/4-in. OD Drill collars.]

Assume 4 3/4-in. OD drill collars flush against the side of a 6 1/4-in. ID hole with two mud-cake thicknesses. Further assume the mud cake is moved from between the 4 3/4-in. drill collar and the wall of the hole to form a pressure seal at the tangential point where the drill collar OD intersects the ID of the hole at the wall cake ID.

With a 1/16-in. mud cake, circumferential length is about 2 in. With a 1/8-in. mud cake, length is about 3.15 in., almost a 60% increase.

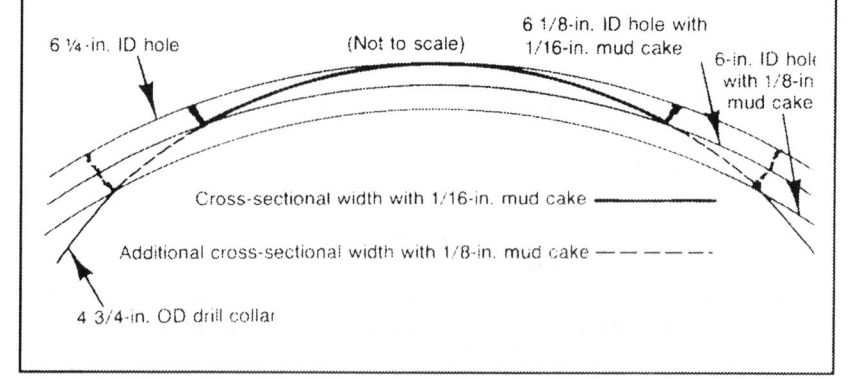

[Lower diagram (Not to scale) showing: 6 1/4-in. ID hole, 6 1/8-in. ID hole with 1/16-in. mud cake, 6-in. ID hole with 1/8-in. mud cake, Cross-sectional width with 1/16-in. mud cake ———, Additional cross-sectional width with 1/8-in. mud cake — — — —, 4 3/4-in. OD drill collar.]

The severity and speed of wall sticking depends upon the size of the contact area, the magnitude of the differential pressure, and related factors. The area of contact depends upon the relative difference in the diameter of the drill collars and the diameter of the hole, the length of the section, and the wall cake thickness. Sticking can occur very rapidly, in some cases within a few minutes. Wall sticking is more common in freshly drilled holes, especially while making connections and using a low-quality mud. The formations are apparently not completely sealed at this time, so the pressure differential is more active. Mud loss is relatively uncommon.

Wall sticking can be prevented in many cases by eliminating the causes listed in Table 2–4.

Table 2–4 Factors conducive to wall sticking

- Overbalanced mud system that increases the differential pressure.
- Increasing the contact area due to large-diameter collars in a small-diameter hole.
- Porous formations that permit the pressure differential to be effective.
- Soft, thick mud cake that effectively increases the contact area and helps provide material for sealing around the contact area.
- Immobile drilling assembly that allows time for sealing and pressure differentials to act.
- Monitor and control mud properties with emphasis on a minimum mud weight.
- Wall sticking in oil mud is less common but does occur.
- Drilling in deviated and crooked holes where the collars lie against the wall with some force.
- Drilling in depleted reservoirs or formations with subnormal pore pressure.
- Excess drag and torque.
- Keyseating conditions are highly conducive to wall sticking.

All of the factors in Table 2–4 are important, but the ones that should be emphasized are using good quality, minimum mud weight; keeping the drill tools moving as much as possible; and taking special precautions in crooked, deviated, and horizontal holes, especially under keyseating conditions. Prevent or alleviate sticking by setting down a maximum safe weight on the drilling assembly immediately after the assembly wall-sticks, or keyseats.

General Rules and Procedures

Add walnut hulls as described in the section "Drag and Torque." Mud additives, such as graphite, to increase mud lubricity are of questionable value. Spiral or fluted-type collars have recessed areas with less contact area, which reduces the risk of sticking. Heavy drillpipe serves the same purpose and also helps reduce drag. Stabilizers help hold drill collar bodies away from the wall of the hole, but are seldom used because they can cause other sticking problems.

Release wall stuck assemblies by working, soaking and other actions, as described in the section "Wall Sticking" in Chapter 5.

KEYSEAT

A keyseat is a slotted or recessed section in the wall of the wellbore, cut by the movement of the downhole assembly (Fig. 2–9), or less commonly by a wireline. A section of the downhole assembly with a larger diameter can wedge into the keyseat, sticking the tools. Keyseats are a major cause of fishing. The severity of keyseats is emphasized by the fact that an estimated 20% of all severe fishing jobs in softer formations and about 40% in harder formations are related to or caused by keyseats. Almost all keyseats can be detected and either eliminated or reduced in severity to permit continued operations.

Keyseats occur in almost all formations and at all depths. They are more common in crooked and deviated holes, especially at higher angles of inclination. The simplest keyseat usually is near the point where the wellbore first deviates appreciably from vertical. They often occur in pairs such as on opposite sides of an S-shaped crooked hole section. Keyseats may occur in deviated sections either as bottom or top side keyseats when the hole angle changes in the upward direction. Keyseats are described in the vertical plane but occur in any orientation.

Keyseats develop at varying rates, normally related to factors that cause them and affect their severity (Fig 2–10). They continue developing until they are deep enough to stick the drilling assembly. They develop slower in hard formations but are more difficult to wipe out and often cause problems for a longer time. The opposite holds true for softer formations. Layers of alternating hard and soft rocks are similar. The combination of hard formations and crooked hole, such as severe doglegs, causes most keyseat problems.

Keyseat growth rate and severity generally increases with time unless remedial action is taken. Drilling deeper increases the weight hanging below the keyseat. This allows more time for growth and

Figure 2–9 Keyseats in layered formations

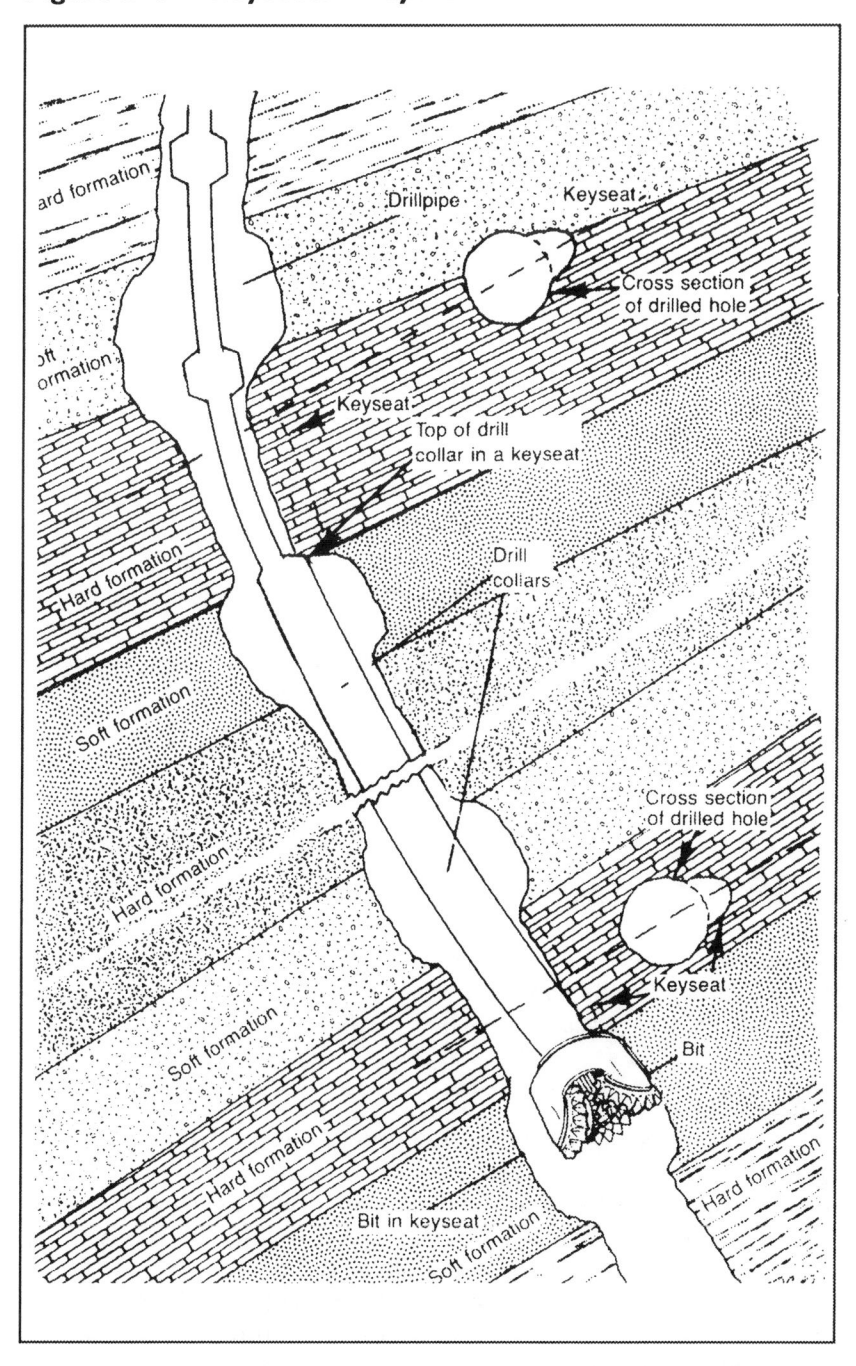

General Rules and Procedures 127

Figure 2-10 Keyseat growth

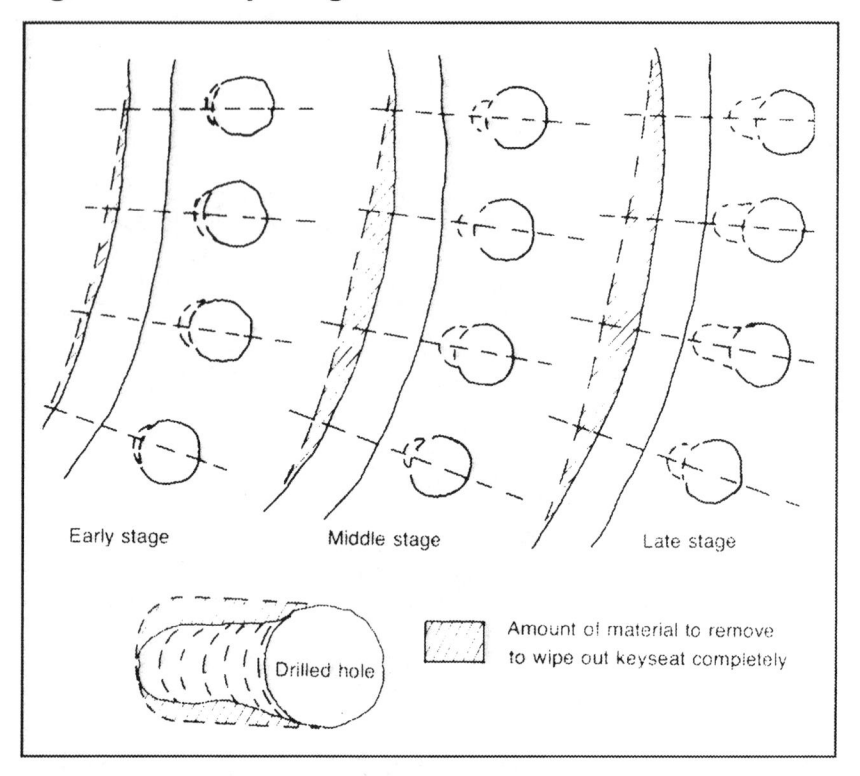

Early stage   Middle stage   Late stage

Drilled hole

Amount of material to remove to wipe out keyseat completely

increases the magnitude of the forces applied to the keyseat area. The types, sizes, and differences in diameters of component parts of the drilling assembly and how it is handled also affect the growth of keyseats. The actual rate of growth after a keyseat starts to develop is unknown. It is believed to develop at a reducing rate with time because of the increasing area that must be cut and the decreased force available per unit area.

Detection

In most cases, keyseats can be detected in time to take preventive action before sticking the pipe. It has been well established that experienced, conscientious personnel often will detect these problems by careful observations. Proper prevention includes observing signs of keyseat development and taking action.

Detect keyseats by observing drill tool action as indicated by torque and weight, including fluctuations, while drilling, making connections, and tripping. Begin trying to determine the exact location and possible severity of the keyseat when it is first detected or suspected. Review deviation and directional surveys and information on crooked-hole sections. Note hole conditions and information on keyseats on the drilling reports and pass them on verbally to the next crew.

Early detection reduces the risk of sticking. Also it is easier to ream out a growing keyseat compared to a mature one. If the keyseat is cleaned out at an early stage, less material must be removed; therefore it takes less time and the keyseat can be removed faster and easier. There is less risk of fatiguing, or possibly having a failure in the drilling assembly, and the risk of sticking is reduced.

All holes have a natural level of drag and torque. This is relatively low for a straight hole and higher for crooked, deviated, and horizontal holes. Drag should be reduced at a constant rate as the pipe is pulled out of the hole. Keyseats may be detected by a periodic increase in drag or pull while moving the assembly upward. If the weight indicator shows a slight drag or extra pull at 31-ft intervals (assuming schedule 2 drillpipe), this is an indication that the tool joints are forming a keyseat. When the drill collars reach this point on the trip out of the hole, there should be an additional pull, even though it may not be excessive.

The top of the drill collars may ride over a developing keyseat and show only a moderate pull. This may occur as a higher drag or pull at the point where the top of the drill collars, stabilizers, reamers, or bit moves upward through the keyseat.

It often is difficult to detect which point on the drilling assembly is hanging up in the keyseat. If the weight indicator shows extra drag at 31-ft intervals, and a heavy drag or pull or actual keyseating occurs immediately after the periodic (31 ft) increases in drag, then there is a high probability that the keyseat is at the top of the drill collars. This can give a relatively accurate depth measurement to the bottom of the keyseat. If the periodic increases in drag stop and the pipe is pulled out of the hole a distance equal to the length of the drill collars before a heavy drag or possible keyseat occurs, this indicates that the bit is hanging up and again provides a definite depth to the keyseat.

A similar drag effect in the reverse order may be observed while running the pipe in the hole. However, this usually is less common and more difficult to detect. Reamers and stabilizers may cause additional drag and torque and may or may not obscure drag and torque indicators. Spiral-type stabilizers frequently cause a torquing effect on the drillpipe when they pass through a tight section.

It is not uncommon to detect minor indications of a keyseat on one pipe trip and make a trip later without any problems. Then the drilling assembly might stick in a keyseat on the next trip, causing a fishing job. Normally, the second trip after a keyseat is detected will give a more accurate indication of keyseat formation, but this is not always the case.

Detection of the keyseat location can be complicated by reamers and stabilizers in the drill-collar assembly, as well as the existence of two or more keyseats. Also, the periodic increases in drag may be wiping out a rough place in the hole and not indicating an actual keyseat. Take all of these into account and use with available data to locate the keyseat(s).

If pipe is rotated out with the rotary on trips out of the hole, then the relative free movement of the rotary can be an indication of hole condition. When the drill-collar assembly is in potential keyseat sections, the rotary movement may be restricted more than when the drill-collar assembly is hanging free. Observing backlash helps locate sections where keyseats are forming. Reamers and stabilizers can also cause extra torque when they pass through a tight section.

Comparison of caliper logs from two- and three-arm tools may indicate a keyseat or hole section subject to keyseating. Most pad-type logging tools have two arms. The two-arm caliper can run with one arm in a keyseat section, thus reading a larger hole diameter than a three-arm caliper tool, as illustrated in the following example.

Two- and Three-arm Calipers Detect Keyseats

A 9½-in. hole was being drilled at 17,000 ft below 10¾-in. casing set at 12,000 ft in a 3° hole. After drilling out, a packed hole assembly was used, and the hole deviation continued to increase at a gradual rate to 6½° at 14,000 ft. A pendulum was run at this point, and the hole angle dropped at 1°/100 ft to 3½° at 14,600 ft. This angle was maintained to 17,000 ft. Some keyseat drag action was observed in the interval from 14,000 to 14,500 ft while drilling deeper.

After drilling to 17,000 ft the hole was logged and the two- and three-arm calipers compared. The intervals from 12,000 to 14,000 ft and from 14,500 to 17,000 ft had an average diameter of 13 in. with the three-arm caliper and an average diameter of 15 in. with the two-arm tool. The section where keyseating was suspected, from 14,000 to 14,500 ft, had a diameter of 12 in. with the three-arm caliper and the two-arm caliper read off scale, greater than 20 in. During subsequent drilling operations, it was necessary to ream this section several times. Similar caliper log results have been observed in keyseat sections on other wells.

Figure 2-11 Limit dogleg angle to prevent keyseats
Note: If hole sections *a* and *b* are not in the same plane, check tables of absolute angles to find the absolute dogleg.

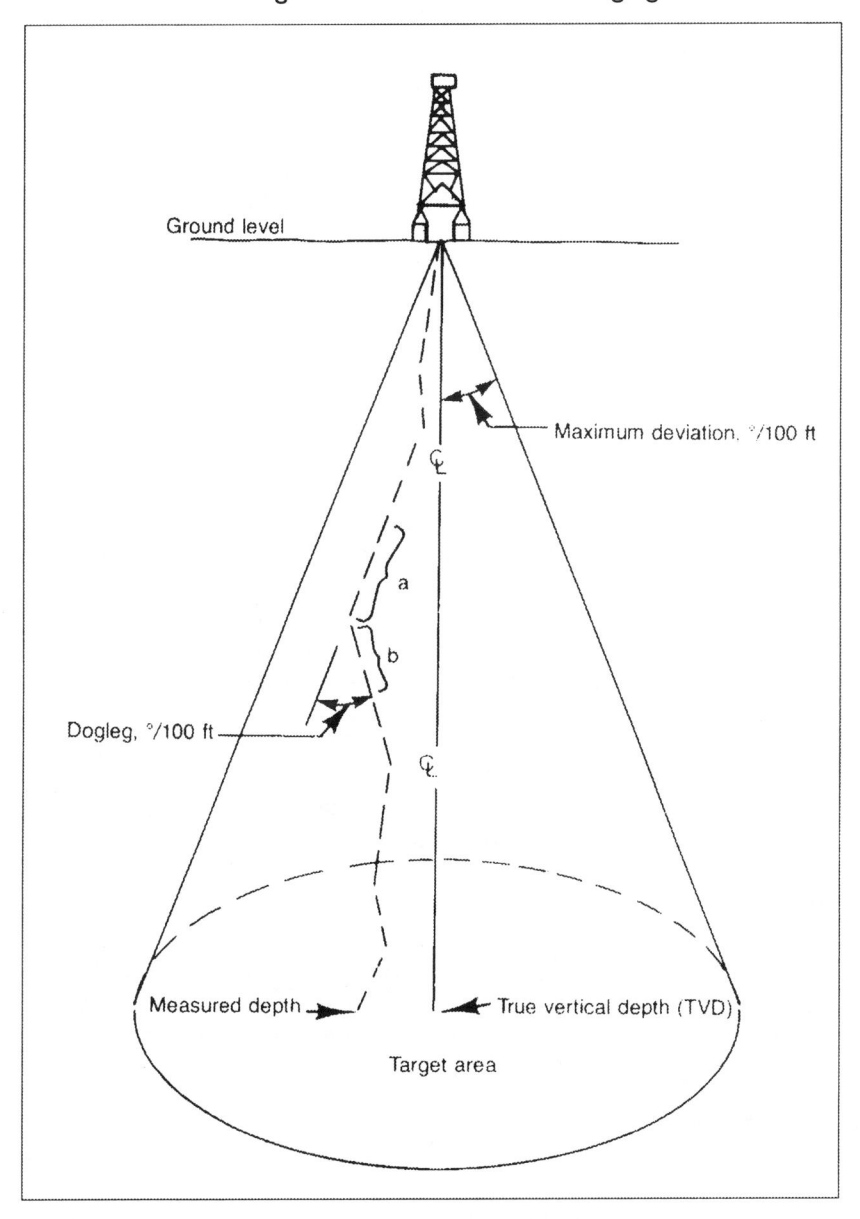

General Rules and Procedures

Tools normally stick in a keyseat while being moved upward and can be circulated freely. In some cases, tools stuck in a keyseat may be moved downward but cannot be moved upward past the point of sticking. Except for this and the direction of movement during sticking, sticking in a keyseat is very similar to wall sticking. Pipe stuck in a keyseat is highly susceptible to wall sticking.

Prevention

Prevention basically includes countering the actions that cause keyseats. It is equally important to recognize the occurrence of keyseats and take action accordingly. One of the best preventive actions is to drill a straight, vertical hole, especially with minimum dogleg (Fig. 2–11). However, sometimes the higher cost of drilling a straight hole through crooked-hole formations is not justified compared to the cost and risk of preventing keyseats. Other holes are drilled directionally and some horizontally for economic and other reasons. Handle these by minimizing the conditions that cause keyseats, such as drilling smooth curved and straight sections. Otherwise watch for keyseat developments and remove them as necessary. However, the straighter and the more vertical the hole is, the less trouble will occur because of keyseats.

The most common point subject to keyseating is at the top of the drill collars. Reduce severity by placing the correct type of keyseat wiper on top of the drill collars. The recommended wiper is the floating, tapered-sleeve type described in the section "Keyseat Wipers" in Chapter 3. It serves a dual purpose. It is moderately effective in wiping out keyseats, especially developing keyseats. More important, it is very effective in releasing an assembly keyseated at the top of the drill collars. Ensure that the keyseat wiper selected has the proper design and adequate strength to perform the intended operations.

Drilling jars and bumper subs do not aid in releasing a keyseated pipe if the top of the drill collars are stuck in a keyseat and these tools are below the sticking point. However, they often are very effective if the assembly is stuck below them, such as at the bit.

Removal

Take action immediately when the keyseat is first detected, or even suspected. The type of action to take ranges from watching the keyseat for added development to removing it by reaming. It depends upon various factors. What is the type and weight of the downhole assembly and does it have a keyseat wiper? What type of formations are being drilled? Is the hole in good condition? Is it crooked or deviated? How severe is the keyseat?

Keyseats can be removed. However, this seldom removes the cause, and the keyseat usually begins to form again when drilling resumes. Remove or partially remove keyseats based upon severity. Remove very minor keyseats by reaming with the assembly in use, especially a stiff assembly, or a keyseat wiper run on top of the drill collars. Otherwise, remove the keyseat by reaming with a keyseat wiping assembly or string reamer as described in the section "Reaming."

The decision of when to clean out a keyseat is based on experience and knowledge of the particular well conditions. It is not an easy decision to shut down drilling and spend one or two days wiping out a keyseat, especially when the operations are relatively hazardous. Unfortunately, in field practice there may be a tendency to wait too long before making the decision.

Releasing keyseated tools is described in the section "Wall Sticking" in Chapter 5.

Wireline Keyseat

Wireline keyseats form when the wireline wears a groove or slot into the side of the open-hole wellbore. Like tubular-type keyseats, they most commonly occur in crooked and deviated holes. In a few cases the wireline may wear a keyseat in a metal obstruction, such as the bottom of the casing. These are less common but can be very severe.

Drilling a straight hole is the main preventive procedure but is not always practical. Otherwise, observe the drag on the tool and ream out the hole if a keyseat begins to develop.

Open-hole logging tools are most commonly keyseated in wireline keyseats. Like tubulars, if the tool is pulled too tightly into the keyseat, it will stick and is difficult to remove. In a few cases it can be worked out of the hole. Otherwise it must be removed by fishing, as described in the section "Wireline Fishing" in Chapter 6.

REFERENCES AND SUGGESTED READING

Abel, Wilson L., "Blowout Risks Cut with Contingency Plan," *Oil & Gas Journal,* June 7, 1993: 30–36.

Adams, Neal, and Joe D. Thompson, "How a Geothermal Blowout Was Controlled," *World Oil,* June 1989: 38–40.

Eric, Erick, and Case Jansen, "A Method for Handling Gas Kicks Safely in High Pressure Wells," *Journal of Petroleum Technology*, June 1993: 570–576.

Flack, Larry, "How Well Control Techniques Were Refined in Kuwait," *World Oil*, May 1992.

Flack, Larry H., and W.C. Goins, Jr., "New Relief Well Technology Is Improving Blowout Control," *World Oil*, Dec 1983: 57–61.

Flack, Larry H., John W. Wright, and John W. Ely, "Blowout Control: Response, Intervention and Management," *World Oil*, Part 1 Nov 1993: 71, Part 2 Dec. 1993: 55–61.

Framnes, Einar, and Joseph R. Roche, "Shear Rams Hold 10,000 Psi After Cutting Pipe," *Oil & Gas Journal*, May 6, 1991: 56–62.

Grace, Robert D., and Bob Cudd, "Fluid Dynamics Used to Kill South Louisiana Blowout," *World Oil*, April 1989: 47–50.

Grace, R.D., A.F. Kuckes, and J. Branton, "Operations at a Deep Relief Well," *World Oil*, May 1990: 44–54.

Gross, Richard, "Apache Blowout Successfully Killed," *Drilling*, March, 1984.

Holbert, Don, "Conventional Tubulars Can Cut Cost of Drilling in Horizontal Holes," *Oil & Gas Journal*, June 3, 1985: 68–73.

Javanmardi, Kazan, and Debbie Gaspard, "Soft Torque Rotary System Reduces Drillstring Failures," *Oil & Gas Journal*, Oct. 12, 1992: 68–70.

Johancsik, C.A., D.B. Friesen, and Rapier Dawson, "Torque and Drag in Directional Wells — Prediction and Measurement," *Journal of Petroleum Technology*, June 1984: 987–992.

Kempt, Gore, *Oilwell Fishing Operations: Tools and Techniques*, Gulf Publishing Company, Houston, TX, 1986.

"Kuwait: The Mother of All Blowouts," *World Oil*, May 1991: 38–41.

Leraand, F., J.W. Wright, M. Zachery, and B. Thompson, "Relief Well Planning and Drilling for a North Sea Underground Blowout," SPE 20420, New Orleans, LA, Sept. 23–26, 1990.

Lester, C.B., "The Basics of Drag Reduction," *Oil & Gas Journal*, Feb. 4, 1985: 51–56.

Montgomery, Michael E., "Testing Improves Surface BOP Equipment Reliability," *Oil & Gas Journal*, June 14, 1993: 44–46.

Moore, Preston L., *Drilling Practices Manual*, PennWell Publishing Company, Tulsa, OK, 1974: 269–296.

Sayers, Bob, "Capping Blowouts from Iran's Eight-Year War," *World Oil*, Part 1, May 1991: 45–48; Part 2, July 1991: 81–82.

Sonnemann, Paul, "Circulate-and-Weight Well Control Method Has Several Advantages," *Oil & Gas Journal,* Jan. 31, 1994: 96–97.

Uzcategul, Humberto, Darrell Hewitt, and Reinaldo Golindano, "Precise Guidance Puts Record-Depth Relief Well On Target," *World Oil,* June 1991: 39.

Varcoe, Brian E., "Shear Ram Use Affected by Accumulator Size," *Oil & Gas Journal*, Aug. 5, 1991: 34–37.

Watson, Davie W., and Preston Moore, "Momentum Kill Procedure Can Quickly Control Blowouts," *Oil & Gas Journal,* Aug. 20, 1993: 74–77.

Wikie, D.I., and W.F. Bernard, "Detecting and Controlling Abnormal Pressure," *World Oil,* July 1981: 129–144.

Williamson, Joseph Stephen, "Casing Wear: The Effect of Contact Pressure," SPE 10236, Society of Petroleum Engineers, San Antonio, TX, Oct 5–7, 1981.

Chapter 3

TOOLS AND ASSEMBLIES

Fishing is a critical operation. Knowing how to select and use fishing tools is almost as important as preventing fishing. The importance of selecting the best, most applicable tool and operating it correctly cannot be overemphasized. Successful operators are familiar with the various tools and how they are operated.

Some fishing tools can be used in either a cased or open hole. Others are designed in such a way that they are more efficient and safer to use in one type of hole. Fishing tools can be run on tubulars, wirelines, or both, depending on the tool. This is a natural division and is commonly accepted method of classification.

- Tubular fishing tools are operated by rotation, reciprocation, or a combination of both motions. They are more common because most fishing occurs with tubulars and they are more efficient.
- Wireline tools are operated by reciprocation only and can be less efficient than tubulars.
- Combination fishing tools are run on either tubulars or wirelines. They operate by rotation and reciprocation when run on tubulars and by reciprocation only when run on wirelines. They generally serve the same purpose in either case, but are modified to function differently, depending on the way in which the tool is to be run and operated.
- Auxiliary tools are not always classified directly as fishing tools, but they are used to aid fishing operations or to help prevent a fishing job.
- Fishing assemblies are used on all fishing operations.

FISHING TOOL SELECTION

The wide variety of fishing operations necessitates an extensive assortment of fishing tools. A larger selection of heavy-duty tools is available in medium and large sizes. Availability of tools diminishes with decreasing hole size. Tools for smaller diameter holes have reduced strengths and selection may be limited.

Fishing tools fall into various categories. Catch tools are the largest group and most used of all fishing tools. They are subdivided into several broad groups.

- The more common outside catch tools are slipped over the fish to catch it on the outside.
- Inside catch tools fit into the fish to catch it on the inside.
- A swallowing tool slips over and encloses the entire fish.

In a few cases, one tool may catch a fish by a combination of two methods. Each operates somewhat differently, either in the manner in which the fish is caught or released or in the way a fish is controlled or manipulated. Some tools operate on the same general principles and could fit into other groups but their importance justifies separate classifications. Cutting, parting, and milling tools part the fish and serve various milling functions.

The type of operation or the condition of a fish normally dictates which fishing tool should be used. Frequently there is only one obvious selection. However, occasions do arise where one of several tools can be used. In such a case, always select the strongest tool. As a secondary consideration, select the tool with the strongest catch if there is an option. Tool strength is an overriding consideration. Sometimes the choice of a less functional tool is fully justified when the tool is appreciably stronger than other available tools. This is especially important when conditions of severe jarring and tension pull exist.

For example, an overshot is selected to catch a 4½-in. drillpipe inside an 8¾-in. hole. A larger overshot with a thin wall in the bowl section can be dressed to catch the drillpipe tool joint and may have the satisfactory strength characteristics. However, a smaller OD overshot with a thicker and stronger wall in the bowl section dressed to catch the drillpipe body may have a higher safety factor in handling tension under the severe conditions of jarring and pulling that are frequently encountered. This extra margin of safety may be the difference between a successful fishing job and the creation of a second fish. Using the smaller tool also reduces the risk of sticking.

A number of tools usually not considered fishing tools can directly aid in fishing or prevention. Standard rig equipment used in a drilling operation normally is suitable for fishing. Pumps keep the hole clean and may help pull stuck drillpipe out of the hole. Coiled or other small-sized tubing can clean inside the fish to allow running back-off tools. Cement plugs can isolate lost circulation or high-pressure zones that may affect fishing operations. Fracture sand or barite plugs may block lost circulation zones or bridge around pipe in an annular space as a way of holding pressure or preventing the pipe from dropping. Pipe with left-hand connectors can back out conventional right-hand connected pipe. Perforating tools help establish circulation. Logging tools may be used to locate the top of the fish, to determine the extent of sticking, or to obtain pipe stretch data in order to locate the free point. Other wireline tools run and detonate a back-off shot or part the pipe by different methods.

When considering the use of a special fishing tool, either become thoroughly familiar with the tool and its operation, or obtain information about the tool from the supplier. Two fishing tools may appear similar; usually there are differences. The supplier of each tool should be considered the final authority on the use and operation of the equipment. Knowing the tool's appropriate use and operation may also help in selecting between two similar pieces of equipment. If this is insufficient, contact a knowledgeable third party for help.

TUBULAR FISHING TOOLS

Tubular fishing tools are run on tubular work strings, such as drillpipe and tubing. In most cases they are run on a fishing assembly that is equipped with a jar and bumper sub. These tools are actuated by rotating and reciprocating the work string.

All references to rotational movement are described as if looking downward in a vertical line. For example, right-hand rotations means turning the tubular clockwise, or in the right-hand direction, as seen from above the tool while looking downward.

Overshot

An overshot is the most important and widely used fishing tool. There are many variations of this outside catch tool. Overshots make a strong catch and are efficient, versatile, and reliable. They probably have caught over 90% of all fish recovered. Sometimes it is difficult to disengage the overshot from the fish. Otherwise, the most common failures are a split shoe, split body, or body connection failure. Always use the strongest bodied overshot possible (Fig. 3–1).

Single Bowl

The single bowl overshot is the most common overshot and is widely used. Its basic parts are a top sub, bowl with taper, slip device, guide, packoff if desired, and various types of shoes. The top sub connects the drillstring with the overshot bowl and also serves as a stop when engaging the fish. Stops can be inserted to shorten

Figure 3–1   Overshots and accessories (courtesy Bowen)

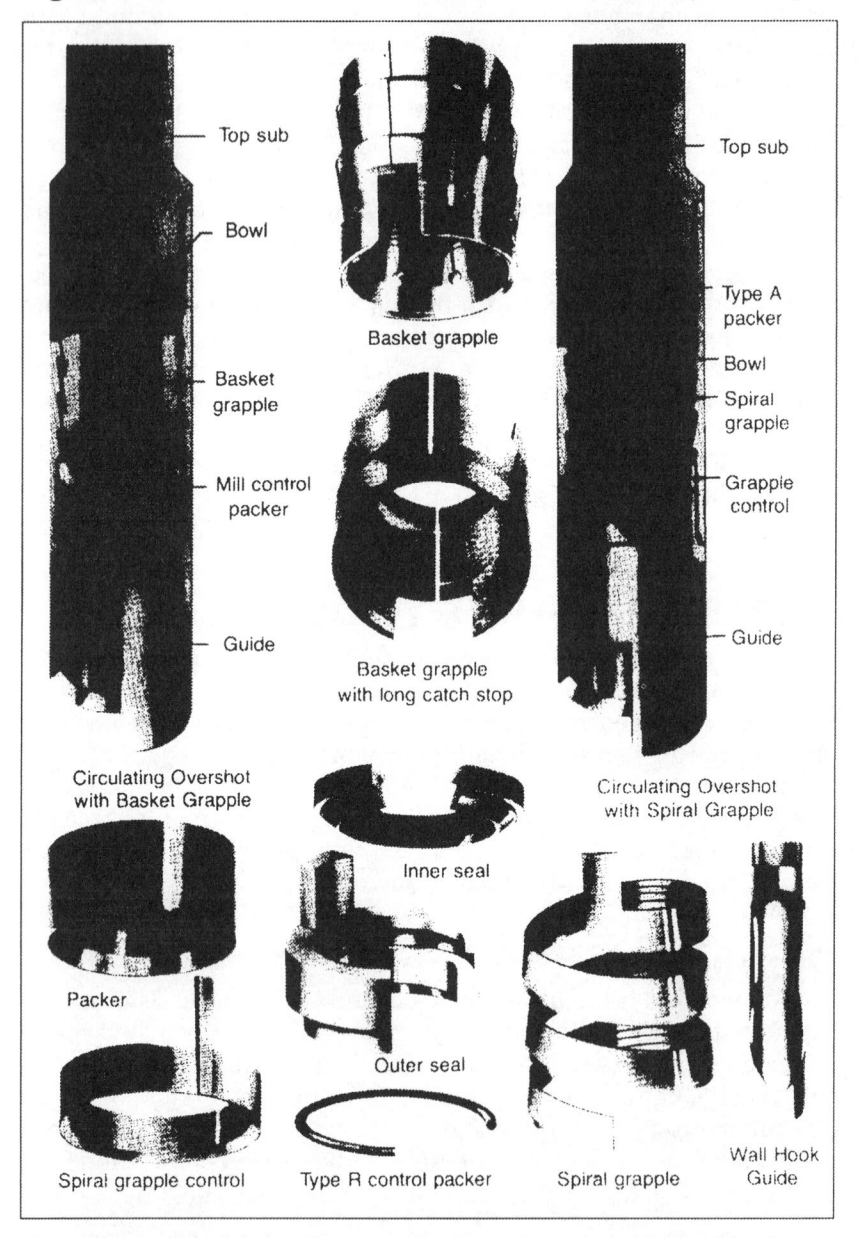

the clearance length of the bowl if it is too long, or as extension joints to increase bowl length. The inside diameter (ID) of the bowl determines the maximum and, to some extent, the minimum diameter of a fish that the overshot will catch.

The overshot is connected to the bottom of the bumper sub on the fishing assembly and is run into the hole. It is rotated and low-

140 *Chapter 3*

Figure 3–1   Overshots and accessories (continued)

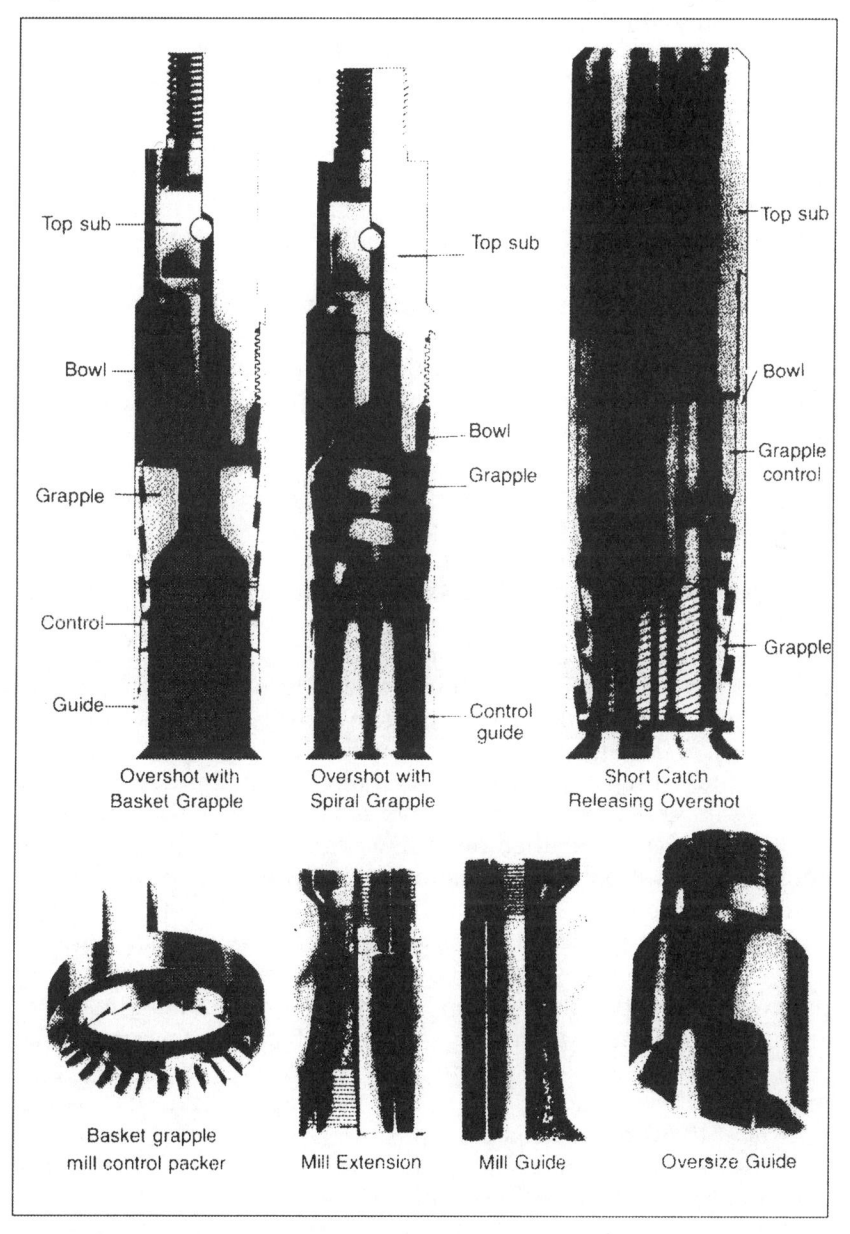

Overshot with Basket Grapple

Overshot with Spiral Grapple

Short Catch Releasing Overshot

Basket grapple mill control packer

Mill Extension

Mill Guide

Oversize Guide

ered over the fish so that the top of the fish passes through the guide shoe, slip, or grapple section, and fits up into the bowl of the overshot. Do not rotate the overshot excessively since this can damage the wickers and may prevent catching the fish. When the fishing assembly is picked up, slips or grapples catch and hold the fish firmly.

Tools and Assemblies

The fish is worked as necessary until it is free, and then pulled. Ordinarily, bumping down during a job is limited because this is a releasing procedure. Tools such as a stuck pipe log or back-off string shot can be run through the overshot into the fish if the top of the fish is open.

The overshot's inner tapered bowl section can be fitted with either slips or basket spiral grapples to catch a fish. These tools have tapered exteriors to conform with bowl tapers. They normally catch a fish with an outside diameter of about $1/16$ in. over or under the nominal catch size of the grapple.

The spiral grapple, or long catch, is a left-hand helix with interior ridges or wickers. A basket grapple is an expandable split cylinder that is helically tapered to conform to the tapered bowl section. The interior has left-hand wickers. The grapple is equipped with a key or spine that allows the grapple to move up and down a short distance inside the bowl and that provides a means of transmitting torque from the bowl to the grapple. The basket grapple can be a short catch tool since it has an upper lip that prevents the fish from traveling up into the bowl. When the basket grapple is used, the fish must have a clean top and be full-gauge since the basket grapple only catches the upper few inches of the fish.

The overshot is released if the fish cannot be recovered or if it is necessary to pull the fishing string. Release the overshot by bumping down and rotating to the right while slowly lifting the fishing string. This releases the slips on the conventional right-hand releasing overshot so the fishing assembly can be pulled. Sometimes the slips or grapples may stick when the overshot is subjected to heavy working loads. The release may then require heavier bumping down and the application of stronger right-hand torque, often with considerable effort. If this does not release the fish, then a string shot may be run and detonated opposite the slips or grapples to jar the fish loose.

When the overshot hangs above the fish, the normal circulating pattern is down the drillpipe and overshot, around the guide shoe, and up the annulus. After taking hold of the fish, use the same circulating pattern to remove it. Do not circulate excessively or with high pressure in this configuration since the mud can erode the catch tool, causing it to release the fish.

Most overshots can be fitted with packoff elements. These seal the annular space between the outside top of a fish and the inside of the lower part of an overshot. This fitting diverts the fluid flow down into the fish. If circulation can be established through the fish, it normally is easier to release and recover the fish (Fig. 3–2). Packoffs usually are not high-pressure devices but will often withstand sufficient pressure to establish circulation through the fish.

Figure 3-2 Overshot circulating patterns

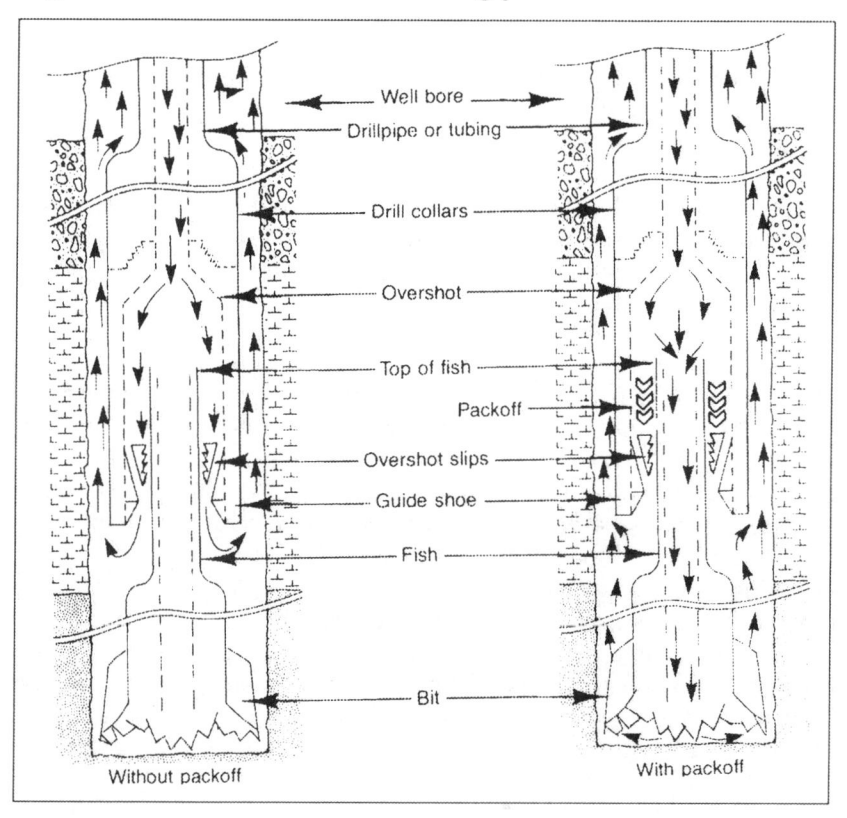

Both the grapples or slips and the circulating packoff can be easily damaged if the top of a fish is ragged, contains splinters of steel, or is out of round, bent, or otherwise damaged. For protection, some overshots can be fitted with a mill control in the lower sub. The mill control removes burrs and repairs minor damage to the fish's top as it guides the fish into the overshot. A guide or mill guide retains the grapple and serves the same purpose. To a limited extent, either may be dressed with tungsten carbide so the tool can serve as a milling shoe. The mill may be run below a mill extension guide that has been fitted to the bottom of an overshot. DO NOT mill excessively with the mill shoe. There is a risk of breaking the mill shoe or damaging the grapple. Use a regular mill if the top of the fish is severely damaged.

It may be best to catch a lower tool joint if it is near the top of the fish, especially if the condition of the top of the fish is questionable. Dress the overshot with slips or grapples to catch the tool joint. Select the bowl length and use bowl extensions so that the grapple,

Tools and Assemblies

and packoff if run, is placed at the correct position on the fish. This larger diameter tool allows the overshot to pass over the damaged fish top and catch the tool joint.

Various types of shoes are available, including guide, mill, rotary, mule, and cut lip guide or wall hook. Use the muleshoe or cut lip guide if the fish is difficult to catch because it is crooked or slightly out of gauge. Use a wall hook if the top of the fish is laid over in a washout. Rotate the wall hook slowly and carefully to prevent breaking it off. If the wall hook is unsuccessful in guiding the top of the fish into the overshot, run a knuckle joint, bent sub, or bent joint to help engage the fish.

Overshots have many operating advantages but some precautions apply. The fish can be torqued to the right or left when the grapples are under tension. Fine thread bowl connections are the weakest "back-off" point, so use lock rings to help prevent backing off. The control finger and grapple tong are subject to failure. Reduced pump pressure while engaging the fish will help prevent hydrostatic lift. Use a minimum circulating rate while the fish is engaged to prevent fluid erosion of the overshot and grapples. A single bowl normally is preferred over a double bowl. A longer catch with spiral grapple is preferred to a short catch with basket grapple.

If there is an option to catch either the drillpipe body or the tool joint, carefully evaluate which type of catch to use. Determining the size of worn tool joints may be difficult but is not usually a problem. There is a higher risk of splitting the slip body by catching the tool joint if a fish must be worked very hard. There is less risk of damaging the tool joint and a higher risk of damaging the pipe body during heavy working, but there is also less risk of damaging the fishing tool. For cases involving similarly sized holes, an overshot to catch the small diameter fish (pipe body) usually is stronger than a tool used to catch a larger diameter fish (tool joint).

Double Bowl

A double bowl or combination overshot is essentially two single bowl overshots run one above the other. Dress the upper bowl with the smaller size catch. For example, dress the upper bowl to catch the pipe and the lower bowl to catch a tool joint. A mill guide can only be used on the lower bowl. A combination tool may be slightly more risky to run because of the longer length and increased number of fine thread connections. Otherwise, a double bowl is operated similarly to the single bowl overshot.

Triple Bowl

The triple bowl is similar to three single bowl overshots run one above the other. It contains three catches. The smallest catch is in the top of the bowl and the largest is in the bottom. It is seldom used because of the increased problems when compared to those of a single or double bowl tool. Otherwise, the triple bowl is operated similarly to the single bowl overshot.

Trap Door
The trap door overshot has a slip block that travels on an inclined carrier. It catches long, small-diameter fish, such as coil tubing or partially collapsed tubing that may not be completely circular in its cross section. The trap door is lowered over a fish, which passes up into the tool and into the drillpipe. The slip block closes against the fish and holds it for recovery when the overshot is picked up.

Side-door
A side-door overshot is used to recover wireline as described in the section "Fishing for Wirelines" in Chapter 6.

Continuous
The continuous overshot, sometimes known as a full-opening overshot, is modified for different uses. One type has a full opening into the running assembly to allow the overshot to slip over a long fish that then passes up into the fishing string. This type is used primarily to fish for small diameter tubing strings and coiled tubing by swallowing long sections.

Casing Bowl
A casing bowl is an outside catch tool sometimes called a casing patch, external patch, or casing overshot. It is similar to a slim-hole type, full opening single bowl overshot (Fig. 3–3) and operated in a similar manner, except it is not designed to be released. It is used to connect or reconnect casing strings and other large-size tubulars.

Normally a casing bowl is run into a hole on casing to connect to a casing fish. It is latched onto the top of the casing fish, tension is applied, and the casing is landed in the casing head and nippled up. A seal type can be used, or the repair can be cemented. This tool and its procedures make a good, full-gauge casing repair.

Basket
Baskets catch small junk such as bit cones, sidewall cores, bullets, tong dies, and remnants from milling by swallowing them. Various types are available.

Regular
A regular basket or junk basket has an upper body, a bowl (or basket), a sub to hold the retainer ring and fingers, and a shoe (Fig. 3–4). The fishing assembly connects to the upper body. The bowl

Figure 3–3 Casing bowl (courtesy Homco)

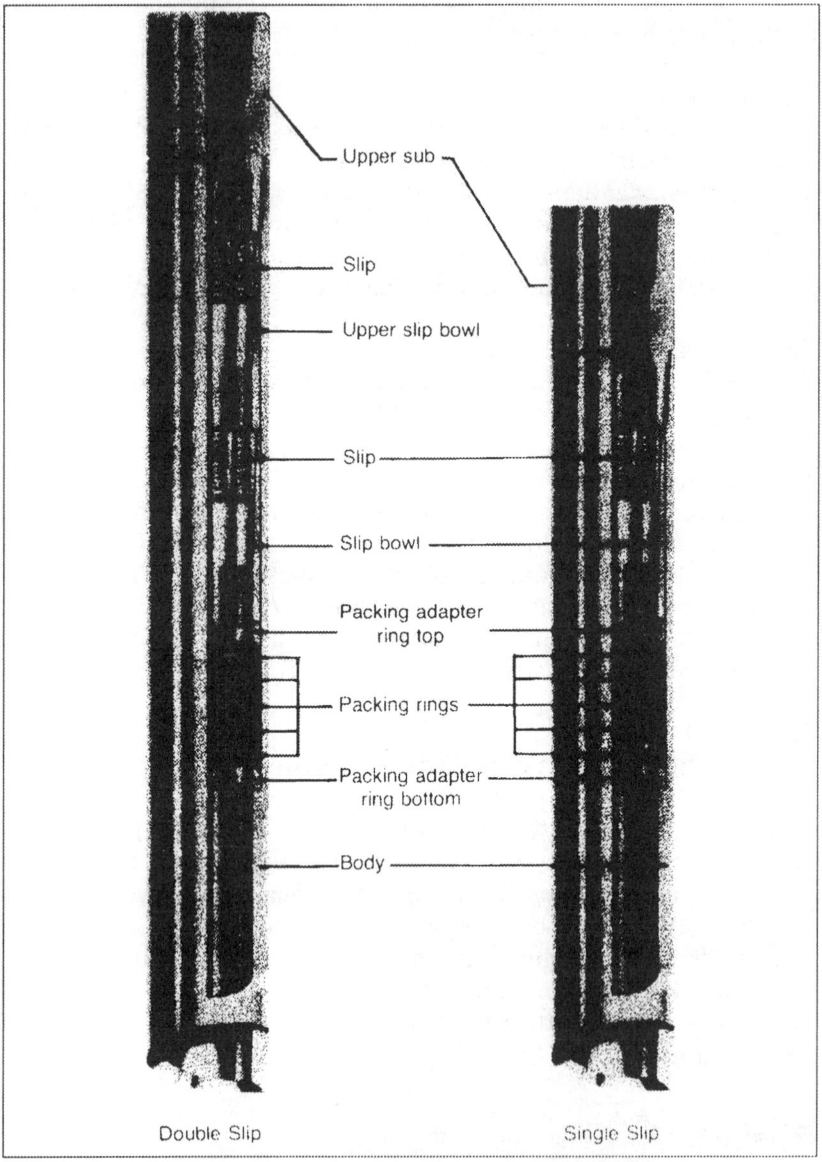

Upper sub
Slip
Upper slip bowl
Slip
Slip bowl
Packing adapter ring top
Packing rings
Packing adapter ring bottom
Body
Double Slip
Single Slip

holds junk caught in the basket. Spring-loaded fingers mount on a ring that is free to rotate inside the sub. The fingers normally are horizontal and swivel upward to let junk pass up into the bowl of the basket. Then the fingers return to the horizontal position to prevent the junk from falling out of the basket.

A magnet insert, usually with a flush guide, may be placed in either the junk basket or the reverse circulating basket in lieu of the

Figure 3–4 Junk baskets (courtesy Bowen)

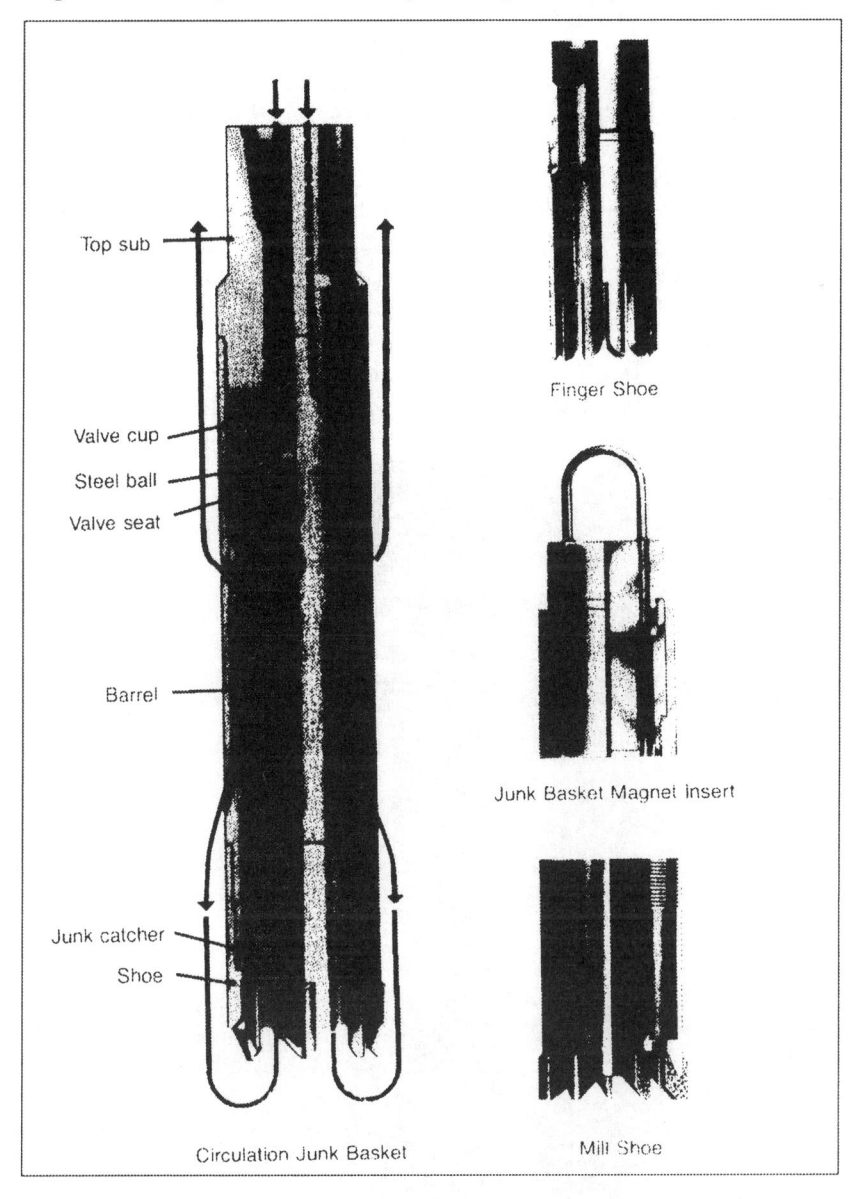

catchers. Several types of shoes are available, although normally some type of drag tooth shoe is used.

The junk basket is durable and strong. It can recover any metal object left in the hole that can pass up through the shoe and into the bowl. Such objects include bit cones, bow springs from logging tools, slip fragments, dropped hand tools, backup pads, and short pieces of wireline.

Tools and Assemblies

Run the junk basket into the hole on the bottom of the fishing assembly. After circulating, rotate the basket down over the pieces of junk, which pass up through the shoe and fingers into the bowl. The fingers retain the junk while the basket is pulled out of the hole.

One disadvantage of a junk basket is that the fingers are inherently weak because they are small. They can break, allowing the junk to fall out of the basket. The fingers are located between 6 inches and 1 foot from the shoe's bottom, so the basket must be drilled down over the fish. They often break, especially when drilling lasts over 30 minutes and when the junk is relatively large. To alleviate breakage, some junk baskets are fitted with double catchers (Fig. 3–5).

Figure 3–5 Junk basket (double catcher) (courtesy of Homco)

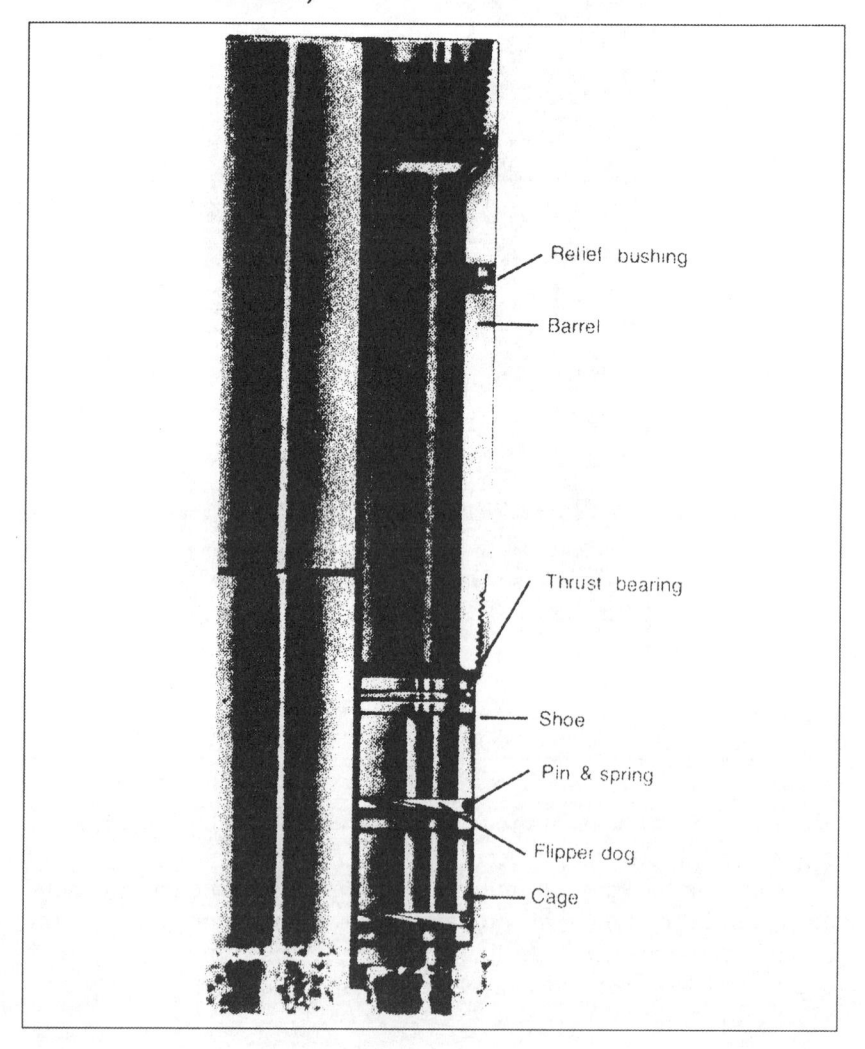

Double catchers partially account for increased tool efficiency in softer formations. To avoid the problem of broken fingers, the tool sometimes is drilled down until the bottom is literally plugged with a core of formation. If the core remains in place while pulling the tool, it normally will recover any junk left in the bottom of the original hole.

To drill a core and keep it in place, start drilling in a normal manner but with a reduced pump rate. After drilling about 1½ ft of formation, slow the pump rate again, increase the drilling weight, and reduce the rotary speed. The net effect of this action is to jam the core into the bottom of the junk basket. This is very effective in soft formations but less so in harder formations.

Reverse Circulating

One popular modification is the reverse-circulating junk basket. It is similar to a regular basket except that it has a reverse circulating action to help sweep or wash junk, especially smaller junk, into the basket. Channels and ports located in or near the shoe direct the fluid flow from the periphery of the shoe toward the center of the basket. As the junk sweeps into the basket, the mud passes through a second set of ports into the annulus.

In operations, run the tool into the hole and circulate normally down through the junk basket, around the shoe, and up the annulus. Then drop a ball to divert the mud flow and help sweep the junk into the basket. For all other steps, the tool is operated conventionally.

Poor Boy

Effectively, a poor-boy basket is a piece of casing strongly welded to a bit sub or tool joint to allow connecting to the bottom of the drill tools (Fig. 3–6). Saw-tooth-type teeth are cut in the bottom of the casing. Each is about as long as half the casing diameter. Various tooth designs are used, and sometimes the outer part of the teeth are tapered to help them bend inward during drilling. Vertical slots may be cut in the top of the basket to let fluids pass while running in the hole.

A poor-boy basket is operated similarly to a regular basket. Run it to the top of the fish, circulate it, and then reduce the circulation and drill over the fish until the basket stops moving downward. In an ideal case, the junk passes up into the basket and the drilling motion, combined with the frictional heat caused by reduced circulation, bends the teeth inward to close the bottom of the basket. Then pull the tool and cut it open to recover the junk.

In one modification of the tool, short fingers of cable are welded on the inside of the bowl in order to help retain the junk. This tool

Figure 3-6 Poor-boy junk basket

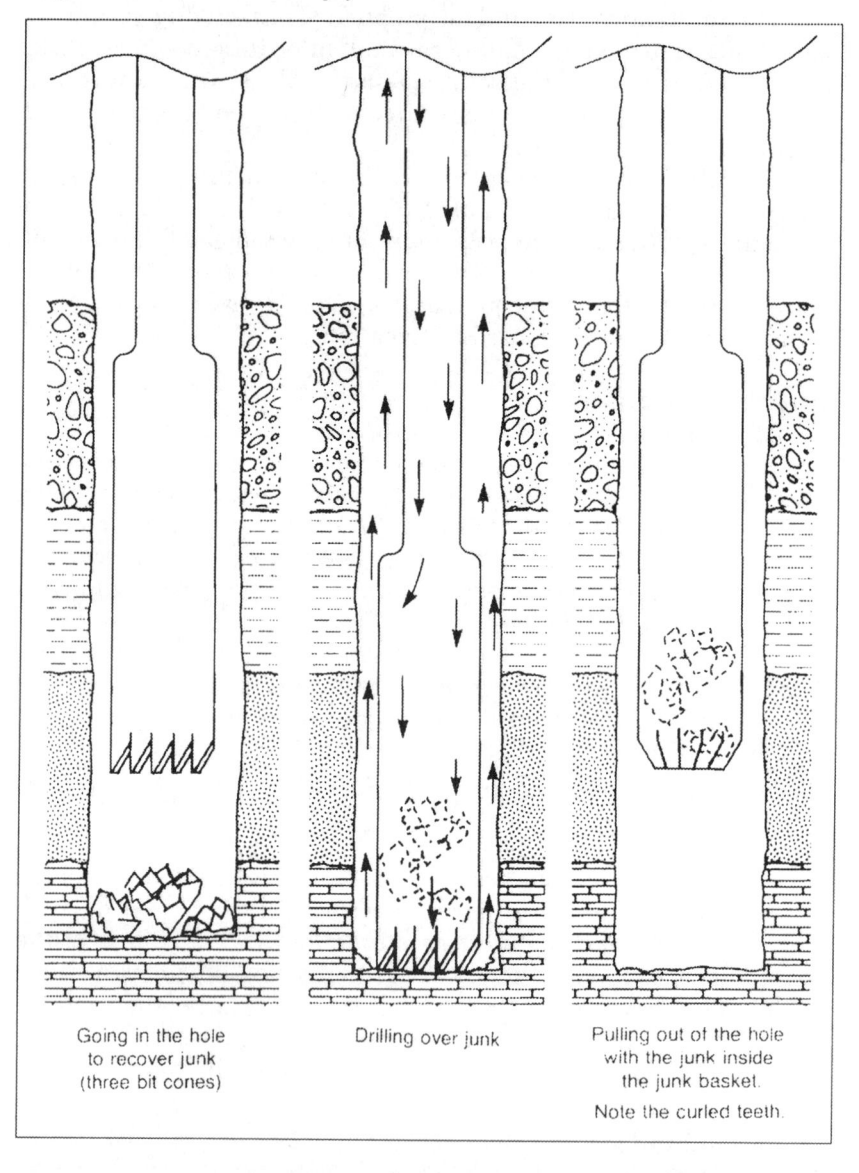

Going in the hole to recover junk (three bit cones)

Drilling over junk

Pulling out of the hole with the junk inside the junk basket. Note the curled teeth.

works well in soft formations but is less efficient in harder formations because of the difficulty in drilling and the danger of teeth breaking off.

Boot

The boot basket, often called a junk sub, is a highly efficient tool for catching and recovering large fragments of cement, metal, and other materials in the mud flow stream near the bit, most commonly

Figure 3–7   Junk sub (courtesy Homco)

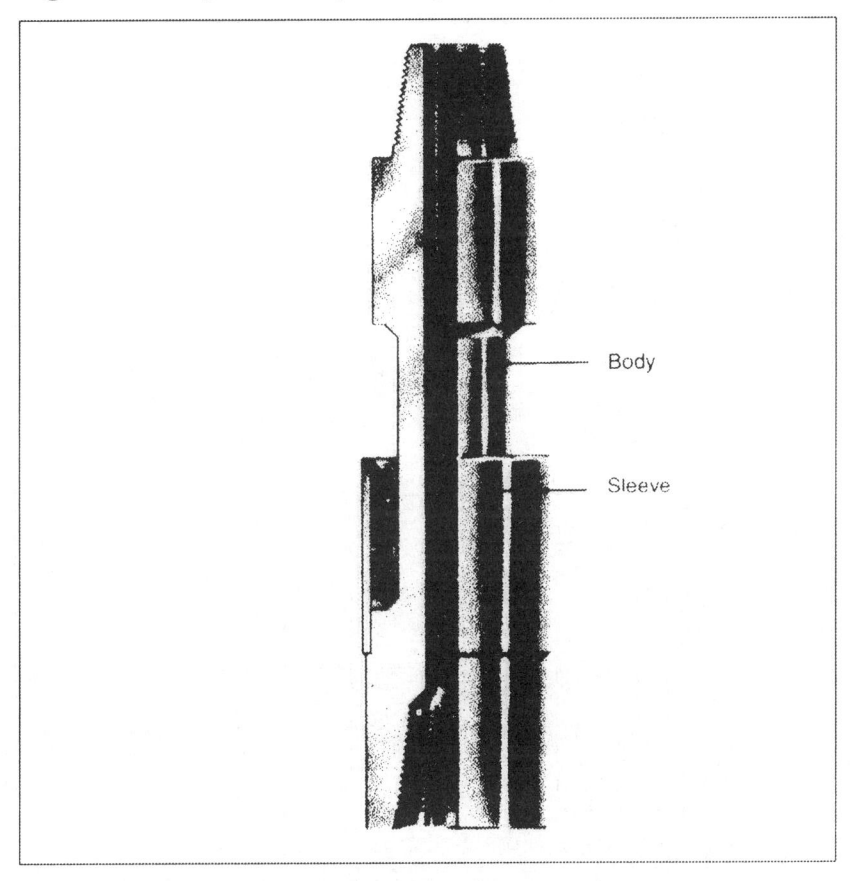

caught while milling. This tool eliminates the need for redrilling the fragments, allows more of them to be carried out of the hole with the circulating mud, reduces drilling time, and extends the life of the bit or mill (Fig. 3–7).

The junk sub has an inner barrel or body with tool joints on both ends. An outer body or bowl is fixed to the lower tool joint and extends upward approximately two-thirds to three-fourths of the distance between the two tool joints. The space between the outer body and the inner body forms a bowl or basket used to collect junk and debris.

During an operation, run the junk sub immediately above a mill or bit. Metal cuttings and cement fragments are milled or drilled into smaller pieces and then carried up the annular space beside the junk sub by the circulating mud (Fig. 3–8). When the mud passes the space between the upper lip of the bowl and the lower part of the upper tool joint, the mud's velocity reduces slightly because of the increased

Tools and Assemblies

Figure 3–8 Junk sub circulating pattern

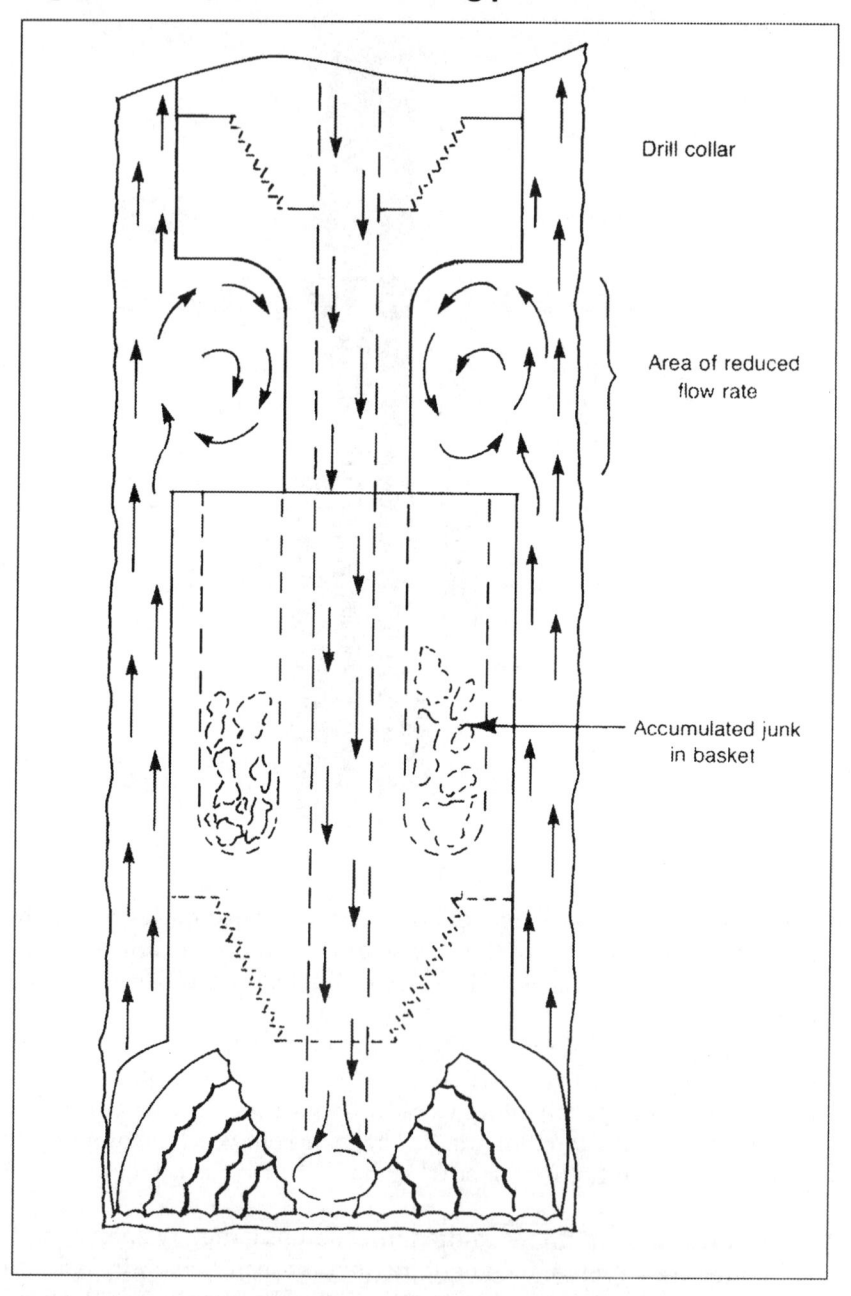

cross-sectional area. The mud flow also rolls or creates eddy currents as it passes by the lip of the bowl into the larger cross- sectional area. These actions allow the heavier junk and debris to drop into the bowl of the junk sub, where it is trapped. The junk remains in the basket until the junk sub is pulled from the hole.

Junk subs should always be run with mills and bits when drilling metal, such as using a bit to break up bit cones. These subs commonly are run when drilling cement, cement diverting tools, float collars, and cementing shoes. Junk subs are especially effective in recovering metal fragments when milling or drilling on bit cones and other metal objects.

Junk subs frequently are run in tandem, one above the other. In one case, when three junk subs were run, the first (lower) junk sub collected 15 lb of junk while about 5 lb were in each of the upper junk subs.

Junk subs are available in various configurations, including long and short barrels, side-door types, and basket top cut on a slant to provide a larger opening on one side. Junk subs are one of the most trouble-free tools that can be run. They are highly efficient and should always be used whenever applicable.

Spear

Spears are probably the second most common type of catch tool. They catch tubular fish with an inside catch. Tool strength normally is less than that for an outside catch used on smaller diameter fish, such as tubing, but often stronger for catching larger fish, such as casing. Otherwise, spears are an efficient tool. There are a wide variety; some have different activating mechanisms for catching different types of fish, such as casing, drillpipe, tubing, and wirelines.

Pipe

The pipe spear, commonly called a drillpipe spear, is the most common spear. Pipe spears generally catch casing and large tubulars, not drillpipe or tubing, although they may be used for these items in special situations. To enter smaller diameter tubulars a spear must also be relatively small, so it may not have the necessary strength to release the small tubular that is stuck. Sometimes, it may be difficult to run the spear into the drillpipe or tubing in larger diameter holes. In these cases an overshot is a better choice. The inside top of a fish must be clean and not out of round for a successful catch.

The conventional spear has an upper body that connects the tool to the running string. It may have a stop or no-go to prevent running the spear too deeply inside the fish. The body contains a tapered section and a cage to hold the slips in a recessed position. The inner face of the slips has a recessed, cone-shaped taper, while the outer face is serrated to catch the inner wall of the fish. The cage is fitted with a "J" slot and drag springs, or friction blocks, that contact the inner wall of a fish (Fig. 3–9). The spear is similar to a tension packer. In fact, in a few cases either tension packers or inverted regular hook-wall packers have been used as spears.

During an operation, connect the pipe spear to the bottom of the fishing assembly, run this into the hole, and position it inside the fish. The slips are released by rotating the fishing string to the left and picking it up slowly. The drag springs or friction blocks resist this movement slightly, so the slips are released to move downward, relative to the fishing string. Then the fishing string is picked up slowly. This moves a taper on the body of the spear up and under the slips. The slips move outward to engage the inner wall of the fish. The fish is then pulled or worked. If the fish cannot be released, the pipe spear can be released by holding right-hand torque and bumping downward.

Large pipe spears are strong and can withstand relatively high forces, but they have disadvantages. A fish cannot be bumped or jarred downward since this motion releases the spear. Like all tools that have upside-down slips, the slips must be released before pulling the tool. Sometimes high working forces cause the slips to stick, making it difficult to release the spear. When the spear cannot be released by bumping down, apply weight, run a string shot positioned opposite the slips, and detonate it. This jarring action often releases the slips. It may not be possible to use a string shot on smaller spears because of their restricted ID.

The no-go may cause a problem. If the fish cannot be released, it is necessary to release the spear by bumping down a short distance as noted. If the tool has been run into the drillpipe and is seated in or near the no-go, there may not be enough distance to move the spear down in order to release it.

There are various modifications of the basic spear. It can be equipped with a packoff, which allows circulation through the fish similar to that of a circulating overshot. In another modification, the spear is operated hydraulically and does not require rotation to set or release.

Washpipe

The washpipe spear, sometimes called a Lebus spear, is a special application. The tool is similar to a regular spear with some modifications. An inverted spear is run inside the washpipe and held in-

Figure 3–9 Circulating and releasing spear (courtesy Homco)

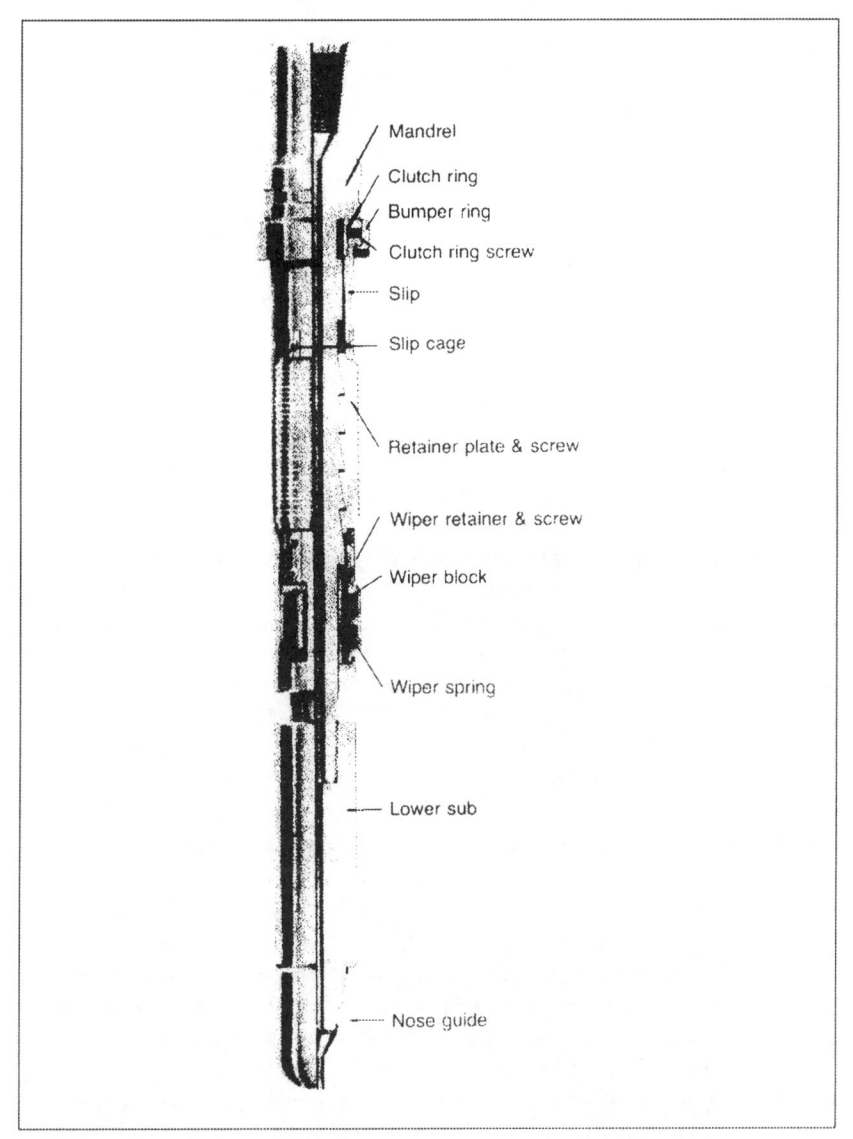

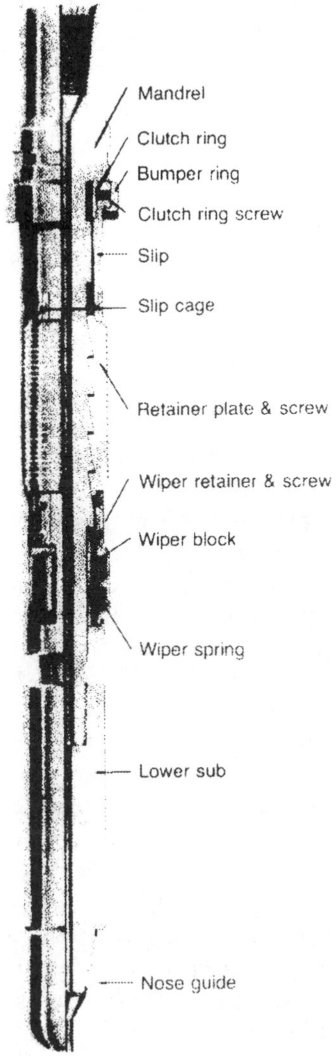
Mandrel
Clutch ring
Bumper ring
Clutch ring screw
Slip
Slip cage
Retainer plate & screw
Wiper retainer & screw
Wiper block
Wiper spring
Lower sub
Nose guide

side by a retainer on the bottom of the washpipe. The slips on the spear catch the washpipe and are controlled by restriction rings. A catch tool is run on the bottom of the spear to catch the fish.

A spear is run inside the washpipe so that a fish can be washed over and recovered in one fishing trip. This is much faster than the conventional procedure of washing over the fish with washpipe, and then tripping to catch and recover the fish. The relatively long time

period between washing over and catching the fish often allows part or all of the fish to stick again. For example, a fish may be washed over for 350 ft, but by the time it is caught and equipment is installed, only 100–200 ft of the fish may still be free.

During an operation, run the spear inside the washpipe assembly. ALWAYS run a back-off or unlatching tool above the washpipe. Run the spear and washpipe into the hole and connect or catch the fish with the spear. Wash over the fish while the washpipe slides over the stationary spear.

If the washover shoe becomes worn, normally the operator can back off and recover the released section of fish. Otherwise, release the spear from the fish, and pull the washpipe and spear. Replace the shoe and rerun it to continue washing over.

If the spear takes excess weight, this indicates the fish is free, so pull it and the washpipe out of the hole. If only a section of a fish is free, back off the free section and recover it. If the entire fish is stuck, disengage the catch on the bottom of the spear and then pull the spear and washpipe. Otherwise, it is necessary to unlatch or back off in the drillpipe above the washpipe.

A washpipe and spear combination frequently is used to recover a keyseated fish. Washing over the fish without a catch allows it to drop, frequently damaging the dropped tools and causing another fishing job. Often the preferred procedure is to back off above the keyseat and use the washpipe and spear combination. Usually, when an assembly is keyseated, only a relatively short length must be washed over so there is less risk of danger.

Running a washpipe and spear combination is a moderate- to high-risk operation. The described operations are straightforward but often difficult to perform. Using the combination involves running long sections of large diameter tools with minimum clearance, so there is a high risk of sticking, even under favorable conditions. Close tolerances cause high torque with a risk of twisting off. It may be necessary to work torque down to the fish to release the inverted spear. This may be very difficult because of the large diameter washpipe. The same difficulty applies to releasing stuck washpipe connected to the fish by a spear, so there is a high risk that the fishing job will not be completed successfully. The washpipe and spear combination is commonly considered as a "last resort" tool. If it is stuck, there is little chance of milling it out of the hole, and the main alternative is to sidetrack.

Packer Retriever

The packer retrieving spear, sometimes called a packer plucker, recovers packers during milling. This tool is similar to a regular spear with modifications. The spear connects to an extension below the bottom of the shoe-type mill. After running it into the hole, the spear passes through packer bore, and catches inside of the mill-out extension below the packer. The top slips are milled off the packer and the spear engages the mill-out extension in order to pull the packer out of the hole (Fig. 3–10).

Conventionally, mill packers by cutting one or both sets of slips, chasing the packer to the bottom of the hole, and then milling it completely. Using a packer retrieving spear often is a faster way of removing a packer. This tool eliminates drilling the lower slips, the extra trip to recover the packer after milling off the slips, and the slow drilling on a loose packer.

Figure 3-10 Packer retriever (courtesy Fishing Tools, A Petrolane Co.)

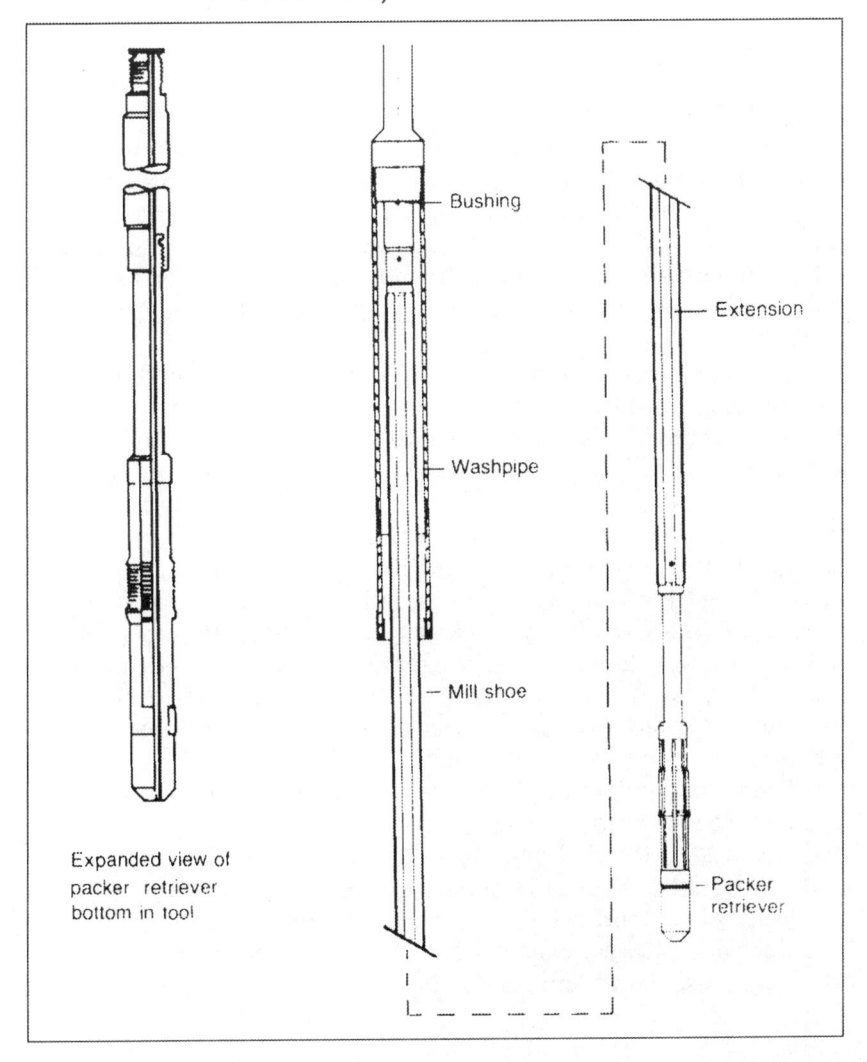

Tools and Assemblies

Screw-in Sub

The screw-in sub sometimes is called a screw-in pin or pin sub. It is one of the strongest catches and is commonly used in catching a backed-off fish if the fish has a good connection on top. Screwing into the fish is not difficult if the hole is clean and vertical. Otherwise the tool joint pin (or box) may sit on the edge of the mating tool joint on top of the fish and spin without engaging the fish. This occurs most often in an inclined hole where either the top of the fish or the bottom of the fishing assembly is slightly above the lower side of the hole.

The screw-in sub has a small notch in the bottom of the pin or box shoulder to eliminate the problem of spinning. If the fishing pin or box spins on the fish, the notch will either catch and kick the fishing pin or box over the fish to catch it, or it will slip off to one side. Normally, it takes only a few attempts to catch the fish.

Taper Tap

The taper tap is an inside catch tool. It has an upper connection and a tapered body that are constructed of hardened tool-type steel (Fig. 3–11). The long, tapered body has machine-type threads that cut and or thread the tap into the inside of the fish. This tool is used when an outside catch cannot be made or the inside of the top of the fish is either too small or irregularly shaped so that a spear cannot be used.

Connect a taper tap to the bottom of the fishing assembly, run it into the hole, and screw it into the fish. It is more difficult to screw in the tap if the fish is loose. The fish is pulled after careful working as necessary.

The taper tap is used mainly in special applications because it has several weaknesses and disadvantages. It is relatively strong in a straight pull. However, it cannot withstand any appreciable jarring or bumping. It effectively cuts a thread in the bore of the fish, but these threads are relatively weak so it is easy to jar the tap loose. A tap with a longer taper cuts a stronger thread than one with a shorter taper. However, the tap with a longer taper breaks more easily. The tool steel is relatively easy to break because of its brittleness and rigidity.

The taper tap does not have a releasing mechanism. Normally, it cannot be backed out since the upper fishing string would probably back off first. If the tool must be released from the fish, try pulling and jarring the tap. Usually, this is successful, but there is a risk of breaking the tap and leaving a piece in the hole. If pulling and jarring does not release the tap, back off the assembly above the tap and use a more efficient fishing assembly.

Figure 3-11 Taper taps (courtesy Hendershot)

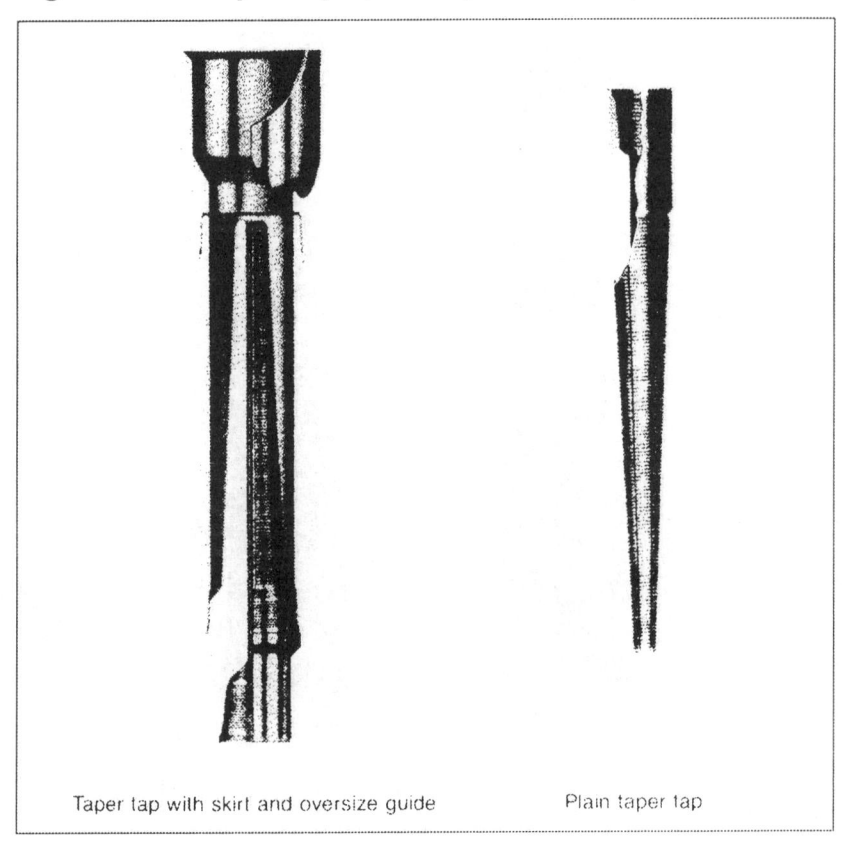

Taper tap with skirt and oversize guide   Plain taper tap

Die Collar

The die collar is similar to a heavy-duty collar with a standard top thread for connecting to a fishing assembly. The bottom has machine-type tool steel threads that cut and/or thread over the outside top of the fish. It is an outside catch tool used to catch a fish that has an irregular top (Fig. 3–12).

A die collar is used primarily to catch smaller or lighter weight fish when the top of a fish is rough and cannot be milled off easily in order to accept an overshot. This tool also is used when the top of a fish is uneven or out of round, such as a slightly flattened tubing top.

One disadvantage of the die collar, and the taper tap, is that it is difficult to determine how much torque to apply during a catch. If the tool is not rotated enough, the threads that are cut on the fish will be too shallow and the die collar may strip off the fish at a relatively low tensional pulling force. If the die collar is rotated too much, the threads will strip and the die collar will not take hold.

Tools and Assemblies

Figure 3-12   Die collars (courtesy Gotco)

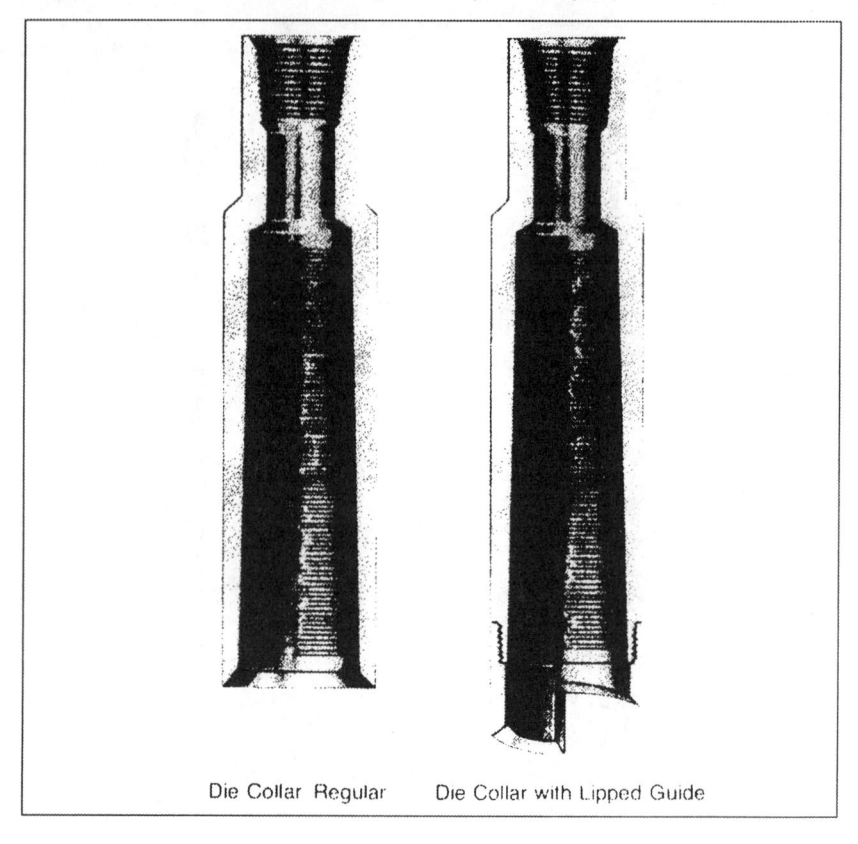

Die Collar Regular   Die Collar with Lipped Guide

The die collar is a strong tool with a strong body. Like the taper tap, it cannot be worked and jarred too heavily because the hold may break. However, there is little risk of breaking the collar. Except for this tendency and the outside catch feature, the die collar generally is similar to the taper tap and operated in a similar manner.

Mills

Mills are widely used, efficient tools that are safe to use when operated correctly. They do not have moving parts so there is little risk of leaving junk in the hole as the mill wears.

Mills are constructed from tubulars, generally with thick walls. Bottom cutting mills have a tool joint connection on the top, normally in standard sizes similar to bit pins, that connect to the bottom of the drill collars. Pilot-type mills have a similar connection on the upper end and a smaller connection, usually a pin on the lower end.

String reamer mills normally have a box-type tool joint on the upper end and a pin-type tool joint on the lower end, cut to standard sizes to run in a drill-collar or drillpipe assembly.

The bottom or cutting face of a mill is designed for its intended use. Flat bottom mills have a flat cutting face with fluid channels so that the fluid cleans the bottom of the hole and removes pieces of metal cuttings. Reamer mills have ribbed surfaces where the face of the ribs serves as the cutting surface and the channels between the ribs as mud courses.

The mill cutting face is coated with a hard facing, usually particles of tungsten carbide, embedded in a matrix material and applied by acetylene welding. Use larger particle sizes for milling larger objects, such as a bit. Coarse particles are often characterized by rough running and high torque. Smaller particle sizes mill a smoother surface for such operations as dressing off the top of a fish. Particle size, particle density in the matrix, and thickness of the combined particle matrix are designed for optimum cutting action.

Since mills wear out, one might ask, "Why not put extra cutting material on the cutting face so the mill can run longer?" If it takes two concave mills, each dressed with a ⅜-in. thick layer of cutting material, to drill or mill three bit cones, then why not use one mill with a ¾-in. layer of cutting material? The tungsten carbide and matrix material combination is tough, wear-resistant, and extremely durable. However, it is not as structurally strong as high-strength steels. Cutting material deposited in an overly thick layer can break off a mill in chunks or pieces, thus reducing the overall efficiency of the tool.

Mills use torque to drill. The amount of torque depends on the size of the mill, the material being milled or drilled, the rate of rotation, and the milling weight. Amounts can vary from light and relatively constant torques to very high, widely fluctuating ones. Excessive torque can damage the running assembly, especially tool joints or connections, and pose a risk of twisting off because of over-torquing or fatigue failure.

Normally, run a mill on an assembly with butt-shouldered, tool-joint-type connections. A mill may be used on tubing but remember care must be taken to prevent damaging the tubing, especially the connectors that can be ruined by over-torquing. In all cases, control the milling rate and resulting torque in order to minimize the risk of work string failure.

Mills serve many purposes, such as drilling (milling) cement, packers, and plugs, and dressing off a damaged fish top. They drill metal junk ranging in size from small tong dies to large drill collars and casing. Mills are commonly used to ream collapsed casing, to remove sections of casing, or to cut casing windows for sidetracking (Fig. 3–13).

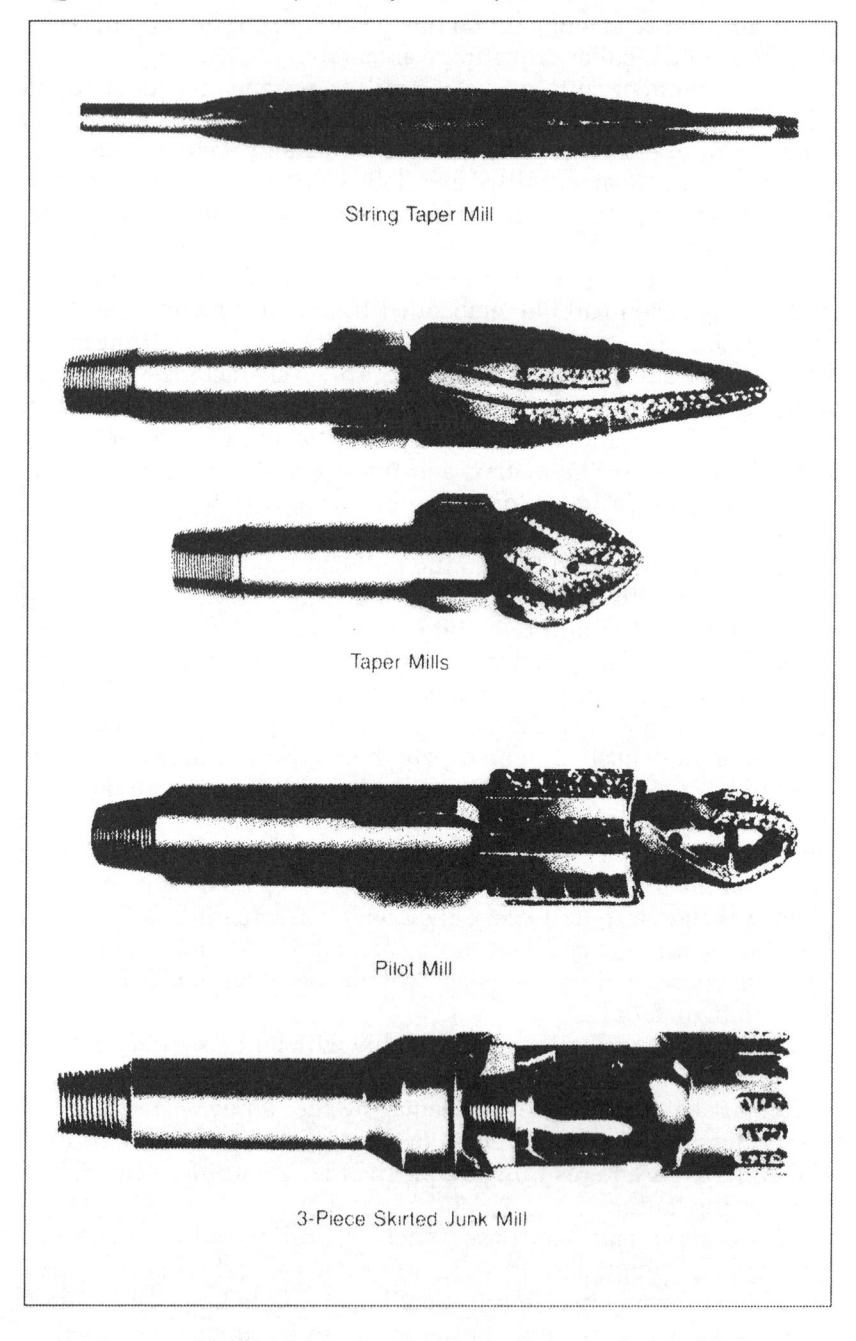

Figure 3-13 Mills (courtesy Servco)

String Taper Mill

Taper Mills

Pilot Mill

3-Piece Skirted Junk Mill

Figure 3-13 Mills (continued)

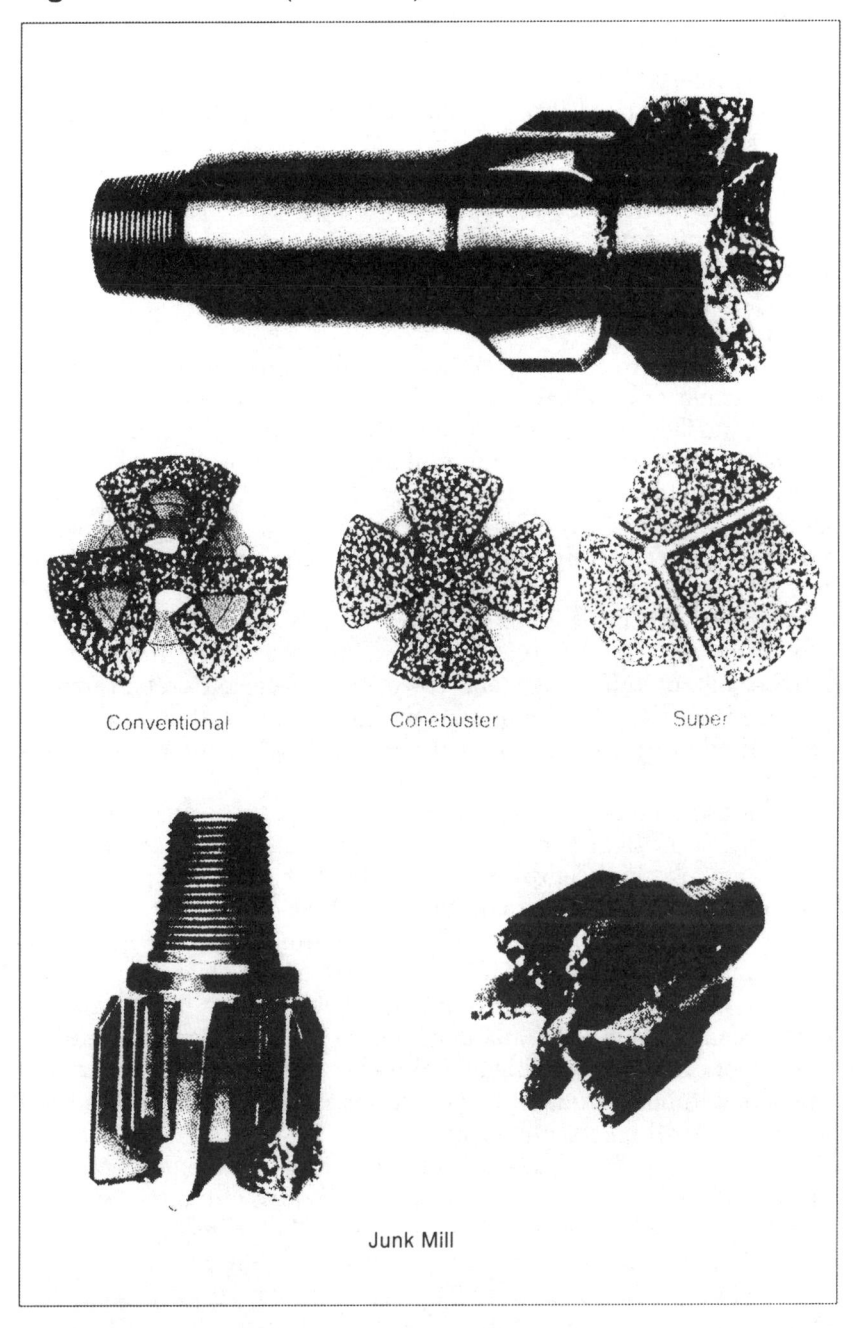

Conventional   Conebuster   Super

Junk Mill

Tools and Assemblies 163

Take special precautions when milling inside of casing. Do not dress the outer edge of a mill with hard facing material since there is a high risk of cutting the casing. It is better to remove about $1/16$ in. of cutting material from the outer edge on the mill's bottom as an additional precaution. Milling at one place in casing may also wear or damage the casing because of mill cuttings wear, even after taking the precautions noted.

Mills are broadly classified according to their uses. Sometimes two different types of mills may serve the same purpose and, frequently, the same mill can be used for different purposes.

Junk Mills

Junk mills are constructed in three basic bottom configurations: flat, concave, and ribbed. These mills expose a maximum amount of cutting surface to the junk to be drilled. Some of the mills have "smooth-type" hard facing material on the lower peripheral edges to allow some side cutting action. Junk mills generally are massive and can drill almost any type of junk. They are also used for other purposes, such as cleaning out cement. It is necessary to run junk subs, or boot baskets, above most junk mills.

Use concave mills when pieces of junk are small. Small pieces of junk often tend to wall off or to be suspended above the edge of a mill because of fluid movement. The concave face helps retain smaller pieces in the center of the hole where they can be rolled, broken up, and milled until they are small enough to circulate out or to recover in a junk basket.

Flat-bottomed mills, especially ribbed types, are dressed on bottom with additional cutting material because of their construction. They may also have additional strength, although this probably is not a significant factor since the bottoms of the other mills are equally strong. These mills are preferred when drilling large pieces of junk.

When milling in casing, it is common to have a clearance of about $3/8$ to $5/8$ in. between the mill and the inside of the casing and no cutting material on the outside of the mill. This clearance reduces the risk of milling the casing and allows the mill to trip in and out of the hole without difficulty. It also creates a sufficient cutting surface in order to drill most junk encountered.

In milling operations, small pieces of junk often bypass or work up around the edge of the mill but are too large to circulate out. If the mill is raised while milling inside casing, these pieces can wedge the tool. With a slightly larger clearance between the mill and the inside of the casing, it is normally possible to work the mill up above the cuttings or junk so that it can be moved back under the mill and milled to a smaller size. About the same tolerances are used when milling in an open hole, although some operators prefer a full-gauge tool, especially in softer formations.

Skirted Mills

Skirted mills are like junk mills except for a short sleeve connected to the lower outer edge of the mill. This shirt or mill guide centers the mill over the fish. For example, when dressing off a small fish, the top may wobble. A regular junk mill could mill down beside the fish, but a skirted mill will stay centered over the fish. The design of a skirted mill is similar to a throated mill. However, the skirt is not covered with cutting material and is used only as a guide.

Cement Mills

Cement mills are similar to junk mills and are often interchangeable with them. Fuller gauge cement mills provide an increased amount of cutting surface. Cement is usually somewhat friable, so there is less risk of sticking a cement mill if a small piece works up past the edge and lodges.

Cement mills normally are not as efficient as roller bits for drilling cement, especially when cement is to be drilled inside larger casing. However, a correspondingly smaller bit must be used in smaller sized casing. The rollers on small bits have a limited life because of their smaller bearings. As a result, there can be a higher frequency of lost cones when drilling cement in small holes; therefore, a mill generally is recommended in these cases.

Pilot Mills

Pilot mills are regular flat-bottom mills with an added centralized bottom extension that guides the mill. A smaller mill or pilot bit extension may be run below some larger pilot mills. The extension enters the tubular to be milled and holds the mill in a centered position. The extension slides inside the object to be milled while the cutting action takes place at the face of the mill.

Pilot mills are designed for milling tubulars such as drill collars, drillpipe, or casing liners and for cleaning the top or upper inside of some fish. These mills can be highly efficient and have a relatively long milling life. Some mills, with cutting material on the extensions, cut and mill the inside of the tubular. This action cleans inside the top of a fish, and in other cases may reduce the amount of material that must be cut by the face of the mill. Pilot mills also recover packers, as described in the section "Packer Retriever." In special cases, a pilot mill can enlarge the ID of a large fish so it can be caught with an inside catch tool.

Reamer Mills

Reamer mills are run on the drill tools, similarly to regular reamers, but are used for cutting metal, usually casing. They enlarge a hole to or near the original diameter, or ream the inside walls of the casing or other tubulars.

String-reamer-type mills have tool joints on both ends and are run in the drilling assembly. Full-gauge vertical or spiral blades taper at each end to the diameter of the body of the reamer. The cutting material is laid on the outer surface of these blades.

Bottom-reaming mills are similar to regular flat or ribbed-bottom junk mills except that they have a longer, full-gauge guide section near the bottom. Cutting material is placed on the bottom and may be placed in vertical rows or ribs on the side depending upon the use of the mill.

Tapered Mills

Tapered mills have spiral or vertical blades on the reamer body. The blades' outer surface tapers from the diameter of the body at the bottom to full gauge at the top. Cutting material is laid over the outer surfaces, and the channels between the blades act as fluid passages.

A tapered mill is run on or near the bottom of the drill-collar assembly. It reams out tight places in casing, severely collapsed casing, casing shoes, liner tops, and the windows used for sidetracking.

Pineapple Mills

The pineapple mill is a variation of the tapered mill but with a sharp point on the bottom. This mill can remove most casing obstructions, even when the entry is very small.

Throated Mills

Throated mills have a long, heavily constructed, tubular body with cutting material on the bottom and inside or throat of the mill. Cutting material may be used on the outer part of the mill, depending upon the conditions under which the mill is used. One main use of a throated mill is cutting a fishing neck. Less frequently, it is used to wash over the top of a fish for a short distance so that there is clearance for a catch-type fishing tool.

Mill Shoes

Mill shoes, sometimes called washovers or rotary or burning shoes, are run on the bottom of overshots, junk baskets, and washover pipe. They wash or mill over a fish. The term "rotary shoe" generally implies a tool used in drilling formation material with a toothed mill (Fig. 3–14). Similarly, mill shoes are dressed with cutting material for cutting metal and, less commonly, cement.

Figure 3-14 Rotary shoes (courtesy Homco)

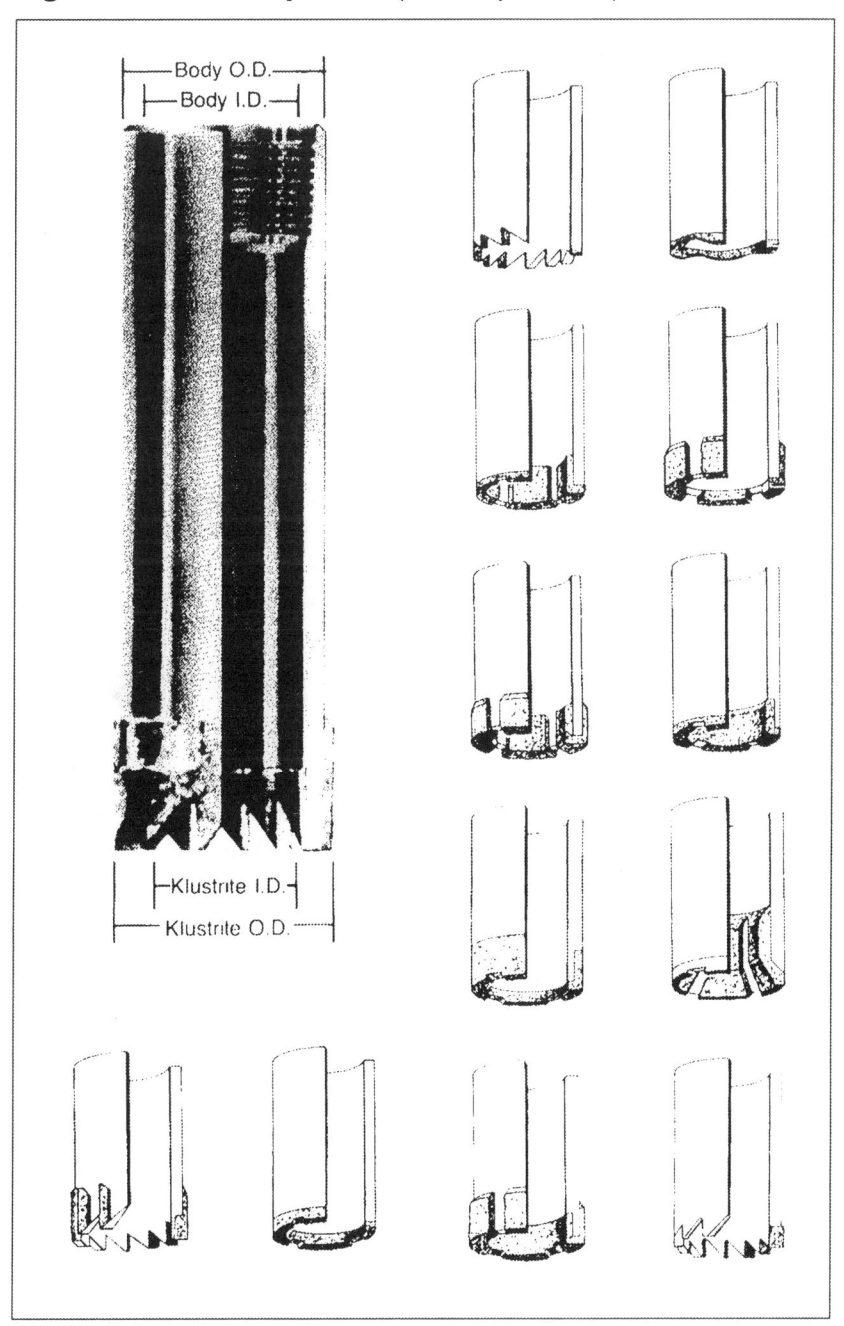

Tools and Assemblies 167

Mill shoes have different bottom designs. A flat to scalloped bottom is more efficient for cutting metal and some very hard formations. Tooth-shaped shoes cut formations. Tooth designs vary from longer teeth on wider spacings for soft formations to shorter teeth with closer spacings for harder formations. Drag-tooth designs have reversed-type teeth that point backward. Teeth may be coated with cutting material for very hard formations and when metal may be encountered.

Mill shoes are dressed in one of four different ways:

1. Bottom only — cuts vertically downward relative to the axis of the bore hole. It is used where no side cutting action is needed.
2. Bottom and inside — frequently run inside casing. There is no cutting material on the outside to damage the casing and all cutting action is on the bottom and inside. This type of mill shoe is commonly used to cut a fishing neck.
3. Bottom and outside — often used when inside cutting action is not needed to cut over a fish.
4. Bottom, inside, and outside — often used for washing over in the open hole. This type of mill is commonly run on tools such as junk baskets when the fish may require cutting before it can be swallowed.

When additional cutting action is needed, the lower part of the shoe may be built up outward or inward and cutting material may be laid over the built-up sections. For example, if overshot slips break while milling over junk, the inside of the mill shoe can be built up, reducing the inside diameter. During the washover operation, the mill shoe cuts part of the outside of the fish, reducing its outside diameter so that it will pass through the overshot slips with less risk of damaging them.

Pipe Cutters

Pipe cutters are sometimes called casing cutters or section mills. They cut tubulars and generally are subdivided into two types. Inside pipe cutters run inside larger pipe and cut from the inside outward. Their use is limited to tubulars with larger inside diameters. Outside pipe cutters run over smaller pipe and cut from the outside inward (Fig. 3–15). First, casing cutters cut through the pipe and then begin cutting and removing the pipe. Inside-type cutters are commonly used to remove a section of casing for sidetracking in cased holes.

Figure 3–15 Pipe cutters (courtesy Homco)

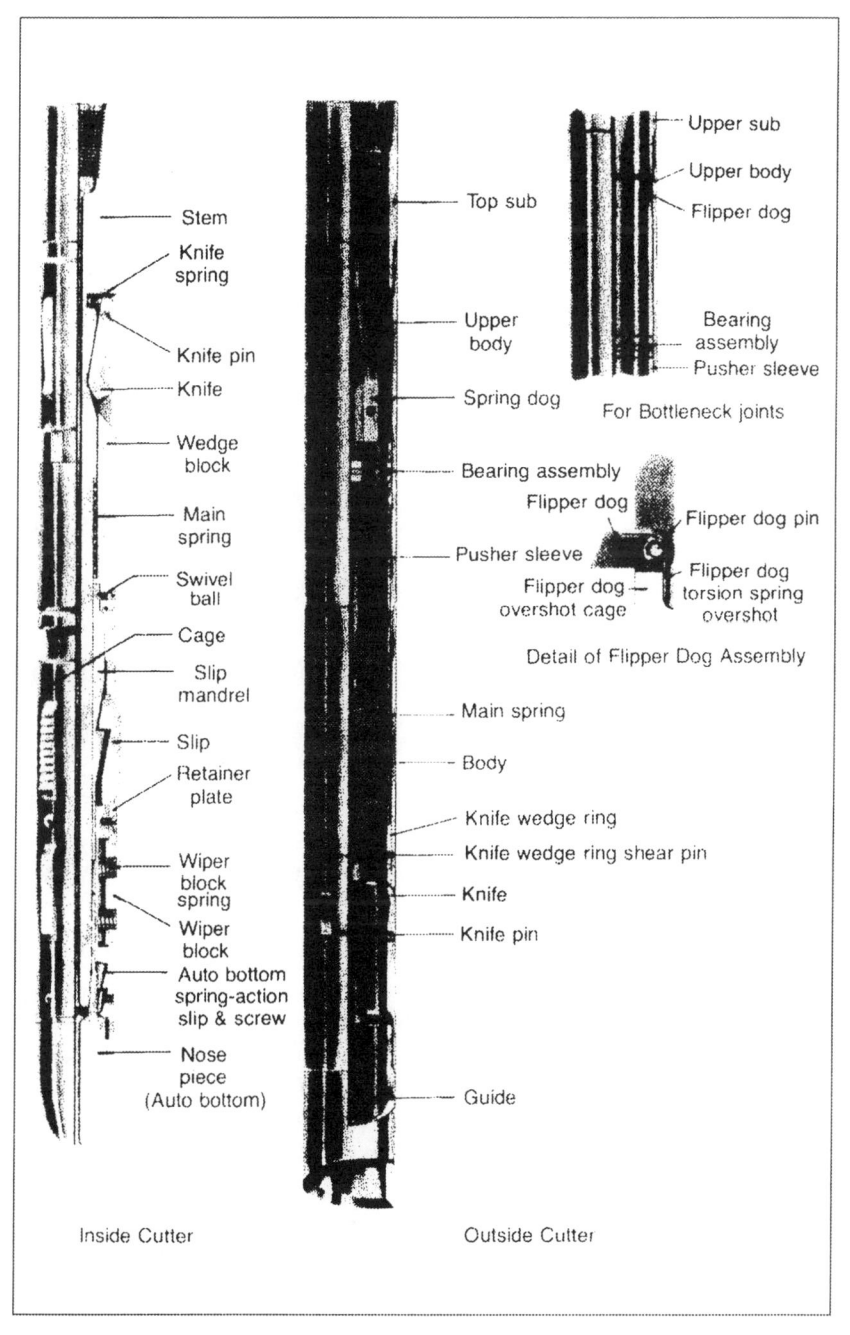

Tools and Assemblies 169

There are various types of cutter blades or knives. A common version has two or three knives dressed on the bottom with tungsten carbide to perform the cutting. One preferred type has two, or preferably three, blades, each with two or three leaves per blade. Each leaf has circular inserts of tungsten carbide. Two or three leaves are bonded together with the tungsten carbide inserts in a staggered configuration. As the leaves cut and the leaf metal wears, there is always cutting material exposed on the cutting edge. These can be highly efficient blades. For example, in one case a tool was cutting a 7-in., 26-lb/ft casing at the rate of 15 ft/hr. This high rate caused a problem with the removal of metal cuttings.

Both mechanical and hydraulically actuated tools are available. There are inside cutters for small tubulars, but they are seldom used because of their reduced strengths. Small tubulars are most often cut with a chemical cutter because this tool is easier to run and there is less risk of problems. Sometimes, a chemical cutter cannot be used, such as when a damaged fish top is encountered, so an outside cutter probably would be used. This cutter will also probably carry the cut section out of the hole on top of the cutter blades.

Jars

Jars are important fishing tools and responsible for releasing most stuck tools. Two main types are the more common hydraulic jar and the less used mechanical jar. They are further subdivided into drilling and fishing jars and may be supplemented with accelerator jars. Understanding when to use jars and their operation, strengths, and weaknesses is an important preventive procedure (Fig. 3–16).

Jars are available in a wide range of sizes. They normally are full opening, allowing for mud circulation and passing small diameter tools, such as a back-off string shot. They are run above any tool that may stick. They deliver a high, upward-impact blow on the tools connected beneath them. Torque can be transmitted through the jar for regular drilling or torquing a fish. They can be used immediately, which is often a major advantage. The chances of releasing stuck tools are improved by starting jarring immediately after sticking; therefore, jars help prevent and also expedite fishing.

Jars are widely flexible in the strength and frequency of the jarring stroke. The force of the jarring blow is determined by various factors, including stroke length, internal construction, weight concentration (usually with drill collars) immediately above the jars, and the amount of overpull when the jars trip. Adjust the jarring frequency by manipulating the speed and distance of upward and downward movement of the work string.

Figure 3-16 Drilling jars (courtesy Dailey Oil Tool Inc.)

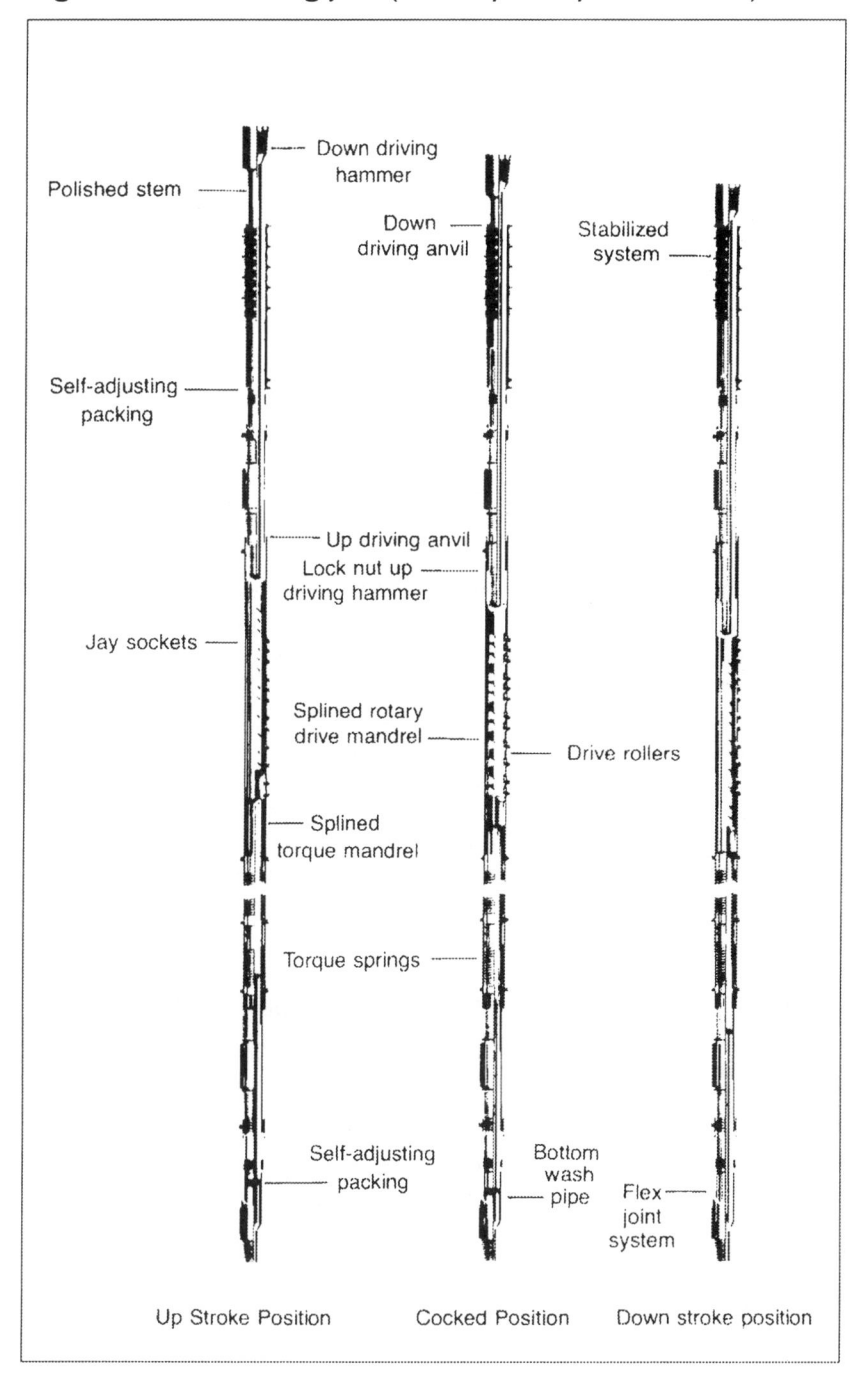

Tools and Assemblies 171

Jars are constructed in different ways. The most common hydraulic jar has an annular-type piston inside an annular-shaped cylinder. The "top" of the piston may be above or below the cylinder — above for this description. Spines prevent internal rotation. The upward-moving work string pushes the piston upward and increases tension on a fish. Then the piston enters an enlarged or grooved section, or a fluid choke bypass, which allows a rapid release of fluid. This causes the drill collars above the jar to deliver a strong upward blow that is transmitted to the tools below the jar. Lowering the fishing assembly lowers the piston, and fluid bypasses rapidly to cock or reload the jar in readiness for the next blow.

During an operation, run regular jars below the drill collars and above the fishing tool that is used to engage the fish. Catch the fish and work it conventionally. If this is unsuccessful, begin jarring the fish. In some cases, start jarring immediately. Drilling jars are similar to regular jars and are operated in a similar manner.

When a stuck assembly is jarred, the forces are dynamic and much larger than average static loads. Consider a stuck assembly that is jarred by tripping the jars at an overpull of 40,000 lb over the free point weight. The static load on the pipe at the stuck point is 40,000 lb.

Much higher forces are obtained by jarring. Picking up or stretching a downhole assembly stores energy in the assembly. The energy is released when the jars trip. The upper section of the jars and the work string move upward rapidly a distance about equal to the stroke length of the jars (drill collar stretch is negligible) and are then stopped abruptly. This dynamic force, which can be very large, is transmitted directly to the top of the stuck tools and indirectly to the point of sticking. The effectiveness of the jars can be increased by adding concentrated weight, such as drill collars, in the work string immediately above the jars. This extra weight provides additional momentum and a higher impact. *Caution:* Heavy-duty jars can be worked to the point that the force is very high and may be capable of parting some drillpipe and tool joints. Always work the jars within the manufacturer's specifications.

Hydraulic Fishing

The hydraulic fishing jar, sometimes called an oil jar, is the basic tool described earlier in this section. Hydraulic fishing jars are highly reliable and simpler to operate than other jars. They have a relatively short stroke length of about 4–6 in. Depending on hole depth and other factors, hydraulic jars can be tripped at a rate of one to four jarring blows per minute. These jars are generally preferred over mechanical jars because of their reliability and faster, easier operation.

Hydraulic Drilling

The hydraulic drilling jar generally is operated similarly to a hydraulic fishing jar and serves the same purpose, with minor modifications to comply with drilling action. It has a longer stroke of about 12–15 in. and it is run while drilling for immediate use if sticking occurs.

Mechanical Fishing

The mechanical fishing jar is one of the first rotary jars. It is generally less efficient than other hydraulic jars, may not last as long without failure, and cannot be stroked as fast. Otherwise, this jar serves the same purpose as the hydraulic fishing jar and has a similar stroke length, but it is operated in a slightly different manner.

Some mechanical jars require placing about one-half turn of torque in the assembly, picking up to the required tension, and then releasing the torque in order to release the jars for the stroke. Depending on operating factors, mechanical jars can be torqued to deliver one to two blows-per-minute. Normally, the torque is placed with a rotary. If the assembly sticks firmly so that the kelly cannot be installed, then place the torque with the slips or tongs, which is more time consuming than using the kelly. The speed of the jarring action is reduced correspondingly, usually to one blow-per-minute or less.

Mechanical Drilling

The mechanical drilling jar is generally similar to the mechanical fishing jar. It has a heavier construction and a stroke length of about 12–15 in. and generally is operated in a similar manner. However, it is not used as often as the hydraulic tool for reasons similar to those for the mechanical fishing jar.

Accelerator Jars

Accelerator jars, sometimes called intensifier jars, may be run along in the fishing string with regular jars. An accelerator jar normally is full opening to the inside diameter of conventional drill collars. Torque can be transmitted through the jar for torquing the fish. It is not used as a drilling tool.

An accelerator jar acts as an air cushion or fluid spring, serving as energy storage and release for regular jars, thereby increasing their efficiency. The cushioning action also lessens the strong jarring and impact forces on the remainder of the downhole assembly as the regular jars are tripped.

Figure 3–17 Bumper subs and jars (courtesy Bowen)

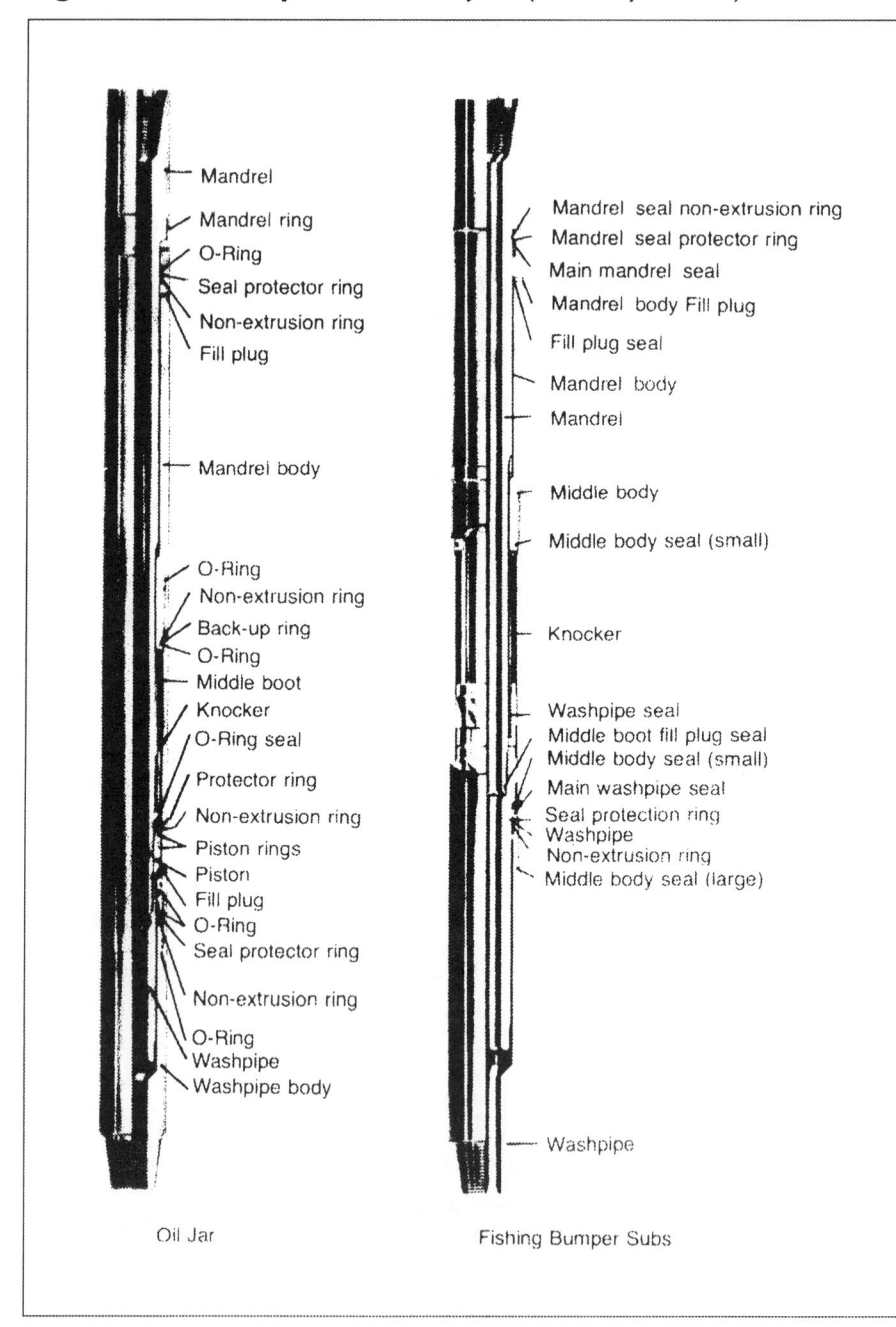

Oil Jar

- Mandrel
- Mandrel ring
- O-Ring
- Seal protector ring
- Non-extrusion ring
- Fill plug
- Mandrel body
- O-Ring
- Non-extrusion ring
- Back-up ring
- O-Ring
- Middle boot
- Knocker
- O-Ring seal
- Protector ring
- Non-extrusion ring
- Piston rings
- Piston
- Fill plug
- O-Ring
- Seal protector ring
- Non-extrusion ring
- O-Ring
- Washpipe
- Washpipe body

Fishing Bumper Subs

- Mandrel seal non-extrusion ring
- Mandrel seal protector ring
- Main mandrel seal
- Mandrel body Fill plug
- Fill plug seal
- Mandrel body
- Mandrel
- Middle body
- Middle body seal (small)
- Knocker
- Washpipe seal
- Middle boot fill plug seal
- Middle body seal (small)
- Main washpipe seal
- Seal protection ring
- Washpipe
- Non-extrusion ring
- Middle body seal (large)
- Washpipe

Chapter 3

Jar Intensifier

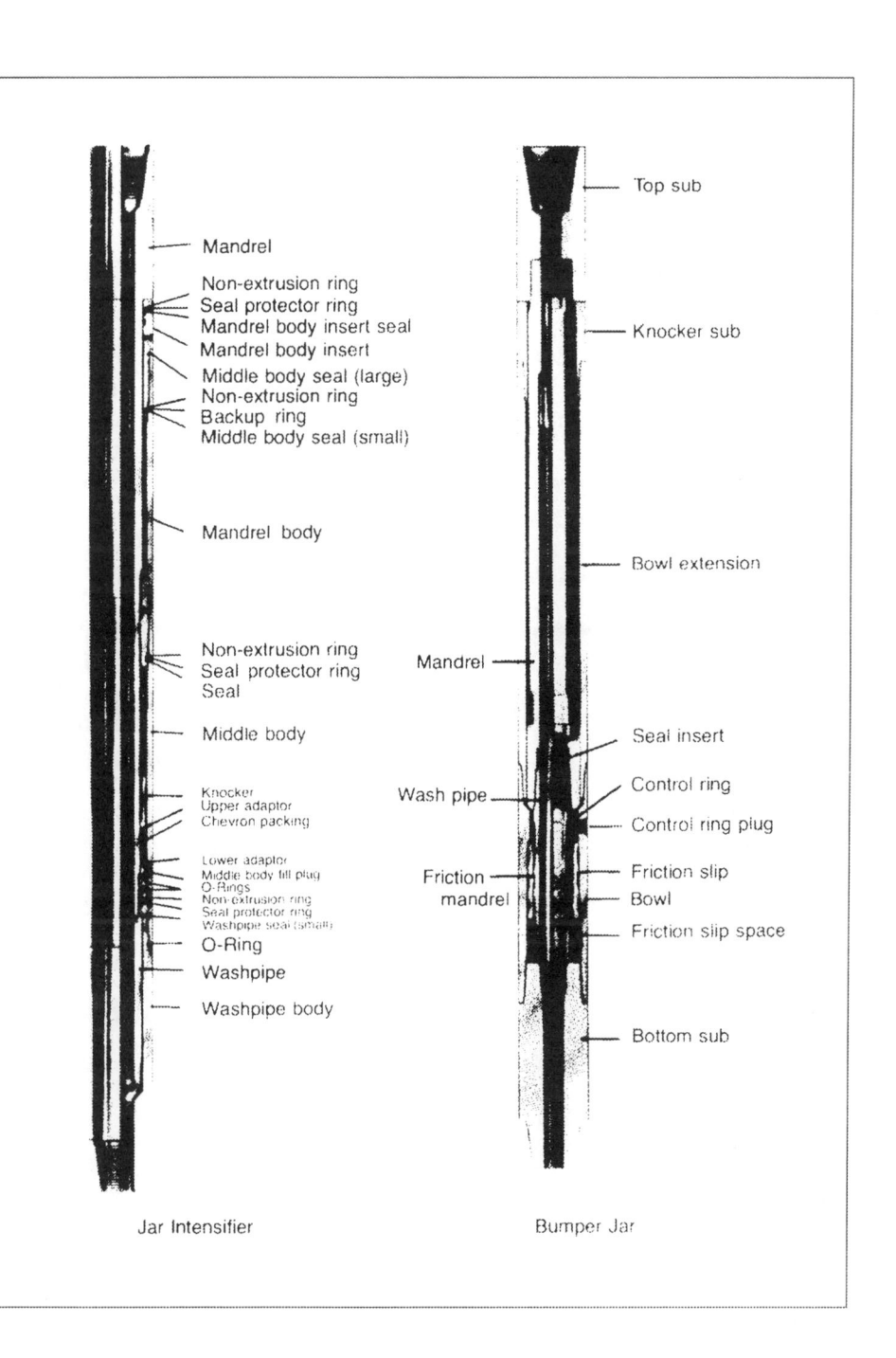

- Mandrel
- Non-extrusion ring
- Seal protector ring
- Mandrel body insert seal
- Mandrel body insert
- Middle body seal (large)
- Non-extrusion ring
- Backup ring
- Middle body seal (small)
- Mandrel body
- Non-extrusion ring
- Seal protector ring
- Seal
- Middle body
- Knocker
- Upper adaptor
- Chevron packing
- Lower adaptor
- Middle body fill plug
- O-Rings
- Non-extrusion ring
- Seal protector ring
- Washpipe seal (small)
- O-Ring
- Washpipe
- Washpipe body

Bumper Jar

- Top sub
- Knocker sub
- Bowl extension
- Mandrel
- Wash pipe
- Friction mandrel
- Seal insert
- Control ring
- Control ring plug
- Friction slip
- Bowl
- Friction slip space
- Bottom sub

Tools and Assemblies 175

Bumper Sub

The bumper sub sometimes is called a mechanical slip joint (Fig. 3–17). It strikes a sharp downward blow directly to the tools below the sub and indirectly to the point where the tools are stuck. Its action complements that of the jars in the fishing assembly. Bumper subs also help to release slip-type tools when slips are stuck on the taper, such as in bumping down to release overshots, and less frequently, to apply a constant weight on lower tools. The bumper sub normally is full opening to the inside diameter of conventional drill collars. Torque is transmitted through the sub for torquing the fish and also for regular drilling. Bumper subs are available in different sizes.

When the fish cannot be released with an upward blow from the jars, it may be necessary to drive the fish down. The bumper sub is designed for this purpose. It has a free-traveling mandrel that provides the stroke length of the stroke. At the end of the stroke, the fishing assembly imparts a sharp blow to the tools located below the bumper sub. Drill collars above the jars increase the force of the blow.

During an operation, the bumper sub normally is located between the jars and the fishing tool. If the bumper sub were located above the jars, especially hydraulic jars, blows would be struck against the jars. This could cushion the blow and lessen its effect.

Bumper subs, like jars, are designed with different stroke lengths for different purposes. They also can be obtained with built-in safety joints. In addition, some types combine the functions of both a bumper sub and a jar in one tool. One type of tool operates as a mechanical jar but also can bump downward by using a similar rotating action with weight set on the tool. When the tool is set in a second position, it can be used as a conventional bumper sub to bump down or bump upward through the free travel of the stroke length.

Combination Hydraulic Jar and Bumper Sub

The combination hydraulic jar and bumper sub sometimes is called a jar bumper. It combines a bumper sub and hydraulic jar in one tool and is available in different sizes. Each part operates similar to and has the same use as the individual tools already described.

This tool is very efficient when working with the alternate actions of a hard pull to jar upward and an immediate slack-off or drop-off bump down with maximum force. A slight hesitancy as the jar closes increases the efficiency of the bumping down stroke.

Surface Bumper Sub

The surface bumper sub sometimes is called a surface jar. It is run at the surface, above the rotary, and delivers a downward blow. Various sizes are available.

In operation this tool is positioned at the surface and connected to the stuck tools. It is stretched with an upward pull and the tripping mechanism releases at preset loading. The released energy causes downward momentum, delivering a downward blow at the bottom of the assembly. This tool is not commonly used.

Washpipe

Washpipe or washover pipe is used to wash over fish and stuck tools. It is made from high-strength casing and fitted with flush-joint, butt-shouldered connectors. These connectors resist continued tightening and possible damage caused by the high torque conditions under which washpipe often is operated. The washpipe is sized to obtain the optimum clearance possible between the fish and the inside diameter of the washpipe, and in the annulus between the outside wall of the washpipe and the inside wall of the hole (Fig. 3–18).

During an operation, run a washover shoe on the bottom of the washpipe. Make up and run the required amount of washpipe, and then connect it to the bottom of the fishing assembly. Run this mechanism to the top of the fish and begin circulating and drilling, or washing over the fish. Up to 500 ft of washpipe is used in big holes that have adequate clearances and lesser lengths in smaller holes.

After washing over, pull the washpipe and run a fishing assembly to catch and recover the fish. As an alternate procedure, run a spear inside the washpipe to catch and recover a section of the fish after it has been washed over. There is a high risk of sticking the washpipe in this process, especially wall sticking, because of the close tolerances. Recovering stuck washpipe is very difficult at best, because of the required sidetracking and/or the increased risk of losing the hole.

Safety Joint

The safety or back-off joint is a disengaging tool. Run it on the bottom of the fishing assembly for a screw-in connection; otherwise, run it above the catch tool. This tool allows backing off and leaving a fish or stuck tools in the hole without using a string shot or cutting

Figure 3-18 Washover pipe (courtesy Gotco)

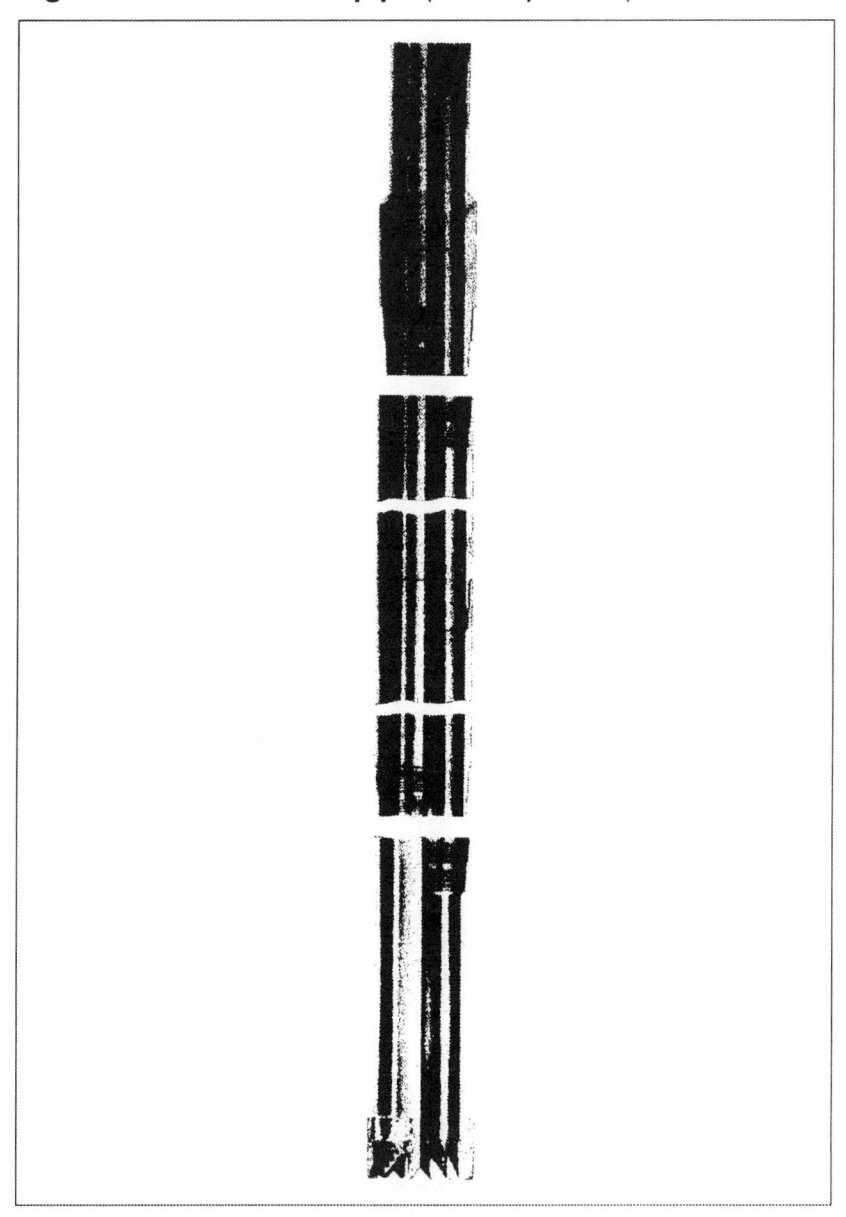

the fishing string (Fig. 3–19). Pull the fishing string for repairs or changes, then run it back into the hole, reengage the fish with the upper part of the safety joint, and resume fishing. The safety joint is widely used and relatively safe to run.

One common use of safety joints is to replace jars that fail, usually because of heavy working, Similarly, if the hole begins to cave

Figure 3-19 Safety joints (courtesy Homco)

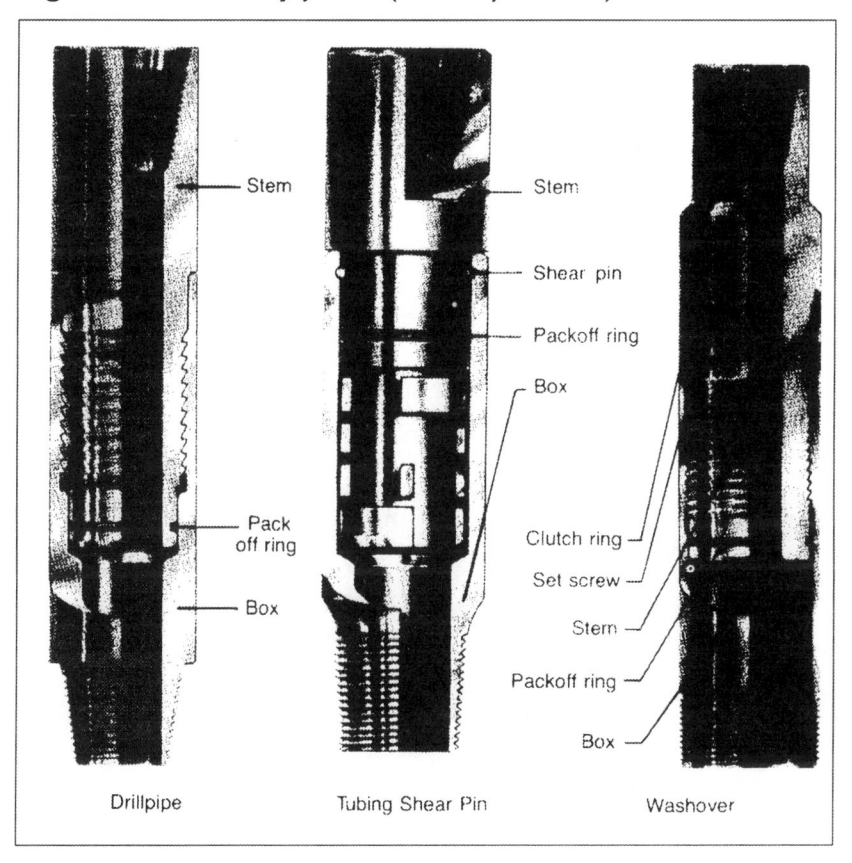

around the fishing string, it may be best to disengage from the fish and pull the assembly above the point where caing has occurred or to allow for a cleanout trip. If the operator must wait for a string shot to back off the fishing asembly, the fishing assembly could stick. With a safety joint, the fishing assembly can be released almost immediately and pulled, or other action taken.

There are various back-off joints; most are full opening for circulation and for running tools such as string shots. One type has a coarse-threaded connection in the middle of the safety joint. This connection requires less torque than the regular tool joint connection. The tool is backed off by rotating the assembly in the left-hand direction. The amount of back-off torque is less than that normally required to back off regular tool joints. However, sometimes applying sufficient torque at the tool may be limited because of drag that increases with depth, higher angle of hole inclination, and dogleg severity. All of these may cause problems in working the torque down to the tool in order to back it off.

Tools and Assemblies 179

Another type of safety joint is backed off and released by a combination of short, reciprocating movements and right-hand torque. Others are released by different manipulations of the fishing assembly. Normally, a safety joint is reconnected by reversing the motions required to back it off. For example, the coarse-thread, left-hand, back-off safety joint is reconnected by reciprocating and rotating to the right. Select a safety joint based on the motions required to connect and disconnect it, and that are compatible with the working motions of the other tools and the way the fish is worked.

Knuckle Joint

Knuckle joints connect fishing tools to the bottom of the fishing assembly and change the direction of the tools suspended below it. The axis of a center line through the fishing tool string changes a few degrees at the knuckle joint so the tools below the knuckle joint will point in a slightly different direction (Fig. 3–20). The knuckle joint serves the same purpose as a bent sub, bent pin, or bent joint.

Figure 3–20 Knuckle joint (courtesy Bowen)

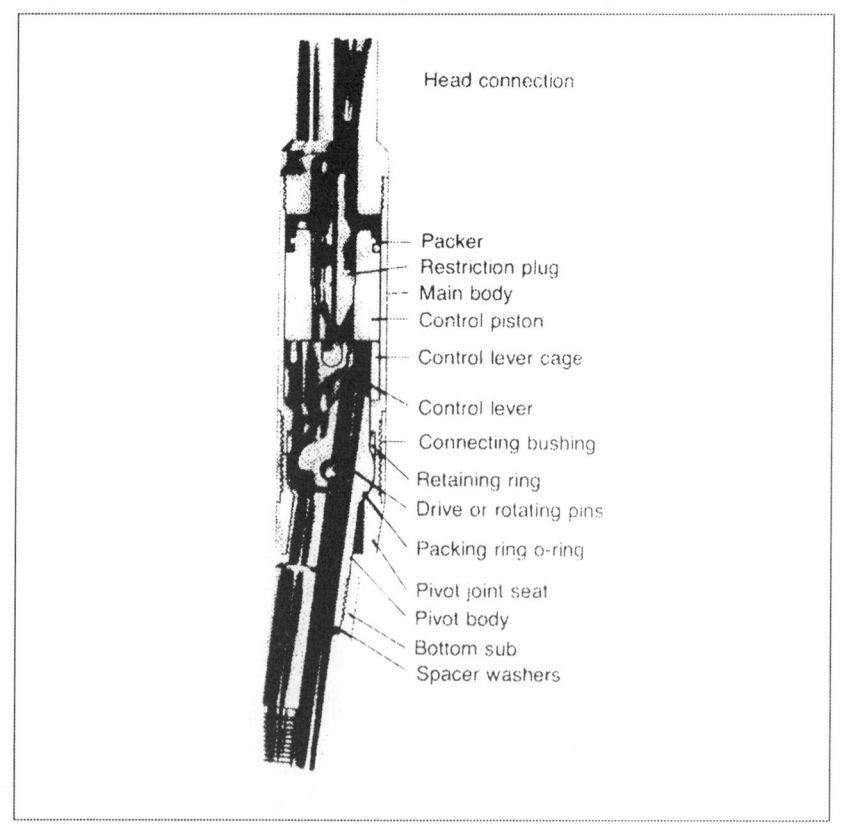

The knuckle joint allows the fishing catch tool to take hold of a fish top that is out of alignment (usually laying to the side in an out-of-gauge hole section). A fish top in this position may be difficult or impossible to catch with a regular fishing assembly, because the tools tend to run down beside the fish.

There are various types of knuckle joints, including both mechanical and hydraulic. A common hydraulic tool aligns with the axis of the work string during running. The lower part of the tool deflects out of alignment, extending to one side when it is actuated by mud pressure. Depending on its size, the knuckle joint may not be as strong as some of the other tools used to deflect the fishing assembly.

Bent Sub

The bent sub, or kick sub, changes the direction of the axis of tools connected below it from the direction of the axis of the running assembly above it (Fig. 3–21). A bent sub normally is constructed from a regular sub. The most common bent sub is a bent pin sub. The pin on the bottom of the sub is cut in the normal manner but is kept out of alignment with the sub body and upper box connection. The angle of misalignment is about 2–5° from the center line of the sub. The sub normally is named by the amount of misalignment, such as a 3° bent pin sub that has pin cut with a 3° misalignment.

A bent box sub is less common. It has the box thread in the top of the sub cut a few degrees off center and the pin is aligned with the body of the sub. Otherwise, this tool is similar to and used in the same way as the bent pin sub.

Bent subs serve the same purpose as, and (other than the fixed angle) operate similarly to, the knuckle joint and bent joint.

Bent Joint

A bent joint is a joint of drillpipe, tubing, or in some cases, a small drill collar that is bent so the lower pin is out of alignment with the upper box connection. This tool is used for the same purpose and generally operated in the same manner as a knuckle joint or bent sub. However, a bent joint usually can move the bottom of the fishing tools a further lateral distance from the center line of the wellbore than the bent sub or knuckle joint. The bent joint, especially a bent drill collar, is stronger than a knuckle joint or a bent sub.

Normally, an overshot with a cut lip guide is run with a bent joint. Ensure that it is connected to the bottom of the bent joint so the open side or lip of the guide is on the right-hand side (looking down) relative to the bend in the bent joint. As the bent joint is ro-

Figure 3-21 Bent sub

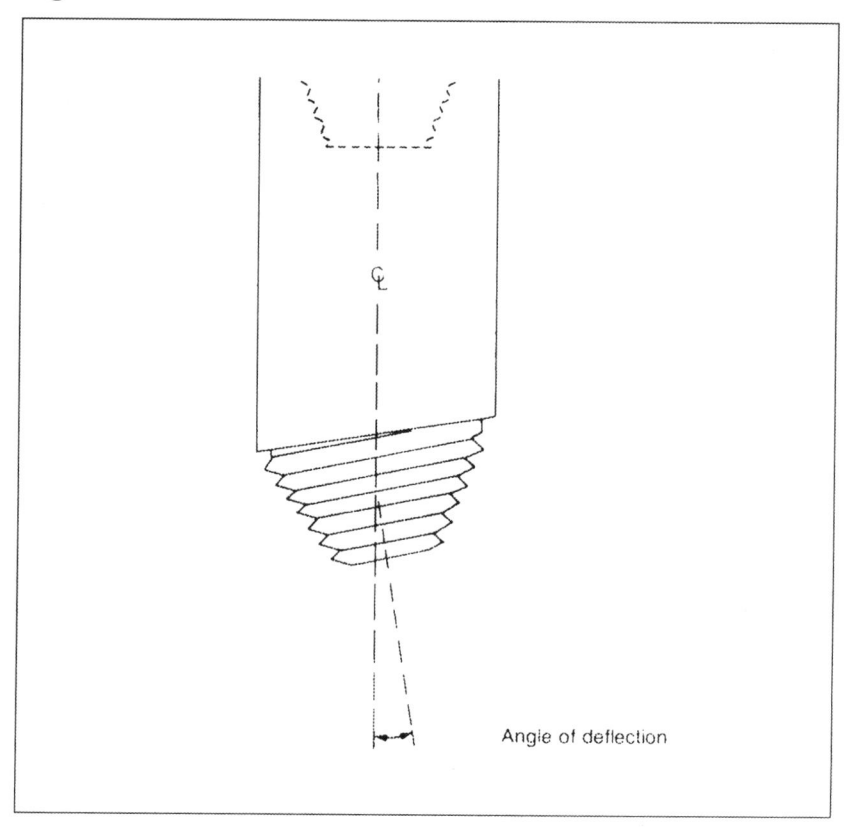

[Figure showing a bent sub with centerline and "Angle of deflection" labeled at the bottom]

tated slowly down over the fish, the open side of the cut lip guide can contact the fish and guide it into the overshot. The joint can be bent at the rig site, but make sure that the overshot and cut lip guide are correctly positioned.

Wall Hook

The wall hook is a hook-shaped tool constructed as part of a washover shoe. It helps catch the fish when the top is off center or out of alignment with the axis of the wellbore, such as when a top is laid over in a washed-out section. This tool also can be used when the overshot is run under a bent joint, knuckle joint, or bent sub.

During an operation, connect the wall hook to the bottom of the overshot. Run the fishing assembly into the hole until the bottom of the wall hook is immediately above the top of the fish, and then rotate and lower it slowly. The rotating wall hook contacts the top of the fish, and hooks it toward the center of the hole and into the shoe of the overshot. The fish enters the overshot, is caught by the slips, worked, and recovered.

The action of the wall hook is like that of a muleshoe or cut lip guide. However, it increases the lateral extension and helps engage a fish that is deeper in a washed-out area and further from the axis of the hole. Handle a wall hook carefully since the long extension makes this tool easy to break.

Circulating Sub

A circulating sub, run near the bottom of the fishing string, allows the option of circulating through the fish or out a circulating port in the side of the sub. Various types of circulating subs are available. One has a circulating port on the side of the sub that can be opened or closed by manipulating the drilling assembly.

Another circulating sub has a ported nipple, allowing circulation though the perforation. A perforation ball sealer is dropped to seal off the perforation, which allows circulation through the fish. If circulation is not established through the fish, release pressure, allow the perforation ball sealer to drop off, and circulate again through the perforation. A main disadvantage of this procedure is that it leaves the perforation ball sealer inside the fish, which could restrict downhole operations inside the fish. In a few cases the ball may be recovered by reverse circulation, but usually there is a high risk of plugging the perforation.

Another sub has side circulating ports closed by rupture discs. When these are used, the first operation is an attempt to establish circulation down through the fish. If this cannot be done, increase the pressure until the discs rupture, then circulate around the fishing assembly.

A circulating sub may not be used when fishing with an overshot without a packoff because it allows circulating around the fishing string. However, other methods, such as screwing in, do not allow this option, so a circulating sub may be needed.

Circulating through a stuck fish whenever possible is a major aid in releasing the fish or preventing additional sticking. It is always a good practice to circulate around the fishing assembly to prevent sticking, especially when fishing through hazardous formations, to keep the hole clean, and to control high-pressure zones. The drillpipe can be perforated near the fishing tool to establish circulation around the fishing assembly. However, the circulating sub often is a better selection because of the time delay and cost of perforating.

Ported "Side-Door" Back-off Sub

The ported side-door back-off sub allows tools, usually wireline tools, to run through the fishing assembly, and out into the annulus through the sub. Normally, the sub is connected immediately above the catch tool. Some tools, such as a string shot, pass through the work string, while the tapered inner section guides the wireline tool through the side-door choke so that it passes down the annular space outside of the fish.

Shock Sub

Shock subs dampen shock loads and vibrations when drilling in high torque and in rough drilling situations. Normally, run these tools immediately above a bit or mill. They can extend the lives of bits and mills, minimize drillstring vibrations, and reduce the risk of fatigue failure.

Shear Rams

Shear rams fit in the blowout preventer. They have a knife-blade-type cutting face. They serve as an outside cutter designed to cut drillpipe and seal the wellbore in severe emergency conditions.

Tools Run Inside Casing

Two tools are run only inside casing because they catch the inside casing wall. In some cases, these tools are set inside casing but extend below the casing. There is a risk of damaging the casing with either tool, and retracting the anchor to release the tool may also cause problems.

Pulling Tool

The pulling tool is used inside casing to exert a very high tension force on a fish. It is a highly efficient tool when applicable. Connect the pulling tool between the bottom of the fishing assembly and the catch tool. Catch the fish. Then slips on the pulling tool will engage the casing wall, anchoring the tool. The tool is rotated, exerting a strong upward tensile force to the fish.

Mechanical pulling tools are actuated by rotating the work string, which acts through a screw mechanism, creating a high lifting force. Mud pressure acts against a piston in the hydraulic pulling tool to exert a strong upward force.

Reversing Tool

The reversing tool is used inside casing to exert left-hand torque in order to unscrew a fish. Like the pulling tool, it is run above the catch tool. Tools below the bottom of the reversing tool have left-hand connectors. After engaging the fish, the reversing tool anchors against the casing wall. Apply right-hand torque and rotation with the work string. The tool converts this force to a higher level of left-hand torque and rotation below the tool as a way of backing off part of the fish.

One problem with the reversing tool is that often only a short section of a fish is recovered. The fish can be backed off at any connection, but the uppermost connection is the most likely one to back off. Additional trips are required to recover the fish.

Roller

Casing rollers open the collapsed casing in place, restoring it to or near the original ID. The casing roller has a tapered metal body with a tool joint connection on top that connects to the drill-collar assembly (Fig. 3–22). Cylindrical rollers are positioned longitudinally around the tapered body section of the tool. Ports provide for mud circulation.

Figure 3–22 Casing roller (courtesy Eastman Whipstock and Fishing Tools)

Tools and Assemblies

The casing roller connects to the drill-collar assembly below a bumper sub and jars. Run this tool into the hole, set it on top of the collapsed casing, and rotate. A combination of weight and rotation restores casing diameter. Depending on the severity of the collapse, several rollers may be used in consecutively larger sizes to open the damaged casing.

Whipstock

The whipstock is used for sidetracking in cased holes. A tapered lower edge directs the drill tools away from the original center line of the wellbore.

TUBULAR OR WIRELINE FISHING TOOLS

Some tools are designed to be run either on tubulars or wirelines. In most cases, these tools are more efficient when run on tubulars.

Casing Patch

The casing patch repairs holes in casing without the need for cementing or pulling the casing. A common patch has a corrugated steel sleeve that is run on a swaging device. When this tool is run and positioned, a swage is pulled though the sleeve, expanding it and pressing the sleeve firmly against the inner wall of the casing. The casing patch restricts the casing ID slightly and is more commonly used in production operations. A similar type internal casing patch tool run on a wireline can seal holes in small tubulars (Fig. 3–23).

Casing and Tubing Swage

The casing or tubing swage restores flattened or partially collapsed casing and tubing in place to or near its original inside diameter. A swage is more commonly used in tubing and a casing roller in casing. Normally, a swage is run on tubulars for swaging casing and on a wireline for swaging tubing.

The swage has a cylindrical body with a tapered lower end and a tool joint connection on the upper end for connecting to the work string (Fig. 3–24). It has ports for circulating and grooves for mud channels. Tubing swages are smaller sized tools of similar design.

Connect a casing swage to the bottom of a bumper sub and jars with drill collars for added weight. Run this tool into the hole down to the point where the inside diameter is restricted. Bump the swage into the collapsed section until it is lightly stuck, and then jar it free.

Figure 3–23 Tubing liner (courtesy J.C. Kinley Co.)

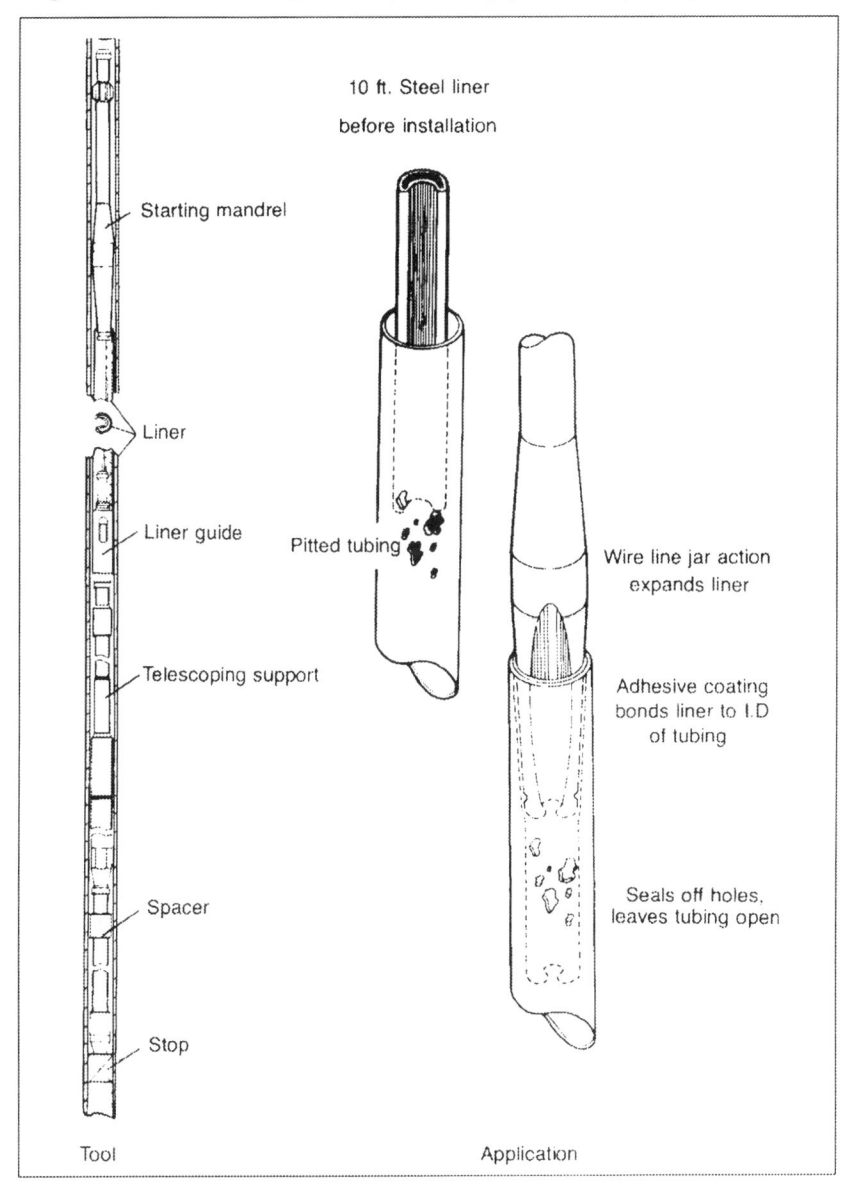

Repeat this procedure until the casing has been swaged out to the diameter of the swage.

The amount of force used to jar down is determined by the force of upward jarring required to release the tool. The outside diameter

Tools and Assemblies

Figure 3-24 Swages

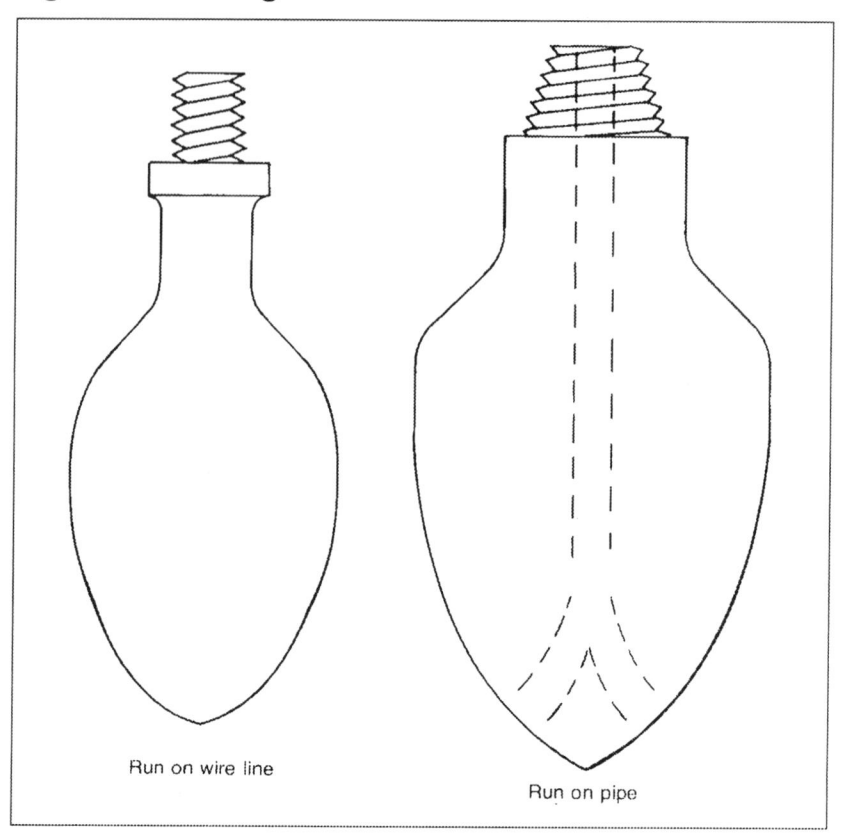

Run on wire line

Run on pipe

of the swage will be slmost that of the inside diameter of the casing for lightly damaged or flattened casing. For more severely collapsed casing, a smaller diameter tool may be used first, followed by successively larger sized tools. Swaging is repeated until the casing is restored to or near the original ID. Wireline operations are completed similarly, but generally with smaller tools and wireline jars and a bumper sub.

Chemical Cutter

The chemical cutter cuts tubing, drillpipe, and small diameter casing from the inside. It may be used to partially cut through the wall of thicker walled tubulars that are then jarred apart.

Run a chemical cutter inside the tubular and position it at the point to be cut. Chemicals are forced under high pressure out of jets in the cutting tool, cutting the tubular. Standoff distance is important for efficient operation.

Grab

The grab, frequently called a wireline grab, recovers wireline fish (Fig. 3–25). This tool consists of a tool joint with 2 or 3 prongs that have pointed ends extending downward from the bottom periphery of the tool joint. The inner sides of the prongs are fitted with barbs. Barbs are spaced alternately on the prongs so there is one barb every 4–10 in. from the base of the tool joint to the bottom of the prongs. Normally, select a grab with a barb diameter that is slightly larger than the diameter of the wireline fish.

The grab is structurally weaker when compared to the spear, but it is used in a similar manner. Generally, the same precautions apply. There is less risk of bypassing a wireline fish when using a grab,

Figure 3–25 Grab
Note: Larger, heavy-duty tools run on drillpipe and tubing; lighter duty tools (illustrated) run on rods and wirelines.

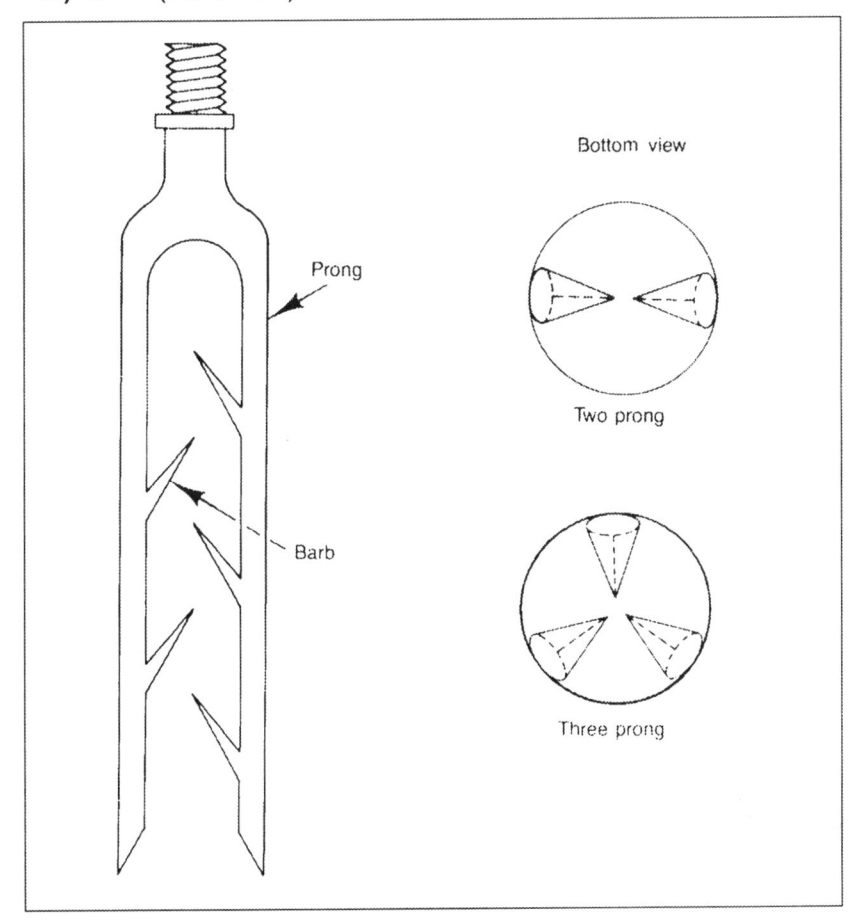

Tools and Assemblies

but it can happen. There is a higher risk of bypassing in an open, out-of-gauge hole. A stiff-pronged or sprung wireline grab may work better in this case.

A sprung grab is used in cased and open holes, especially in out-of-gauge open holes below casing. This tool is the same as a regular grab except that the prongs are modified slightly. The prongs are bent radially outward so the equivalent fixed, unsprung diameter of the bottom of the prongs is about 25% to 50% greater than the casing ID. They are sprung together at the bottom of the grab to a smaller, sprung diameter and tied with strands of soft rope or soft wire. This allows the prongs to pass through the preventers and other surface equipment and to enter the casing. The retaining wire wears and falls off as the spear runs through the upper casing. The the prongs slide against the wall of the casing or open hole. Any wireline fish are gathered inside of the prongs and caught with the barbs. This reduces the risk of bypassing the fish and increases the chance of a successful catch.

Jet Cutter

Jet cutters cut tubing, drillpipe, and small diameter casing with a specially shaped, explosive jet charge. They cut tubulars from the inside and are operated like a chemical cutter.

Junk Catcher

The junk catcher screens the mud to remove small floating debris, such as pieces of packer and pump piston rubbers, perforation ball sealers, and other small particles. This junk collects in the bowl of the catcher or is pushed to the bottom of the hole. Otherwise, close-tolerance tools, especially those run on wirelines such as gauge rings and packers, can stick next to this debris by wedging against the side of the hole.

The junk catcher has a tool joint or other screw-type connection on top for connecting to the work string or wireline. It has an inverted basket with perforated or slotted sides. The diameter at the bottom of the basket approaches the casing ID. The slots or perforations are sized to catch large pieces of suspended debris as mud passes through them. Debris that passes through the slots usually is small enough so that it can pass around close-tolerance tools without sticking them.

Run the junk catcher on a wireline or work string to screen mud. It also may be run below a packer or other close-tolerance tools. It screens mud efficiently so other tools can operate in clean fluid and minimizes the risk of sticking.

Junk Shot

Junk shots break up large junk into smaller pieces so they can be easily drilled, milled, walled off, or caught in a junk basket (Fig. 3–26). The junk shot is a large, shaped jet charge that fires vertically downward.

Figure 3–26 Wireline junk shot

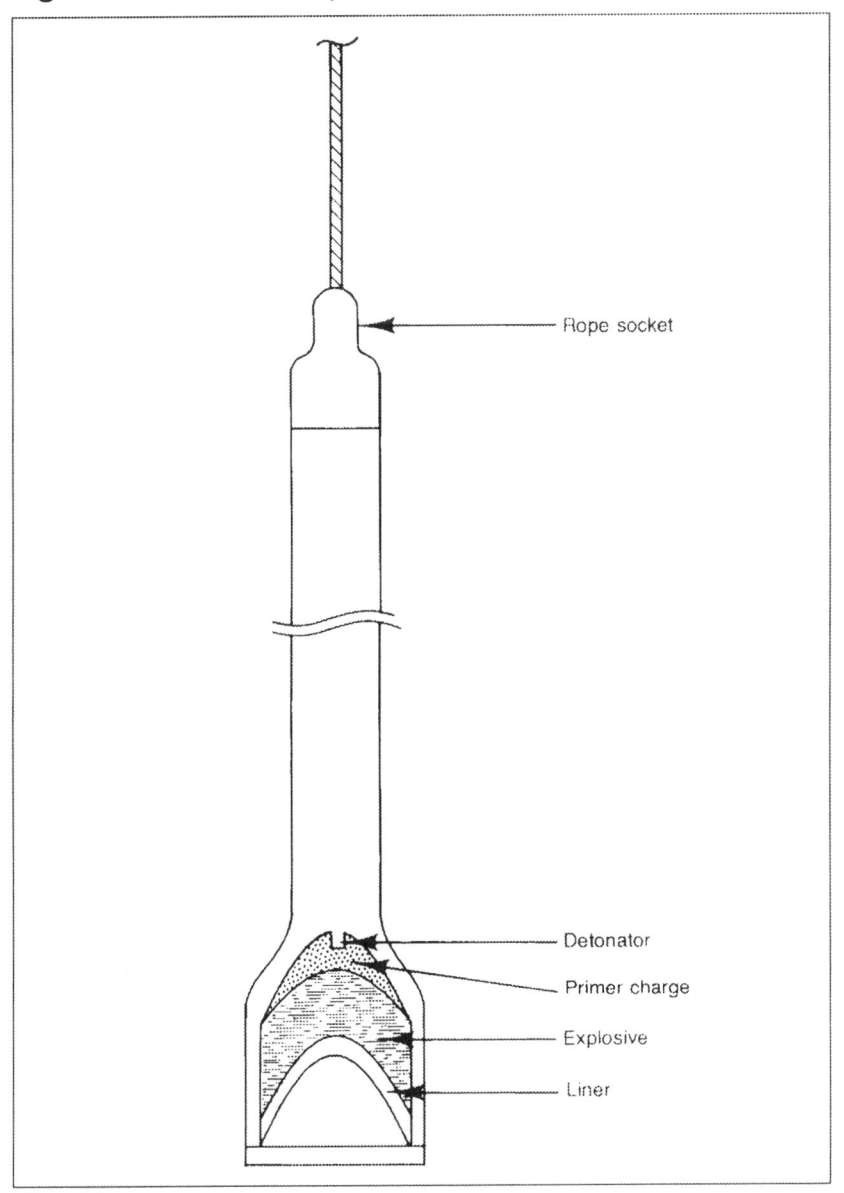

Tools and Assemblies

Junk shots must be very close to or almost in contact with the junk to be effective. If the fish is in the open hole, it frequently is covered with a small amount of cuttings or other fill material. This prevents the junk shot from being detonated close enough to the fish to be effective. Because of this, junk shots are less effective when run in the open hole on a wireline shielded conduit. An effective junk shot run on tubulars has circulating ports to clean off the top of the fish. Then the shot can be positioned correctly for effective detonation.

Magnet

A strong permanent magnet recovers small junk. In the common configuration the magnet is encased and attached to a tool joint. It is a close-contact tool. The magnetic insert must touch or be very close to the fish to be effective (Fig. 3–27). Magnets run on wirelines are seldom effective in open holes because of the small amount of cut-

Figure 3–27 Fishing magnets (courtesy Gotco)
Note: Similar type, smaller tools run on wirelines.

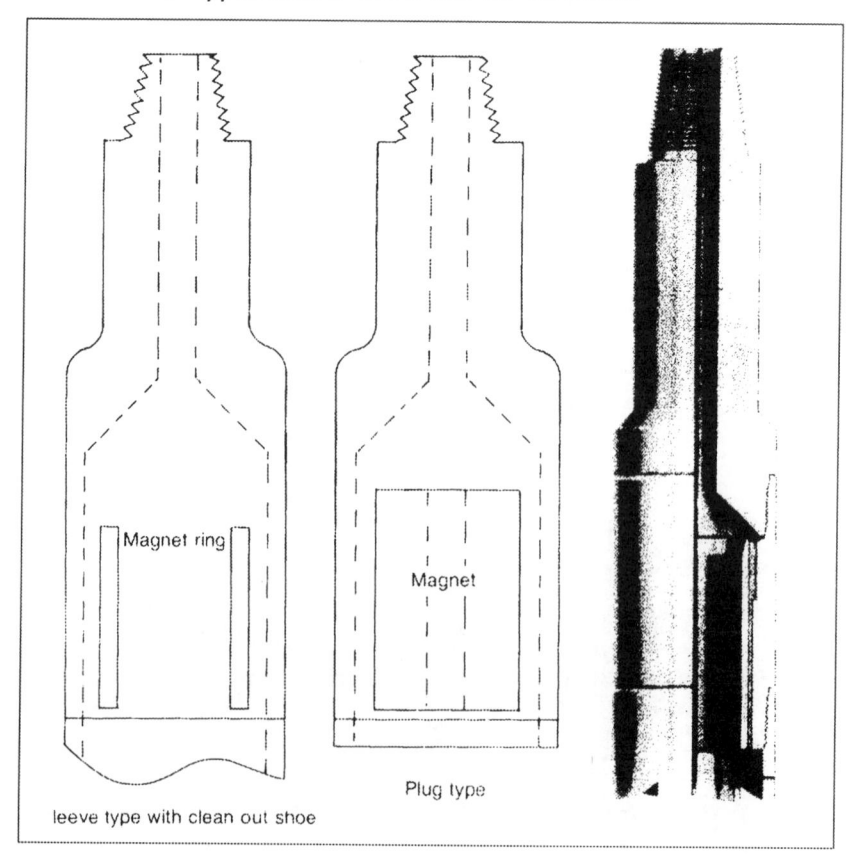

tings or other fill material that commonly cover the junk. Magnets can be more effective in a cased hole if it is clean.

A magnet designed to wash and clean the bottom of the hole can be run on tubulars so the magnet can contact the junk. A magnet may be fitted with a skirt to keep the junk from being knocked off during tripping. Toothed skirts may be used for minor drilling and cleaning out in order to better expose the junk to the magnet. In most cases it is better to take such positive actions as running a junk basket or short mill-tooth bit to break up the junk and catch it in a boot basket or circulate it out of the hole.

Socket

The socket or rod socket is like an overshot, except that it is smaller so that it can catch smaller fish. It has slip teeth aligned horizontally on slip-type inserts. The socket usually is run on a wireline or pump rods. It is highly efficient for engaging and recovering broken pump rods, weight bars, and fishing necks on rope sockets. If run on tubulars and if a fish cannot be recovered, then the slips generally can be either released or broken and released by jarring and bumping.

The wireline version of the socket usually has a shear-pin release. If the fish cannot be released and recovered, bump down and shear the releasing pin, which then releases the slips from the fish. The shear-pin release may also be used when the socket is run on pump rods, but this is less common.

Wireline Spear

Wireline spears recover lost wirelines. A wireline spear has a slender, elongated metal body or prong with a tool joint connection on the upper end (Fig. 3–28). The average tool has three to six barbs connected to the prong and oriented to cover 360°. The diameter of the prong should be approximately two times the diameter of the barb. Select a spear with a barb diameter slightly larger than the diameter of the wireline fish. Run larger spears on tubing or drillpipe and smaller spears under sinker bars on wirelines.

Generally, run a guide above the spear. This prevents the wireline from getting above the tool and causing it to stick. The guide also pushes or bunches the wireline below the spear so the barbs can engage it. To prevent sticking, it is important to use a guide when running this tool in casing in order to ensure that the clearance between the guide and the ID of the casing is less than the diameter of the wireline fish.

Figure 3-28 Wireline spears

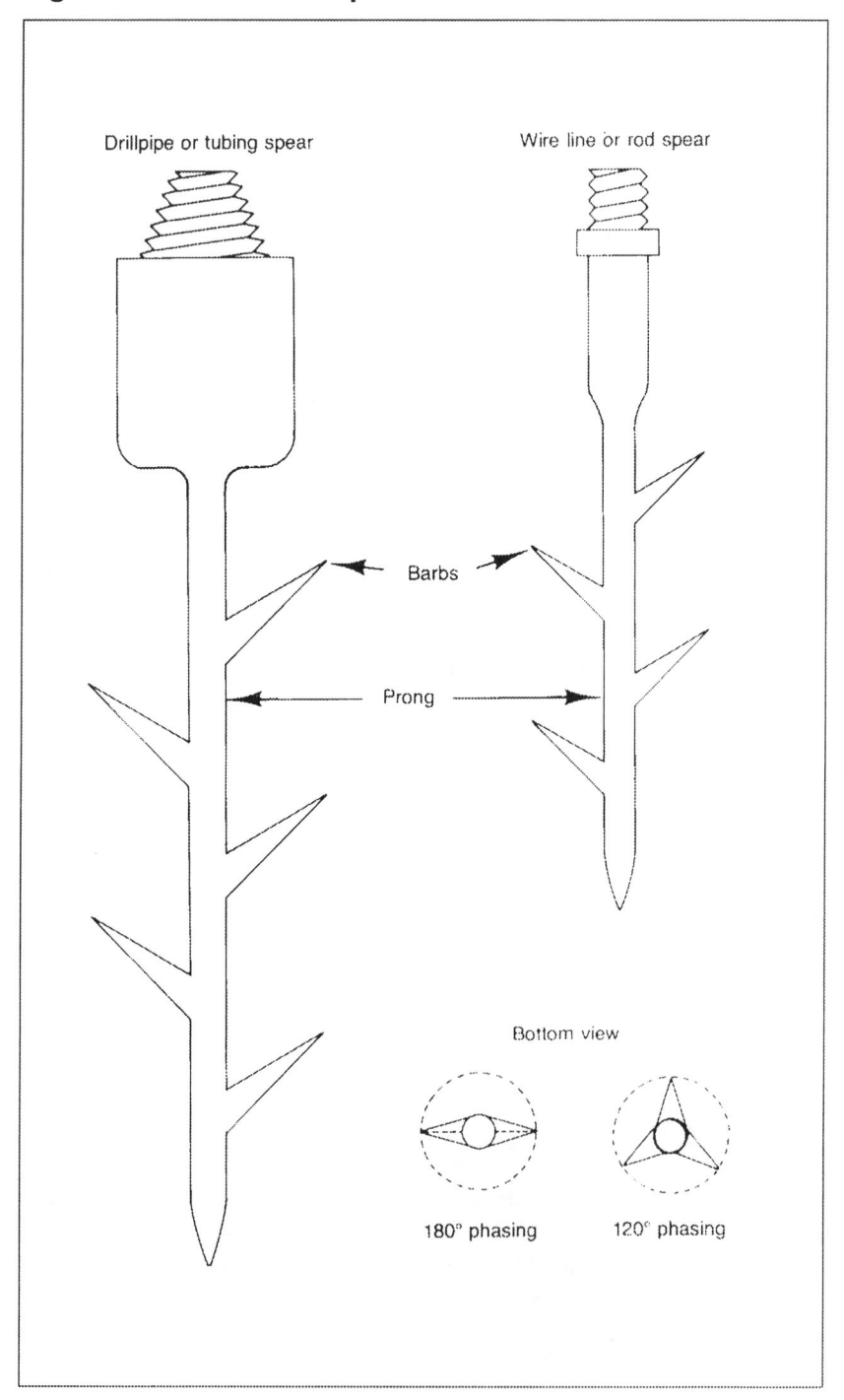

194 Chapter 3

For operation with tubulars, connect a wireline spear below a bumper sub and jars. Run this tool to engage the fish with minimum weight. It is best to recover the fish without bumping or jarring since the fish can break where it is entangled in the barbs. The tool is run in a similar manner on a wireline. The same precautions apply to tubular and wireline operation. Handle the wireline spear carefully to prevent breaking, especially when fishing for large lines with large tubulars. The spear is a stronger tool when compared to a similar sized grab. In most cases a "sprung grab" is preferred, at least to bunch up the line. Then use the stronger spear to catch and pull the bunched line out of the hole.

As a very rough rule of thumb, the top of a wireline fish will be at a distance from the bottom of the hole equal to about ½ to ⅗ of the stretched length of the wireline fish. Run the spear on tubulars in the open hole with care. Use slick tools as much as possible. When tagging the top of the fish, DO NOT rely on the weight indicator's set-down weight. That indication may come too late, and the tool may already be stuck. Take special precautions, such as double pulling stands, and pull an extra stand if there is any question about the fish being caught.

When pulling a fish in the open hole, there is a high risk of sticking as it is pulled into the bottom of casing. If the fish sticks, do not pull too hard. Work down and back up to wear out the bunched wire fish, or birdnest, so it can enter the casing. Use minimal jarring and bumping action (Fig. 3–29).

WIRELINE FISHING TOOLS

Wireline fishing tools are run on wirelines and are primarily operated by reciprocating motion. In most cases, this limits their efficiency as compared to other types of fishing tools. Wireline operating efficiency decreases with depth because of line stretch caused by increased weight. It then becomes more difficult to determine specific tool action. This adversely affects the operation and increases the risk of additional fishing jobs.

To illustrate this point, the jarring action is crisp and well-defined when jarring on a fish with wireline tools at 5,000–8,000 ft. Tool action is less distinct at 14,000–16,000 ft and more difficult to determine. If the line is slacked off too much on the downward jarring stroke, excessive line flexing at the top of the rope socket will result in line failure. When a tool is stuck and the wireline cannot be cut or broken, it can be worked in this manner to break it near the rope socket.

Most suppliers of wireline equipment also supply fishing tools specially designed to recover their own equipment.

Tools and Assemblies

Figure 3-29 Balled wireline below a casing shoe

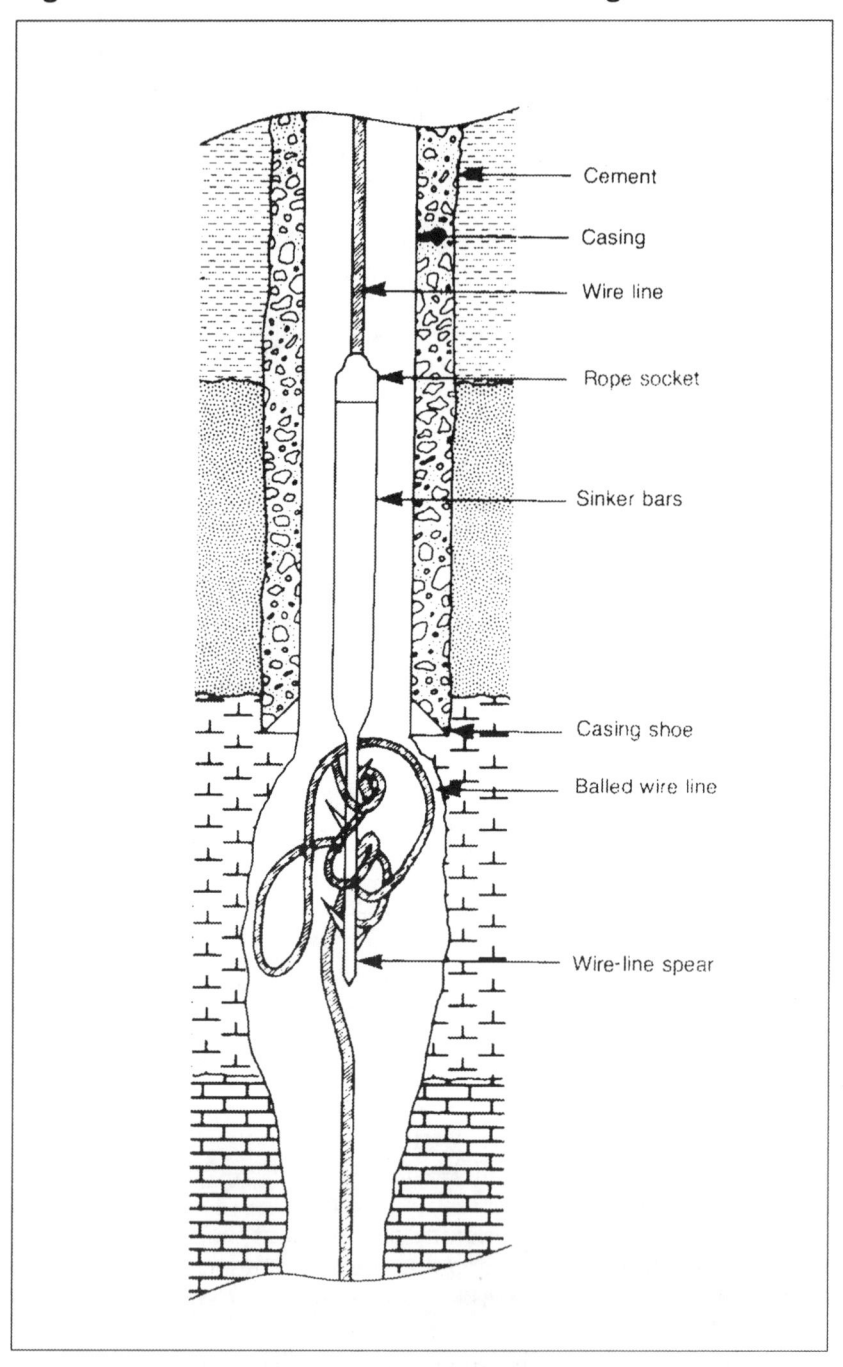

Bailers

Bailers remove fluids from the hole. They also may drill or clean out the hole to a limited extent and be used to dump sand or cement slurry on plugs or at the bottom of the hole. Bailers are most commonly run on wirelines (Fig. 3–30).

Figure 3–30 Bailers (courtesy Hendershot Tool Co.)

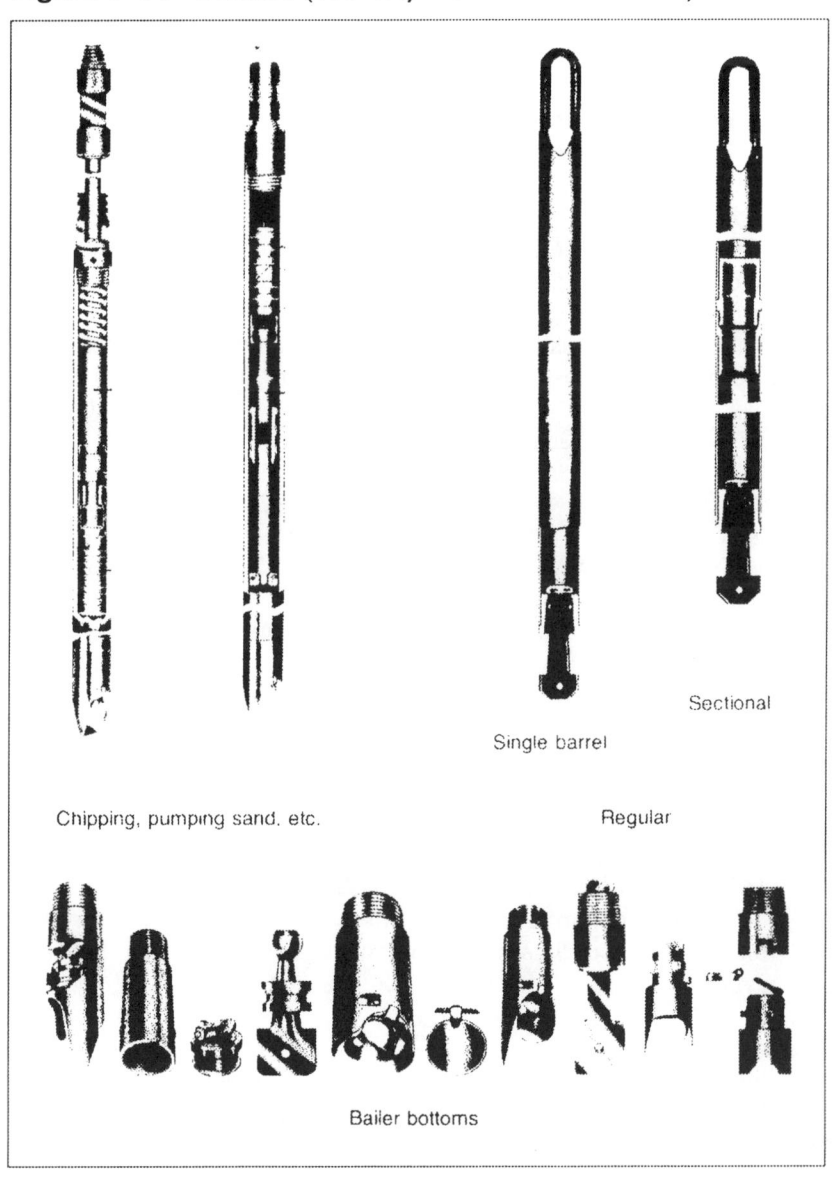

Chipping, pumping sand. etc.

Single barrel

Regular

Sectional

Bailer bottoms

Tools and Assemblies

Bucket Bailer

The bucket bailer removes fluids from the hole. It is a long tube with a sealed bottom and a threaded connection at the top so it can be attached to the wireline rope socket. Older versions had the wireline attached to a bail connected to the open top.

In operation, the bailer is run into the hole below the fluid level. Fluid runs into the bailer through the upper ports. After pulling, the bailer is emptied at the surface by inverting it and pouring the fluid out.

Dart-Bottom Bailer

An improved version of the bucket bailer has a check valve in the bottom. The check extends out the bottom of the valve in a dart shape. Fluid enters the bailer through the upper ports and through the check valve in the bottom. After the bailer is pulled, the fluid can be dumped through the dart bottom more conveniently than by inverting the bailer.

When the hole has a fluid column, the dart-bottom bailer picks up fluid from its lowest point. Fluid passes through the bailer as it is lowered. When the tool is stopped, it contains fluid from that point. This can be used to locate a water-oil interface.

Dump Bailer

The dump bailer carries sand or cement slurry to the point in the hole where it is to be deposited or dumped. One of the most common applications is dumping cement slurry or sand on top of plugs. The mechanical type has a modified dart on the bottom of the bailer that remains closed until the dart touches the bottom of the hole. The bailer is filled with sand, cement slurry, etc., and run slowly to the bottom of the hole. When the bailer is set on bottom, the dart valve locks open and the material in the bailer is dumped out by a combination of gravity and the upward movement of the bailer.

Another modification of the dump bailer is run on a shielded electric line. It is filled with material and run to the bottom of the hole, where the material inside the bailer is dumped by an electrically actuated mechanism. These bailers are more commonly run in smaller tubulars, such as when cement is dumped on top of a plug set in tubing.

Sand Pump

The sand pump has a piston that travels the entire length of the bailer. It is connected to the bottom of a rod that extends out through the top of the bailer and has a connection to attach it to the running assembly. The lower part of the piston and the bottom of the bailer are each fitted with a check valve.

Run the sand pump to the bottom of the hole. Work the piston by listing and lowering the running line a distance approximately equal to the length of the bailer. As the plunger moves upward, it creates a suction action that helps pull sand and fluid into the bailer barrel. On the downstroke, fluid passes through a check valve and into the plunger. As the plunger is lifted again, the suction stroke moves more fluid, with sand and other debris, into the bailer barrel. When the bailer is full, it is pulled to the surface, emptied, and rerun.

Chipping

The chipping or cleanout bailer is a modification of the sand pump in which the bottom of the bailer is replaced with a cutting face. Various designs of cutting structures are available. On the upstroke, the entire bailer is lifted off the bottom. It is lowered rapidly and the cutting structure on the bottom of the bailer cuts and loosens debris at the bottom of the hole. On the upstroke, the debris cut loose by the downstroke is pulled into the bailer barrel. Depending on the skill of the operator and the running equipment, a chipping bailer can be an efficient tool. For example, a 4½-in. drillable bridge plug was drilled with a chipping bailer until it released and was then pushed to bottom in about 6 hours.

Blind Box

The blind box cleans the top of a fish by removing metal burrs that prevent the fishing tool from making a solid catch. It also cuts off short strands of wireline remaining in a rope socket after the line parts (Fig. 3–31). If the top of the fish is burred or swelled, such as a swelled sinker bar tool joint shoulder, the blind box may cut the swelled section off so the fish can be caught.

The blind box has a screw-top connection on top of a steel cylinder with a sharp, lower inner edge. The inside diameter is very close to the outside diameter of the top of the fish. The socket is set on top of the fish and driven down over it with jars. This cuts and smooths the side of the top of the fish so it can be caught with a socket or other type of catch tool.

For example, one of the most common wireline fishing jobs is fishing for rope sockets lost when the wireline parts, usually at the top of the socket. Frequently, several short strands of wireline remain on the rope socket. These prevent the rod socket from going over and taking hold of the rope socket fishing neck. In this case, run a blind box below the jars. Jar the tool down over the rope socket to cut the wire strands. Then pull the blind box and recover the fish with a rod socket.

Tools and Assemblies

Figure 3-31 Blind box
Note: *a*, cut long strands off rope socket, trim swelled joints, etc.; *b*, cut short wires off rope socket, burrs on the edge of a box or pin, etc.

[Deep cut] [Shallow cut]

Multiple-sized blind boxes are similar but have one or more cutting recesses set in steps. The lower part of the tool has the largest inside diameter with stepwise decreases higher inside the tool. These tools are used in the same manner as a regular blind box and serve the same purpose, except in cutting over one or more diameters on the top of the fish.

Friction Socket

The friction socket uses a friction hold to catch small fish that are not stuck too tightly. It cannot be jarred or worked very hard without losing the friction hold. It has a screw-top connection on top of a steel cylinder. The inside of the cylinder is cone-shaped, apex upward, with a relatively long taper.

Run the friction socket on a wireline. Set it on top of the fish and drive it down over the fish with the link jars. This forces the fish into the taper, where it is caught by a friction hold and then pulled out of the hole.

Cutters and Perforators

There are various wireline tools used to cut wirelines, tubing, and casing. A few perforate tubing.

Wireline Cutter

The wireline cutter cuts a wireline inside the hole, usually as near the bottom as possible (Fig. 3–32). When a wireline tool becomes stuck downhole and cannot be released, one alternative is to pull the line apart. If the rope socket has not been crippled, the line often will part near the surface, usually at the crown running sheave where the maximum tension in the line occurs or at a shallow weak point. Then both the wireline and tools must be recovered. Eliminate fishing for the wireline by cutting the wireline near the rope socket and pulling it out of the hole. Recover the remaining tools by fishing.

Various types of wireline cutting tools are available. One of the most efficient is fitted over a taut wireline and dropped in the hole. When the cutter reaches bottom, a charge detonates and activates a wedge. This in turn activates a knife edge that cuts the wireline. The cutter is recovered as the wireline is pulled. The tool also has a fishing neck so it can be recovered by fishing if necessary. Mechanical tools that operate similarly are also available.

Some downhole wireline cutters are run on a second wireline and activated electrically from the surface. The tool operates satisfactorily and cuts the line efficiently. However, after the line is cut, there is a strong tendency for the two wirelines to become twisted and tangled, often causing a problem recovering the cut line.

Wireline Tubular Cutter

The wireline tubular cutter, sometimes called a casing ripper, is used to part casing. It now is used only for special situations.

The inside pipe cutter has an upper connection and a carrier body containing spring-loaded cutter blades or knives. The knives swivel from a vertical upward-running position to horizontal for cutting and are stopped by a backing device. The length of the knives is adjusted so they can extend horizontally outward inside a pipe collar (Fig. 3–33).

Figure 3–32 Wireline cutter (courtesy J. C. Kinley Co.)

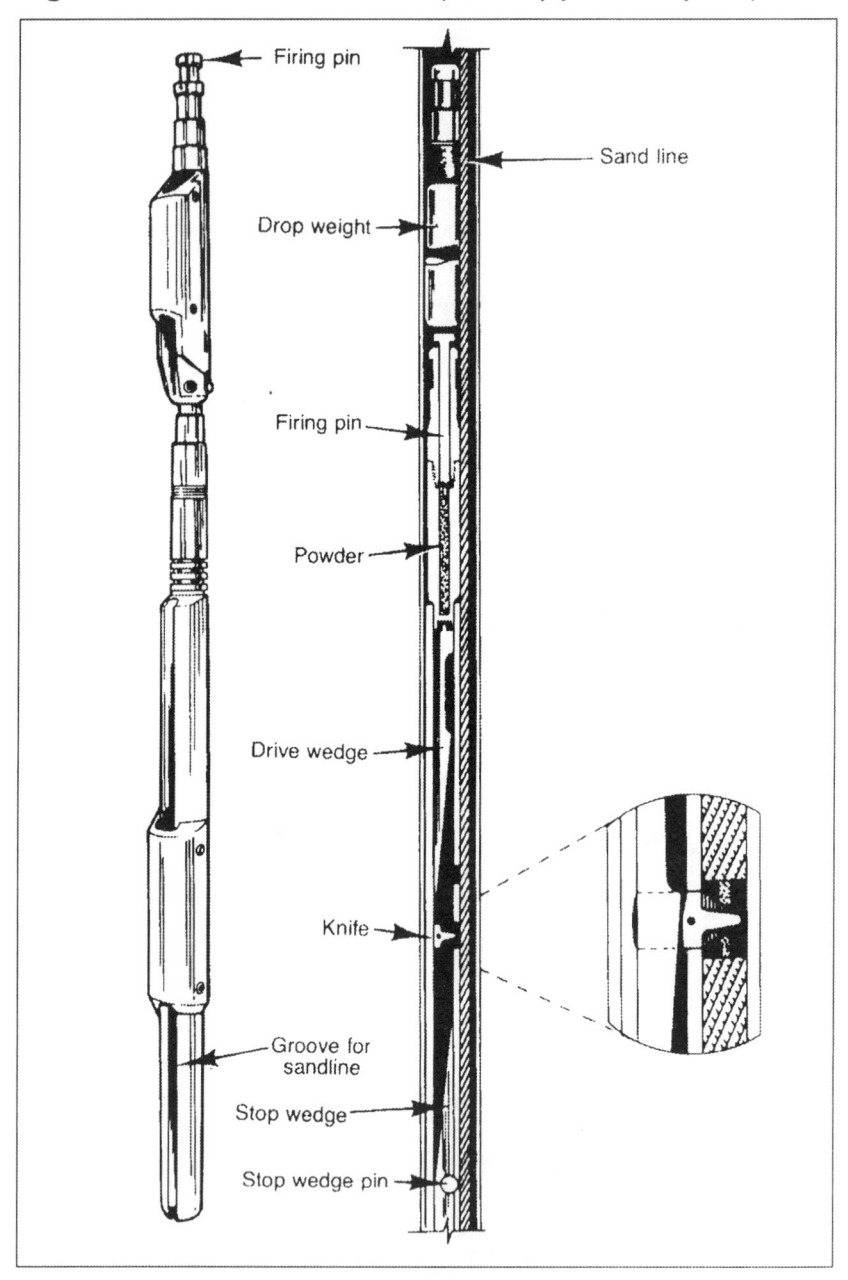

Run the cutter into the hole with the spring-loaded knives riding against the edge of the casing. When the tool reaches the point where the pipe will be cut, raise the cutter slowly until the knives enter a casing collar and stop on the bottom end of the pin of the joint. Then

Figure 3-33 Wireline pipe cutter (ripper)

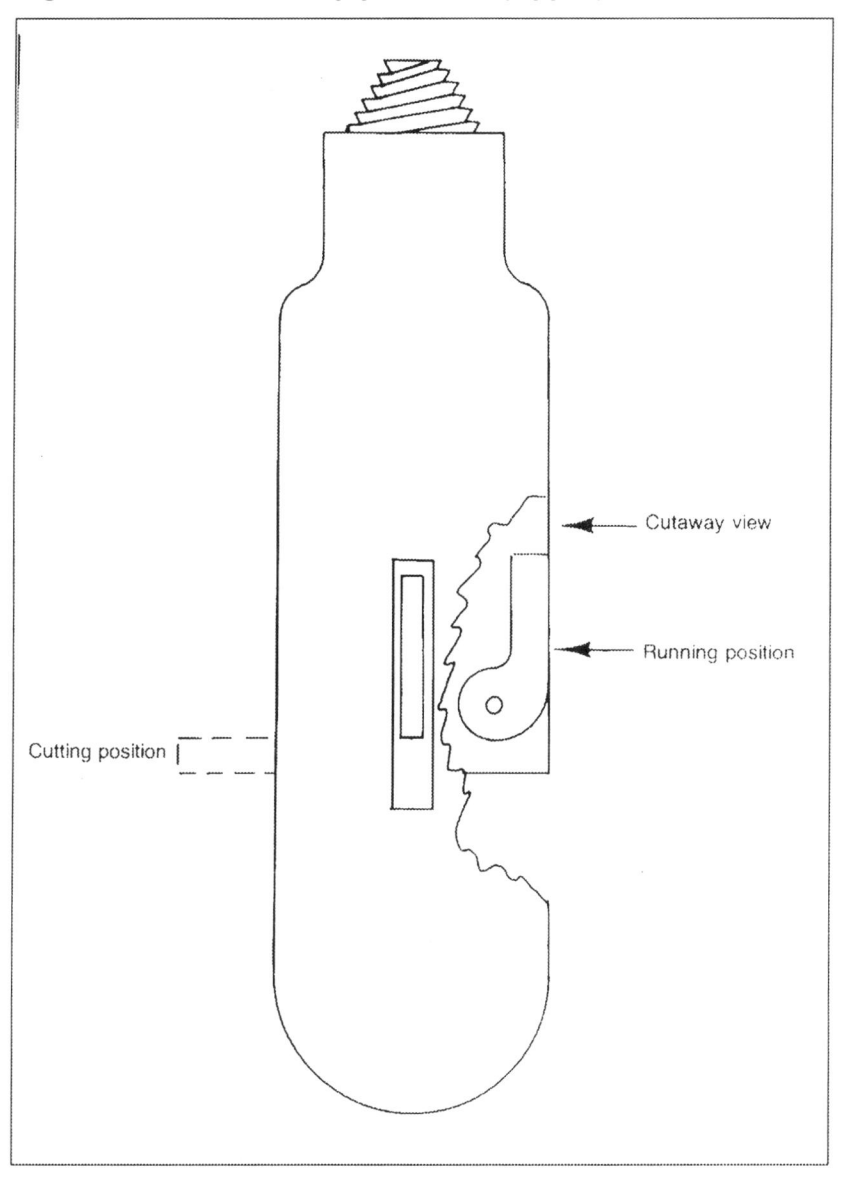

jar the cutter upward. Each knife makes a slit in the threaded portion of the pipe pin. If this does not release the pipe, which normally is held in tension, lower the cutter and raise it again to make another cut at a different point. Usually, a second cut will part the pipe. The knives are released by a tripping action, the cutter is pulled, and then the pipe is cut. Short lengths of casing may be recovered on the knife blades.

Tools and Assemblies

Normally, rotary methods of cutting the pipe, such as with regular rotary pipe cutters and chemical cutters, are more efficient. The pipe cutter is used primarily in cable tool operations.

Mechanical Wireline Pipe Perforator

The mechanical wireline pipe perforator perforates holes in tubing. It is somewhat similar to the wireline cutter except that the wedge actuates a punch to puncture the tubing wall (Fig. 3–34).

Figure 3–34 Mechanical wireline perforator (courtesy J. C. Kinley Co.)

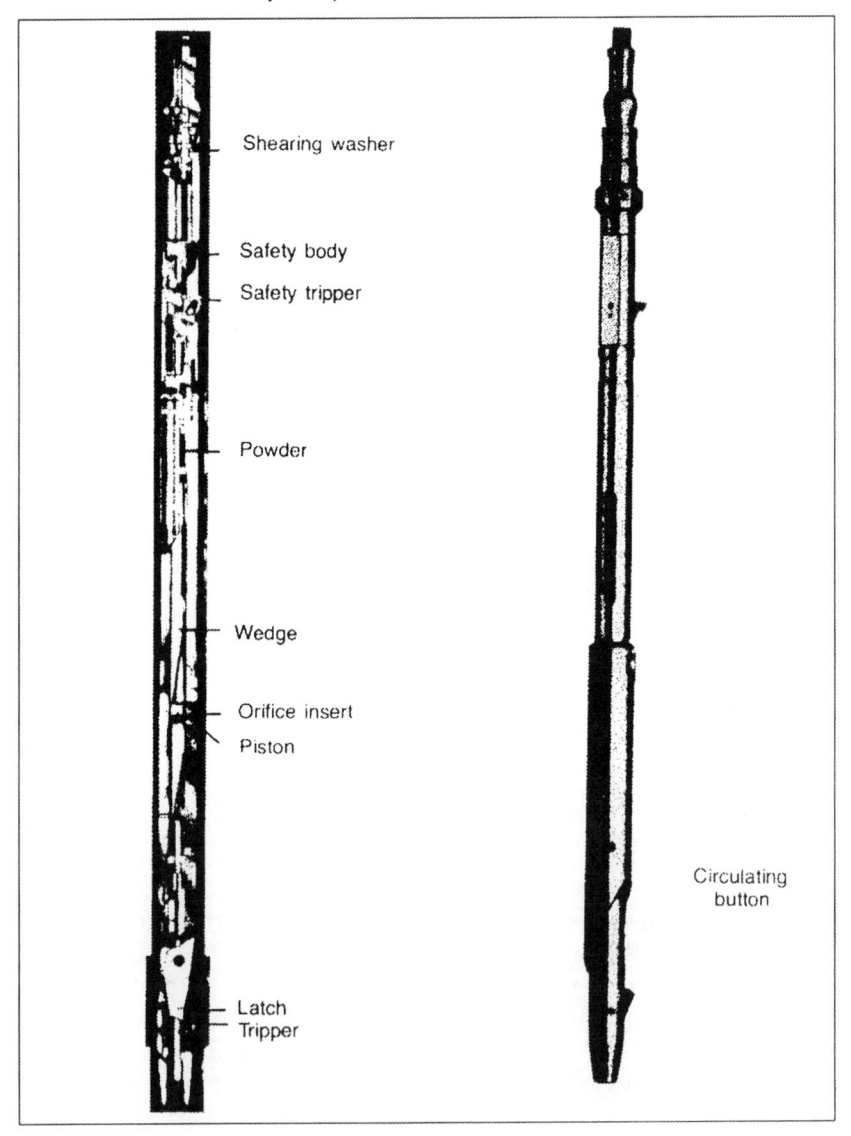

204   Chapter 3

Impression Block

Impression blocks obtain an imprint of the top of the fish (Fig. 3–35). This information may help determine what type of fishing tool to run or how to conduct the fishing operation.

The impression block has a short sub, open at the bottom, with a tool joint on top. The bottom of the sub has welded pins on the inside or recesses. These pins help retain lead or babbit that is melted and poured into the recess. This, or other means of ensuring the lead is retained in the impression block, is extremely important. If the lead is lost, it is very difficult to drill out. The bottom part of the sub is filled with several inches of lead, which is smooth on the bottom.

Figure 3–35 Impression block
Note: Various types of circulation ports (side, bottom, none, etc.) are used on impression blocks run on pipe.

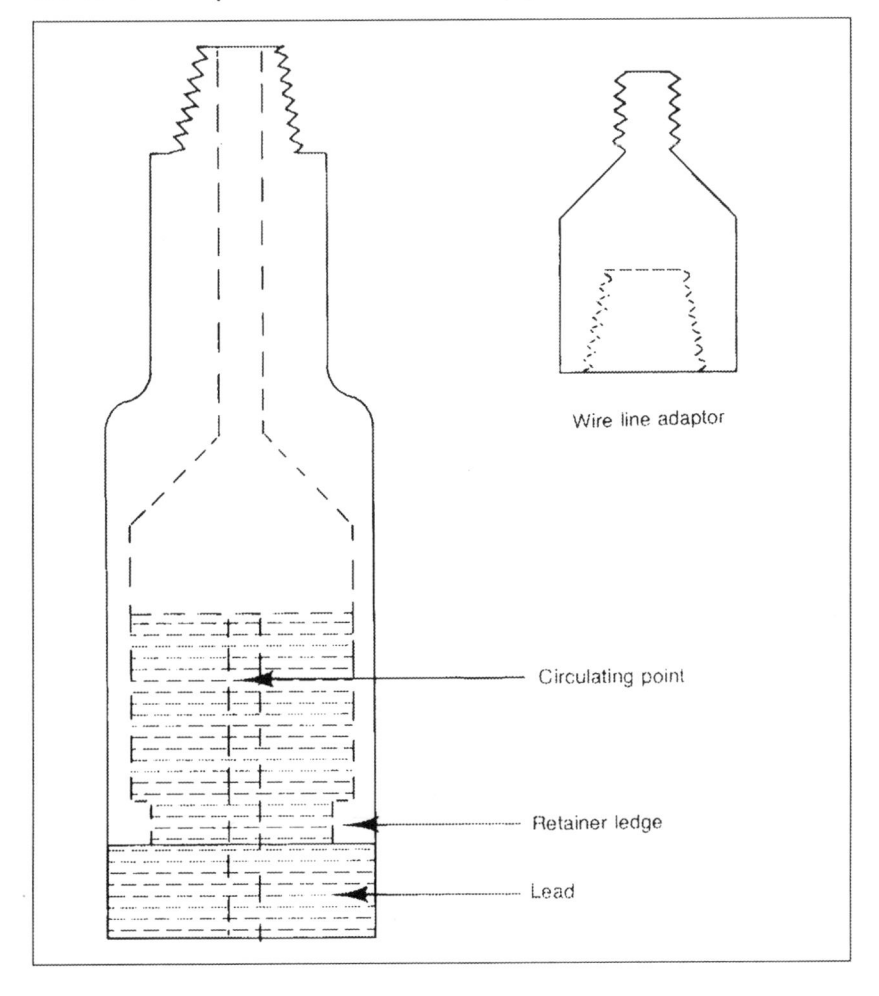

Tools and Assemblies

Some impression blocks have a small port through the center of the lead for circulation. Others have circulation ports in the walls of the subs above the lead.

Connect the impression block to the downhole assembly or a wireline and run it into the hole. Usually, set it on top of the fish with a sharp, relatively light force. Some operators prefer striking the fish twice. Then pull the block and examine the indentation on the face. Interpret the impression by visualizing the configuration on the top of the fish that will give the impression obtained on the impression block.

Fair to good impressions often are obtained inside casing. Results may be less favorable in the open hole because the impression frequently is obscured by cuttings that accumulate on top of the fish.

Hydraulic Jars

Wireline hydraulic jars serve the same purpose as tubular jars and operate similarly. Weight bars are used to increase their efficiency.

Mechanical Bumper Jars

Wireline mechanical bumper jars deliver a jarring blow in the upward direction and a bumping blow in the downward direction to the top of the fish. These jars have an upper connection for attachment to the running assembly, while the catch tool connects to the bottom connection. Sinker bars provide weight and increase efficiency. Either link or mandrel (cylindrical) jars are used. The links provide travel for the jarring or bumping action. The plunger tool has a free-moving mandrel with a stop to provide the travel necessary for jarring and bumping (Fig. 3–36).

Sinker Bars

Sinker or weight bars pull the wireline into the hole and provide rigidity for the tools below the bars and additional weight to increase jar efficiency. One sinker bar normally is sufficient for catching a fish but two or more may be needed to provide tool rigidity and weight for working tools. Conventionally, sinker bars are made of steel rods with connectors on each end. Special heavyweight or leaded sinker bars are steel tubes filled with lead.

Casing and Tubing Swage

The wireline casing and tubing swage is similar to the one run on tubulars, but it has a lighter construction and is used on smaller

Figure 3–36 Wireline mechanical jars (courtesy Camco)

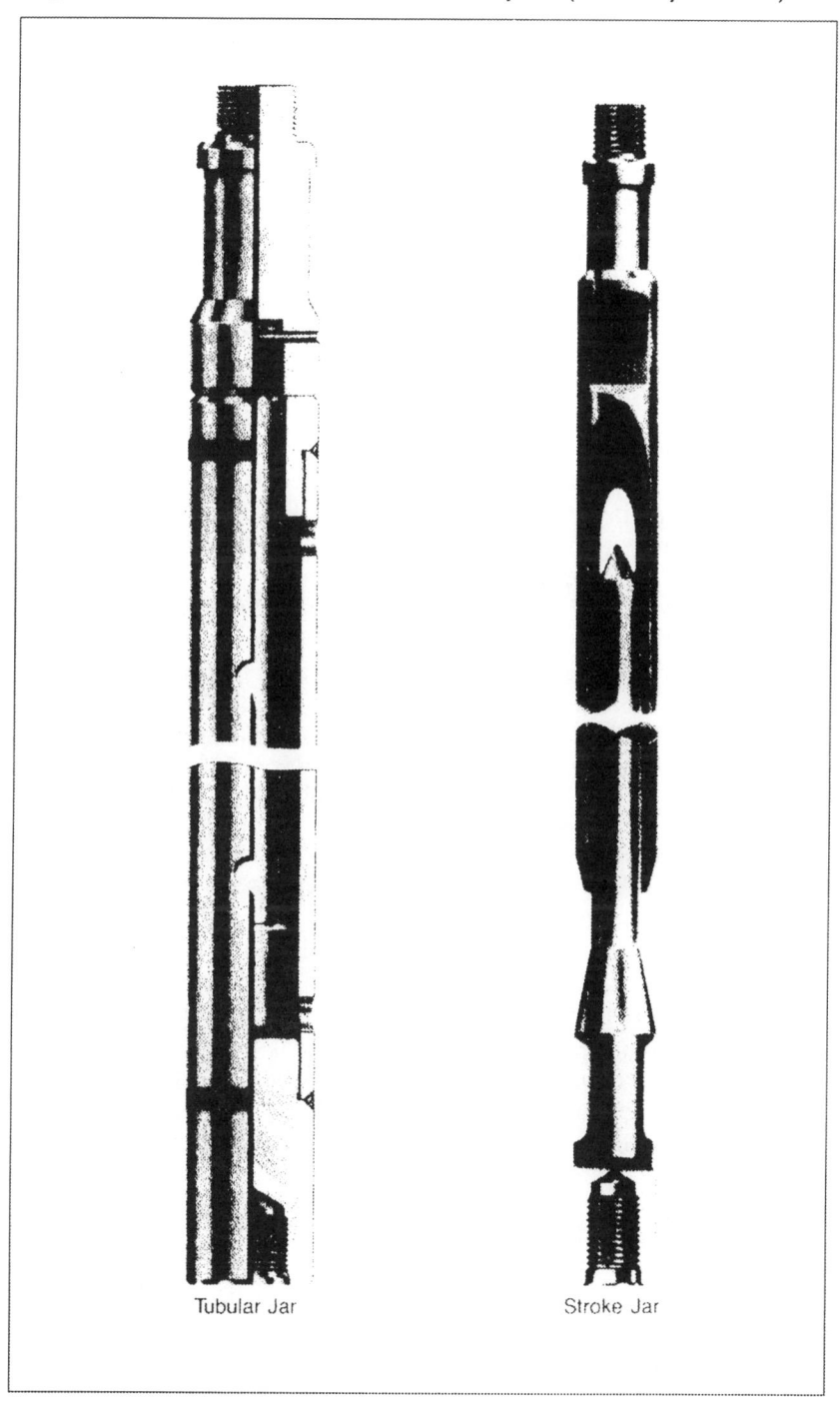

Tubular Jar   Stroke Jar

tubulars. It is also operated in a similar manner but is less effective because of the limitations of running on a wireline.

Explosives

Primer cord or dynamite are the two main explosives used in fishing, excluding jet charges. Primer cord is used in string shots. Dynamite is used for cutting and recovering casing for abandonment. Explosives such as nitroglycerine and ammonium nitrate, usually pelletized and mixed with 5–8% diesel oil, AMFO, were originally used for stimulation but are seldom used now. Large-shaped charges may be used to cut off wellheads and casing of blowout wells. Nitroglycerine or plastic explosives are used to blow out well fires.

Rope Socket

Rope sockets provide a connection between wireline tools and the wireline itself. The bottom of the rope socket is threaded to receive tools. The top of the rope socket will have one of various types of wireline connections.

Single-strand and small multi-strand lines commonly have wire twisted around a small interlocking bolt, lug, or screw-type friction device. Larger lines are fastened with lead or babbit, either into the hollow rope socket or with a small sleeve or thimble that fits into the rope socket. Shielded electric lines use special connectors. Bevel the top of the rope socket and cut part of the wires in the head in order to cripple it, as described in "Wireline Fishing" in Chapter 6.

Overshot

The wireline overshot or rod socket catches small fish with slip-type grapples. Otherwise, this tool is similar to and operated in a similar manner as the tubular overshot, with allowances made for the reduced efficiency of wireline operations. It may be necessary to bump the overshot down over a close-tolerance fish. Release the fish by jarring upward, and then recover it. The tool is strong but may split or fail under heavy jarring action. Run a shear-pin release if needed, as described in the section "Wireline Assemblies."

ASSEMBLIES AND EQUIPMENT

Assemblies and equipment include all of the surface and downhole equipment. This section emphasizes those used for preventing fishing and conducting fishing operations when they do occur. Exceptionally severe fishing jobs may place additional demand upon these tools, so be prepared to make adjustments and changes accordingly.

Surface Equipment

Surface equipment used in fishing is generally the same as that used in regular operations. Strength ratings of the lifting equipment is normally in excess of the drillstring and is satisfactory for most fishing operations. Top drive systems have many advantages, especially in operations such as cleaning out.

Pay particular attention when using a lifting system because of the heavy working necessary to free stuck tools. Be sure the drilling line is in good condition, and provide for slipping and cutting regularly. Jarring and bumping can cause a flattened drilling line, so it is necessary to take special precautions, such as slipping the line a small distance frequently. Worn crown and traveling block sheaves accelerate this problem, so ensure that they are in good condition. Always keep the brake system in good condition.

Special inspections are seldom necessary for only a fishing job. However, for severe fishing jobs, consider inspecting the brake bands, crown, traveling block component parts, etc., as is necessary before running very heavy casing loads. Periodic visual inspections are a good preventive procedure and highly recommended.

The circulating system can be a source of problems. There is often a high stress on mud-handling equipment, so check it carefully and frequently. Likewise, ensure that all blowout safety equipment is in place and in good operating condition and that crew efficiency is verified by blowout drills. Most rigs have a safety valve between the swivel and kelly. Consider converting to a double-valve system with a second safety valve located on the bottom of the kelly. This is highly recommended for deep wells and wells drilled in very high-pressured formations. ALWAYS position the opening and closing wrench in the same conveniently available place. This also applies to the inside preventer. Be sure it is open, and its opening and closing wrench is stored properly. Ensure that all personnel are aware of where this equipment is located and how to use it.

Drill and Work Strings

Drillstrings are tubulars such as drillpipe or tubing that connect the bottomhole assembly to the surface. The term "work string" is sometimes used but generally refers to the entire assembly, including the drillstring and bottomhole assembly. In most cases, the work string used for normal drilling operations is satisfactory for fishing. Smaller tubulars, such as tubing, are often used for workovers and completions. These may be inadequate if sticking or fishing occurs.

Severe sticking and fishing may place considerable stress on the drillstring. Have it inspected and replaced if there is any question about its condition. Generally, do not use tapered strings except for special situations, such as increasing overpull. Use slips designed for the correct loading.

Inspect and maintain, tighten and operate each type of connection within design specifications as a major preventive action. This is very important, as emphasized by the fact that a 10,000-ft, range 2 drillstring will have about 340 connections, each a potential point of failure. Connectors loosen during heavy jarring, so it is a good practice to tighten all of the work string connectors after a long period of heavy working. Fishing may require using tools with fine thread connections, so take special precautions since these are liable to fail.

Tool joint selection is important, especially when drilling in deeper holes with long strings of relatively small intermediate casing. As an example, consider drilling below a combination string of 7-in., 26 and 29 lb/ft casing set at 12,000–14,000 ft. Conventionally use 3½-in. drillpipe with regular or internal flush tool joints. However, 4-in. drillpipe, with the same weight as the 3½-in. drillpipe, fitted with slim-hole tool joints, has about the same tensile strength as the 3½-in. drillpipe and appreciably better mud hydraulics. The higher flow rate can be a major help in hole cleaning, especially in highly deviated and horizontal holes.

Tool joint failures probably are the most common cause of fishing after fishing for bit cones.

Drillpipe

Drillpipe used in regular operations is normally satisfactory for most fishing jobs. This is subject to the condition of the pipe, including wear, which is normally measured in total number of feet drilled. Other items include slip and tong cuts, fatigue, use in corrosive environment, and tool joint condition. Consider replacing the pipe, or at least inspecting it, especially after severe fishing jobs.

Drillpipe can be replaced with the same size pipe that is in better condition or with pipe of a larger size or higher strength. Another alternative is to replace part of the drillpipe with higher grade pipe, usually in the upper part of the string. Some very high grades of pipe may require using double elevators and other precautions, but these can increase the risk of problems, so in general lower grades are preferred.

Hydrogen Sulfide (H_2S) creates special problems because it is highly poisonous, even in low concentrations. It also causes embrittlement that can lead to tubular failure. Tempered X-95 and Grade E drillpipe are most common for H_2S service. This may require using tapered strings to attain the desired overpull. Use of plastic-coated drillpipe may be questionable. Using the correct mud may help. Take all precautions and maximum well-control measures to prevent taking a kick that contains H_2S.

Hard banding on tool joints does not necessarily affect fishing but in some cases can cause fishing. Incorrect hard banding causes excess casing wear and even keyseats in certain conditions. Do not use coarse tungsten carbide particles. Normally, use particles sizes smaller than 60 to 80 mesh with a smooth surface in a band about 3 in. wide.

Aluminum Drillpipe

Aluminum drillpipe has been tested in various drilling and a few fishing situations. Theoretically, it should make a good drillstring because of its favorable strength-to-weight ratio. However, field results indicate that operational problems outweighed the benefits. Some problems include corrosion in most mud pH ranges (except for a limited few), excess wear of the pipe body and tool joint, and high flexibility resulting in buckling and impact damage. Aluminum drillpipe requires special tapered slips, and applying torque with the slips or rig tongs may cause severe pipe damage.

It is also difficult to fish for or with aluminum pipe. It is very difficult to back off, requiring about two rounds of back torque per 1,000 ft, compared to about one round for steel drillpipe. There is a risk of accidentally backing off, and it is difficult working the torque down. Aluminum drillpipe may be parted with a shaped charge or chemical cutter, but string shots may shatter the pipe body adjacent to the tool joint. Catch tools, such as an overshot grapple, may damage the pipe body; in one case, aluminum pipe failed at about one-half of the tensile rating. The best alternative is to mill or cut off the extra pipe and fish for the tool joint. Another alternative is replacing the top joint of the fish with a steel joint if possible.

Tubing

Tubing normally is not recommended for fishing except for less severe operations. Its strength characteristics, springiness, flexibility, and connector design are unsatisfactory for most fishing situations. It is especially susceptible to damage during milling and similar high-torque applications.

Threaded tubing connections, either plain end or external upset couplings, are not designed for extensive tripping and the use of high torque, even if the body strength is satisfactory. The connections may become excessively tight or tighten until they become damaged and fail. Disconnecting overly tight joints may cause damage. Extensive connecting and disconnecting causes wear and premature failure. Tubing may be fitted with butt-shoulder-type tool joints, but this does not eliminate the problem of damage due to excessive torque on the screw-type thread. It is best to replace the tubing with small diameter drillpipe or to drill tubing designed for lighter weight work strings. The same reasoning applies to macaroni tubing, normally in the range of 1½ in. or less in diameter. Regardless of these problems, tubing is used successfully in many cases by careful, skilled handling.

Coil Tubing

Coil tubing is a high-strength, ductile steel tube. Common sizes are 1 and 1¼ in., but other sizes are available. Common strength ratings are 70,000 psi yield strength. In operation, an injection head seals the tubing and lifts and lowers it. Coil tubing is run into a hole at about 100 ft/min or even faster with special units.

Coil tubing is used to pump fluids at rates of 1–2 bbl/min. Common usage is light-duty cleanouts, sometimes with a small mud motor and bit. Other uses include low-volume acid cleanouts, low-pressure kills, drillstem testing, running as siphon string to unload low-pressure gas wells, displacing nitrogen to lift fluid out of the hole, and carrying logging tools and perforating guns in high-directional and horizontal holes. Coil tubing has been used for light-duty fishing such as cleaning sand off the top of a plug and recovering a plug and squeeze cementing.

Assembly Equipment and Tools

Various equipment and tools are used to construct bottomhole assemblies.

Heavyweight Pipe

Heavyweight pipe is not only heavier than normal drillpipe but also generally lighter than drill collars for the equivalent hole size. It has one or two extrusions distributed along the body. These extrusions are shaped like tool joints for lateral support.

Heavyweight pipe may be used to modify bottomhole assemblies by replacing some of the drill collars to reduce overall tool weight and related drag and torque. It also reduces assembly rigidity and provides a more balanced connection between smaller drillpipe and larger drill collars. It reduces the risk of sticking, especially wall sticking, because there is less contact area. Compared to regular size drill collars, heavyweight pipe is easier to recover by fishing.

Heavyweight pipe is most commonly used to replace drill collars run above drilling jars and bumper subs. It may be used to replace all of the drill collars in highly directional and horizontal drilling that use a positive displacement, mud motor, or turbine. Heavyweight drillpipe does not affect the basic action of the assembly since the configuration of tools on the lower part of the assembly primarily determines performance.

Drill Collars

Drill collars are the main component of the bottomhole assembly. They are constructed of a steel alloy and heat treated to increase hardness. API specifications include a minimum-yield strength of 110,000 psi for drill collars with an outside diameter less than 7 in. and 100,000 psi for larger diameter collars. Standard practice is to run larger or full-gauge collars on the bottom and to use worn, smaller collars in the upper assembly. Always use drill collar clamps when handling drill collars, even the bottleneck type.

There are a wide variety of drill collars (Fig. 3–37). All serve the same purpose of providing bit weight and the degree of flexibility or rigidity necessary for different types of bottomhole assemblies to perform their designed functions. Some differences in drill collars are noted in the following list.

- Bottleneck collars have a recessed section near the top or box to aid in holding them with the slips and to prevent dropping.
- Spiral or fluted collars have continuous, recessed flutes or mud courses cut in a spiral on the entire length of the outer surface, excluding the handling sections at the ends. These collars minimize the risk of pressure differential sticking by reducing the contact area. Compared to round collars, spirals reduce the weight of the collar about four percent.
- Nonmagnetic collars are similar to regular drill collars but their nonmagnetic state permits recording compass directions with instruments that are run inside the collars.
- Drill collar weights, donuts, or drill collar rings concentrate weight on the bit for drilling very large diameter, "big holes," normally in excess of 6–10 ft in diameter. They are short steel cylinders that slip over regular large-diameter collars. The inside diameter of these tools is slightly larger than the outside diameter of the drill collar they are fitted over. Outside diameter depends on hole size.
- Pony collars are short drill collars 6–15 ft long and are used for spacing tools, such as stabilizers, on a drill-collar assembly.
- Square drill collars aid drilling a straight hole in severe crooked-hole conditions. They are seldom used since the development of good stabilization equipment and practices.

Figure 3–37 Drill collars

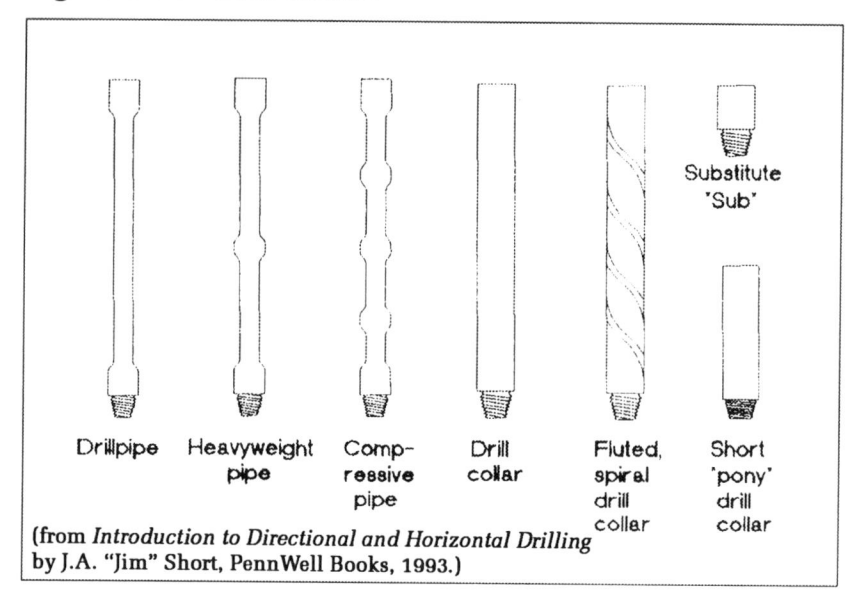

| Drillpipe | Heavyweight pipe | Compressive pipe | Drill collar | Fluted, spiral drill collar | Short 'pony' drill collar |

Substitute 'Sub'

(from *Introduction to Directional and Horizontal Drilling* by J.A. "Jim" Short, PennWell Books, 1993.)

- Eccentric or off-center drill collars have an unbalanced weight distribution around the center line axis. They reportedly increase drilling rates but either are applicable only in special situations or are not widely accepted in the industry.
- Bit or crossover collars or double box collars run on the bottom of the drill-collar assembly. The bottom of the collar has a box connector that fits the bit pen.

Selecting the number and size of drill collars depends upon various factors. Smaller drill collars are safer to run, while larger drill collars are generally more efficient. A major consideration is when an operator should run larger diameter collars that cannot be washed over.

Larger drill collars efficiently concentrate weight near the bit. They allow drilling straighter holes through crooked-hole formations. Fewer collars are needed for a given assembly weight. When using larger drill collars, there is a higher risk of sticking, especially differential sticking.

Smaller diameter collars may be stabilized to give the same or better results than larger diameter collars. Normally, stabilizers can be cut off by milling if washing over is required. However, some reamers used as stabilizers, such as roller reamers, are very difficult to mill off.

Evaluate all of the preceding factors. Generally, if there is an option of using large drill collars that cannot be washed over or using stabilization, then stabilization probably should be used. If it is necessary to use large drill collars that cannot be washed over, they may be released by circulation, pulling, or jarring. If this does not work, it is not excessively difficult to sidetrack them. An example of this is dropping angle with a pendulum while drilling very difficult formations. One alternative is using very large collars to drop angle fast enough and at a reasonable drilling rate. Another alternative is to control inclination by continuing drilling at a rate that is 20–30% of that possible with the large drill collars. Rig time and total cost savings justify the risk of a possible fishing job using the larger collars.

Keyseat Wiper

Keyseat wipers wipe out keyseats. The preferred type has a solid body and a floating tapered sleeve with right-hand, ribbed spirals and is fitted with a bottom clutch. The outside diameter at the top is approximately equal to the diameter of the drillpipe tool joints. The OD at the bottom should be equal to the diameter of the drill collars. During drilling, the sleeve remains stationary and the mandrel rotates inside the sleeve, thus preventing wear on the ribs (Fig. 3–38).

A keyseat wiper is moderately efficient at reaming keyseats, but its primary purpose is as a safety device to release a keyseated drilling assembly, especially if the assembly sticks when drilling jars and a bumper sub are not used, or if the assembly sticks above them.

When the keyseat wiper is pulled up into a keyseat, the sleeve sticks or keyseats. Release it by bumping DOWN and out of the keyseat, utilizing the mandrel travel of 12–18 in. Apply right-hand torque to engage the bottom clutch and to help rotate the keyseated wiper sleeve DOWN and out of the keyseat. After the pipe is free, remove the keyseat by jarring and working the keyseat wiper into and out of the keyseat. Repeating this procedure helps to wear out the keyseat.

Figure 3–38 Keyseat wiper (courtesy Hendershot Tool Co.)

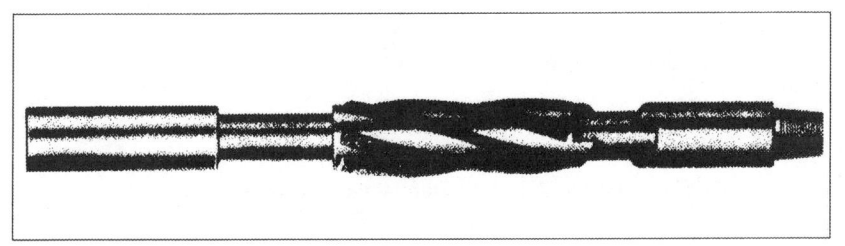

Reamer

Reamers primarily enlarge under-gauge hole and restore it to full-gauge. Run reamers in the drill-collar assembly or drillstring as described in the section "Reaming" in Chapter 2. They commonly are run immediately above roller bits as a "near-bit" reamer to prevent a tapered hole and to ensure a full-gauge hole is maintained as the bit becomes worn and under-gauge. Reamers are frequently used as stabilizers and in some cases, the name is interchangeable with "stabilizers" (Fig. 3–39).

Stabilizer

Stabilizers are used with drill collars to make up various assemblies. It is important to understand the principles of stabilization since stabilizer placement in the assembly primarily determines the function of the assembly.

Figure 3–39 Reamers (courtesy of Eastman Christensen, a Baker-Hughes company)

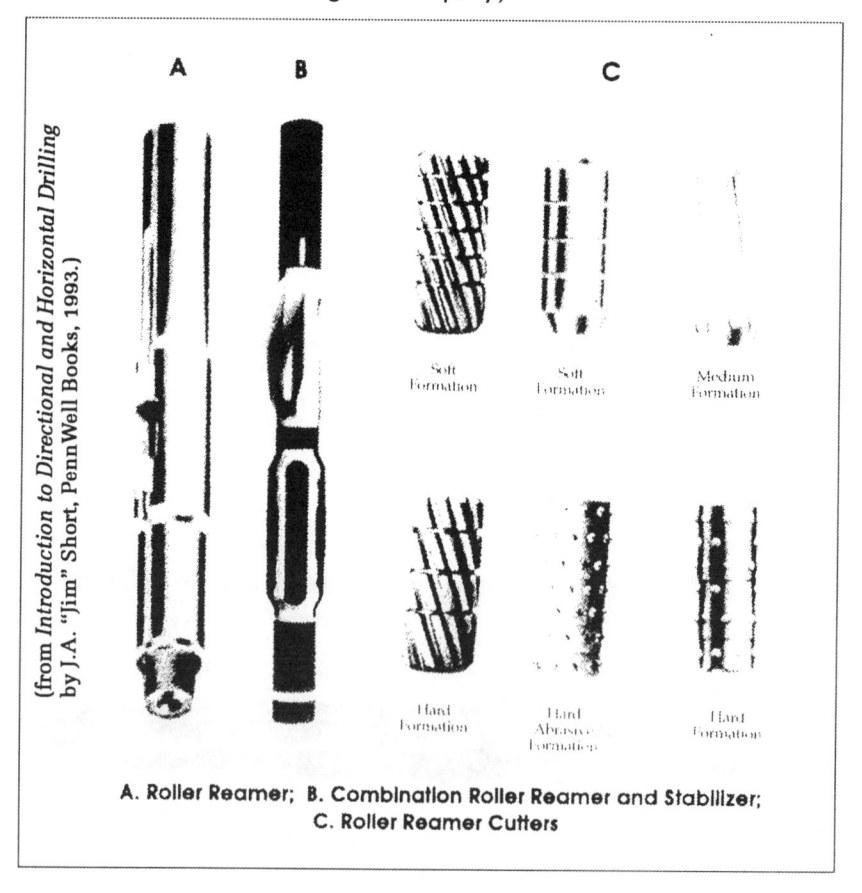

A. Roller Reamer; B. Combination Roller Reamer and Stabilizer; C. Roller Reamer Cutters

Stabilizers steady the drill-collar assembly by providing lateral support, such as for a stiff assembly. They serve as a fulcrum in pendulum and angle-build assemblies. Run them immediately above drag-type bits, diamond bits, and polycrystalline diamond bits, as near-bit stabilizers to steady the bit for improved performance and longer life. Normally, these tools must be full-gauge to be effective. Full gauge is within $1/16$ in. of the bit diameter and never less than $1/8$ in. Use roller reamers as stabilizers for hard drillability and abrasive formations.

Different types of stabilizers are available. Primary selection considerations are blade contact area, fixed or replaceable blade, vertical or spiral types, and whether the blade can be milled off if washing over is required. List stabilizer placement on drill-collar assembly by position (Fig. 3–40).

Reamers and stabilizers are additional points of potential failure in a drilling assembly. They are subject to extra fatigue and bending movement stresses. They also increase the difficulty of recovering the fish when a failure occurs. Prevent these problems by using the minimum number of stabilizers required. For example, do not run a dual-stabilized packed-hole pendulum when one stabilizer in a common pendulum is sufficient.

Do not stabilize the upper bottomhole assembly to prevent wall sticking. Consider using fluted drill collars or heavy drillpipe for severe conditions. Normally DO NOT cover more than the lower 150 ft for $6\frac{3}{4}$-in. and smaller diameter limber drill collars and 180 ft or less for larger diameters. Packed pendulum assemblies may require a slightly longer stabilized length.

Double-Drilling Bumper Jars

Double-drilling bumper jars are used on some split horizontal and high-angle drilling assemblies. Run drilling bumper jars on top of almost all directional assemblies. Some high-angle and horizontal assemblies are divided into two parts to reduce high torque and drag. A lower drill-collar assembly containing the bit, motor, orientation equipment, and nonmagnetic drill collars is run in the horizontal or high-angle hole. Additional drill collars are run in the vertical or low-angle hole to provide weight to push the lower tools through the horizontal section.

Drilling bumper jars are placed in the upper drill-collar assembly and set to begin jarring at a normal level of overpull. The jarring action is effective to aid in releasing a stuck upper assembly and some length of the pipe below it. However, it is not effective in releasing a stuck lower assembly located a long distance (in the range of 500 ft or more) from the jars in the upper assembly.

Figure 3–40 Stabilizers (courtesy of Eastman Christensen, a Baker-Hughes company)

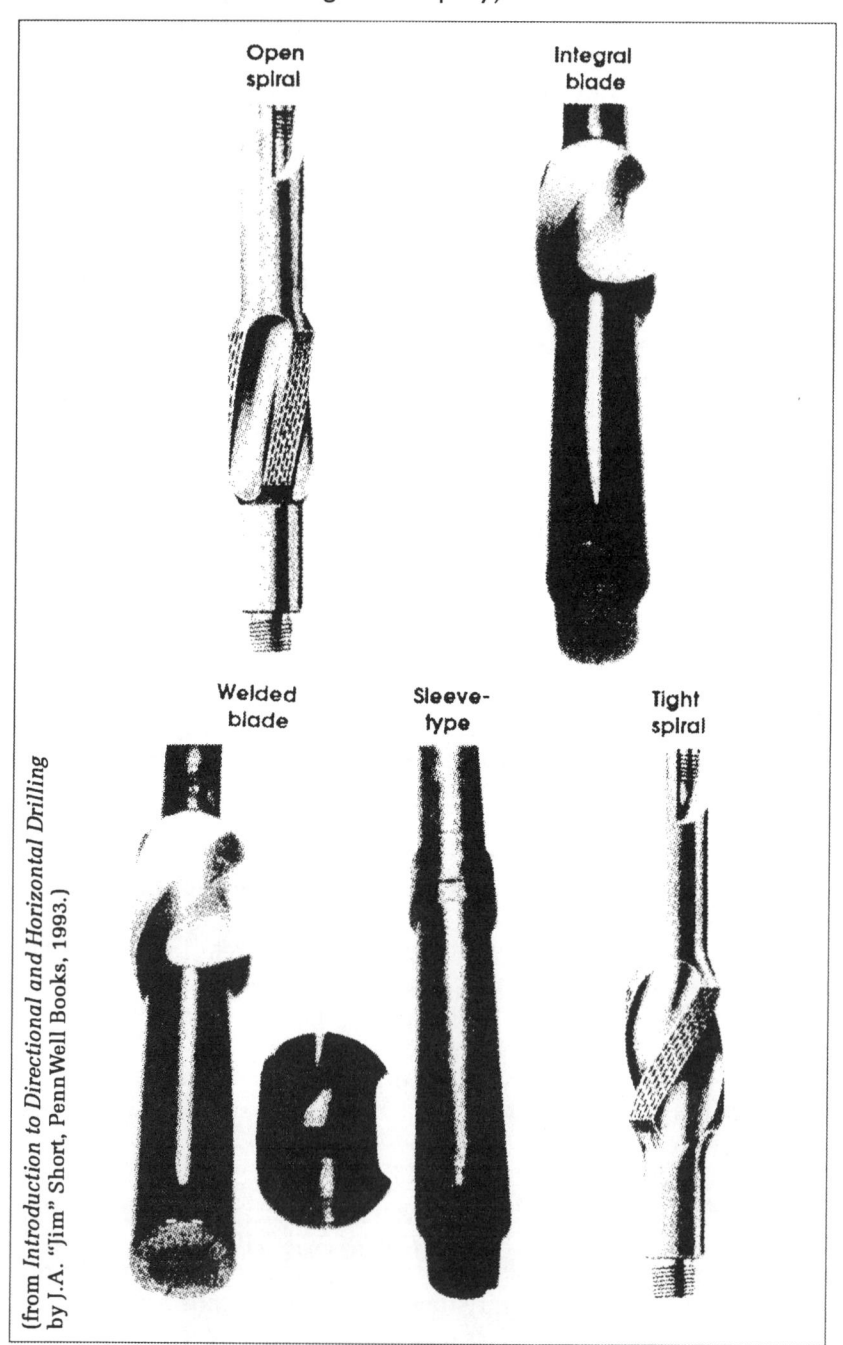

(from *Introduction to Directional and Horizontal Drilling* by J.A. "Jim" Short, PennWell Books, 1993.)

Figure 3-41 Split assembly

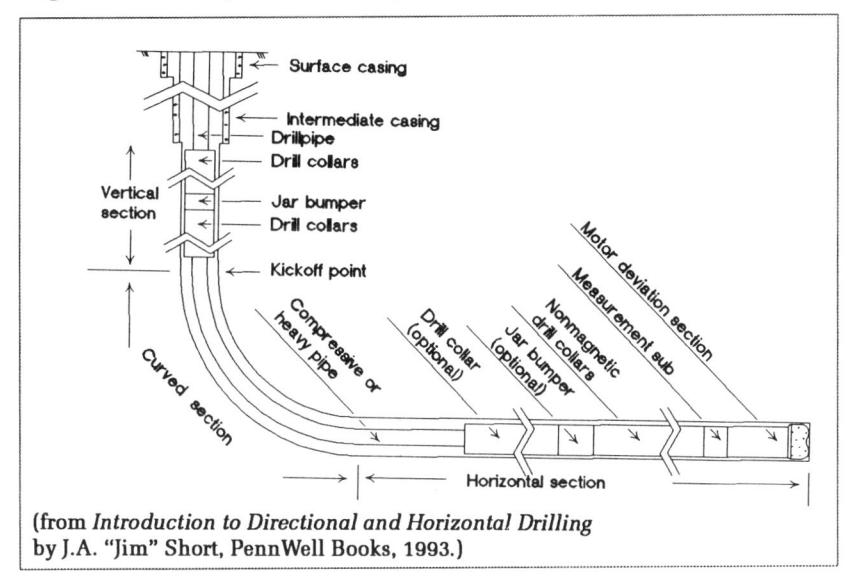

[Figure showing directional drilling assembly with labels: Surface casing, Intermediate casing, Drillpipe, Drill collars, Jar bumper, Drill collars, Kickoff point, Vertical section, Curved section, Compressive or heavy pipe, Drill collar (optional), Jar bumper (optional), Nonmagnetic drill collars, Measurement sub, Motor deviation section, Horizontal section]

(from *Introduction to Directional and Horizontal Drilling* by J.A. "Jim" Short, PennWell Books, 1993.)

Place a second set of double-drilling bumper jars near the top of the lower assembly (Fig. 3–41). Set them to begin jarring at a lower level of overpull compared to that of the upper jars. The jarring action will have a lower force but is more effective since the jars are located near the lower assembly. There frequently is a higher chance of success in releasing stuck tools with double-drilling bumper jars.

Assemblies

Assemblies are made up of various tools, including drillpipe, heavyweight pipe, tubing, drill collars, reamers, stabilizers, and other special tools. The design purpose of the assembly is determined by the type and placement of equipment. Stabilizers, diameter of the drill collars, and method of operation determine the flexibility and rigidity of the assembly.

In most cases, the number and spacing of stabilizers, and sometimes reamers, usually on about the bottom 150–200 ft of the assembly, determines the type of assembly. Subs and pony collars are used for fine-tuning stabilizers' spacing adjustments. Nonmagnetic collars may replace part of the regular drill collars. Additional equipment, such as a keyseat wiper, jars, and a bumper sub are frequently used in the upper section of the assembly.

Tools and Assemblies

The basic purpose is to provide a downhole assembly that has sufficient strength to perform its design function, such as drilling, reaming, and fishing. Allow for adequate safety factors. Compute strength characteristics on tensile strength, but also consider torsion, collapse, burst, corrosive conditions, and mud hydraulics. Allow for downhole failures, and provide for proper sizes, whenever possible, to permit fishing operations. When there is an option, use the least-rigid assembly that will accomplish the job.

"API RP 7G: Drillstring Design and Operating Limits" is a common guideline for designing and operating drill tools. Use the correct equipment and only that needed to perform a job. Eliminate extra connections such as crossover subs, since each connection is a point of weakness and potential failure. Also, provide for fishing for the assembly if it fails. However, this should not deter an operator from running the required equipment so the assembly can perform correctly and accomplish its design purpose.

Drilling

Various assemblies are used in overall drilling operations, including fishing and vertical, directional, and horizontal drilling. These tools are broadly divided into two general classes. Rotary assemblies, such as a limber drilling assembly, are driven by the rotary or top drive. Motor assemblies, such as a directional drilling assembly, have a positive displacement mud motor (PDM) or a turbine. Combination-type assemblies, such as some high-angle directional drilling assemblies, may use either power source.

Various types of assemblies are included as a reference in the following list.

- Drilling — Regular drilling and miscellaneous operations
- Cleanout — Special hole cleanout situations
- Fishing — Performs all fishing and various other fishing-related operations
- Reaming — Enlarges and smooths the wellbore
- Packed Hole — Maintains current hole direction and drills a straight hole
- Pilot and Hole Opening — Drills a small diameter hole and opens it to full size
- Pendulum — Reduces hole angle
- Angle Build-up — Increases the angle of inclination
- Angle Drop — Decreases the angle of inclination
- Deviation — Sidetracks and changes hole direction
- Hooligan — High-rate angle building assembly

A classification may include more than one type of assembly, such as the pendulum that includes both regular and packed-hole pendulums. The same assembly may be used in different operations, such as using a limber drilling assembly for both drilling and cleanout. Some assemblies may be more difficult to recover by fishing if they become stuck, such as a packed hole and various directional drilling assemblies.

Competent operators are familiar with the design, maintenance, and operations of all of these assemblies. Knowing how and where to use each assembly is an important preventive procedure.

Overpull

Overpull is the difference between the maximum allowable design load and the actual or hook load on the tubulars. Common strengths used are:

- Safe Strength — Allows for wear and includes a safety factor often 75% of new, or less for extra wear
- Tensional Strength — Based on tensile strength before deformation
- Ultimate Strength — Based on tensile strength at failure

In most cases, overpull is the tensile strength less the actual buoyant weight of the work string or tubulars in the hole. In some cases, it may be necessary to pull the maximum amount possible, which would be up to the ultimate strength.

Use the ultimate strength with caution and almost as an emergency or "last resort" procedure since it may damage the metal even if failure does not occur. Maximum or breaking strengths are not reliably predictable above the tensile strength because of stretching and elongation. Ultimate strength normally is in the range of 10% to 20% over tensile strength.

A long-standing industry practice is to design deeper drilling assemblies with a minimum of 100,000 lb of overpull. A more applicable recommendation is listed in Table 3–1.

Table 3–1 Recommended overpull

Overpull, lbs	Recommended Depth Application
100,000 lbs	Shallow holes to 8,000 ft
150,000 lbs	Medium depth holes 8,000 to 15,000 ft
200,000 lbs	Holes deeper than 15,000 ft

A straight pull will not always release stuck pipe and is not necessarily recommended. However, sometimes this is the only alternative and is frequently successful, especially with working. The additional pull (between 100,000 and 200,000 lb) could be sufficient to release the stuck pipe. This pull is important and can be the difference between a successful or failed fishing job.

Handling and Inspection

Handling all downhole tools in the correct manner is a major preventive measure. Do not overlook small actions. For example, use recommended slips. Longer slips with a larger contact area may be needed for heavier loads. Run slip-type elevators for very heavy loads and double elevators for tools built with higher grade steel. Seat them carefully without excessive shock. Always use drill collar clamps on all tools with small differences in diameter, even bottleneck drill collars. There is at least one case where bottleneck drill collars were set on slips and the elevators released rapidly. The drilling jars opened, causing the collars to vibrate, work downward through the slips, and drop.

Watch for equipment wear. Always make up tool joints using the correct techniques and torque. If a freshly disconnected tool joint has mud in it, check for joint damage. Watch for worn bits in order to detect them before loosing the cones. Leaks may be difficult to detect early, especially at higher pressures. But if they are not detected, a fishing job may be the result. During fishing, replace the jars and bumper sub after any trip in which they have been used. Observe the precautions listed in the section on tripping. Do not overlook blowout preventer drills and testing.

All of the above precautions and others are mostly common practices but may be overlooked in the distraction of fishing operations. Don't commit a *pilot error,* since it can lead to fishing or a fish on top of a fish.

Inspection is another important preventive action. Normally, an operator should have a schedule of inspection times for all tubulars. Use qualified, experienced inspection personnel. Consider using both sonic and ultraviolet light and know the advantages and disadvantage of each. Do not overlook the importance of good visual inspections, since they can detect a number of potential failures. Perform periodic inspections based upon the experience operators have had in that particular area.

Ensure that all fishing tools are inspected. Always inspect assemblies after severe, rough service, including cases of drilling on bit cones, reaming, and similar operations. Fishing, especially heavy jarring and bumping for long periods, can be hard on the entire fishing assembly. Check the assembly carefully, and replace damaged equipment.

The second most common downhole equipment failure is in the tool joint. Eliminate this failure or reduce its frequency through proper selection, operation, and periodic inspections. The main causes of tool joint failures are excessive wear or exposure to failure during extremely severe operations, such as fishing.

There is a high incidence of tool joint failure in a lower drillpipe and drill-collar assembly during, or shortly after, severe services. This also applies to the latter part of a long bit run, especially if the bit has bad cones and bearings. These tool joint failures are caused by the high, fluctuating torque that occurs during milling or drilling.

One corrective procedure is to inspect the drill collar tool joints at shorter time periods. If a milling job occurs shortly before a scheduled tool joint inspection, then inspect the tools earlier than the scheduled time and with reference to the milling job. An alternate procedure might be to break, clean, visually inspect, dope, and make up each tool joint, including the drill-collar assembly and five or ten stands of drillpipe above the collars. Although this involves some extra work, it often is justified and may prevent a more severe fishing job than just recovering or milling up the bit cones. A similar failure is illustrated in the following example.

Drill Collar Pin Failure

A wildcat well had $7^5/_8$-in. intermediate casing set at 13,500 ft and was drilling in a 6½-in. hole at 15,000 ft in hard, abrasive formations with a premium grade button bit. The bit had drilled 325 ft in 71 hours (4.6 ft/hr) when torque became excessive, indicating a cone failure. The bit was pulled and two cones were missing.

A junk basket with a hard-faced combination milling and drilling shoe was run. After drilling over the bit cones for a short period, the drilling assembly parted. Pipe weight indicated parting at the top of the drill collars. The pin failed on the bottom joint of drillpipe, leaving all drill collars in the hole with a Dutchman, or broken pin, in the box of the top drill collar.

An overshot was run on a conventional fishing assembly and caught the fish without difficulty. The fish came free with a 50,000 lb overpull after working for about 2 hrs. The assembly was chained out of the hole, recovering the entire fish, including the two bit cones in the junk basket.

Overall, this was a successful fishing job. Junk baskets often are less efficient at recovering cones in hard formations, but the overshot recovered the drill collar fish efficiently. It is more difficult to detect bit torque at this depth with the relatively limber and springy 3½-in. drillpipe, especially when compared to a larger assembly. The failure indicated an appreciable amount of downhole torque. A shock sub may have helped.

Tool joint inspection records were not available, but the assembly could have been approaching an inspection period in view of the long bit run. A good visual inspection might have prevented the fishing job.

Fishing

The fishing assembly or fishing string is the complete downhole assembly used in a fishing operation. Compared to most other operations, there is a higher risk of a drillstring failure in fishing. The assembly should be as strong as possible and have the minimum number of connections. Extra strength ensures safer operation under high-stress conditions.

A limber drilling type assembly is used as the basis for a fishing assembly. Normally, it is made up with the drillpipe and collars used in regular drilling. The assembly should be slick, without any stabilizers or other external obstructions. This reduces the risk of hanging up caused by caving or sloughing and facilitates washing over if required.

Both drilling jars and bumper subs are very effective if the assembly is stuck below them. Normally, they are positioned below the top three to five drill collars. Always locate jars above the bumper sub. The location of the bumper sub relative to the jars will not significantly affect the action of the jars. However, a bumper sub closes faster than the jars when bumping down. If a free travel bumper sub is positioned above the jars, the jars may cushion the action of the sub and reduce bumping efficiency.

Common fishing assemblies include the following equipment, listed from the top down:

- Drillpipe or tubing — From the surface to the bottomhole assembly
- Keyseat wiper — Use only if needed. Run a heavy-duty, spiral blade, solid-bodied keyseat wiper if there is any question about forming keyseats.
- Accelerator jar or jar intensifier — Seldom used
- 4 to 6 drill collars — These provide concentrated weight for bumping and jarring. Do not run excess drill collars in the fishing assembly, only the number needed. The number of drill collars is determined by the manufacturer's recommendations for the weight to be used above the fishing jars. A rule of thumb is one standard drill collar for each

inch of jar diameter. Standard collars have about the same OD as the bumper sub and jars.
- Fishing jars
- Bumper sub — Combination fishing jars and bumper sub may replace the individual tools.
- One or two drill collars for assembly rigidity if needed
- Back-off sub — Run a back-off or similar type sub in most cases when the rapid release of a fish is needed. However, back-off tools may not be needed with some types of fishing tools, especially those that have built-in back-off equipment, such as some overshots.
- Tool to catch the fish

Milling

The standard milling assembly is similar to the fishing assembly except that the catch tool is replaced with a mill and 4 to 6 drill collars are run above the mill for weight and stability. Run one or two, and not over three, junk subs above the mill. The bumper sub may be used to provide a constant weight for milling operations. In this case, adjust the number of drill collars between the mill and bumper sub to provide the required weight.

Wash Over

A washover assembly is similar to the milling assembly except that the mill is replaced with washpipe and a washover or mill shoe is run on bottom of the washpipe. Always run the back-off sub. The collars below the back-off sub may be eliminated, or two or more may be used for short lengths of washpipe. The washpipe also may contain a spear for catching the fish.

Reaming

A drill collar reaming assembly is similar to a regular drilling assembly but has one or more reamers and sometimes a stabilizer. The number of reamers and possibly stabilizers depends upon hole and formation conditions and the depth location and length of the section to be reamed. A hole opener with a pilot bit is the safest tool to run on the bottom of a reaming assembly in order to prevent sidetracking.

A conventional assembly for reaming on or near bottom may have reamers in the drill-collar assembly, sometimes with stabilizers in some positions as listed below.

- Position 30 — Minimal risk of sidetracking
- Position 30 and 60
- Position 30, 60, and 90 — Moderately aggressive

Tools and Assemblies

- Position 0, 30, 60, and 90 — Very aggressive
- Position 0, 10, 30, 60, and 90 — Extremely aggressive and seldom used

Drill collar reaming assemblies for reaming off-bottom are less aggressive. Again, a hole opener with a pilot bit is the safest tool to run on the bottom of a reaming assembly in order to prevent sidetracking.

A drill collar reaming assembly with a reamer at position 30 is the least aggressive assembly, but it is only moderately effective. A slightly more aggressive assembly that is also more effective has a reamer between a stabilizer and the bit, such as the reamer at position 30 and the stabilizer at position 60. A more aggressive, more efficient assembly has a reamer between the stabilizers, such as stabilizers at positions 30 and 90 with a reamer at position 60. Two reamers may be used at positions 60 and 90 with stabilizers at positions 30 and 120 as a way of increasing aggressiveness and efficiency.

Other more aggressive assemblies containing more reamers and stabilizers are seldom used. The increased number of reamers and stabilizers increases the risk of twisting off or sticking and also increases the difficult of recovery by fishing.

String reamer assemblies have one or more reamers positioned in the drillstring above the bottomhole assembly. Their placement and use is described in the section "Reaming" in Chapter 2.

Clean Out

The cleanout assembly is a regular fishing assembly, but it has one or two stands of drillcollars and a bit below the jars and bumper sub with their associated drill collars. The assembly is slick and used without extra collars because they are not needed and they represent a risk of sticking.

The choice of bit depends upon the situation. For cleaning an open hole, run a rollar bit or a fixed-blade type to eliminate the risk of losing rollers. A dull regular or new button-type roller bit normally will not damage the top of a fish. The correct shoe will not damage the fish and can also clean around the top of the fish for an outside catch. The cleanout assembly is sometimes not recognized as a separate assembly. However, it is included here because it often is needed and using it can be an important preventive action when it is needed and used.

The normal procedure is to clean out with a regular limber assembly or whichever assembly happens to be in use at the time. In most cases this is a satisfactory practice, but it is not the best because about twice as many drill collars are being run as needed. Usually, a hole needs cleaning because of cuttings accumulation, sloughing, formation swelling, and similar problems. These often are not high-risk situations, but they should be considered as a warning of pending problems. Therefore, use the correct cleanout assembly, and do not run unnecessary tools into the hole.

The most common reason for not running a cleanout assembly is the time required to replace part of the stands of drill collars with stands of drillpipe. However, this is not a valid reason because it is easy to leave several stands of drillpipe standing back in the mast so they can be picked up rapidly when needed. Another reason for not using a cleanout assembly is the belief that a longer assembly is needed for cleaning out. This is incorrect in most cases.

Wireline Assemblies

Wireline assemblies are used to conduct various operations, including fishing with a wireline. The basic wireline fishing assembly from the surface down includes the following:

- Wireline from the surface to the rope socket
  It provides for lowering and recovering tools and transmitting reciprocating action while working the tools.
- Rope socket to connect the wireline to the tools
- One or more sinker bars as needed for stability and weight
- Hydraulic or mechanical bumper jars
- Shear release tool if needed
- Catch tool

One or more sinker bars may be run below the jars in cleanout operations for additional weight.

It may be necessary to disconnect the fishing tool assembly from the stuck fish if the fish cannot be released. Use a shear-pin release for this. Many fishing tools are equipped with a shear-pin release that is designed to release the fish by jarring either up or down, depending upon the tool selected.

The type of shear release to run depends upon the situation. A shear-down release may prevent hard bumping to catch the fish, such as driving an overshot over the fish because bumping down would shear the pin and release the overshot hold. The shear-up release may prevent recovering a fish that requires heavy jarring. For example, a shear-down release would be run above an overshot. If jarring up does not release the fish, then bump down to shear the pin and release the overshot hold on the fish. In some cases, a tool is run without the shear-pin release, thus relying on cutting the wireline or parting it at the rope socket if the stuck tools cannot be released.

REFERENCES AND SUGGESTED READING

American Petroleum Institute, RP 7G, "Drill String Design and Operating Limits."

Hornbrook, P. Rosemary, and Clifton M. Mason, "Improved Coiled-Tubing Squeeze-Cementing Techniques at Prudhoe Bay," *Journal of Petroleum Technology,* April 1991: 455–458.

Lane, J. Bret, and J. Parks Wesson, "Magnetic Ranging Tool Accurately Guides Replacement Well," *Oil & Gas Journal,* Dec. 21, 1992: 96–99.

Lubinski, A., "Factors Affecting the Angle of Inclination and Dog-Legging in Rotary Boreholes," *Oil & Gas Journal,* March 23, 1953.

Millheim, Keith, "Behavior of Multiple-Stabilizer Bottom-Hoe Assemblies," Part 5, *Oil & Gas Journal,* Jan. 1, 1979: 50–64.

Millheim, Keith, Steven Jordan, and C.J. Ritter, "Bottom-Hole Assembly Analysis Using the Finite-Element Method," *Journal of Petroleum Technology,* Feb. 1975: 265–274.

Mullin, J.A., S.H. McCarty, and M.E. Plante, "Fishing with 1.5 and 1.75 in. Coiled Tubing at West Prudhoe Bay, Alaska," SPE paper 20679, New Orleans, LA, Sept. 23–26, 1990.

Moore, Steven D., "The Coil Tubing Boom," *Petroleum Engineer International,* April 1991: 16–20.

Tarr, Brian A. and Iain A. Graham, "North Sea Experience with Aluminum Drill pipe", SPE Drilling Engineering, Dec. 1990: 317–324.

Woods, H.B., and Arthur Lubinski, "How to Determine Best Hole and Drill Collar Size," *Oil & Gas Journal,* June 7, 1954.

Woods, H.B., "Use of Stabilizers in Controlling Hole Deviation," *Drilling and Production Practices,* 1955.

Chapter 4

FISHING PROCEDURES

Basic fishing operations are commonly used in many sticking and fishing situations. They expedite operations and prevent fishing and/or aid in recovering the fish. When the average fishing job occurs, conduct one or more basic procedures. These include; milling, drilling and walling off, repairing leaks, working stuck tools, parting and catching the fish, sidetracking, and other procedures in special situations. Fishing jobs vary widely, occur under a broad range of conditions, and require various tools and different operational procedures. A variety of problems is encountered.

Overall, fishing operations are similar for both cased and open holes and generally the same types of fishing tools are used, with a few exceptions. Most formation problems are eliminated in cased holes. There is a risk of a higher loss in an equivalent cased hole because the well is nearer completion, and a higher percentage of the allocated funds has been spent. Cased holes often use smaller tools that may restrict selection and increase operating difficulty. Wireline fishing often is more effective in a cased hole.

Successfully conducting fishing procedures requires a good knowledge of general rules on fishing and equipment used.

MILLING, DRILLING AND WALLING OFF

Milling, drilling, and walling off are common operations for removing or recovering bit cones and other small pieces of junk. These steps are usually a minor operation but can lead to other fishing operations if not conducted correctly. Mills can be used on almost any type of junk and are very efficient in many cases for milling or drilling cement. Observe all precautions described in the section "Milling, Casing and Tubulars."

Mills, or short-tooth, hard formation-type bits are used to drill on junk. Drilling usually is confined to very small pieces of junk or larger pieces, such as bit cones, to break them into smaller pieces. Drilling or milling on junk often is relatively rough with high and fluctuating torque so there is a risk of leaving bit cones in the hole. There is an increasing tendency to use mills because they do not have moving parts that can be lost in the hole, and they normally mill efficiently for a longer period. The development of better hard-facing material such as tungsten carbide and matrix material further increases the use of mills.

Selecting the Drill or Mill

Select a short, mill-toothed roller bit because longer teeth tend to break off. Bits are somewhat more efficient than mills at walling off small pieces of junk by pushing and burying them into the wall of the hole in soft and medium formations. Bits often are used on large pieces of junk, such as bits cones, that can be broken into smaller pieces, especially in an open hole. They are also more effective when the junk spins. Bit life is short, ranging from up to 30 minutes for 6¼-in. or less for smaller bits, and about 1 hour or less for larger bits. All bits are subject to bearing failure and broken cones, so do not run them too long and risk leaving additional cones in the hole. Use torque and bit action as a partial guide to bit condition.

Select a mill designed for the specific job. Concave mills tend to hold junk under the mill and grind it up more, and flat-bottomed mills tend to wall it off. The mill should have ports or channels that allow unrestricted fluid flow so that cuttings pass up the side of the mill into the junk subs or out the hole. Mills are dressed with various sizes of tungsten carbide particles, generally smaller sizes for metal and larger sizes for cement mills. Do not have cutting material on the outer side of the mill, especially when milling inside casing.

Generally, run full-size bits or mills, seldom if ever undersized. Always run one or more junk subs above the mill or bit except for special conditions.

Milling Operations

Milling or drilling action on small junk usually is rough. The random-shaped metal pieces roll and slide under the grinding, cutting action of the mill or bit. This causes a fluctuating torque which, in turn, causes a fluctuating rotary speed. The action becomes smoother as the pieces break up. The action is similar when milling large pieces of fixed metal, except the torque increases as the metal surface becomes smoother. Higher, constant torque indicates milling at optimum efficiency.

Start milling with minimum weight and rotary speed and increase it with time. Do not over-torque or use excess weight. Normally circulate at slightly reduced rates. Have sufficient circulation to cool and flush the milling surface and to carry metal cuttings to the surface to prevent redrilling them. Higher circulation rates may carry large pieces of junk to the top of the drill collars where there is a risk of sticking. Milling weights vary from 500 to 2,000 lbs per inch of mill diameter. Have sufficient drill collar weight available.

Milling speeds vary, and Table 4–1 is only a guide.

Table 4-1 Suggested rotary speeds

OD of Mill, in	Speed, rpm	OD of Mill, in	Speed, rpm
$3^7/_8 - 4^1/_4$	175	$8 - 8^7/_8$	80
$4^3/_8 - 4^7/_8$	150	$9 - 9^7/_8$	75
$5 - 5^7/_8$	125	$10 - 10^7/_8$	70
$6 - 6^7/_8$	100	$11 - 11^7/_8$	65
$7 - 7^7/_8$	90	$12 - 12^7/_8$	60

The mill may be worn or the junk may be rotating with the mill if downward progress is restricted and the assembly runs smoothly at a relatively constant rotary speed with low torque, even at higher weights. Knowing how long the mill has been running and the type of junk being milled helps determine which action is occurring. Replace worn mills.

Rotating junk or a fish causes a problem. The fish or junk spins or turns with the mill. The lower part of the object may become smooth and rotate on the formation or on other junk, somewhat like an inefficient bearing, and continue this way for an extended period. Actual milling is almost negligible in this case, so change the operation. Bumping down sharply may stop the junk from spinning. Alternately reduce milling weight and vary the rotary speed. Another procedure that must be used cautiously is reducing the pump rate. This may let some cuttings settle downhole, wedge around the fish, and prevent it from turning. Rubblizing it with a junk shot may be justifiable. If the junk is relatively free try catching it with a junk basket. It may be possible to fix the junk in place by dumping cement as a last resort.

Always try to understand the milling action, what is being milled and how it is occurring. For example, milling a clean top on the fish is a standard procedure. Nevertheless, it is important to know that the mill is running on top of the fish. The top could be against the edge of the hole in a washed out section, so the mill may be cutting on the edge of the fish below the top (Fig. 4–1). Continued milling in

Figure 4–1   Milling on the side of a fish
Note: a, fish with the top laid over in an enlarged section of the hole. The top of the fish is ragged and must be milled before it can be caught; b, mill bypasses the top of the fish and cuts it below the top; c, cut-off piece of fish falls downhole alongside the remainder, increasing the complexity of the overall fishing job.

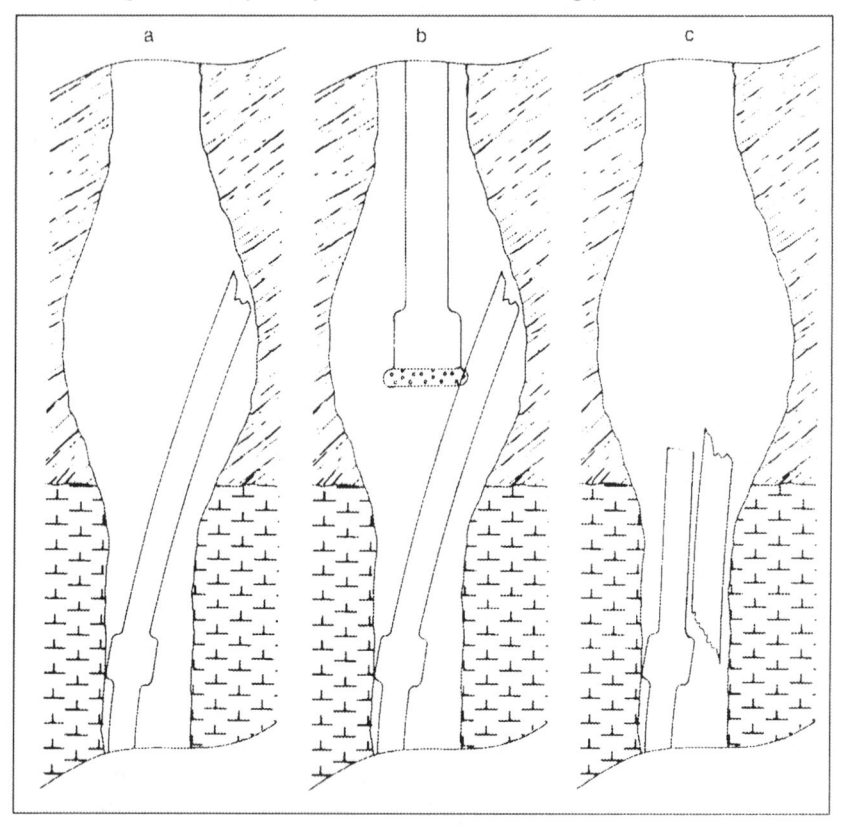

this case may leave an additional fish in the hole. Eliminate the risk by using a skirted mill and be very careful to have correct depth measurements.

Mills can be extremely efficient tools as illustrated in the following example.

Milling up a Bit and Junk

A 17,000 ft on-structure wildcat had 13 3/8-in. conductor casing set at 100 ft, 9 5/8-in. surface casing set at 2,500 ft, and 7-in. intermediate casing set at 13,000 ft. When the 7-in. casing was cemented, the casing pressured up during final displacement, leaving about 1,000 ft of cement inside the casing. The operator suspected cuttings had bridged in the open hole in the spearheaded water column ahead of the cement. Nevertheless, the casing was adequately cemented, as verified by subsequent logs.

The 7-in. casing was landed, the BOPs changed, and a drill-collar assembly and 3½-in. drillpipe picked up, while the crew installed rubbers on alternating joints. Precautions to prevent running into soft, contaminated cement had been emphasized. After picking up the bottomhole assembly and about 9,000 ft of drillpipe, the pipe began taking on a small amount of weight. This was much higher than the projected cement top, but a stand was pulled to ensure that the pipe was free. Some drag was encountered, so the first stand was set back and a second pulled. This had to be worked part-way out of the hole, and the kelly was picked up to circulate and help free the pipe.

At this point, missing drillpipe rubbers were noted. They had been installed correctly, so there was no explanation of why they were gone. The pipe stuck several times, requiring circulating, rotating and further working. Many broken drillpipe rubbers were also circulated out. The few recovered on the drillpipe had slipped, and some were partially damaged. When the bit was pulled to the bottom of the BOP, it hung up and had to be worked free. The assembly was worked out of the hole in about 12 hours.

After pulling the assembly out, the preventers were checked. The wear bushing was not seated properly and locked correctly (Fig. 4–2). The rubber protectors sheared off, sliding past the cocked wear bushing. The weight needed to cause this was not noticed because it was small compared to the overall pipe weight.

The chisel-tooth bit originally run was unscrewed and set on the edge of the rotary. Later it was knocked over by the last stand of drill collars and fell into the hole. The hole was uncovered and the BOP was open. It was unknown whether the bit fell pin up or upside-down.

A long, mill-tooth bit was run to push all junk as deeply as possible and clean out any floating rubber. It was made up with two junk subs and drilling jars and a bumper sub . This assembly was run and the hole was cleaned out to about 10,000 ft. The assembly was pulled and two unsuccessful runs made with impression blocks. Several unsuccessful runs were made with short skirted overshots to catch the pin above the bit. Finally, another impression block verified that the bit was upside-down. Junk baskets with heavily dressed shoes on the bottom and inside edges were run to cut over the bit shanks and swallow the bit, but were unsuccessful. These activities pushed the bit down to 10,500 ft.

Heavily dressed concave mills with triple junk subs to catch metal fragments were run to mill the bit. The average life was 10 hours per mill. Luckily, the bit turned very little under the mill. The metal fragments from the drillpipe protector rubbers may have helped hold it in place. The bit

Fishing Procedures

Figure 4-2 Wear bushing

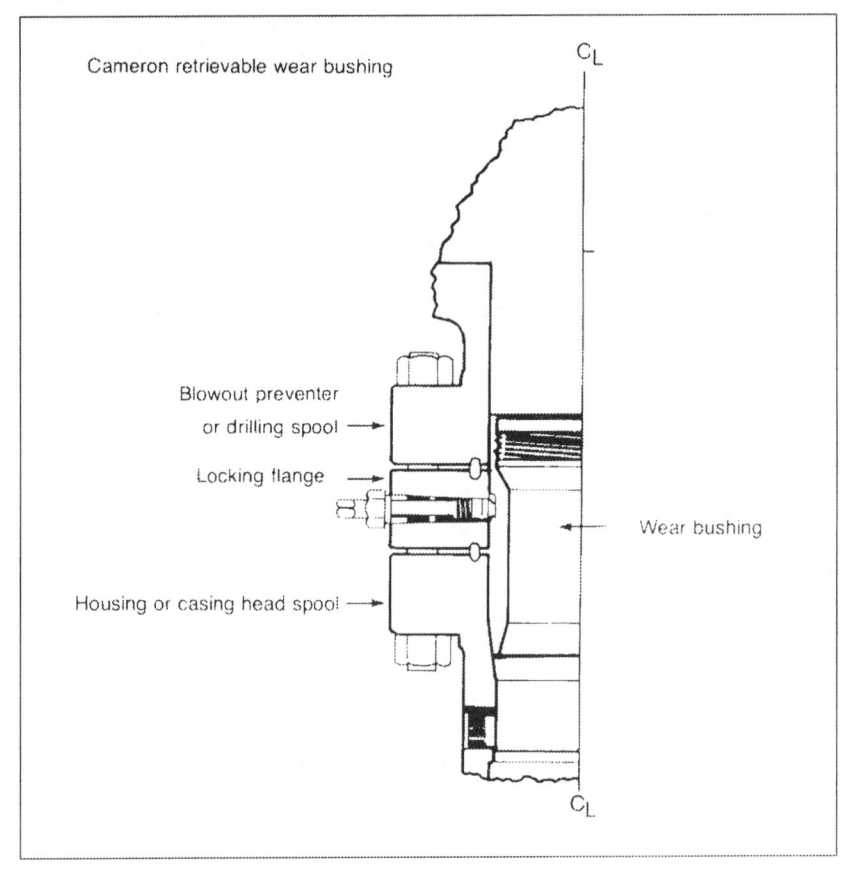

Cameron retrievable wear bushing

- Blowout preventer or drilling spool
- Locking flange
- Housing or casing head spool
- Wear bushing

was pushed downhole to about 11,000 ft while milling, and 5 to 10 lbs of metal per junk sub per mill run were recovered.

A fourth mill was run. and after about 6 hours of milling, 36 hours total milling time, it broke through. This same mill washed and reamed to the top of the cement at 12,000 ft and then drilled 1,000 ft of cement at 20 ft/hr. The casing was pressure tested, and the same mill then drilled the float shoe, and normal operations were resumed.

Small Fish

A "small fish" here most commonly is bit cones but may be tong dies, nuts, bolts, hand tools, and other small pieces of junk. Recovery procedures depend on the size, shape, depth, formation hardness, hole size, etc. Bit-cone size is determined from the bit size and how long they were drilled. The recovered bit may provide information. Did the cones drop off, and did the bit show wear from running on junk? Impression blocks are seldom used because results are very questionable. In most cases the time can be used more effectively with other procedures.

Magnets are a close contact tool and less efficient on larger size junk and junk with a small contact area, such as the end of a wrench. They are less efficient in soft formations compared to harder formations since smaller pieces of metal tend to bury in the soft caving material. Caving and cuttings may be removed by circulating with a skirted magnet run on tubulars for improved recovery. Wireline magnets are less effective compared to running on tubulars. However, run the magnet on a wireline at shallow depths, if it is conveniently available.

Junk shots are seldom effective on small junk. They are a close contact tool, and the junk frequently is covered with at least a small amount of drill cuttings and caving material in the open hole. A circulating-type junk shot may be available for running on tubulars, but its use is questionable. Junk shots are not recommended in cased holes, except special situations such as very large casing and a small shot, well centralized to prevent casing damage.

Junk baskets can be effective for recovering small junk, especially in softer formations. Reverse circulating baskets can be very effective on very small junk. In many cases, run a junk basket as the first try unless the junk is very small. Run a junk basket for two or three cones, wrenches, short pieces of pipe and similar items. Handle the junk basket carefully since the retaining fingers are fragile and easily broken. When possible, cut over the junk and then cut a few feet of formation as a plug to hold junk in the basket. Run a junk sub on top of the basket to catch small pieces of metal.

In many cases the junk cannot be recovered with a basket or magnet and must be milled away or walled off. Small pieces of metal conventionally are drilled and milled up with flat-bottom or concave mills or a short-toothed, mill-toothed bit as described earlier in this section. Run one or two junk subs above either a mill or bit.

Small junk can be walled off relatively easily in soft and medium formations (Fig. 4–3). Any walled-off junk could fall back into the hole and possibly stick the downhole assembly. Generally walled off junk either remains walled off or if it is small enough to wall off and later falls into the hole, it will not affect the drilling assembly severely. There are very few, if any, reports of walled-off junk falling back into the hole at a later date and obstructing operations.

Figure 4–3   Walling off junk
Note: Run junk sub above bit in these operations.

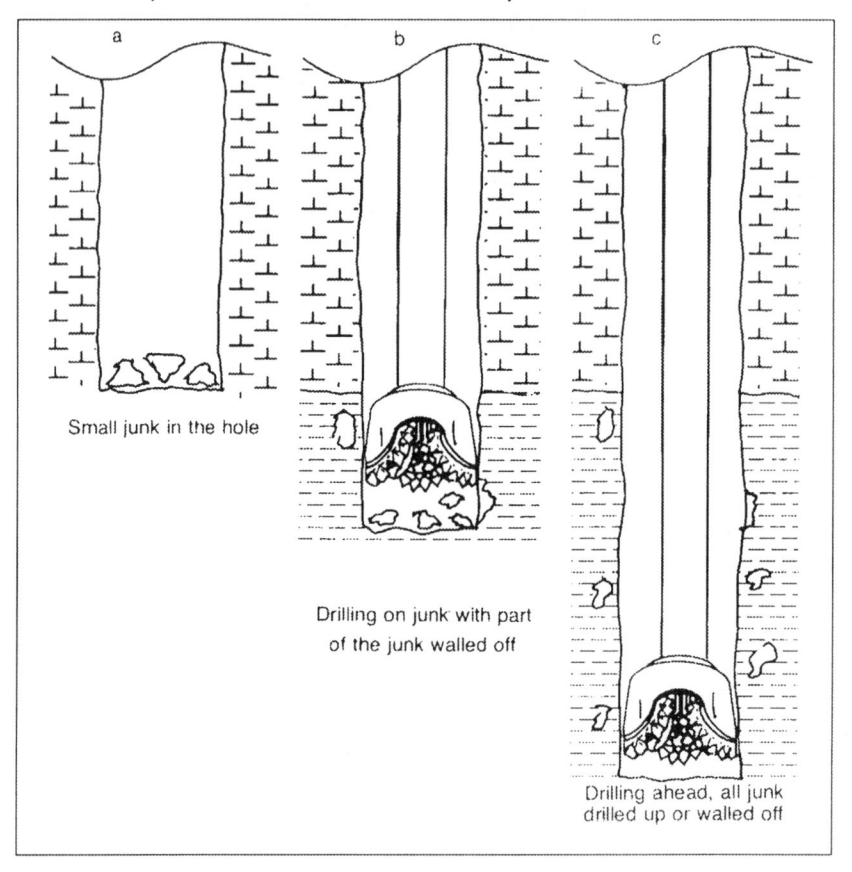

Small junk in the hole

Drilling on junk with part of the junk walled off

Drilling ahead, all junk drilled up or walled off

Large Fish

"Large fish" as used here includes complete bits, short subs, and similar-sized pieces of metal. The type of fishing tool depends upon the shape and dimensions of the fish.

Screw into bits or large junk with a threaded connection on top. If there is a neck of any type, try to catch it with an overshot. Full-gauge bits normally will not turn over if dropped in casing. They may or may not turn over in an open hole depending on hole diameter and washouts. Careful clean-out should not damage the connection. Short-grab overshots seldom catch a bit pin because of the thread taper.

Make at least one attempt with a junk basket if it will cover the fish. Junk shots may be effective subject to precautions regarding their use. It is better to mill large metal objects that are hard to wall off. The preferred method is to mill and cut over the object so it can be recovered as a larger piece. This reduces the amount of metal to be milled. Use a mill shoe below either a heavy-duty junk basket or short length of washpipe.

If the fish cannot be caught, generally the best approach is to break the large junk into smaller pieces and then fish for it by the procedures described for small junk. If the entire object must be milled, normally select a concave or flat-bottomed mill. Run junk subs. Mills can be very efficient.

Dressing Off the Top of the Fish

The top of the fish often is bent, twisted, flattened, or broken. These occur when the fish is twisted off or pulled apart. The top of the fish must be dressed-off to present a clean top to the fishing tool before it can be caught and recovered (Fig. 4–4).

The best procedure, when applicable, is to cut off the top of the fish with an outside cutter so there is a clean fishing top. It can seldom be cut from the inside because of entering the damaged top or the small diameter of the fish. Otherwise mill a clean top for one of several catches, depending upon the situation.

Milling off a short section of the top of the fish with a flat-bottom mill often is easiest and gives a satisfactory top in many cases. Run a skirted mill if there is sufficient clearance between the fish and the inside of the casing or the wall of the hole. This helps hold the top of the fish centralized and against the face of the mill and reduces the risk of milling down the side of a small-diameter fish.

There may not be enough clearance on the outside diameter to catch the fish after it is milled, such as a larger drill collar inside the casing. Then, after milling a flat top if necessary, cut a fishing neck with a mill shoe dressed with cutting material on the bottom and inside, and catch the fish with an overshot. Otherwise, mill a larger hole through the inside top of the fish with a combination flat-bottomed and centered, tapered mill and catch it with a spear or taper tap.

LEAKS

Leaks are a relatively common problem. The frequency of leaks is probably second only to losing bit cones and generally is more severe. Normally detect surface leaks with a good visual inspection. Downhole leaks can cause a fishing job if they are not detected and repaired in a short time. Leaks are more common in drillpipe tool joints and less common in drill collar tool joints. The most common

Figure 4–4   Dressing off the top of the fish

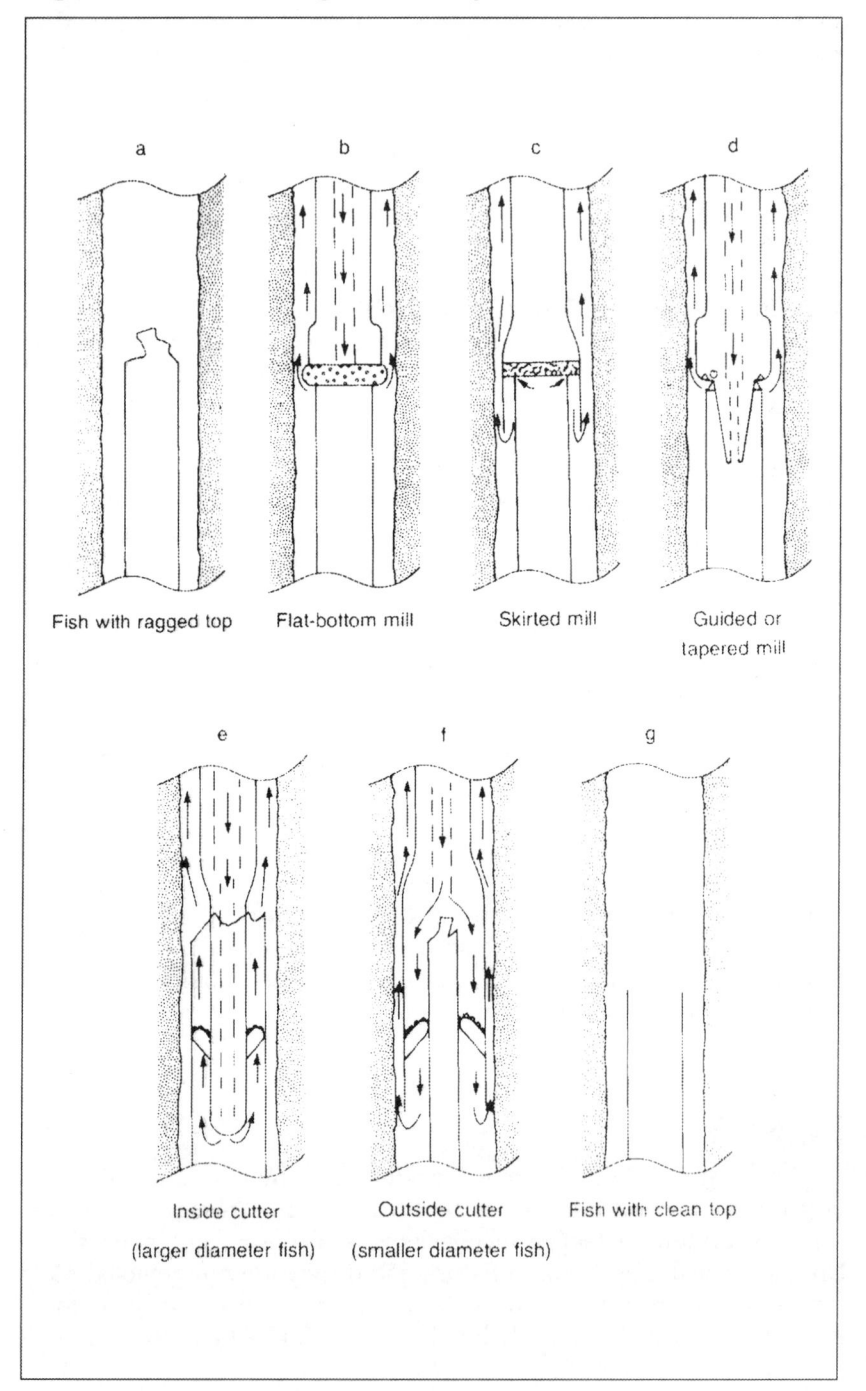

a — Fish with ragged top
b — Flat-bottom mill
c — Skirted mill
d — Guided or tapered mill
e — Inside cutter (larger diameter fish)
f — Outside cutter (smaller diameter fish)
g — Fish with clean top

cause is improper tool-joint makeup. The abrasive mud frequently combines with a loose tool joint to erode and wash out the metal, causing a washout. These are more common in tool joints but may occur in the body of the pipe.

Leaks also may occur in some high-pressure operations such as squeezing, circulating out a kick and testing. Some of these may create a dangerous situation or cause a severe fishing situation. An example is cement leaking past a retrievable packer during squeezing. A common preventive action is to pressure-test the pipe before conducting some high-pressure operations.

Most fluids, especially drilling mud, are abrasive. The circulating fluid erodes the metal surfaces. The rate of erosion is a function of fluid velocity, pressure, metal hardness, and the fluid abrasiveness. The inside diameter of tool joints often differs from the ID of the pipe body. This changes the flow rate and may create eddy currents, increasing erosion effects.

Most leaks caused by fluid erosion develop relatively slowly. Leaks in damaged tool joints or those not made up according to specifications may occur faster. Regardless of the cause, fluid passing through the leak erodes metal. The leak grows at an increasing rate with increasing fluid volume through the leak. This removes metal by fluid erosion and leads to a tension failure, resulting in fishing.

Leaks increase in frequency and severity with increasing well depth. There are more connections in a deep well than a shallow well, so there are more points subject to failure. Deeper wells generally operate at higher pressures, which increases the risk of leaking correspondingly. The higher pressure differential across the leak passes more fluid compared to a more shallow well operating at a lower pressure. The net effect is that erosion and leaks occur faster in deeper holes.

Tubulars in deeper wells normally operate at higher levels of tension. If 10% of the metal is eroded, then theoretically, the pipe would fail at 90% of the new strength rating. Normally the leak would be detected well before this. With operations at higher stress, a failure may occur faster in a deeper well compared to a more shallow well.

Leak detection is a strong fishing preventive measure. Normally detect leaks by changes in pump pressure. The amount of change depends upon depth of the leak, where it occurs in the pipestring, mud pressure and rheology, and related factors. A pressure reduction of 5% in shallow wells and less in deeper wells can be significant. The time between the first occurrence of a leak and when it develops to the point of equipment failure can range from a few minutes to several hours.

Most leaks can either be prevented or detected and repaired before they cause a fishing job. The first step is to insure that all tool joints are in good condition and tightened to the correct torque. A smooth-operating pump, with good pulsation dampeners, operating at a constant pressure is very important. Pump stroke and pressure recorders provide a means of checking prior performance against current performance. A balanced mud system helps prevent small pressure fluctuations. Normally, this should not be confused with a leak since the operator should be aware of mud weight variations. Watch for tool joints that are hard to make up or break on trips. Boxes and pins coated with mud indicate a developing or an actual leak.

It is more difficult to detect small leaks at the higher pressures associated with deeper wells. For example, if a shallow well is operating at 1,500 psi, it is relatively easy to detect a 100 psi drop or 6.6% change. For a deeper well operating at 3,000 psi the 100 psi pressure drop is a 3.3% change. This is more difficult to detect and could be confused with normal pump fluctuations. Also the leak in the deeper well is eroding at a much faster rate because of the higher pressure differential.

Pull the drillstring immediately after a leak is suspected or detected. Do not rotate the drillpipe with the rotary. Always rotate the stand with the top drive or drillpipe tongs, or chain out.

Drillstring Leaks

An 8¾-in. hole was drilling at 12,000 ft with a standard 4½-in. drillstring. Over a 30-minute period, the mud pressure dropped from a normal 1,500 psi to 1,350 psi with a slight increase in the pump strokes. The pressure dropped relatively constantly over the period, indicating a developing leak. A mechanical malfunction of the pump was less likely because of the slight increase in pump strokes. The second pump, a twin to the first, was placed on-line as a check and also gave 1,350 psi. In the next 30 minutes, the pressure dropped another 150 psi to 1,200 psi. At that time the driller picked up the assembly, and it had a normal weight, indicating that it was intact.

The pressure drop indicated a severe leak. A trip was started to pull the drilling assembly from the hole to check for a leak. The pipe was rotated out at the surface, chained out in order to prevent excessive downhole movement. This was justified because of the apparent severity of the leak and the possibility of a large washout.

After pulling 8,000 ft of drillpipe, a tool joint in the middle of a stand appeared to be leaking as indicated by mud on the side of the pipe. The pin broke when the tool joint was unscrewed. About half the broken area was a new break made when the tool joint was unscrewed. The rest of the break was eroded. Rotating the entire drillstring while pulling it would almost certainly have caused the failed pin to break and would have dropped the bottom section of the drilling assembly. In this case, the problem was recognized and the correct action taken, preventing a fishing job. It might have been advisable to pull the pipe after verifying that the pump was operating correctly.

This can be compared with another case of drilling a $6\frac{1}{8}$-in. hole at 15,000 ft with a $3\frac{1}{2}$-in. drillpipe. Declining pump pressures indicated a leak. The pipe was started out of the hole, rotating out normally. After pulling five stands, the pipe parted, dropping about 9,000 ft of drillpipe and the drill collars. The fish was recovered with an overshot after three runs.

There is no record of how far the pressure dropped, and chaining out of the hole, a longer procedure than the regular method of rotating out, may not have been justified. Also, there is no insurance that the lower part of the drilling assembly would not have dropped while chaining out of the hole. However, there is less chance of dropping a partially failed drilling assembly when chaining out compared to pulling out normally.

In another case, while drilling a 15,000 ft hole, a leak developed in the drilling assembly, resulting in an unsuccessful fishing job, and the lower hole was sidetracked.

Sometimes locating a leak, especially a small leak, is difficult and time-consuming. A leak can almost always be found by plugging the bit. Use perforation ball sealers on a jet bit and rags or a plugged swab rubber on a conventional bit. Then pull a wet string. However, this is slow and sometimes dangerous, and it can mean losing a considerable volume of expensive mud.

Another method is to dump a gallon of paint, usually aluminum and mixed with several gallons of diesel fuel, down the drillpipe before pulling out. In the operation of pulling the drillstring, the paint mixture floats on top of the mud column. A small amount will run out of the hole as the leaking stand is picked up above the rotary. The paint can be seen on the side of the pipe, indicating the leak, usually in a tool joint. Do not run the pipe wiper; it may wipe the paint off. Wash the pipe with a small stream of water.

The coring or sand line may be used to help locate a leak. Use one sinker bar and tie several long strands of soft line just above the rope socket. Circulate at a reduced rate and lower the line slowly through a pressure pack-off while measuring it. When the soft line is opposite the leak, it is pumped into the leak and will stop the downward movement of the sinker bar, thus locating the leak to within a few feet.

Locate very small leaks by pressure testing. Run the assembly in the hole with a jet bit plugged with perforation ball sealers. Pressure test the pipe every 10 to 15 stands with the kelly. When the leak is detected, pull stands or singles as needed, pressure testing as necessary until the leak is located. Replace the leaking joints and reverse the perforation ball sealers out of the hole. This procedure works well in a cased hole, but the bit may plug while reversing out in the open hole.

Various methods of measuring lag time can also determine the approximate depth to the leak. For example, a little carbide or diesel fuel can be circulated downhole slowly and detected with a gas detector as it returns to the surface. Depending on the size of the leak, most of the material will circulate down through the entire downhole assembly. Nevertheless, a small amount will pass out through the leak and return to the surface. The first returns help calculate the approximate depth to the leak.

Other methods of locating leaks have been used or proposed. Detect the leak by pumping several long strands of soft line down the hole with one pipe volume. Then pull out of hole. The strands on the side of the pipe show the position of the leak. A wireline production flow meter, spinner survey, might locate the leak by measuring the difference in flow rates above and below the leak. This probably is questionable since, if the leak is bad enough to detect in this manner, the pipe probably has already parted.

Normally circulate as little as possible after a leak is detected, to prevent additional damage. Sometimes, though, there is an advantage in knowing the location of the leak while the drilling assembly is still in the hole. For example, the leak could be so severe that there would be a risk of dropping the assembly if the washed-out point occurred high in the assembly where it was subject to maximum forces. In this case, depending upon specific conditions, consider setting the assembly on bottom, part it below the leak, recover the released portion, repair the leak, and then recover the pipe left downhole. This is a relatively straightforward operation, again depending on conditions. There is a risk of sticking the pipe left on bottom but the risk of dropping and damaging the downhole assembly is reduced.

WORKING STUCK TOOLS

Working stuck downhole tools is the procedure of moving the workstring various ways to apply releasing forces to the stuck assembly. Working includes both reciprocation and rotation with emphasis on specific methods. Most working begins when the tools stick during drilling or other operations. Otherwise, begin working after catching a stuck fish. The same general working principles apply in both cases and in open and cased holes. Working is both a preventive action and common fishing procedure.

Work stuck downhole assemblies by one or a combination of the following actions: working to establish movement or circulation, working up, working down, and a final working procedure. Working creates stress reversals on the formation at the stuck point which can be effective in releasing stuck tools. Knowing where, how, and why the assembly is stuck helps determine the correct working procedure. Working the assembly correctly greatly increases the chance of releasing it.

Working downhole assemblies often is considered as much of an art as it is a science. This is not correct. Working is the application of known procedures that have been successful many times in the past and will be equally successful in the future. Unfortunately it is all too common in the industry to either work stuck assemblies for too short a time and or work them incorrectly. The tendency to "Call a fisherman!" is too often considered the only solution to stuck pipe.

Working requires various concepts to be successful. Have a mental picture of what is sticking the assembly and how the working method will release it. Begin working as soon as the assembly sticks; this often releases it. Work steadily in a deliberate, methodical manner. Be patient, an early release is the exception rather than the rule, but it occurs. Work using one procedure long enough to determine whether it will be successful. Then try another procedure depending upon the circumstances. The working procedures may not be successful, so have an alternate working procedure or other plan.

Free Pointing and Free Pipe

The free point is the point in the assembly where the pipe is neither in tension nor compression in the direction parallel to the axis of the pipe. The free point moves depending upon the pull at the surface. For example the free point would be at the bottom of the pipe for a string hanging free in the hole. It would be at the top of the string, at the surface if all of the pipe weight was set down. Normally the free point cannot go below the point where a string is stuck. The stuck point is important for working stuck pipe since the pipe above it is free. Determine the amount of free pipe by the fast, convenient pipe-stretch method or the more accurate free-point tool run on an electric line. Locate various intervals where pipe may be stuck with special electric logs.

Pipe-Stretch Method

The pipe-stretch method is a fast, convenient procedure for determining the free point. The data for the calculation is easily obtainable with normal rig equipment. It is not used for calculating the stuck point in the drill collars. If the free point is calculated near the top or below the top of the drill collars, then assume it is at the top of the drill collars for working purposes. Calculating the free point by the pipe-stretch method is less accurate compared to the free-point log, but generally satisfactory for working stuck pipe. It can be affected by drag in crooked, deviated or horizontal holes. It is very good for detecting freepoint movement while working pipe (Fig. 4–5).

Figure 4–5 Pipe-stretch data (courtesy Dailey Oil Tools Inc.)

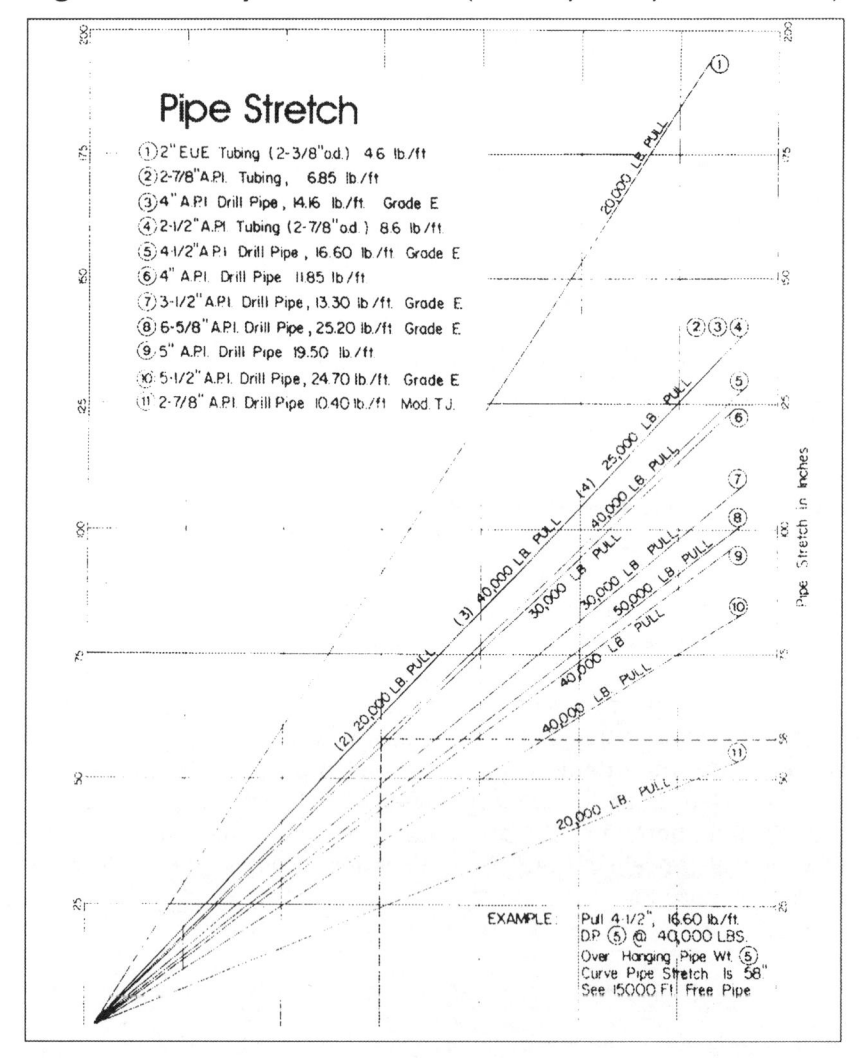

Pipe Stretch

(1) 2" E.U.E. Tubing (2-3/8"od.) 4.6 lb./ft
(2) 2-7/8" A.P.I. Tubing, 6.85 lb/ft
(3) 4" A.P.I. Drill Pipe, 14.16 lb./ft. Grade E
(4) 2-1/2" A.P.I. Tubing (2-7/8"od.) 8.6 lb/ft
(5) 4-1/2" A.P.I. Drill Pipe, 16.60 lb./ft Grade E
(6) 4" A.P.I. Drill Pipe 11.85 lb/ft
(7) 3-1/2" A.P.I. Drill Pipe, 13.30 lb./ft. Grade E
(8) 6-5/8" A.P.I. Drill Pipe, 25.20 lb/ft Grade E
(9) 5" A.P.I. Drill Pipe 19.50 lb./ft
(10) 5-1/2" A.P.I. Drill Pipe, 24.70 lb./ft. Grade E
(11) 2-7/8" A.P.I. Drill Pipe 10.40 lb./ft Mod. T.J.

EXAMPLE: Pull 4-1/2", 16.60 lb./ft.
D.P. (5) @ 40,000 LBS.
Over Hanging Pipe Wt. (5)
Curve Pipe Stretch Is 58"
See 15000 Ft. Free Pipe

The formula used to calculate the length of free pipe is:

$$L = (735.294 \times S \times W) / (T)$$
Where: L = length of free pipe, ft
S = pipe stretch (elongation), in.
W = nominal pipe weight, lb/ft
T = total load, 1,000 lbs

Work the pipe for a short time to minimize the effect of drag, then pick it up to remove all slack and take a small overpull of a few percent of the pipe weight. Mark the pipe opposite a reference point such as the top of the rotary or kelly-drive-bushing. Record the weight indicator reading. Pick up more and stretch the pipe slightly, but do not exceed the elastic limit. Mark the pipe again opposite the same reference point, measure the amount of stretch between the two marks, and record it as S. Deduct the first weight indicator reading taken from the reading at this point and record as T.

Enter T, S and W into the formula and calculate the length of free pipe or depth to the stuck point. This formula is applicable for different grades of steel in various types of pipe, regardless of the mud weight. Free pipe length may also be calculated from charts and calculator programs.

Generally monitor the free point by the stretch method when working stuck pipe. It is a good guide to working and to the types of working procedures. A free point that appears to move downhole on successive measurements, "moving free point," indicates an improving condition with an increasing length of free pipe. Conversely, a free-point movement uphole indicates deteriorating hole conditions and a decreasing length of stuck pipe. This is analogous to the hole caving, causing sticking at a higher point.

Electrical Free-Point Pipe Logs

Various electric-type logs are run through the work string on an electrical line to provide additional information on the stuck condition of the pipe. These give more precise free points as compared to the stretch method.

One of the most common is the free-point logs. They measure the degree of sticking by measuring either torsional or longitudinal strain, and they locate the lowest point where pipe is free. Normally torsional strain is more accurate for directional and horizontal holes and floating operations. Often the tool is run with a string shot to back off the pipe after locating the free point. They are especially accurate when backing off in the drill-collar assembly (Fig. 4–6).

Figure 4–6 Freepoint tool (courtesy Dia-Log)

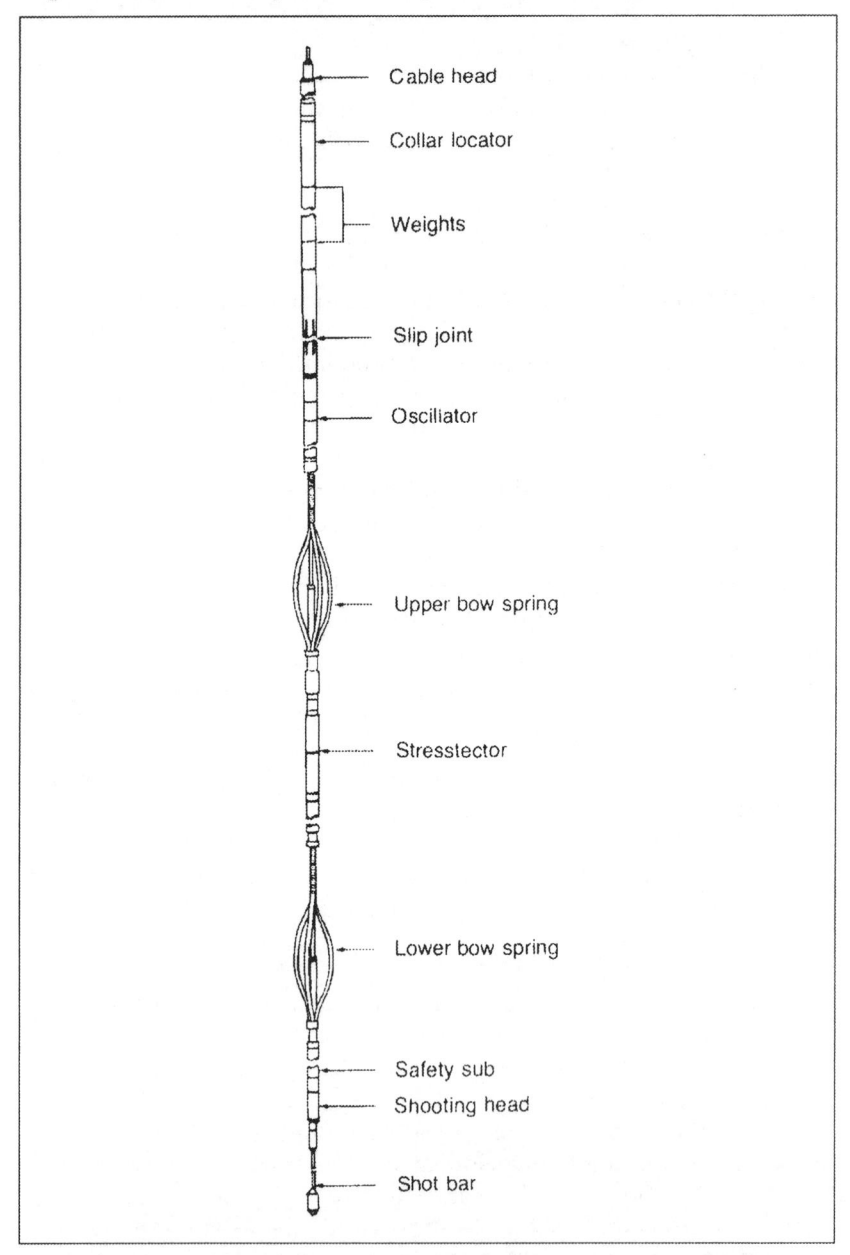

- Cable head
- Collar locator
- Weights
- Slip joint
- Osciliator
- Upper bow spring
- Stresstector
- Lower bow spring
- Safety sub
- Shooting head
- Shot bar

Drillpipe recovery logs locate free or stuck sections of pipe below the free point by means of a sonic signal. These aid in situations such as planning the length of washpipe needed and where to back off.

Begin Working Immediately

Begin working stuck tools immediately after sticking and continue working constantly. There is a tendency to work the pipe and then shut down for a few minutes, often to "talk it over." Obtain the best results by continued working.

Some sticking, such as wall sticking, occurs immediately. In other cases the process is more gradual and indicated by increasing drag and/or torque. Immediate working is facilitated if the kelly is connected. If sticking occurs while tripping, try to move the pipe so a tool joint is at or near the rotary so the kelly can be connected. Otherwise, try to connect a flexible-joint metal hose if possible for circulation. It is very important to be able to circulate both for releasing the stuck pipe and for good well control. Always have an inside blowout preventer available.

Working the stuck assembly immediately can be the difference between minor stuck pipe and a severe fishing job. For example, a small amount of caving material may stick the assembly. Immediate, constant working action could release the assembly, often by using relatively low overpull and sit down working weights. Otherwise, the caving may accumulate and stick the assembly severely if working is delayed. Likewise if an assembly is keyseated, early working may release it before it becomes wall-stuck.

Know how to work the assembly immediately, but in all cases start and continue working it. If this does not release the assembly, it may at least reduce the risk of additional sticking. Beginning to work the assembly does not necessarily mean taking an immediate, maximum pull. In fact, this may be the wrong action. However, working immediately at maximum force levels is justified in some cases. For example if the pipe wall sticks and this is known, then immediately bump down and jar up at maximum force levels. This may release the assembly before the contact area is completely sealed.

How Hard to Work

How hard stuck pipe can be safely worked is a primary consideration. Most stuck fish are not worked initially at maximum tension. However in some cases this is instrumental in releasing them. Types of loading include pull, set down, impact and torquing. Maximum allowable force depends upon the type of loading, pipe weight, grade and condition, tool joint strength, type and magnitude of the applied forces, and the length of time the assembly is subjected to the forces. Related factors include hole size, gauge, depth, deviation and dogleg, and hole condition such as keyseats and caving.

The workstring, usually drillpipe or tubing, normally is the weakest component. It is the longest part of the assembly, so it is more exposed. It suspends all downhole tools, so it is subject to higher tensile loading, especially the upper section. The lower section is more subject to bending due to compressional forces. Drill collars and other tools normally are stronger. They are deeper in the hole and subject to lesser forces except for jarring and bumping.

The assembly is subject to tension, compression and torque. High forces can be involved due to heavy jarring and bumping. Picking up and lowering and the use of jars and bumper subs cause impact and cyclical loading. These various forces can cause tool joint failures, cracked pins, split boxes, loosened and washed out tool joints and pipe body failures. The main controllable items are how long and how hard to work the assembly, assuming it is designed and maintained correctly.

For example, drag in open holes dampens forces applied at the surface so they are reduced at the point of sticking. This lessens the efficiency of the working action. Jarring and bumping is less affected by this. Caving can cause additional sticking, indicated by the free point moving uphole. Working may create keyseats and cause caving or unstable formations. It may be better in these cases to back off the pipe above the stuck point rather than allowing more of the pipe to become stuck.

Excluding formation effect, the amount of pulling force that can be applied as a straight pull on stuck tools is the lesser of the strength of the tube, tool joint or connection with an appropriate safety factor for condition. These data are available from standard tables. Strengths usually are based on the minimum yield, where the pipe may stretch but returns to its regular length when the force is removed. The pipe begins to stretch permanently at pulls above this point but normally will not fail until it reaches the ultimate yield, which is about 10–15% greater, sometimes more than the minimum yield. Pipe performance is less predictable when stresses exceed the minimum yield.

Common safety factors are 15–25% so use a maximum working tensile load of 75–85% of the minimum yield. However, the difference between the minimum yield and the ultimate yield is, in a sense, a safety factor. This does not imply the assembly should be worked at force levels above the minimum yield. Nevertheless, it is important to use practical safety factors.

Normally tubulars are not deliberately damaged, but there are cases where other factors may be more important than the risk of pipe damage. For example, the cost of replacing a drillstring may be justified to successfully complete the fishing job on a deep, expensive well. When these cases occur, consider the risks in determining how hard to work the assembly, especially if working above the minimum yield.

Working pipe downward often is one of the most effective ways of releasing stuck tools, especially since two of the most common causes of sticking are keyseats and caving. In both cases, usually the best releasing procedure is working downward. How much weight can be set down on the assembly before it is damaged? The pipe is a column, and the actual weight that can be supported safely by the drillpipe can be calculated by column formulas if all data are known. It usually is assumed that a fully supported vertical pipe column will support as much weight in compression as it will in tension. This implies that the pipe can be set down by approximately the equivalent of the amount of allowable pull.

However, the downhole assembly is not a fully supported column. At best, it may be partially supported at the tool joints. If the hole is in-gauge, the pipe body between the tool joints could bend until it touches the wall of the hole and some minor indention into the wall, depending upon formation hardness. However, there are crooked or deviated holes, washouts, and other out-of-gauge sections where additional bending can occur. So trying to calculate the precise allowable weight that can be set down, except for special cases, is impractical.

The net result is that the maximum allowable set-down weight must be estimated based on the strength characteristics of the downhole assembly, knowledge of the specific downhole conditions, and general experience. More weight can be set down safely if the hole is in-gauge and tool-joint diameter approaches the hole diameter. Again this is difficult to quantify.

Compression caused by setting down weight normally will not damage medium- or large- sized drill collars under average conditions. Therefore, the main area of possible damage is in the lower section of the workstring (Fig. 4–7). Most operators are very cautious about setting down pipe weight. This is partly due to the natural caution of sticking or damaging the drillstring or plugging the bit. There also is a natural tendency to pull stuck pipe upward, rather than slack off.

Fishing Procedures

Figure 4–7 Pipe under compression

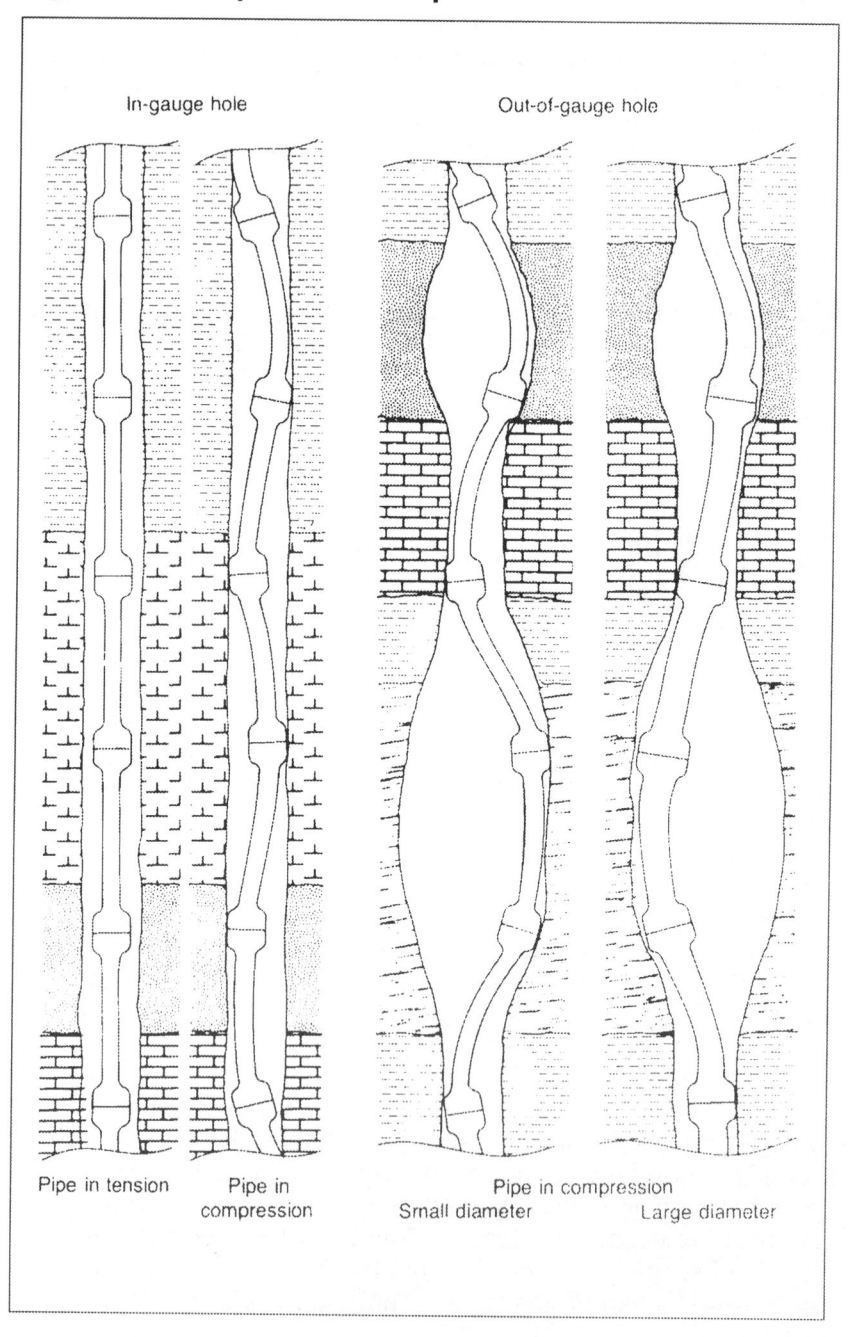

In-gauge hole

Out-of-gauge hole

Pipe in tension

Pipe in compression

Pipe in compression
Small diameter

Pipe in compression
Large diameter

Consider the following field results. 6,000 ft of 2⅜-in. tubing was set on bottom in 5½-in. casing. 8,000 ft of 2⅞-in. tubing was set on bottom in 7-in. casing. 6,000 to 8,000 ft of 3½-in. drillpipe and a standard drill-collar assembly was set on bottom in a 6¾-in. open hole. 8,000 to 10,000 ft of 4½-in. drillpipe and a standard drill-collar assembly was set on bottom in an 8¾-in. open hole. All were set down slowly and picked up later without damage. This information helps establish set-down weights under slow, smooth, pseudo-static conditions.

Next consider a common drilling assembly with 60,000 lb of drill collars and 10,000 ft of 4½-in., 16.60 lb/ft drillpipe that weighs 166,000 lb (air weight). Total string weight is 226,000 lbs in air or 192,000 lb in 10 lb/gal mud with a buoyancy factor of 0.85. There should not be a problem working the assembly down to a surface weight of 141,000 lb, since all of the drillpipe would be in tension. Excluding the risk of sticking, the entire assembly probably could be completely set down slowly in an average hole without damaging the pipe based on field experience.

Therefore it appears that one relatively safe way to work a stuck drilling assembly downward is to slack off slowly the equivalent of at least 5,000, up to 10,000 ft of drillpipe weight above the free point in the average hole. If the assembly is stuck in the drill collars, the free point is assumed to be at the top of the drill collars for purposes of determining how much weight to slack off without damaging the drillpipe. The force acts downward on the formation at the point of sticking. Picking the pipe up with an overpull creates an upward force on the formation at the point of sticking. Repeating this stress reversal over time is a very effective procedure for releasing stuck pipe. This general guide applies to an average hole and is subject to modification based on experience and knowledge in the area and the condition of the specific hole.

When setting weight down, only a portion of the total weight actually is applied to the stuck point. As the assembly is lowered, it tends to coil around inside the hole and drag over inclined and crooked hole sections. The friction at these points supports some pipe weight. The actual amount of weight supported by friction cannot be determined because of the many unknown factors. It can be estimated by observing the weight indicator as the assembly is lowered. If there is no friction, the weight will be transferred to the stuck point over the relatively short travel distance equivalent to the stretch and compression in the free pipe. If the assembly weight is transferred over a longer distance of travel, then the added distance reflects a combination of bending or coiling inside the wellbore and

Fishing Procedures

the effect of deviated and crooked-hole sections. The result is that less weight is applied to the point where the assembly is stuck than is reflected by the surface indications. Consequently, more weight must be applied at the surface to apply a given weight at that stuck point.

Downward impact or shock loading are dynamic forces that are dependent on the rate at which the weight is set down. They are larger but more difficult to analyze. One method is observing extremes or limits. The lower safe limit is slowly setting down at least 5,000 to 10,000 ft of pipe in an average hole.

The upper limits of impact loading can be evaluated to a limited extent by considering several operations. In many cases, drilling and other downhole assemblies have been dropped several thousand feet without parting, although they usually are bent. The lower section of the assembly is exposed to a very high impact force. In a specific case, while tripping in the hole at about 110 fpm, 5½-in. drillpipe hit a solid obstruction at 11,000 ft in the open 12¼-in. hole below 13⅜-in. casing set at 5,000 ft. In the subsequent fishing operations, the drillpipe was found bent and broken immediately above the drill collars. The hole was out-of-gauge at the point where the pipe broke because the fishing tools bypassed the fish on the first few runs, requiring the use of a wall hook. Bits have hit obstructions while tripping in the hole at depths of 5,000 to 10,000 ft, at rates of 90 to 150 fpm without damaging the drillstring.

For impact purposes, an assembly of 5,000 ft, and possibly longer, can strike the stuck point in an average hole at pipe speeds of at least 50 fpm and possibly up to 100 fpm or more without damaging the pipe. These may be reduced for washed out or out-of-gauge holes.

The impact loading on stuck pipe depends upon how fast the work string travels when it is being worked downward. If the pipe is stuck solidly, the only movement is the small amount due to pipe stretch and compression in the upper assembly above the stuck point. A rapid movement in this manner transmits a momentum type force down through the assembly to the stuck point. When working a completely stuck assembly this is the principal downward force applied to the stuck point. If there is some pipe movement at the stuck point, higher forces may be applied. However, these probably are not as high as normally envisioned due to the increasing cushioning effect of hole drag as the assembly is lowered. A bumper sub increases the downward force by an appreciable amount by allowing more distance for downward travel. Concentrated drill collar weight above the sub further increases the efficiency of the downward stroke.

The downward momentum procedure described above may be used in an alternate manner. Lower the workstring rapidly and stop it a short distance above bottom. The pipe will continue traveling downward a short distance, stretching and elongating, to create a strong downward force. Lower the pipe and catch it a little higher. It continues traveling downward a short distance, stretching and elongating. Then it springs upward, imparting an upward force on the stuck point. Repeating the procedure creates stress reversals against the stuck point. This action is sometimes described as a snap-and-bounce procedure and can be effective in releasing stuck assemblies. Efficiency decrease with depth and heavier workstrings.

Pipe sizes have not been specifically used in how hard to work the tools because of unknown data. Normally, the smaller pipe would be in a smaller hole and larger pipe in a larger hole, so to some extent both sizes have approximately equal lateral support. Larger pipe can withstand more total force than smaller pipe, but the larger pipe also weighs more. Therefore, using pipe weights in terms of feet of pipe is probably adequate.

How Long to Work

How long the stuck assembly should be worked depends on a number of different factors. Some are straightforward, but others are intangible and require exercising good judgment. However, do not underestimate the importance of working the assembly. Work it constantly and in the correct manner, usually for an extended time. Work it until one of the following occurs.

- The reason for working is accomplished, stuck pipe is released, or the fish recovered.
- The assembly has been worked to the point where there is a high risk of failure; it needs to be pulled and serviced.
- Hole conditions have deteriorated, so additional working increases the risk and reduces the chance of releasing and recovering the assembly.
- The pipe has been worked for an extended period without obtaining any favorable results, so it is time to change actions. This normally is at least 6 to 12 hours in cased holes and about 8 to 16 hours in open holes, subject to other limitations.

Good judgment, experience, and knowledge determine how long a drilling assembly can be worked before it fails. If the drilling assembly is worked too long and fails, then the resulting fishing job can be more severe than some of the available alternatives, but not always. Also, there are times when the risk of sticking a longer section due to deteriorating hole conditions can more than offset the potential advantage of continuing to work the drilling assembly.

However, the most common situation is not working the pipe long enough to free it. One reason is not knowing how to work the stuck pipe or not being aware of the advantages. Another is fear of a failure in the workstring. If the tools are in good condition, and they should be, then working can continue for an extended period. Otherwise, a fishing job is assured if the stuck tools are worked for only a short time before free-pointing and backing off.

Working time may depend upon whether there is a jar and bumper sub in the work string, if they are working, and if the assembly is stuck above or below them. Also, working action is more efficient if the jar and bumper sub are positioned a short distance above the stuck point. Sometimes it is necessary to suspend working to either run, replace, or reposition them.

The importance of continued working of stuck pipe is illustrated in the following.

Working Stuck Tubing

A Gulf Coast Vicksburg gas well had 9⅝-in. casing set at 8,800 ft, a 5½-in. liner at 14,000 ft and 2⅞-in. tubing latched into a permanent packer at 13,500 ft inside the liner. Packer fluid was 16.8 ppg water-base mud. The well was completed in multiple sands and production had declined abnormally. Sinker bars run on a slick line indicated that most of the perforations were covered with sand. This is characteristic of abnormally pressured Vicksburg gas sands which flow sand into the wellbore under high pressure drawdown.

A work-over rig was moved in and rigged up. The tubing was found stuck near the surface with less than 4 in of stretch at maximum pull. It was worked for 12 hrs with the free point moving downhole. A free point indicator found the tubing 100% free at 11,500 ft and 10% free 50 ft above the packer. The tubing was cut with a chemical cutter, leaving a 20 ft stub above the packer. It was worked free in 6 hours and pulled.

The fish was caught with an overshot, but the latch-in could not be released with torque and jarring. A reversing tool backed the tubing stub out of the latch in the packer. A 1-in. tubing stinger, run on the bottom of the workstring, passed through the packer and washed the sand out of the liner. Another packer was set above the first packer and the well was completed, leaving regular weighted packer fluid behind the tubing.

This illustrates the importance of working stuck pipe for an extensive period to release it before using other fishing procedures. Trying to wash over the tubing would have been very difficult.

Deteriorating hole conditions such as sloughing, caving and bridging are the most common cause for limiting the safe working time, especially without circulation. This emphasizes the importance of circulating while working stuck assemblies because it reduces the risk of additional sticking.

Additional sticking may often be detected by small jumps or erratic movement of the weight indicator needle while increasing or decreasing tension. Another indication is apparent reduced pipe-stretch. Take periodic free points by the pipe-stretch method while working. A deteriorating hole condition is indicated by free points found at successively shallower depths with time or an upward moving free point. As a very rough rule, do not allow the free point to move uphole over 300 to 600 feet before suspending working and backing off.

Working Heavy Stuck Drill Tools

A packed-hole assembly was used to drill a deep well in very hard and abrasive, crooked-hole formations. The assembly included drilling jars, bumper sub and keyseat wiper with 5-in. X-95, 25.60 lb/ft drillpipe, new tensile of 671,000 lb, premium used of 535,000 lb. The lower hole had an average deviation of about 5½°/100 ft and an average dogleg of less than 1½°/100 ft.

The assembly began to hang up at 18,000 ft on a trip to change bits. It was hanging up in the lower part of the drill collars since the drilling jars were working satisfactorily. It was worked carefully to an overpull of 50,000 lb to prevent pulling into a keyseat. Four singles were worked out of the hole in 8 hours by working while circulating. Then the assembly stuck and would not move. The hook weight was about 390,000 lb. The jars were tripped at 425,000 to 450,000 lb and then the pipe pulled up to 475,000 lb. The assembly was bumped downward to a weight of 300,000 lb. It was worked for 6 hours without movement.

Oil was spotted around the drill collars, and the assembly soaked while working for 12 hours without releasing. A free point showed that the bottom eight drill collars were stuck. The first back-off shot was unsuccessful, and the drill collars backed off above the drilling jars on the second shot. The string was screwed back into the fish and the assembly circulated and worked. After 2 hours the pipe began to move slowly. Five singles were worked out in 8 hours. The basic method was to pull up to about 420,000 lb (30,000 lb overpull), slack off to 350,000 lb, and be sure there was some downward movement of the pipe before it was picked up

Fishing Procedures

again. Each time the pipe was pulled up to 420,000 lb it would move upward a small amount. If the pipe would not go down after being picked up, it was worked downward until it could move, even for a small amount. After working for 4 hours, the assembly again began to move upward slowly. Two hours later, it released and was pulled out of the hole.

This example illustrates the reliability of well-maintained tools and the importance of continued working as long as there is a reasonable possibility of releasing them without fishing.

Procedures and Techniques

Various working procedures and techniques are used to release stuck assemblies. Working procedures include one or more of reciprocating, rotating, and circulating. These are used with various techniques such as initial working direction and working force levels. Both are combined to release the stuck tools either by establishing movement and circulation, and working up or down or with a final working procedure.

A common observation is that drill tools are not stuck if they can either reciprocate, rotate, or circulate. As long as one of these actions is possible, even to a limited extent, there is a reasonably good chance the assembly can be worked and manipulated to perform the other functions. Establishing circulation usually has first priority, followed by movement. These are important, both to prevent additional sticking and to release stuck tools.

Working procedures and techniques generally are similar for open and cased holes with allowances based on specific conditions. Assemblies stuck in open holes generally are larger and stronger and often are worked at higher force levels. Open holes usually have more hole problems requiring additional precautions to prevent additional sticking.

Smaller assemblies with reduced strength characteristics are more common in cased holes. They often are worked more efficiently because of fewer hole problems, reduced drag and torque, increased flexibility and reduced weight. Bumping-down operations are more efficient in the cased, in-gauge hole. Establishing circulation in cased holes often is less critical, except if stuck in sand or similar material that can be circulated out with mud.

Working Results Indicate Working Method

Various indicators such as pump pressure and torque may help determine the method and intensity of working stuck tools. They may have slight movement with little or no circulation. The optimum circulation rate will occur with the free point at the stuck point. Pulling slight tension on the pipe may cause a pressure increase as the pipe pulls tightly into the stuck point. This is the approximate upward limit the assembly can be pulled while maintaining circulation. So pick up the pipe until the circulation is restricted and then slacked off. Lowering the pipe usually gives similar results in the opposite direction.

Work the pipe in a slow, easy manner, using the pump pressure as an indicator, and insuring continued circulation. Repeating this often releases the assembly by washing and eroding the caving bridge or other material causing sticking. This method of releasing the assembly would be less effective without circulation. Normally run the pump slowly at first, since there are cases where increased pump rate will stop circulation, and gradually increase it. Using the pump pressure in this manner while working the stuck pipe at times can be more sensitive than using the weight indicator.

Torque or rotation also may guide the working direction. The stuck pipe may rotate slightly while the pipe is in the free point position. Torque may increase with additional pull or setdown. Then work the pipe in the direction to cause increased torque, but not so hard as to stop rotation.

Working Force Levels

Generally start with low working force levels, increasing with time. The rate at which the working forces are increased depends on the type of assembly, how it is stuck, whether it occurs in an open or cased hole, and whether the hole can be circulated. If the pipe is stuck by caving and cannot be circulated, the pipe should be worked at low force levels with gradual increases while trying to establish movement and circulation.

Treat sanded-up pipe inside casing similarly. Both can be difficult fishing jobs. Pipe stuck in keyseats, wall-stuck pipe, and tools stuck inside casing, not stuck with sand, generally can be worked at maximum force levels within a short time. Tools stuck in green cement or highly viscous mud should be worked at a maximum rate immediately. Pipe stuck by caving during a connection should not be worked until after starting the pump and then worked easily with a low pump pressure. Each case must be evaluated to determine the manner and direction of working the stuck assembly and how long and how hard it should be worked.

Initial Working Direction

The initial working direction in almost all cases is in the opposite direction to pipe movement immediately prior to sticking. So if the assembly sticks while being moved in one direction, work it in the opposite direction.

If the bit sticks on bottom, start working it upward. Work in the upward direction after running a bit or other tools into an obstruc-

Fishing Procedures

tion such as a bridge or tight hole on a trip in the hole. Work keyseated drill collars downward. Work packers, retrievable plugs and similar tools in the releasing direction. Initially work a stuck hookwall packer upward and most tension packers downward. Work a balled wireline fish downward that was pulled into the bottom of the casing and stuck.

Sometimes the working direction is difficult to determine. If the bit is at or near the bottom of the hole and pulled into a tight hole or caving, then working downward is restricted, so work it upward. Sticking in a crooked hole may be a similar problem, but generally work in the direction opposite the direction of movement prior to sticking. Normally work drill collars downward after pulling them into a crooked hole. If the bit is forced into the wall of crooked hole at the same time, then work upward or both up and down.

Work to Establish Movement and Circulation

Working to establish movement and circulation is the first action if the assembly is completely stuck and cannot be circulated. If stuck tools can be circulated, there is a much better chance of releasing them. Establishing circulation may be easier if the stuck assembly can be rotated or moved, even a small amount.

The procedures for establishing movement and circulation and releasing the downhole assembly described here generally apply to those stuck in the open hole. They also do not generally apply to releasing keyseated or wall-stuck pipe as covered in Chapter 5. When tools are stuck in the cased hole, establishing circulation is not usually as critical, or as helpful, in releasing the assembly. The exception is an assembly stuck in sand or other material that can be circulated out.

Plugged bits frequently are associated with stuck drilling assemblies, so determine whether a plugged bit is restricting circulation, since it is treated differently than annular restrictions. Detect a plugged bit by the small amount of fluid needed to pressurize the system, a sharp increase in mud pressure and a very limited flowback. When the pipe is open, a larger volume of mud often is required to pressurize the system, and the pressure increases more slowly. Unplug bits in this situation by blowing the nozzles out of the bit with a string shot (Fig. 4–8).

To establish circulation, first move the assembly so the free point is at the stuck point. Then pressure up to about 150 to 250 psi. This is about the minimum accurate reading with the average mud-pressure gauge. Use low pressure when establishing circulation. Higher pressures frequently have a piston effect and may seal the annulus. Low pressures apparently let the mud seep through caving material until circulation is established. Low pressures also minimize the risk of pumping into the formation.

Figure 4–8 Blow plugged jet nozzles out of bit

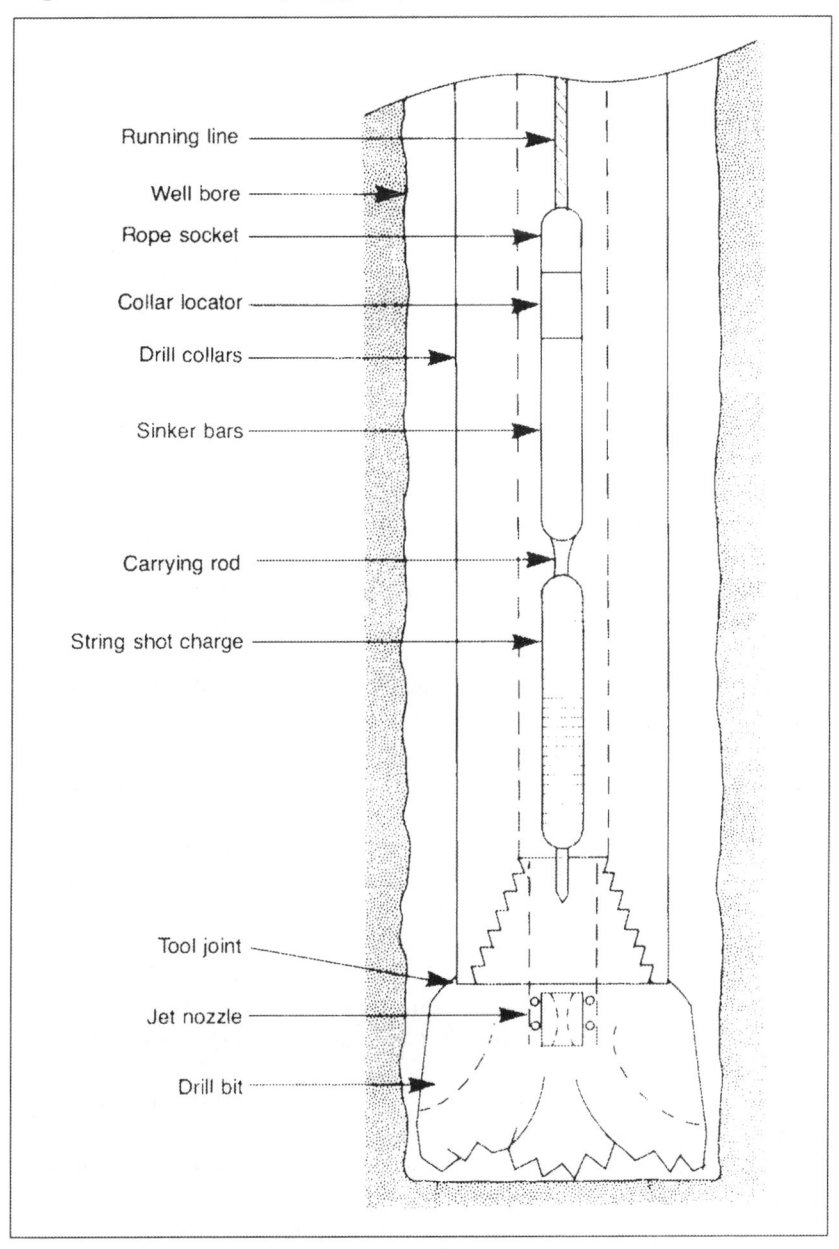

Labels: Running line; Well bore; Rope socket; Collar locator; Drill collars; Sinker bars; Carrying rod; String shot charge; Tool joint; Jet nozzle; Drill bit

After pressurizing, work the stuck tools with about 3,000 to 5,000 lbs, first by overpull and then setting down the same amount. If there is some pipe travel, continue to pull and set down with about the same amount of weight. Continue this for an hour or longer while monitoring the pump pressure. The start of circulation is detected

Fishing Procedures 259

first by a decline in pressure. If it declines, restore it to the same low level. If it can be determined which working technique caused the pressure decline, then repeat that procedure. Continue working and restoring the pump pressure. Then gradually increase the pump pressure while continuing to work the pipe until the pump runs almost continually. Continue until full circulation is established. Use caution to prevent pumping into the formation.

If the procedure is ineffective to this point, leave the system pressurized to a low pressure and apply right-hand torque to obtain rotation. If the kelly is connected, apply torque with the rotary. Otherwise use the drillpipe tongs. Normally do not apply torque with the slips and rotary.

First try to rotate the stuck tools with a neutral pipe weight, free point at the stuck point. If the tools do not rotate, apply a low level of right-hand torque, usually about 20% of the maximum allowable. Hold the torque and work the pipe at 3,000–5,000 lb of overpull and set down. Then check the torque and continue monitoring circulation. If the torque decreases, add torque to the original level and work the assembly again. If the torque does not decrease, apply additional torque and work again. Gradually increase the torque to the maximum safe operating limit. If any procedure appears to be effective in establishing pipe movement or circulation, then continue it. For example, if applying torque indicates the assembly is rotating to some extent, then continue applying torque.

If these procedures are ineffective after about one hour of working, increase the pump pressure to 400 to 500 psi while continuing to work the pipe, and torquing periodically. After a short period, increase the working tensile forces to about 10,000 lb. Usually the maximum efficient pump pressure is about 500 psi for establishing circulation while working the pipe in this manner. If this is not effective, try higher pressure levels for a short time, providing they do not exceed the leak-off pressure. This will be indicated by leak-off without either circulation or a small flow at the flow line. When these higher pressures are not effective, the rest of the pipe-working procedure should be conducted with a pressure of about 500 psi or just below the leak-off pressure, whichever is less. Restore the pump pressure if it reduces.

After working the pipe at 10,000 lb for about one hour, increase the working force to about 20,000 lb and continue working and torquing. At this point the pipe should have been worked for a total of 2–4 hours. If the procedure is unsuccessful in establishing movement or circulation, then increase the working force in steps of about 10,000 lb (to the maximum allowable force) at intervals of about 30 minutes to one hour while torquing periodically.

This procedure generally applies when the assembly is stuck with the bit off-bottom, and indications are that the assembly should be worked both upward and downward. When the assembly sticks on bottom, most of the working will be upward. If the assembly has been pulled up into a tight hole, most of the working will be downward.

The overall objective is to free the assembly, but the immediate goal is to obtain movement and circulation. When the stuck pipe begins to move, work to expand the movement until the assembly is released. Do this by pulling the assembly upward to the end of the free travel and then apply additional pull. Insure that this extra pull is low enough so the pipe does not stick but can be bumped downward. Then slack off until the assembly reaches the bottom of the free-travel section, and apply additional weight in the downward direction. Insure that the weight does not stick the assembly and it can be pulled or jarred upward to the free-travel area. Repeat the procedure and increase the working force as needed.

Concentrate on moving the pipe upward if it begins to move at the upper limit of travel. Conversely, work it in the downward direction if it begins to move downward, depending on how much hole is open below the assembly and other related factors. If free travel can be obtained, the assembly usually can be released.

If the assembly is equipped with jars and a bumper sub and is stuck at a point above these tools, they are ineffective. If the assembly is stuck below the tools, they can be used to help release it.

There are several ways of working with the jars. Pull the pipe upward against the stuck assembly to the tripping weight and wait for the jars to trip, delivering a sharp upward blow on the stuck assembly. Pipe weight reduces slightly after the jars trip due to the elongation of the jars. Another method is to pull the pipe to a higher level of overpull after the jars trip. A third method is to pull the pipe steadily to the maximum weight. Adjust the rate of pull so the jars trip at the desired tripping weight. Then lower the pipe to reset the jars and pick up again to repeat the jarring action.

Use a similar procedure when bumping down. After the bumping stroke, set additional weight down to release the fish. The jarring and bumping action may be modified slightly, depending on the type of tools used and their specific tripping mechanism.

Normally start jarring by tripping the jars at low levels of overpull in the range of about 10,000 lbs of overpull and bumping down about the same amount. Work the stuck pipe at this level for about an hour. If this does not establish movement or circulation, then trip the jars at a higher level by increasing the overpull by 5,000–10,000 lb. Increase the tripping level gradually by 5,000–10,000 lb increments until the stuck assembly moves, or circulation is established, or the jars are being tripped at the maximum safe recommended force (Table 4–2). Increase the level of bumping down similarly.

Table 4–2 Impact values

PULL ABOVE PIPE WEIGHT IN POUNDS

Collar wt. in lbs. above jars	30,000	40,000	50,000	60,000	70,000	80,000
3,000	33,660	44,880	56,100	67,320	78,540	89,760
4,000	44,880	59,840	74,800	89,760	104,720	119,680
5,000	56,100	74,800	93,500	112,200	130,900	149,600
6,000	67,320	89,760	112,200	134,640	157,080	179,520
7,000	78,540	104,720	130,900	157,080	183,260	209,440
8,000	89,760	119,680	149,600	179,520	209,440	239,360
9,000	100,980	134,640	168,300	201,960	235,620	269,280
10,000	112,200	149,600	187,000	224,400	261,800	299,200
12,000	134,640	179,520	224,400	269,280	314,160	359,040
14,000	157,080	209,440	261,800	314,160	366,520	418,880
16,000	179,520	239,360	299,200	359,040	418,880	478,720
18,000	201,960	269,280	336,600	403,920	471,240	538,560
20,000	224,400	299,200	374,000	448,800	523,600	598,400
22,000	246,840	329,120	411,400	493,680	575,960	658,240
24,000	269,280	359,040	448,800	538,560	628,320	718,080
26,000	291,720	388,960	486,200	583,440	680,680	777,920
28,000	314,160	418,880	523,600	628,320	733,040	837,760
30,000	336,600	448,800	561,000	673,200	785,400	897,600
32,000	359,040	478,720	598,400	718,080	837,760	957,440
34,000	381,480	508,640	635,800	762,960	890,120	1,017,120
36,000	403,920	538,560	673,200	807,840	942,480	1,077,120

COLLAR WEIGHT

These values are purely theoretical. They are based on damages to recovered tools in fishing operations and are to be used as a guide only. Any well condition that causes a drag on the drillpipe would alter these figures. Dailey Oil Tools Inc. does not recommend impact valves to exceed the following : for 4⅛ OD jars, 250,000; for 4¾ OD jars, 300,000; for 6 OD jars, 460,000; for 6⅛ OD jars, 460,000; for 6¼ OD jars, 460,000; for 67/8 OD jars, 604,000; for 7¾ OD jars, 770,805.

Source: Dailey Oil Tools Inc.

Note: $0.000374 \times \dfrac{\text{weight of collars}}{\text{above jars}} \times \dfrac{\text{amount of pull}}{\text{above pipe wt.}} = \text{amount of impact in lbs.}$

Table 4-2 Impact values (continued)
PULL ABOVE PIPE WEIGHT IN POUNDS

90,000	100,000	110,000	120,000	130,000	140,000	150,000	160,000
100,980	112,200	123,420	134,640	145,860	157,080	168,300	179,520
134,640	149,600	164,560	179,520	194,480	209,440	224,400	239,360
168,300	187,000	205,700	224,400	243,100	261,800	280,500	299,200
201,960	224,400	246,840	269,280	291,720	314,160	336,600	359,040
235,620	261,800	287,980	314,160	340,340	366,520	392,700	418,880
269,280	299,200	329,120	359,040	388,960	418,880	448,800	478,720
302,940	336,600	370,260	403,920	437,580	471,240	504,900	538,560
336,600	374,000	411,400	448,800	486,200	523,600	561,100	598,400
403,920	448,800	493,680	538,560	583,440	628,320	673,200	718,080
471,240	523,600	575,960	628,320	680,680	733,040	785,400	837,760
538,560	598,400	658,240	718,080	777,920	837,760	897,600	957,440
605,880	673,200	740,520	807,840	875,160	942,480	1,009,800	1,077,120
673,200	748,000	822,800	897,600	972,400	1,047,200	1,122,000	1,196,800
740,520	822,800	905,080	987,360	1,069,640	1,151,920	1,234,200	1,316,480
807,840	897,600	987,360	1,077,120	1,166,880	1,256,640	1,346,400	
875,160	972,400	1,069,640	1,166,880	1,264,120	1,361,360		
942,480	1,047,200	1,151,920	1,256,640	1,361,360			
1,009,800	1,122,000	1,234,200	1,346,400				
1,077,120	1,196,800	1,316,480					
1,144,480	1,271,600	1,398,760					
1,211,760	1,346,400						

Work the stuck assembly in a steady, methodical manner. Each working procedure should be tried for at least 30 minutes to one hour before another method is used. In many cases, stuck assemblies have been worked for a long period before releasing.

Work Up

Working up is one of the more important working methods. Procedures for working upward are broadly similar to those for working to establish movement, with a few specific differences. Always maintain the stuck point in tension. This ensures a constant upward force against the point where the assembly is stuck. Often use a constant minimum upward force in the range of 5,000 to 10,000 lb of overpull. Combine torquing with working up in many cases such as working up to release a tension packer or wedged bit.

Start working upward at a force level of about 10–20% of the weight of the the pipe, calculated to the free point. For openhole stuck situations such as sticking in a crooked hole or pulling into a bridge, work at this force level for about 30 minutes to 1 hour. Then increase the force level by about 10–20% of the free-pipe weight and continue working. Repeat this, periodically increasing the working force level until it is at the maximum allowable strength of the workstring. In the normal case, work at the maximum force level for at least four hours, subject to hole and workstring condition. Then either change the working procedure, such as alternating with working down, or take another course of action.

In some sticking situations, increase the force level at a faster rate, in the order of 10 min intervals. Some of these include a bit stuck in a tapered hole, wall sticking, wash pipe stuck on bottom, stuck openhole test tools, and stuck retrievable hook-wall packers and plugs. However, continue working for at least four hours after reaching the maximum force level.

When working with jars, increase the tripping level correspondingly, using the same jarring procedures as described. When jarring, it may be necessary to set down a small amount of weight in order to cock or reset certain types of jars. Set the weight down slowly, avoiding heavy set-down weights or bumping-down action.

As noted, it is important to maintain the assembly in tension at the stuck point. The main reason for this is that the fish may be moved upward only a very small amount, such as a fraction of an inch. If weight is set on the fish, then it may be pushed back down so all of the benefits gained working the assembly upward to that point are lost. This occurs most commonly when sticking is caused by friction between two surfaces, sometimes tapered, such as a bit wedged in a tapered hole or stuck packer slips.

This can be illustrated by working a hook-wall packer stuck in a cased hole. The seated packer is prevented from traveling downward when the slips engage the casing wall. The slip segments are forced outward by the wedging action of a taper on the packer body. Weight on the packer body by the running assembly increases the wedging action. To release the packer, pull the taper upward so the slip segments retract (Fig. 4–9). Normally, if the taper is moved upward a short distance, the packer will begin to release. It will release completely when the taper moves upward a distance equal to the slip retainer travel.

The significant point is that the wedging taper must move upward a short distance to disengage the slips before the packer is released. This is done by working the assembly in the upward direction. If the taper and the mating face on the back of the slip segments are stuck together, they must be jarred and pulled apart to release the packer. If the assembly has been worked upward for a period of time so that the taper is almost ready to release, and then weight is applied in the downward direction, this can wedge the taper against the slip segments again. This can require working the assembly upward for another long period to release the packer.

Figure 4–9 Working stuck packers
Note: Packers must be jayed while working.

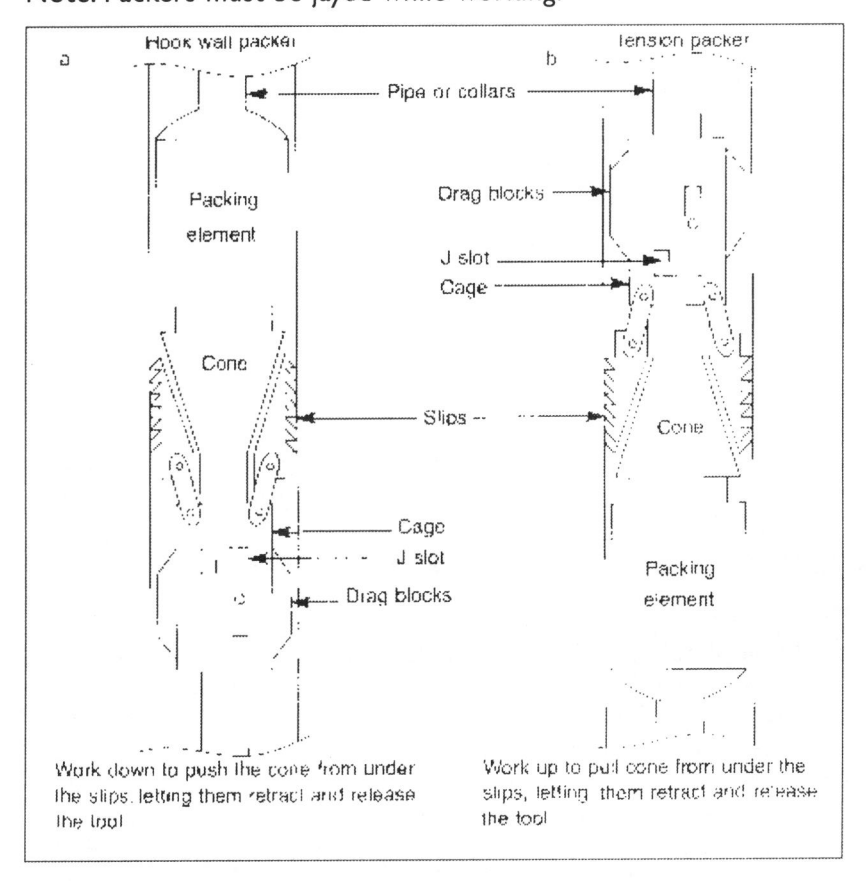

When the assembly is worked upward, the process should be continued for at least one and preferably two or more hours before trying a different method. If it is still stuck, a good alternative is to work it upward for two hours and then work it either downward or to obtain movement for one hour. Sometimes alternate by working up for two hours and working down for one hour. However, it usually is best to spend more time working the assembly upward, especially if the sticking situation indicated the need to work upward.

Work Down

It often is necessary to work the assembly down, such as the initial working procedure when sticking occurs while moving up. Some reasons for working down include an assembly pulled into a keyseat or into a caving or sloughing zone, releasing tension packers, and an alternate procedure when working up is unsuccessful. Combine torquing with working down in many cases, such as working down to release keyseated tools.

Working stuck pipe down often is an effective way to release stuck tools, especially since two of the most common causes of stuck assemblies are keyseats and caving. In both cases, usually the best releasing procedure is working downward. Always maintain the stuck point in compression when working stuck tools downward. Do not pick up more than the pipe weight to the free point; preferably stay about 10,000 lb below it. The reasoning is the opposite of maintaining the stuck point in tension when working upward.

The working procedure is similar to working up except all forces are applied in the downward direction. Use similar force levels, increasing them periodically at similar time periods, again with the forces directed downward. Working forces may be increased at a faster rate when working some stuck tools, such as tension packers. Use impact-type loading if regular set-down weight is unsuccessful. Bumper subs increase the working efficiency, but caution must be exercised to prevent pipe damage.

If the pipe was worked by another method, such as working up, then less time may be required working the pipe at a lower force when starting to work down. Working force and torque may be increased more rapidly to the safe operating maximum.

Final Working Procedure

Final working procedures generally apply when other working procedures have been unsuccessful. First, insure that the fish has been worked at the maximum force level for a sustained period in the direction normally required to release it, such as up for a stuck bit and down for a tension packer.

Then reverse working directions. In this case, work in the new direction for a longer period or a combination of longer and harder compared to working in the prior direction. This helps overcome the effects of working in the other direction. As an example, consider an assembly stuck in suspected caving while picking up after a connection. Assume no movement or circulation, even after attempting to move it and gain circulation; no circulation partially eliminates wall sticking. The assembly was worked down for several hours without releasing.

Next, work it up. In this case, initially working the assembly down could have wedged it. It is necessary to work it up for some time to overcome the additional sticking effect that may have been caused by working it down earlier. Then additional working up is needed to regain the effect of working it upward.

If this is unsuccessful, and jars and a bumper sub are in the assembly and working, then work it both up and down at maximum force levels for an extended period. Consider perforating to establish circulation, but continue working.

If the assembly does not have jars and a bumper sub, then consider backing off and running a fishing string with jars and a bumper for continued working in a more efficient manner. Generally do not consider other fishing operations until stuck tools have been worked extensively with jars and a bumper sub. Exceptions to this include deteriorating hole conditions and similar situations.

In the final case, work up at maximum force levels. Work constantly, alternating by working down about 10% of the time. Work constantly for an extended period, at least four hours or more. If this is unsuccessful, then take other actions such as backing off and running different fishing tools.

Continued working, including the final working procedure, is illustrated in the following example.

Stuck Packer and Final Working Procedure

A flowing oil well with 5½-in. casing was producing through a hook-wall production packer set at 6,000 ft on 2⅜-in., 4.60 lb/ft, EUE, N80 tubing, minimum tensile strength of 104,000 lb. Productivity had declined, and the well was to be placed on pump. A workover rig was moved in and rigged up. The packer circulating valve was opened and the hole circulated with 10 lb/gal salt water. The packer was jayed and found stuck when pulling started.

The mandrel was free to travel about 2½ ft, so it was easy to determine the free point of 22,000 lb, which was in agreement with the calculated free point. The tubing circulated freely, so the packer was stuck in the packer elements or slip-taper section; it did not have a hydraulic holddown. The packer had been in the hole for about three years under relatively light-duty conditions. The most likely cause of sticking was either in the packer elements, possibly due to swelling, or sticking slips in the slip-taper section. There was a possibility of junk and debris on top of the packer, but this was unlikely because of the free mandrel travel. Conditions indicated an initial procedure of working upward.

While rigging up and circulating, the packer was pulled to 37,000 lb, 15,000 lb overpull. The assembly was pulled to 47,000 lb, 25,000 lb overpull, in a slow constant manner without releasing. It was worked between 22,000–37,000 lb, and overpull of 0–15,000 lb for one hour, and the pull increased to 47,000 lb, 25,000-lb overpull for another hour without releasing. The packer was worked with right-hand torque at different times but was ineffective. Since the packer jayed in the right-hand direction, this would leave the packer in a position to release if it could be moved.

At this point the packer had been worked about four hours. Equipment to back off the tubing and run a fishing assembly was ordered, while work continued on the packer until the equipment arrived on location,

Fishing Procedures

due in about eight hours. The plan was to work the assembly up and down for about one hour to establish movement, followed by two hours concentrated on working up, and then repeating the procedure. It was worked in this manner for another four hours.

The tubing string was not equipped with jars or a bumper sub. The only way the tubing could be jarred and bumped was through the 2½ ft of free mandrel travel. In view of these conditions, the working force level was increased gradually to 82,00 lb, 60,000 lb overpull, or about 80% minimum yield. The setdown weight was limited by the weight of the tubing, and the tubing was bumped downward by slacking off at a moderately fast rate and setting the full weight of the tubing on the packer. Right-hand torque also was used intermittently. All of these actions were unsuccessful.

Then the pipe was worked up steadily. After working about 2 hours, about 8 hours' total working time, the packer began to move very slowly. It was worked upward about 2 ft in the next hour and about 10 ft in 30 minutes when upward movement stopped. The packer was worked about 1 hour when it released and pulled freely.

The packer slips probably were in a casing collar when it first stopped its upward movement. Then the slips retracted and the tool was pulled without difficulty. The recovered packer was in surprisingly good condition considering the working time. The rubbers were in almost-new condition, no indication of swelling, and definitely not the cause of sticking. The packer body was unmarked, almost eliminating junk such as a bolt or tong die on top of it. The taper on the packer body and the mating face on the slip segments had a few striations and small indentations about $1/64$ in deep, such as small pieces of hard metal crushing between the two matching faces. Apparently this froze or locked the taper and mating face on the slip segments, preventing the slips from retracting, and causing the stuck packer.

The packer probably could have been released in a shorter working time with a conventional fishing assembly. However, it was recovered in an efficient manner, and the overall time spent was less than would have occurred using the conventional back off procedure.

There are many cases of extended working releasing stuck tools in both open and cased holes, and the procedure is highly recommended.

Working Wirelines

Work wirelines in a similar manner as that for tubulars except that working options are limited to reciprocating action. Wirelines without a wireline bumper sub or jars are worked by stretching and slacking off the line. As with tubulars, the working action is more effective with a bumper sub or jars so that they can be jarred up and bumped down. Weight bars increase their efficiency.

Working generally starts with steady upward pulls followed by slacking off. The usual manner is to begin working the stuck fish at a lower force level and gradually increase to the maximum safe operating limit of the wireline. If the fish cannot be worked free, then release it and try other fishing methods. Shear loose if a shearpin release is used. Most rope sockets are crippled, so pull the wireline out of the rope socket. Wireline cutters are also available. Then catch the fish with a wireline overshot run under wireline jars and a bumper sub, preferably hydraulic, with heavyweight sinker bars. Work to release the stuck tools. If wireline jars and a bumper were run initially, work and release the stuck tools. If the stuck tools are not released, then release from the fish and fish by other methods.

Electrical conduits have a limited resistance to working. Jars and bumper subs cannot be run below them. In this case, working is limited to an upward pull and a rapid slack-off. The most common procedure is to work the stuck fish lightly. If it is not released, then the wireline is released by pulling it out of the crippled rope socket. Then the fish is recovered with wireline or tubular fishing assembly.

One method of working wirelines is bumping up or down by manipulating the wireline in a manner similar to the snap and bounce procedure for tubulars. First determine the neutral point and mark it with a flag on the line. Pull the wireline upward, then slack it off rapidly, and catch it sharply. The elasticity of the line lets the tools continue traveling downward for a short distance and then snap back upward. This action jerks the stuck tools either up or down, depending on where the line flag is stopped in relation to the original reference point. The action is very efficient at shallow depths of 6,000–10,000 ft and efficiency decreases with increasing depth because of the difficulty of catching and releasing the line correctly.

PARTING THE FISH

Frequently, the downhole assembly cannot be released by working, spotting releasing fluids, or similar procedures. The next step is to part the stuck assembly so other actions can be taken. The best method of parting permits catching the fish in the easiest manner, allows using the fishing tool that will make the strongest connection, and can be performed in the least time with the least risk.

This is the actual start of the fishing job. Plan the general fishing procedure before parting the assembly. The plan should include where, why, and how to part the assembly. Often the main reason for parting downhole equipment is to run a regular fishing assembly for more efficient working action.

Before parting the assembly, consider all alternatives to recover it intact. First and foremost, has the stuck string been worked sufficiently and correctly? A parted assembly guarantees a fishing job. If it can be released by additional working under safe conditions, do it and possibly prevent fishing. Before parting the pipe, always consider other alternatives, as described in the section "When and How to Quit" in Chapter 2.

Where to Part

Normally part stuck downhole assemblies as deep as possible to save as much hole as possible if the fishing job is unsuccessful, to recover the maximum amount of pipe, and to leave a minimum amount of fish in the hole (Fig. 4–10). This also allows positioning jars and other tools near the stuck point where they are more efficient. Excluding this, part the fish so that the top is accessible and can be caught by the catch tool.

Often it is preferable to cut the fish in a section with a smaller diameter, such as the pipe body. The smaller diameter may be easier to catch, although there are exceptions. Often, it is easier to cut. Overshots used to catch smaller diameters generally are stronger than those used to catch larger diameter fish.

The position of the top of the fish can be important in catching it. Whenever possible, part the fish in a straight, in-gauge hole section where there is less risk of bypassing the fish. It is more difficult to catch a fish top that is against the wall of a washed out or out-of-gauge hole section. Do not part immediately below a casing shoe for similar reasons. Part the pipe either up in the casing or 50–100 ft below the bottom of the casing shoe.

As a general rule, part stuck drillpipe at least one or two joints above the stuck point. Sometimes it is difficult to start outside catchtools or washpipe over a fish top parted at, or immediately above the stuck point. Normally part drillpipe in the open hole leaving 1 joint, 90% free in tension above the stuck point, or 2 joints, 75% free in tension above the stuck point. Part drill collars in the open hole leaving about one free collar above the stuck point.

Back off stuck test tools, packers, and plugs at the backoff joint, if available, in both cased and open holes. Pull off at the shear joint of a stuck tool so equipped. If necessary, cut the pipe about 2–4 ft above the top of tools fitted with fishing necks. Cut the pipe 8–15 ft above tools that do not have fishing necks. This leaves space to cut another clean fish top if the old top becomes worn or damaged. If the pipe above these tools is stuck, leave free pipe as described above.

Figure 4-10 Where to part the assembly

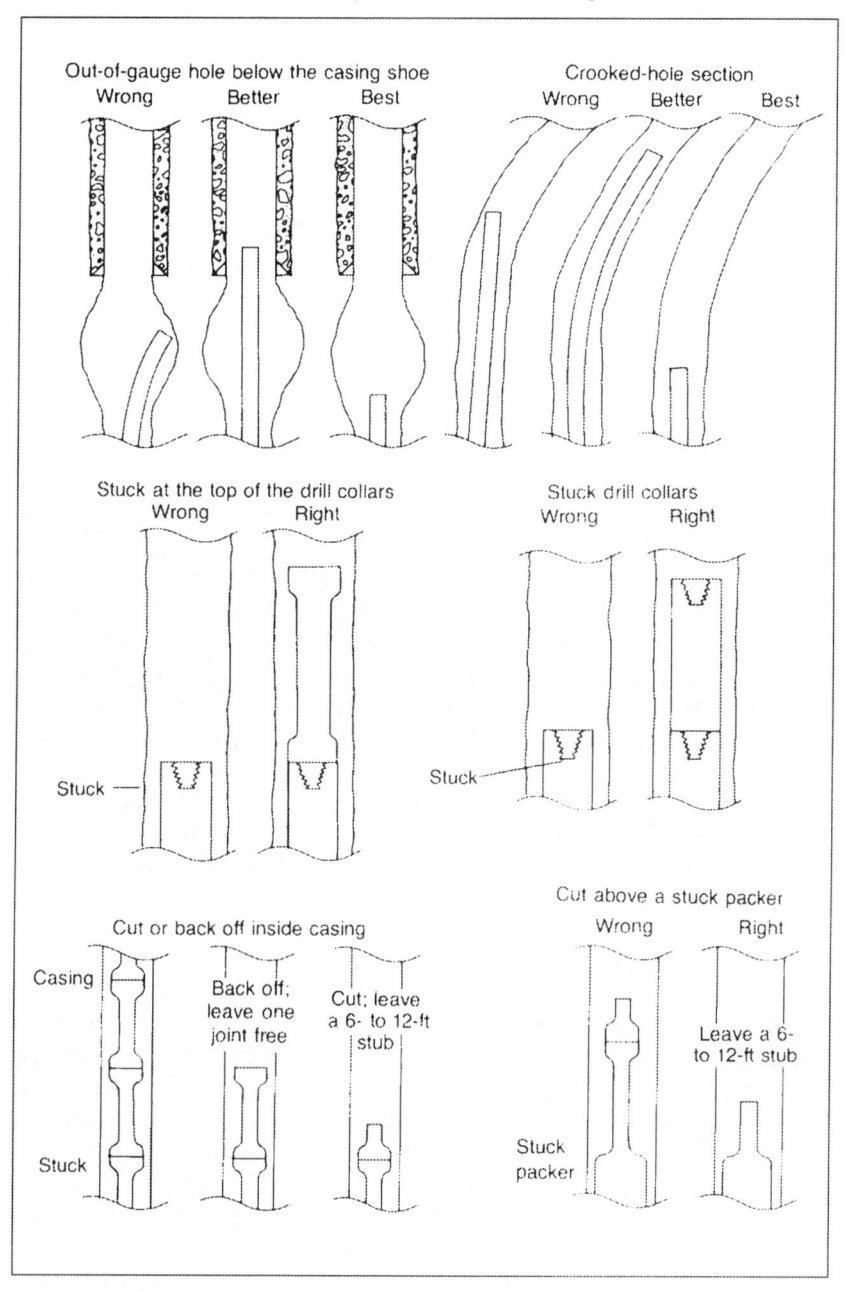

Fishing Procedures

Pipe stuck by sand inside of casing is a special and difficult fishing job. Work longer to free it. It may have to be released by perforating to establish circulation and wash out the sand. If unsuccessful, back off or cut, leaving two or more free joints for washing or milling over.

Backing Off

Backing off is the most common and preferred method of parting stuck assemblies. This usually leaves three options of catching the fish listed below in the normal order of preference.

- Screwing in — very strong catch
- Overshot — strong catch
- Spear — moderate-strength catch

Regular Back-off

The regular backoff using a string shot is the most common back-off procedure. It is very effective on shouldered tool joints. The impact of the explosion jars the mating tool joint shoulders loose so they can be rotated to the left and separated. The back-off procedure also works well with tubing but is less successful with casing. This is attributed to the type of connection, difficultly in placing left-hand torque, and the wide distribution of the detonating force. Using larger charges could be more effective, but there is a higher risk of damaging the casing.

In the ideal case, the tool joint being backed off should be completely free. This is not always the case, but tool joints have been backed off that were less than 50% free. This also may account for the fact that sometimes the assembly will back off from one to three tool joints above the point where the string shot was detonated.

Run a back-off assembly on a shielded electric conduit. It includes a color locator, free-point indicator, and the string shot, usually run on a small-diameter cable or pigtail 10 to 15 ft below the collar locator. This prevents damaging electronics in the collar locator and could prevent sticking the string shot tool, which sometimes happens when the string shot is detonated. In some cases the pigtail may stick when it is run through the top of the fish because the pigtail is too flexible. In this case, run the string shot on a small rod below the collar locator and bars. If the top of the fish is burred or uneven, there may be a problem running this string shot into the fish. One alternative is to use a smaller diameter, longer string shot with fewer wraps of primer chord. Almost the same amount of jarring force can be obtained. It reduces the diameter of the string shot and the risk of sticking or hanging it up during running.

The back-off operation includes running the back-off assembly and locating the free point. Usually determine the free point by changing tension or torsion in the pipe (Fig. 4–11). Place left-hand torque in the pipe and work it down. This is more difficult in crooked, deviated and deeper holes due to high drag and torque. However, the torque must be at the tool joint that is to be backed off to be effective.

The amount of left-hand torque depends upon pipe size and depth and strength of the back-off charge. It is available from standard tables or the company supplying the back-off service. Then locate the next tool joint or collar above the free point with the collar locator (Fig. 4–12). Position the string shot opposite the tool joint to be backed off, detonate it, and pull the string shot assembly carefully out of the hole.

Pull the string shot out of the hole, then rotate the pipe to the left to unscrew the connection, and pull the pipe out of the hole. If there is any tendency for the string shot tool to hang up, then unscrew the pipe and pick it up about 15 ft. Pull the string shot slowly into the upper pipe section and out of the hole. Picking the pipe up about 15 ft leaves clearance for the string-shot tool to enter the upper pipe section if the tool is in the fish or lower open hole. Sometimes the pipe unscrews and releases the fish when the string-shot is detonated. Pull the pipe up about 15 ft when this occurs. Then pull the string-shot assembly carefully, followed by pulling the pipe. Sometimes when the proper equipment is installed, pick up the pipe about 15 ft to allow for circulation before pulling the string shot.

In any case, do not set the pipe back down with the string shot in the hole. If the pipe parts several joints above the backed-off point and is picked up and later lowered, the end of the pipe could sit on top of the fish or go down the side of the fish. The string shot running line would probably be cut in either case, leaving another fish in the hole (Fig. 4–13).

It is not uncommon to back off uphole while trying to work left-hand torque down the hole. This often occurs in crooked and highly deviated holes with high drag and torque and after working stuck pipe at higher force levels. The same tool joint may back off again while placing left-hand torque, even after reconnecting and tightening the assembly. Therefore, prior to backing off, it often is advisable to tighten the drillstring using the procedures described in the section "Blind Backoff." After tightening all tool joints, work the extra right-hand torque out by working the pipe before torquing the assembly to the left for the back-off.

Fishing Procedures

Figure 4-11 Freepointing (courtesy Dia-Log)

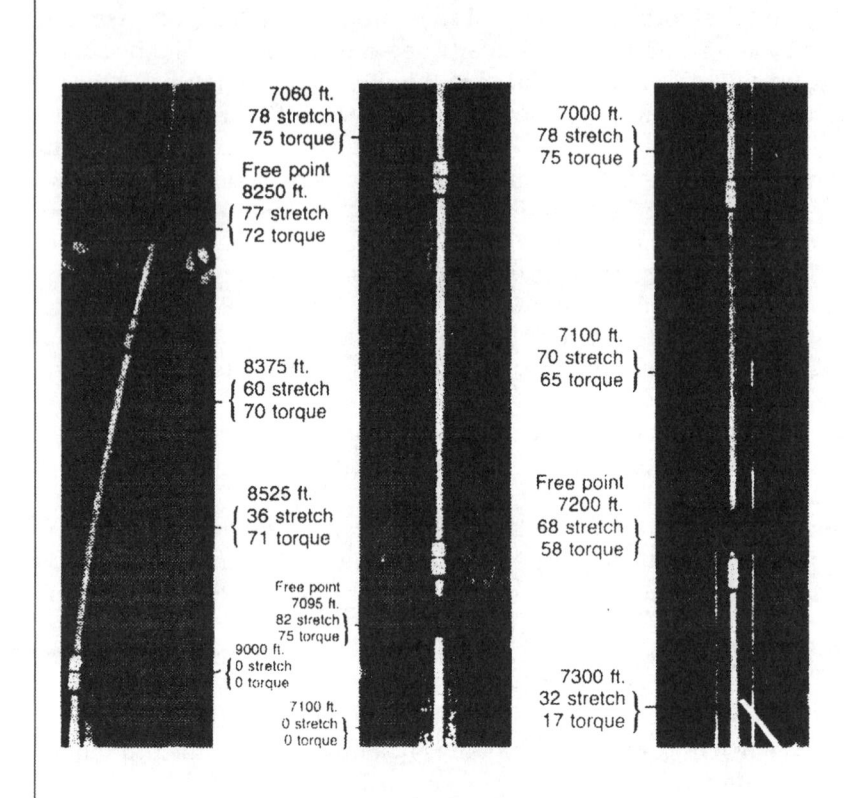

```
                7060 ft.                                    7000 ft.
                78 stretch ⎱                                78 stretch ⎱
                75 torque  ⎰                                75 torque  ⎰
                Free point
                8250 ft.
                ⎰ 77 stretch
                ⎱ 72 torque


                                                            7100 ft.
                                                            70 stretch ⎱
                                                            65 torque  ⎰
                8375 ft.
                ⎰ 60 stretch
                ⎱ 70 torque


                8525 ft.                                    Free point
                ⎰ 36 stretch                                7200 ft.
                ⎱ 71 torque                                 68 stretch ⎱
                                                            58 torque  ⎰

                Free point
                7095 ft.
                82 stretch ⎱
                75 torque  ⎰
         9000 ft.
         ⎰ 0 stretch                                        7300 ft.
         ⎱ 0 torque                                         32 stretch ⎱
                                                            17 torque  ⎰
                7100 ft.
                0 stretch ⎱
                0 torque  ⎰
```

Extremely Crooked Hole	Pipe Cemented or "Sanded Up" in Hole	Mud Stuck Pipe
Normal stretch and torque readings for this string of drill pipe were established at 77 stretch and 72 torque. It should be noted that at 8375 feet and 8525 feet, torque movement was relatively normal but a decrease in stretch movement had occurred. Free Point depth was determined at 8250 feet where pipe was free both in stretch and torque.	Normal stretch and torque readings for this string were established at 78 stretch and 75 torque. Free Point Indicator readings found the stretch and torque movement decreased abruptly to 0 at 7100 feet, thereby establishing the Free Point depth at 7095 feet.	Normal stretch and torque readings for this tubing were established at 78 stretch and 75 torque. Due to congealed mud in the annulus, Free Point Indicator readings gradually decreased with well depth. Long experience has shown under these conditions the tubing can be Backed Off at reduced torque and stretch readings with the Free Point depth being established at 7200 feet.

6000 ft.
{ 78 stretch
  68 torque

6250 ft.
{ 77 stretch
  70 torque

Free point
{ 79 stretch
  69 torque

7000 ft.
{ 0 stretch
  0 torque

7100 ft.
{ 0 stretch
  0 torque

6450 ft.
{ 78 stretch
  75 torque

6510 ft.
{ 76 stretch
  75 torque

Free point
6625 ft.
{ 77 stretch
  75 torque

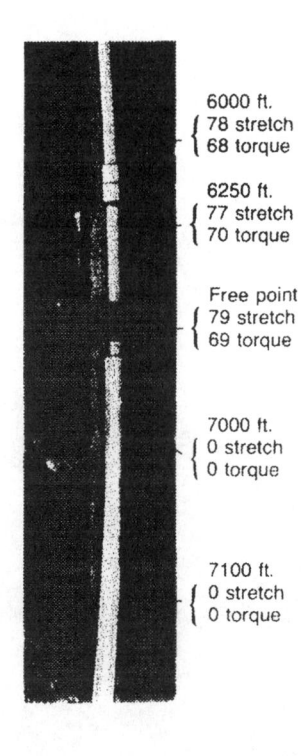

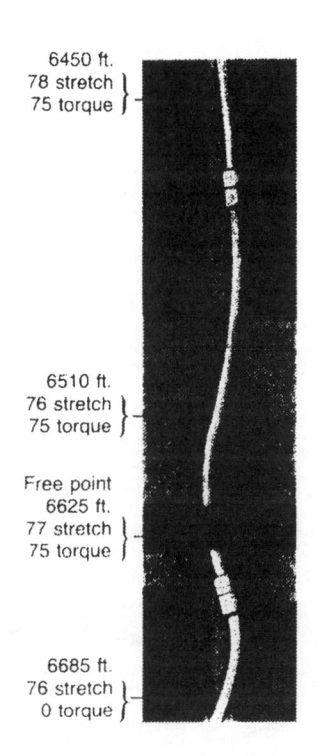

6685 ft.
{ 76 stretch
  0 torque

Drill Collars Differential Wall Stuck

Normal stretch and torque readings for this drill pipe were established at 79 stretch, 69 torque. Free Point Indicator readings at the top of the drill collars were 0 stretch and 0 torque, thereby establishing the Free Point depth at 6995 feet, top of the collars.

Extremely Crooked Drill Pipe

Normal stretch and torque readings for this drill pipe were established at 77 stretch and 75 torque. Due to the crooked pipe, torque movement was highly restricted below 6625 feet with no torque movement at 6685 feet although stretch movement was normal. Free Point depth was determined at 6625 feet where torque movement was sufficient to accomplish String Shot Back-Off.

Fishing Procedures

Figure 4–12 Back-off string shot

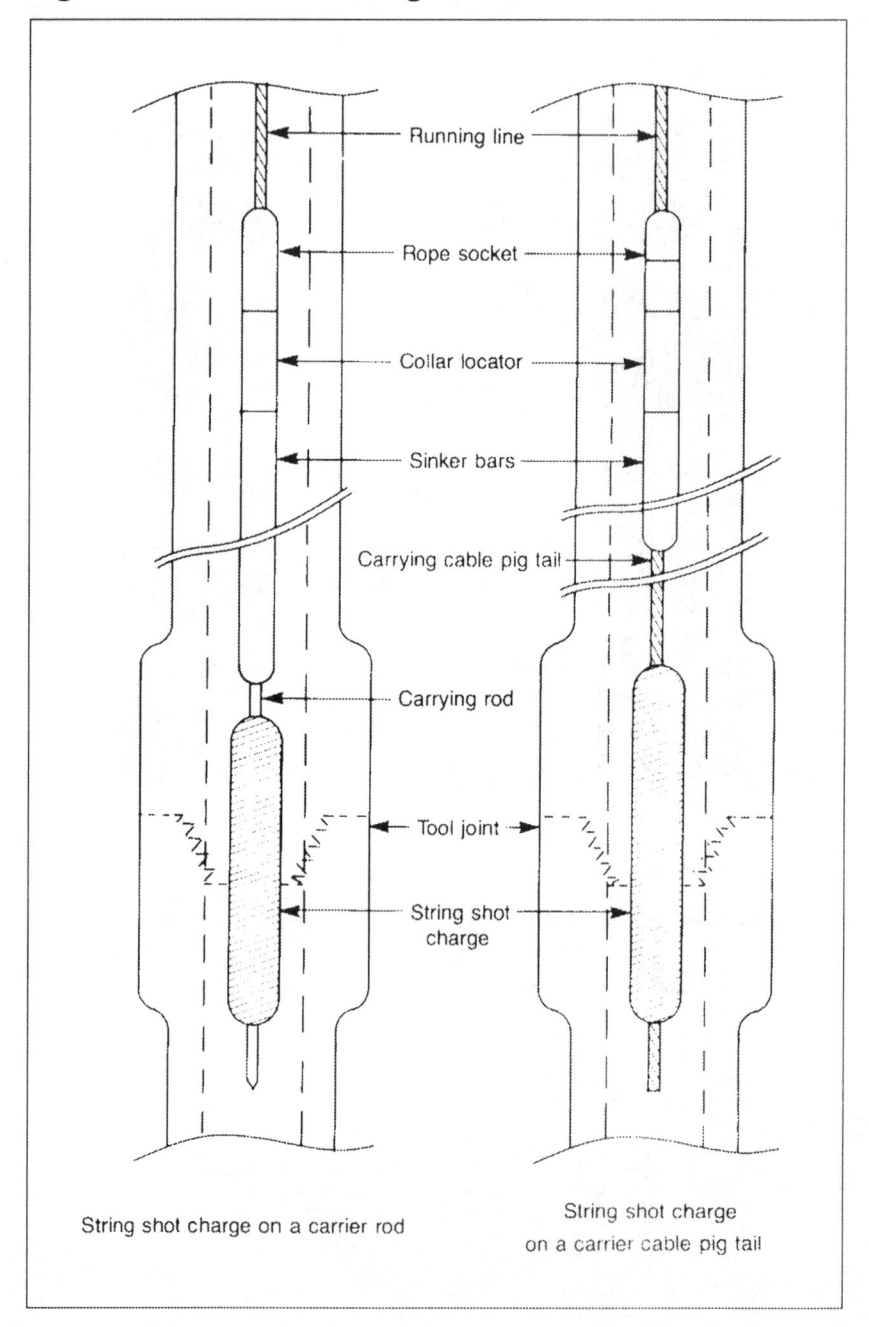

String shot charge on a carrier rod

String shot charge on a carrier cable pig tail

276 Chapter 4

Figure 4–13 Recovering a string shot
Note: After back-off, if tool is stuck, it may be caught as in figure a. Pick up pipe about 15 ft as in figure b so that the tool is free to enter it. Pipe back-offs uphole can occur before or after detonating the string shot and frequently occur while placing left-hand torque. Pull the string shot tool from the hole before lowering the pipe. If the pipe is lowered, it can sit on the box as in figure e or bypass the box as in figure f. The running line will probably be cut in either case, leaving another fish in the hole (inside pipe).

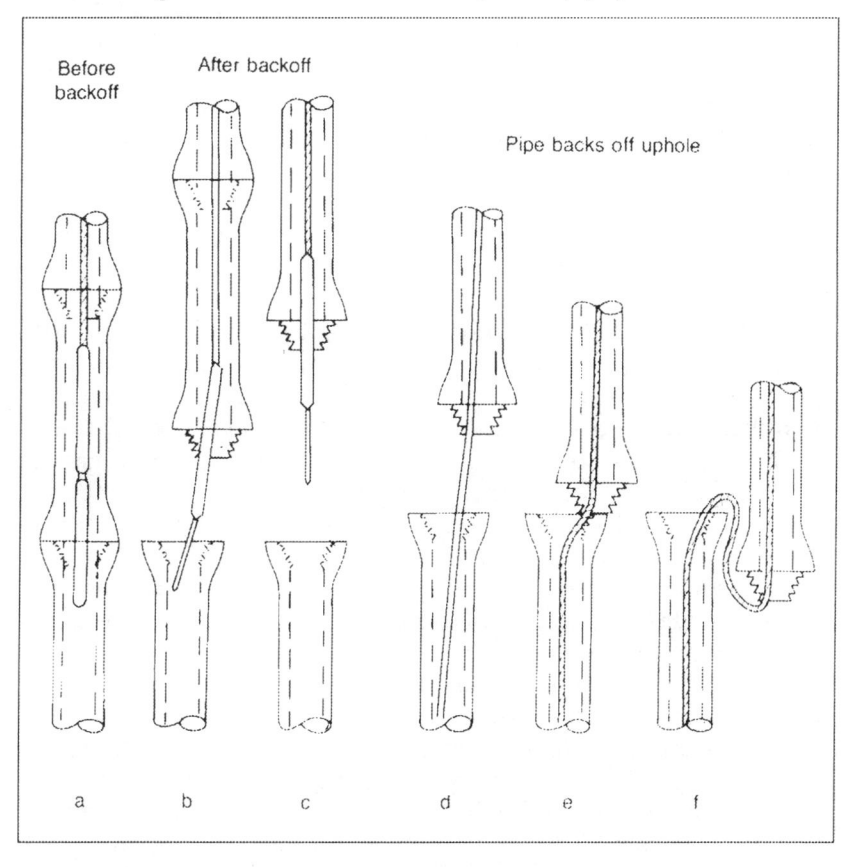

Infrequently, the assembly may rotate while tightening the tool joints. If this happens, rotate the assembly to the right and work it. If it rotates, there is a good chance of releasing it.

String shots may be effective in releasing tools stuck in casing. In one case, a mill stuck while cleaning inside the top of a liner. It was worked several hours without being released. A string shot was detonated opposite the mill but did not jar it loose. Later, while torquing to the left preparatory to backing off, the mill released and was recovered.

Fishing Procedures

Outside Back-off

Back-off shots can be used on the outside of the tool joint or connection to be released, but are less effective. Nevertheless, they have been used externally to back off tubing, drillpipe, and casing sizes up to 5½-in. The overall procedure is like an inside back-off except the string shot is positioned on the outside of the tool joint or connection.

Outside back-offs are seldom used because of the many disadvantages. Charge distribution is not as effective because of the nonconcentric configuration, especially in larger tubulars. It is difficult to run and position the string shots. It usually is difficult to run into and down the annular space because of space limitations, the string shot will not fall freely, and there is a high risk of sticking the tool. It usually cannot be sealed at the surface because of the nonconcentric configuration. There also is a risk of collapsing the casing. The outside back-off generally is restricted to shallow depths and low pressure conditions.

Blind Back-off

The blind back-off is a procedure for backing off the pipe without a string shot and by manipulating the pipe from the surface. The procedure is applicable to drillpipe and tubing and has been used with sucker rods. It is seldom used except in special circumstances.

There are occasions when the conventional methods of parting the downhole assembly cannot be used. The most common situation is when the inside of the pipe is plugged and an outside cutter cannot be run because of the lack of clearance. Common situations include plugged drillpipe or tubing, and annular space too small to run an outside cutter. Blind back-offs are moderately effective under favorable conditions and when applied correctly.

Basically the assembly is rotated to the left until it unscrews and releases. Normally the assembly would be expected to back off at any point, or at least high in the hole. However, if the technique is applied correctly, the pipe can probably be parted within a few tool joints or connections of the desired point. The procedure takes advantage of the fact that a tool joint in either tension or compression is more difficult to rotate as compared to one at the neutral point.

First, find the free point with conventional pipe-stretch data. Then tighten the tool joints to ensure that all the connections are tight. Do this by applying maximum right-hand torque while gradually moving the free point of the pipe from the surface, or high in the hole, downward. This assumes obtaining optimum tightening with the tool joint at the free-point position. Later back-off operations are conducted with all joints in tension, except those in the section to be backed off. The tightening procedure IS VERY IMPORTANT, so take time and do it correctly. Several hours may be required for a moderate depth workstring.

Start tightening the tool joints by setting down the maximum safe set-down weight. In cased holes this may be the full weight of the pipe to the free point and should approach that for open holes. To some extent, do not worry about damaging the lower pipe. If the back-off is not successful, the pipe will probably require milling. Apply right-hand torque to the maximum safe limit. Normally apply and hold this torque with pipe tongs. Pull the pipe upward about 10,000 lb, or about 10–15% of the initial free pipe weight, over the initial set-down pipe weight. Repeat this five or ten times. If the torque reduces, indicating the joints are tightening, then restore it to the maximum level, and repeat the working procedure until there is no indication of additional tightening of the tool joints.

Next pull the pipe upward an additional 10,000 lb, or another 10–15% of the initial free-pipe weight, and repeat the working procedure, replacing lost torque as necessary. It may be necessary to change the position of the pipe tong to allow for picking the pipe up higher in successive working. Use another tong such as an inverted back-up tong. It is best not to release the torque until all the pipestring has been tightened.

Continue the working procedure, increasing the pipe pick-up weight in similar increments and restoring torque as needed until all of the tool joints in the pipestring have been tightened. This is at a pick-up weight of about 10,000 lb over that necessary to move the free point to the desired back-off point. Then release the right-hand torque and work the pipe, with the hook unlatched, to remove all excess right-hand torque. Work the pipe from the set-down weight to an overpull of at least 50,000 lb above the free-point weight.

Next, place and work left-hand torque down. The surface torque will decrease as the left-hand torque is worked down. Use this as a guide of how long to work the pipe before adding more torque. There is a risk of backing off high with associated pipe movement or jump. Tie the elevator latch securely closed so it will not snap open, allowing the pipe to drop.

Work left-hand torque down by picking up until there is about 10,000–20,000 lb of overpull at the point to be backed off. Turn the workstring about one turn to the left. Work the pipe from about 10,000 lb of overpull up to 30,000–50,000 lb for about 15 minutes, or longer if the hole is deviated or crooked. Add another turn of left-hand torque and continue working. Continue adding torque and working the pipe until about 1½ rounds of torque per 1,000 ft of depth, more for smaller tubulars, has been added. The actual number of rounds depends on pipe weight, depth, hole deviation, and dogleg. Keep the free point in tension while working the left-hand torque down.

Fishing Procedures

Start the back-off by continued working, gradually decreasing the overpull from about 30,000 lb of overpull down to 5,000 lb or less. In extreme cases it may be necessary to work the pipe at lower weights, down to setting about 5,000 lb of weight on the stuck point. If the pipe does not back off, add left-hand torque and work the pipe until it backs off. The pipe may back off too high, as indicated by the weight indicator. Then reconnect by screwing back in, and repeat the entire procedure, spending more time tightening, even including bumping, to ensure that the tool joints are fast.

This procedure works well with small, 3½-in. drillpipe tool joints and tubing connections. It may even damage some tubing connections. Larger tool joints such as for 4½-in. drillpipe may require bumping, especially if the pipe failed to back off correctly after the regular torquing procedure.

Bumping or jamming is a procedure for manipulating the pipe at the surface in order to tighten loose tool joints in the hole. Use it carefully and with caution. Torque the pipe to the right to or near the maximum and hold it with the tongs. Pull the handle of the tongs through an arch of about 100–130° with a tong or jerk line. Release the tongs rapidly so they swing backward and bump against a backup post or tong safety line. If a safety line is used to catch the tongs when bumping, install an extra line as a safety measure. The bumping action coupled with the twisting momentum of the assembly imparts a sharp right-hand torque that further tightens the tool joints. Repeat the procedure multiple times and at different weights as necessary to increase right-hand torque. Do not rush the procedure, and ensure that all tool joints are tight. Bumping may not jam the tool joints as tightly as the regular connection procedure, but it is a good substitute under these conditions.

A rotary "jam" procedure is similar but somewhat less effective. Torque the drillpipe tightly with the rotary, normally with slips set and handles tied for safety, and replace the "slip cut" joint later if necessary. Release the torque rapidly in short steps, catching the pipe with the rotary clutch or brake. Repeat the procedure multiple times, often 15 to 30 or more, especially if the pipe tightens. Use MAXIMUM torque.

Sometimes a joint that is not located correctly will back off several times. Clean the lubricant out of the tool joint with mud by connecting and disconnecting it several times to increase the friction. Screw the joints back together lightly and circulate slowly for 5–10 minutes. If the pipe does not circulate, work the pipe lightly and pressurize from 300 to 500 psi; restore pressure as needed. Maintain pressure for 5–15 minutes to remove all pipe dope. Repeat one of the jamming procedures.

Another aid is "burring" the tool-joint threads slightly to increase the tightening effect at the tool joint. The procedure is subject to depth and drag. Rotate drillpipe slowly in the left-hand direction while setting a light weight of about 5,000 lbs on the tool joint. Rotate for 10 to 20 turns; then repeat one of the jamming procedures.

Cutting

In many situations the most practical method of parting the downhole assembly is to cut it. Select an applicable cutting tool as described in the section "Cutters and Perforators" in Chapter 3.

Chemical Cutter

Chemical cutters are preferred over mechanical cutters because they are faster to run, normally provide a positive cutting action, and often involve less risk. Use the correct standoff to cut efficiently. Chemical cutters normally are used to cut smaller tubulars such as tubing. They can cut larger tubulars such as drillpipe in special cases. The chemical cutter usually makes a clean cut, so the fish can be caught with an overshot without dressing off the top.

Observe the same precautions when running a chemical cutter as for all wireline tools. The tool has a relatively full-gauged diameter. Run a screen if necessary to prevent sticking the tool. Equalize the pressure between the inside and the outside of the pipe to be cut. Otherwise, the cutter may be blown uphole and tangle the running line, causing a fishing job. In a few cases it may be blown downhole, parting the running line.

Cutting Tubing with a Chemical Cutter

A submersible pump run on $2^{7}/_{8}$-in. tubing was pumping at 10,000 ft inside 7-in. casing. The pump failed and could not be circulated out of the hole. It was stuck and could not be recovered with a wireline overshot. The packer was stuck and could not be pulled. The tubing could not be backed off conveniently because of the electrical conduit clamped to the side of the tubing. A chemical cutter cut the tubing about 3 ft above the first connection above the pump. After pulling the tubing with the electrical conduit, the pump, cavity, and packer were recovered without difficulty by catching the tubing stub with an overshot run on tubing.

Tubing plugs frequently are run in packer mill-out extensions to isolate a lower zone during completion. Normally open the lower zone later by pulling the blanking plug. It may be difficult to pull due to sticking or sand accumulation on top of the plug. A tubing cutter can cut the tubing immediately above the plug to open the lower zone if perforating the extension is either not possible or undesirable. Ensure that the pressures are equalized before cutting.

In one case similar to that described above, the blanking plug was set immediately below the mill-out extension, which had a slightly larger inside diameter than the regular tubing. The chemical cutter could not cut the mill-out extension efficiently. The small inside diameter in the regular tubing prevented running a larger outside diameter cutter to give the correct standoff in the extension. Therefore it was necessary to kill the well and pull the packer. Prevent this problem by setting the blanking plug below a tubing nipple at least 6 ft long that has the same inside diameter as the tubing in the running string.

The chemical cutter also can be used in special cases to cut drillpipe in a fishing operation, as illustrated in the following example.

Cut Drillpipe with a Chemical Cutter

An on-structure development well had 9 5/8-in. surface casing set at 1,500 ft and was drilling in an 8 3/4-in. hole. The area was known for problem formations, lost circulation, and shallow gas flows. The hole caved, sticking the drilling assembly and shutting off circulation while making a connection at 9,000 ft.

The pipe was worked for 8 hours without obtaining any movement. A free-point indicator was run and the pipe was found to be free to 5,000 ft. The assembly was worked another 2 hours. During this period, an additional 1,000 ft of drillpipe became stuck, indicating continued caving. The assembly was backed off at 4,000 ft and pulled. The fish was caught and worked for about 12 hours, and another 2,000 ft of fish were worked free, backed off, and recovered. Three back-off shots were needed because it was difficult to work sufficient left-hand torque to the free point due to caving.

The fish was caught with a screw-in sub and worked free to about 7,000 ft by stretch measurements. A free-point indicator was run but could not enter the fish. The top was probably plugged with caving that occurred on the cleanout trip. The top of the fish could not be opened with weighted sinker bars and pressure surging. It was opened with a string shot, but circulation could not be established because of caving. The pipe was worked for another 2 hours. A free-point indicator found the fish 25–50% free above 7,000 ft and 25% free at 7,000 ft. The fish was perforated at 7,000 ft, and circulation was established slowly while working the pipe. It was worked free to 7,000 ft when it was necessary to shut down and repair surface flow lines.

After a 6-hour shutdown, the fish was worked and circulated. A free point indicated that the fish was 50% free in stretch at 7,000 ft and 25% free at 7,500 ft, no movement in torsion. The fish was worked an additional 16 hours and became 50% free in stretch at 8,000 ft, with no rotation and about 75% free in rotation at 7,000 ft.

A badly caving shale from 6,000 to 8,000 ft prevented backing off below about 7,000 ft because caving prevented working down left-hand torque. If the pipe could be cut deeper, there was a good chance of recovering that part of the fish through the caving shale section.

A chemical cutter was run. The chemical cutter would probably not make a complete cut through the pipe because of the incorrect stand-off, but a partial cut was expected. Several cuts were to be made if necessary. The cuts could not be positioned precisely over each other, but an effort would be made to position them as closely as possible. In this manner the pipe could be parted at the weak point by tension.

The first cut was made at 8,000 ft, and the pipe was worked for 2 hours without being released. A second cut was made with the tool positioned as near the first cut as possible. The pipe was then worked an additional 2 hours without being released. A third cut was made in a similar manner and, after the pipe was worked for 1 hour, movement was observed. About 2 hours were required to work the first 1,000 ft of pipe out of the caving hole section before it could be pulled freely. A cleanout trip was made, the mud and hole were conditioned, and the remainder of the fish was recovered in a conventional manner.

Using the chemical cutter in this case expedited the fishing operation.

Jet Cutter

Jet cutters are inside cutters run on a wireline. They cut with a specially shaped charge. They are positioned and used much like chemical cutters. Chemical cutters usually are preferred over jet cutters because they make a smoother cut and there is less risk of casing damage in a cased hole.

Mechanical Cutter

Mechanical cutting tools cut and release a section of tubulars and mill or cut out a section of casing. They are used most frequently as inside cutters to cut casing and as outside cutters to cut tubing.

Most tools are self-starting in that the knives, cutting bars, or blades are forced outward, away from the cutting tool body while it is being rotated, inward for an outside cutter. When these blades contact a metal surface, they begin cutting and continue until the blade is horizontal. Depending on the type of tool, it is then moved either up or down by moving the downhole assembly to continue the cutting action. Most cutters are designed so the pipe is cut completely when the cutter blades are horizontal.

Mechanical cutters are relatively simple to operate. One of the main problems is that the knives or blades may be easy to break and must be handled carefully. Another problem is the size of the tools. Close tolerances are often encountered, and this frequently limits use of these tools.

Explosives

Sometimes downhole assemblies are parted with explosives such as dynamite. This is most common when parting and pulling casing during abandonment operations. The explosive often damages one or two joints. However, any attempt to recover the remaining pipe below the shot point requires extensive milling and fishing.

Shear Rams

Shear rams can part the pipestring when the pipe must be dropped deliberately, usually under emergency conditions. They are installed as part of the regular blowout preventers.

Shear rams are seldom used except in emergency conditions. One example might be a situation where a well blew out during a trip. Further assume the drillpipe rams and bag-type preventer were either inoperative, could not hold pressure, or were leaking and could not be changed. If there were no way to pump into the well and control or kill it, the shear rams could be closed, cutting and dropping the pipe and sealing the hole.

If the preventer stack did not include shear rams, and the blind rams were operative, the elevator latch could be knocked open with a sledgehammer to drop the assembly so the blind rams could be closed.

Dropping the pipe is a drastic action with severe consequences. It might be justified under pending blowout conditions where there is no other method of shutting-in the well. Almost any condition or situation is better than an uncontrolled blowout.

CATCHING THE FISH

Catching the fish normally involves selecting the appropriate method and tool. Four basic methods of catching a fish are swallowing, screwing in, outside catch, and inside catch (Fig. 4–14). In most cases the same catch tools can be used in the open or cased hole. However, there is a higher risk of problems when catching the fish in open holes due to out-of-gauge or crooked holes.

Figure 4–14 Catching the fish

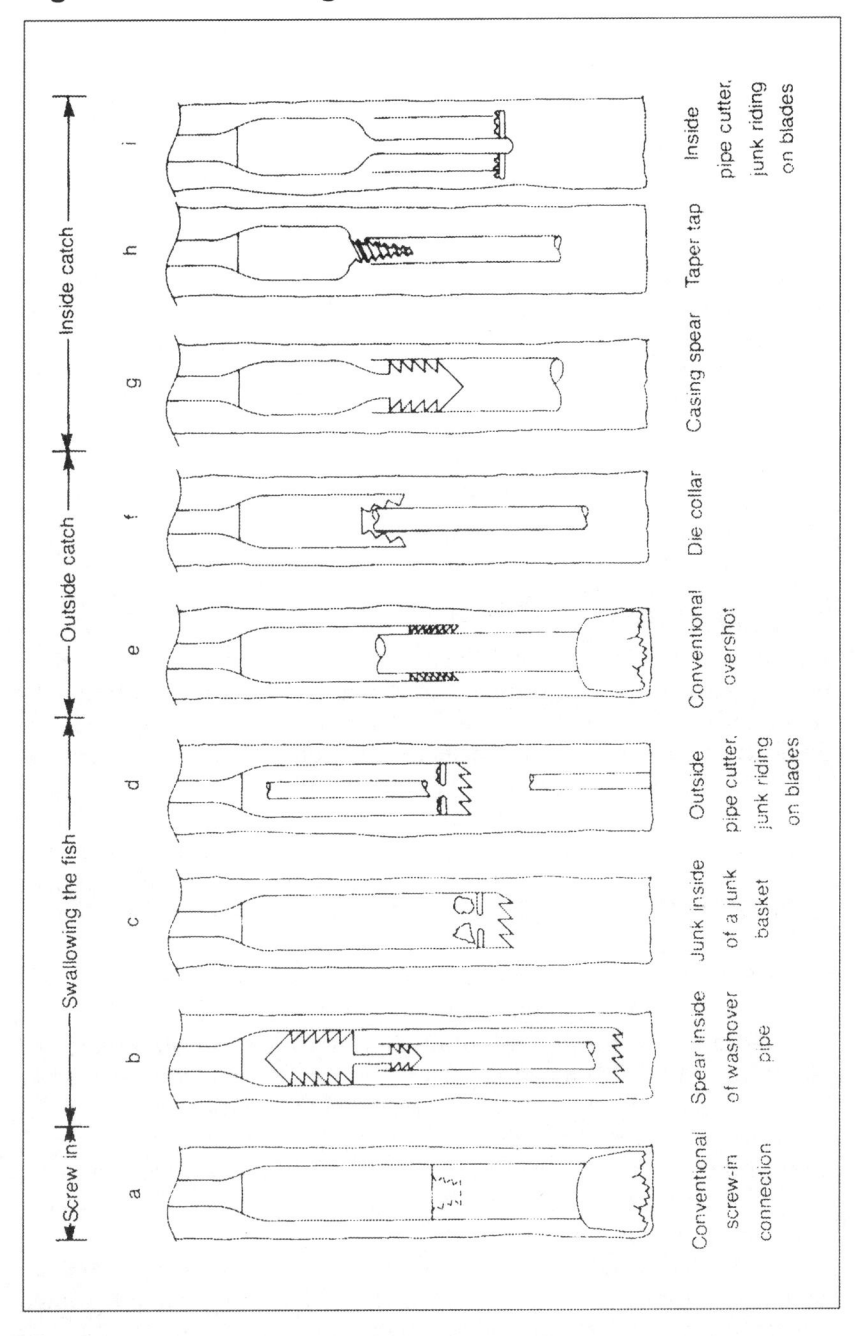

Fishing Procedures 285

If there is a choice of which method to use, the main criteria are ease of taking hold of the fish, strength of the connection to the fish, and tool strength. Always select the method that will give the strongest connection between the fish and the fishing string and fulfill the requirements of the particular job. This usually, but not always, is the strongest tool available. For example, a fish that can be caught with an overshot can often be caught with a die collar. Both tools are equally strong, but the overshot connection to the fish is much stronger than the die collar connection.

Have accurate measurements of the length of the fishing string and the depth to the top of the fish. This ensures going over the fish carefully and prevents running the tools down beside the fish for an excessive distance if the fish top is laid over in an enlarged hole section. This normally does not occur in cased holes except when the diameter of the fish and fishing assembly combined are smaller than the ID of the casing. Check the depth to the top of the fish by running an induction log in the open hole if there is any question about the measurement to the top of the fish.

Keep the top of the fish clean. Minimize rotation in the 30–45 ft interval above the fish. Try to circulate the pipe down the last 5–10 ft at a reduced rate. Tag the fish slowly with about 5,000 lbs, which is the average maximum weight used to catch it. If the fish is going to move, learn it now. Circulate 2 or 3 minutes, continue circulating while pulling up slowly about 30–45 ft above the fish, and then circulate normally. After circulating the hole clean, reduce the circulation rate, lower and tag the top of the fish lightly, and pull out of the hole.

Various tools may be added to the catching assembly. These are not used too often, except when there are direct indications that they are needed. They increase the number of connections and, like all tools with working parts, are subject to failure. Also insure that the mechanical motions to operate the tools are compatible with the motions expected to be needed to catch, work and release the fish.

A safety joint may be run immediately above the catch tool. Under optimum conditions, this permits releasing and catching the fish again as necessary. This can be an advantage when the fish must be released quickly due to conditions such as caving formations and high-pressure zones. Other reasons are to pull the workstring to replace the jars and bumper sub if they fail and to check the connections on the assembly.

A circulating sub may be run above the catch tool for circulating near the bottom of the fishing assembly. This is a special advantage if formations cave around the fishing string. Circulating fluids usually carry caving material out of the hole and prevent sticking the tools.

Cleaning inside the Fish

Cleaning inside the fish is a higher risk procedure. The fish must be open to conduct operations inside it. These include running a free point, backing off, and perforating for circulation (Fig. 4–15). Common causes of a plugged fish-top are drill cuttings, caving, or sloughing. These frequently bridge in the top of the fish. Prevent this by limiting cleanout procedures immediately above the fish. Long bridges can be very difficult to clean out.

Damaged Fish Top

Damaged fish-tops range from minor burring to severe damage, such as a twist-off. The condition may be indicated by the appearance of the break or parted section on the bottom of the recovered pipe. This may also indicate the type of catch to use. It may be verified with an impression block, but these often give questionable results.

Normally cut a clean fish-top if there is an obstruction inside caused by the condition of the metal. This often occurs if the top of the fish is slightly flattened or a small piece of the metal wall is rolled inward. If the top of the fish was originally cut or milled off correctly, it can be damaged by setting a bit or mill down with too much weight. It may be partially plugged by cleaning out too close to the top of the fish.

An overshot may pass over a partially damaged fish-top and catch it below the top. However, the top of the fish could be restricted and prevent running close-tolerance tools inside the fish. If this occurs, cut or mill a clean top so that tools can pass through the fish top.

Cleaning

Plugging in the top of the fish prevents running stuck pipe logs, back-off shots, perforating to establish circulation, and other releasing procedures. An important factor in cleaning out the top of the fish is its accessibility, such as the top of the fish laid over in a washed-out hole section.

The most common blockage that occurs in the top of the fish is a cuttings bridge. The thickness of the bridge inside the fish can range from a few inches to a long interval. Shorter bridges 1 to 3 ft long sometimes can be cleaned out, but longer bridges are more difficult. This emphasizes the importance of circulating the hole clean and using a minimum number of trips to prevent knocking additional material off the walls of the hole.

There are more options for cleaning out the bridge if the fish is caught with a full-opening catch. The bridge may be removed by surging or pumping it out with low pump pressure if the lower part

Figure 4–15 Cleaning out inside a fish
Note: Other methods that apply, subject to general well and operating conditions, include drilling and washing out with slim-hole tubing or washing out with coil tubing. Alternative procedures may include making an outside cut to remove the bridged section or mill off the top of the fish.

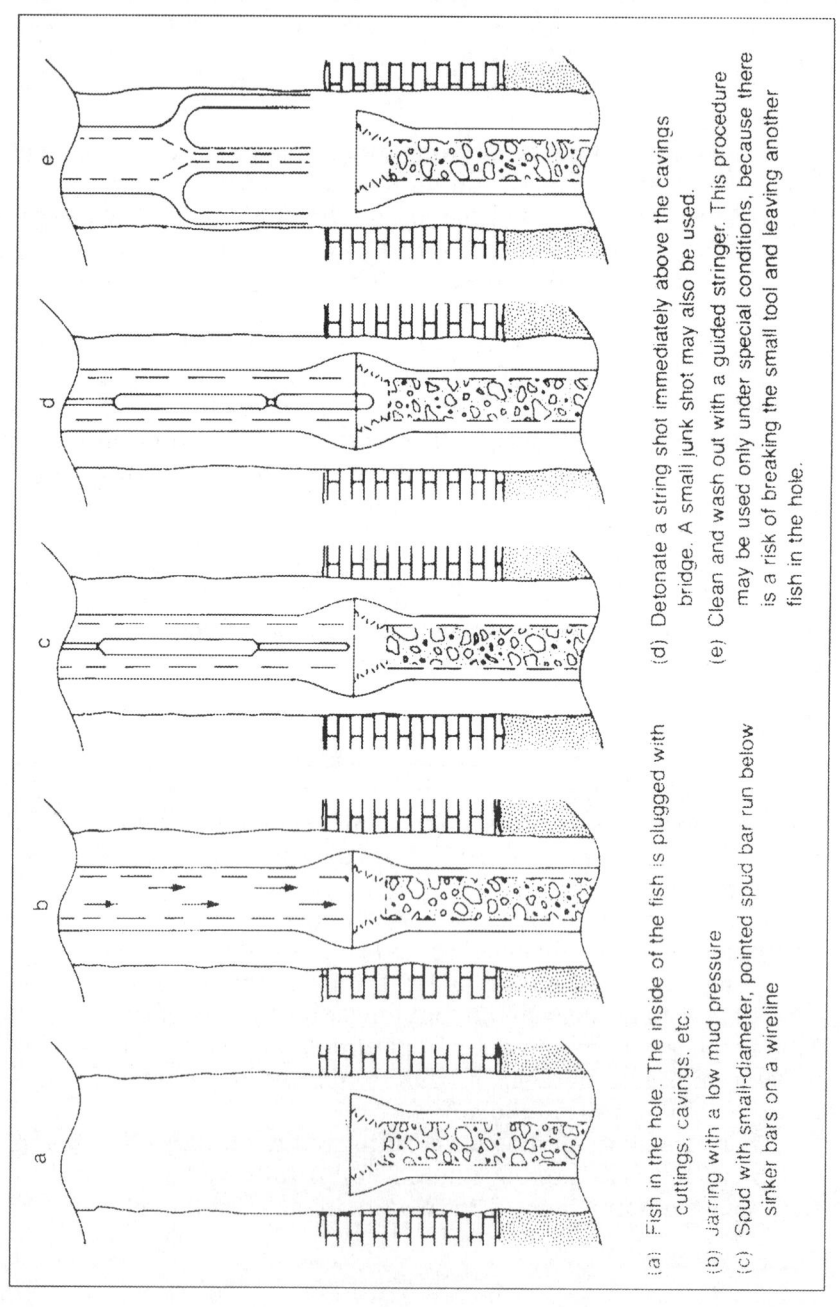

(a) Fish in the hole. The inside of the fish is plugged with cuttings, cavings, etc.
(b) Jarring with a low mud pressure
(c) Spud with small-diameter, pointed spud bar run below sinker bars on a wireline
(d) Detonate a string shot immediately above the cavings bridge. A small junk shot may also be used.
(e) Clean and wash out with a guided stringer. This procedure may be used only under special conditions, because there is a risk of breaking the small tool and leaving another fish in the hole.

Chapter 4

of the fish is open and free to circulate. Sometimes a combination of low pump pressure and jarring may reduce the stability of the bridge so the cuttings either fall, settle, or can be pumped out. The bridge may be removed by spudding with sinker bars run on a wireline. The bottom sinker bar should be pointed and as small in diameter as possible.

String shots may be used and are more effective with shorter bridges. A relatively large string shot or less commonly, a small junk shot can be used. If the fish is caught with an overshot, a smaller string shot must be used since the method is similar to the procedure sometimes used to release an overshot stuck on the fish. Maintain full tension on the overshot, and use a smaller string-shot detonated immediately above the bridge. With the overshot in full tension, the shot usually will not break or shatter the slips or grapples, causing the overshot to lose its hold on the fish. Note that plugged jet bit nozzles can be blown out with a string shot in a similar manner.

Other methods that apply, subject to general well and operating conditions, include drilling and washing out with small diameter tubing. Coiled tubing has been used but its use is limited because of the difficulty of running the tubing and the possibility of sticking.

Another method of cleaning the fish top is to run a short joint of washpipe with a stinger centered inside and connected for circulation. The washpipe guides over the fish, and the stinger enters the fish to wash out the bridge. The main disadvantage of this procedure is that the relatively weak, small-diameter stinger breaks easily.

One of the final alternatives for removing bridges is to cut off the top of the fish. Depending upon the condition of the top of the fish, it may be necessary to mill a smooth top so the cutters can be run over it without damaging them. Another difficulty is that the length of bridged section is unknown. It commonly is 10–20 ft long but can be much longer. There also are the normal risks and problems of making an outside cut. A major consideration is that the excised section may fall downhole and beside the first fish. A blind back-off may be applicable.

Screw In

Screwing in is the most common method of catching a fish. It provides a very strong connection and engages the fish relatively easily. Normally catch the fish by screwing in if it was released by backing off.

In operation, connect a tool joint by mating the one on top of the fish to a fishing assembly and run it into the hole. Stop about 25 ft above the fish and circulate and condition the hole. Then lower the fishing assembly slowly, usually with slow rotation and reduced circulation, thereby engaging the fish. There are several surface indications when the fish is engaged. The weight indicator shows a small weight loss as the tool sits on top of the fish. Increased torque indicates that the fish has been engaged. The pump pressure may increase, sometimes to a high pressure if the fish is plugged.

Use a notched screw-in sub if there is a problem catching the fish. Tighten the connection with a safe, maximum right-hand torque. In some cases the connection may be bumped. If the connection is made tightly and must be released later, normally back it off with a string-shot. Running a back-off sub may help in this case.

A correctly made screw-in connection has about the same strength as a regular connection and normally holds relatively high pressure. This allows using pressure to establish circulation through the fish and may help release it. After catching the fish and tightening the connection, test the catch by pulling 10,000–30,000 lb over the weight of the fishing string. Then begin working to recover the fish.

Swallowing

Swallowing the fish is a good method of catching and recovering small fish. The fishing tool passes over and completely encloses the fish. In this manner the fish is caught and retrieved with a minimum risk of losing it while tripping out.

The most common tool for swallowing the fish is a junk basket, usually employed for catching smaller fish. The outside cut can be another swallowing operation. Run the cutter down over the fish, cut it and recover the cut section inside the cutter barrel, held by the cutter knives. However, the cutter knives are relatively fragile and may break under a heavy load. Another type of swallowing tool is washover pipe fitted with an inside spear, as described in the section "Washing Over."

Outside Catch

The two main methods of making an outside catch are the die collar and overshot. Die collars are a specialized outside catch and seldom used because of their disadvantages, as described in Chapter 3.

Overshots are widely flexible. They can be dressed with slips, or spiral or basket grapples. All are commonly called slips or overshot slips. They are available in a wide range of sizes and lengths, and multiple tools can catch one of several sizes of fish in one run. They may be fitted with a packoff for pressurization, which may help control well pressures and be instrumental in releasing the fish. There are different types of shoes.

The overshot is the most common, widely used outside catch tool. Excluding slim-hole overshots, the standard overshot is approximately as strong as the equivalent tool joint strength for the screw-in connection. The fine thread connections in the upper overshot body are possible weak points, but overall the tool is strong.

The overshot connection to the fish is slightly less reliable and somewhat less flexible for working the fish than the screw-in connection. For example, the overshot cannot be bumped or jarred down, especially while holding right-hand torque, since this is the conventional manner of releasing it. There is little risk of braking the slips and releasing the fish during severe operations and under heavy loads. Normally, though, the overshot will hold as much or more than the remainder of the fishing assembly.

The overshot slips must be sized properly. Several trips may be required to change the slips for other sizes if the size of the fish is unknown. When in doubt about sizes, usually run a slightly larger-size slip. If the slip is too small, it can break while trying to run the overshot over the top of the fish, leaving additional junk in the hole. There also is a risk of breaking the slips with a rough fish top or by not catching it correctly.

When the top of the fish is rough, it may be necessary to mill or cut a clean, smooth top so the overshot can go over and engage the fish. This is not a problem inside casing or in a gauge hole. But, it may be more difficult in crooked or out-of-gauge holes because of the difficulty of going over the fish.

Normal circulation is through a regular overshot body, around the shoe and uphole. This reduces the risk of sticking the fishing assembly and provides better control of exposed high-pressure zones. Limit circulating for long periods or reduce the rate to prevent fluid erosion in the overshot slip section. This can weaken the slips and release the fish but normally is not a major problem. The circulating overshot allows circulating through the fish, which may help release it. The pressure seal can be damaged by a rough fish top (Fig. 4–16).

Normally run the overshot on the bottom of a fishing assembly. Stop it about 25 ft above the fish, and circulate and condition the hole as needed. Then lower the overshot slowly down over the top of the fish, rotating slowly. After catching the fish, test the catch by pulling 10,000–30,000 lb over the weight of the fishing string. This also sets the slips. Then work the fish to release it.

Normally release the overshot by holding right-hand torque while bumping down. It may be hard to release, especially after heavy pulling and jarring. If the overshot cannot be bumped and torqued loose, then try releasing it by detonating a string-shot opposite the slips with some weight set down on the overshot.

Fishing Procedures

Figure 4–16 Circulating vs. regular overshot
Note: A similar type of circulating pattern is obtained in pipe spears run with and without packoffs.

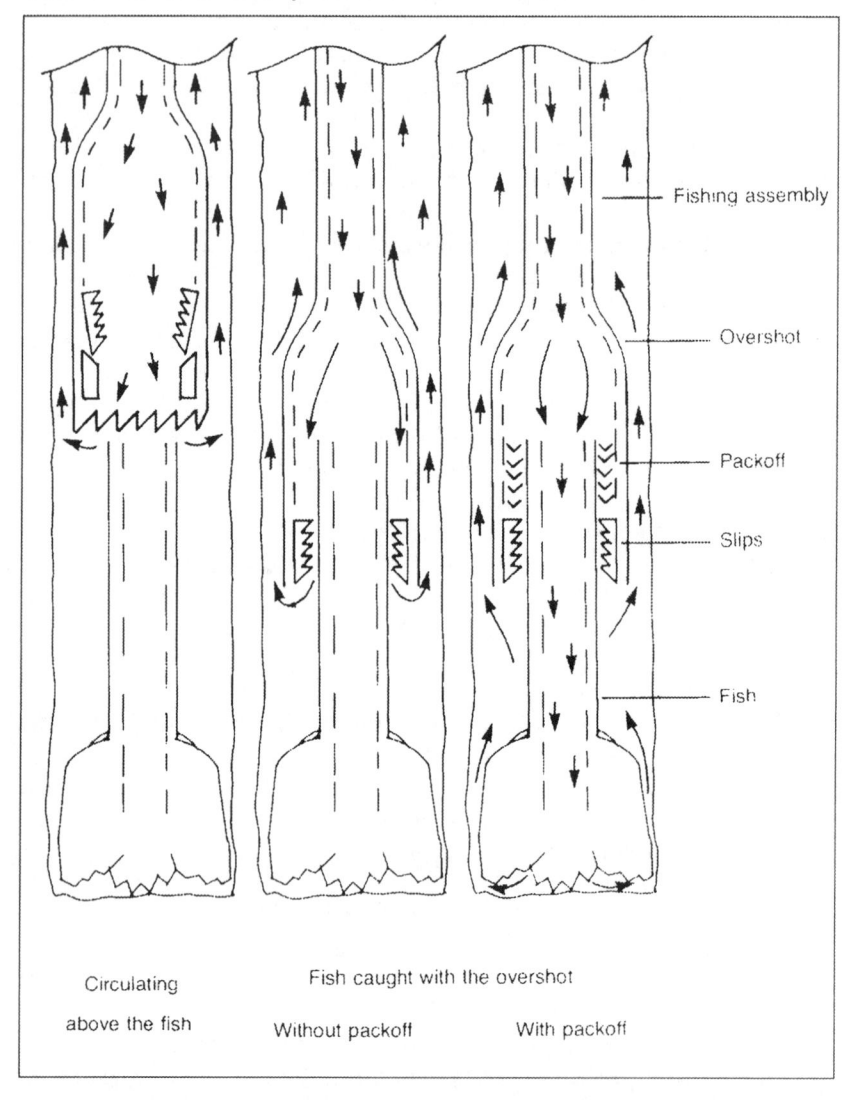

Circulating above the fish

Fish caught with the overshot
Without packoff   With packoff

Fishing assembly
Overshot
Packoff
Slips
Fish

The methods of catching a fish generally are the same in both open and cased holes under normal conditions. However, open holes may be crooked or out-of-gauge, which can increase the difficulty of catching the fish (Fig. 4–17). For a small misalignment, use a mule shoe or a cut lip guide run below the overshot to help catch the fish. Use a wall hook for a greater misalignment. For a relatively high degree of misalignment, run the overshot below a bent joint, bent sub, or knuckle joint. In extreme cases of misalignment, run the overshot below a highly bent joint or place a pup joint extension between the overshot and the knuckle joint or bent sub.

Figure 4-17   Horizontal extension of fishing tools

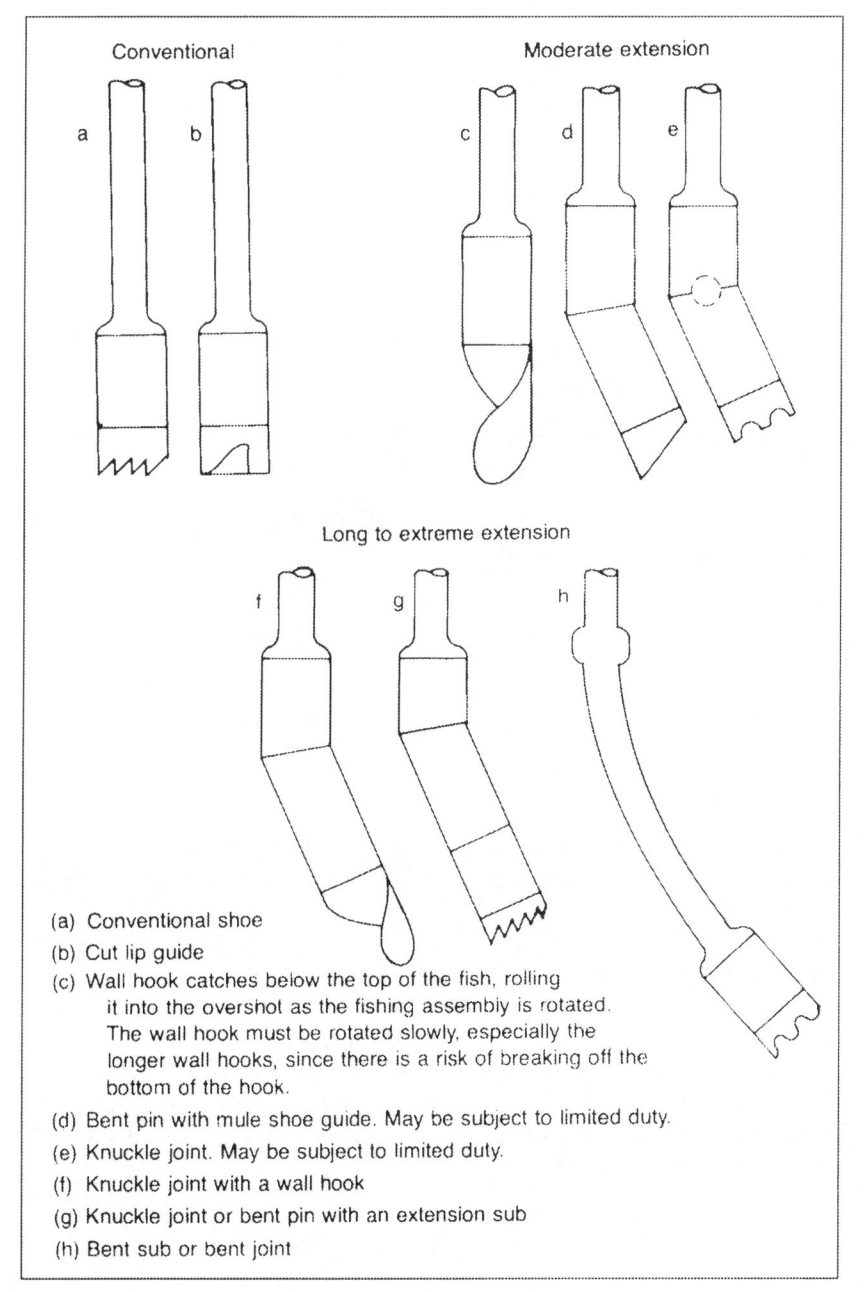

(a) Conventional shoe
(b) Cut lip guide
(c) Wall hook catches below the top of the fish, rolling it into the overshot as the fishing assembly is rotated. The wall hook must be rotated slowly, especially the longer wall hooks, since there is a risk of breaking off the bottom of the hook.
(d) Bent pin with mule shoe guide. May be subject to limited duty.
(e) Knuckle joint. May be subject to limited duty.
(f) Knuckle joint with a wall hook
(g) Knuckle joint or bent pin with an extension sub
(h) Bent sub or bent joint

Fishing Procedures

The fish is difficult to catch when there is misalignment as that just described (Fig. 4–18). However, in some cases it may be difficult to determine whether this is due to a crooked or out-of-gauge hole or because of a damaged fish top. The corrective procedures for these two situations are decidedly different.

Figure 4–18 Recovering a fish in a washout
Note: In the correct procedure, the operator recognized that the fishing tool was bypassing the top of the fish. If the top of the fish is ragged and must be dressed off, use a mill that will not have a side cutting action.

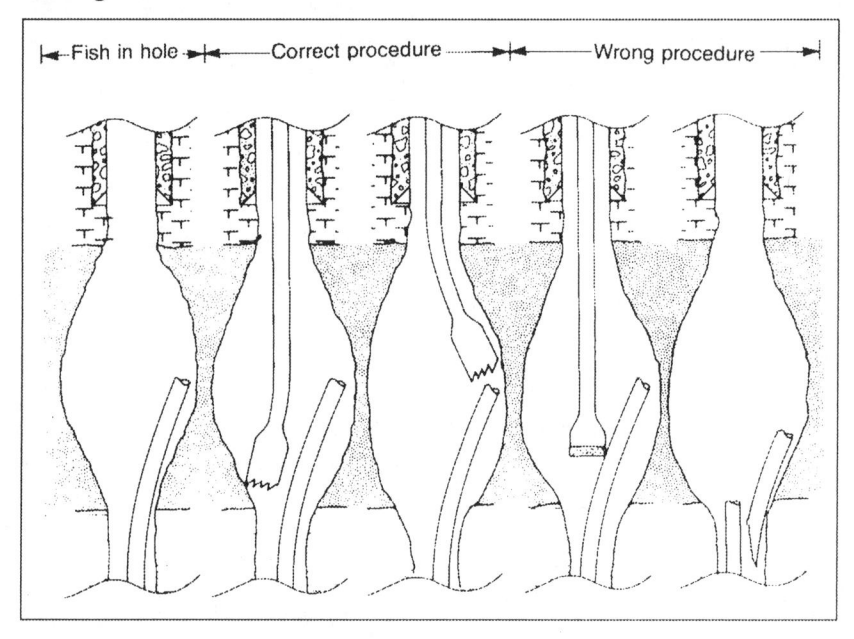

Some fish are difficult to catch because of the position of the top of the fish, such as the top laid over in an enlarged section. At other times, catching the fish is difficult due to the condition of the top of the fish. If the top is bent or flattened, it may not enter the fishing tool. Either cut it off or mill a new top. When the fish is hard to catch, especially when the reason is unknown, it is important to try to catch the fish in the normal manner by using a wall hook, bent joint, bent sub, etc. The action of the fishing assembly, impression blocks, and the appearance of the recovered piece of pipe above the break or sometimes helps determine whether the fish can be caught with regular fishing tools.

Consider a situation where the fish top is laid over in a washed-out or crooked-hole section, and this is either unknown or not recognized. Several unsuccessful trips may be made to try and catch the fish, based on the assumption that the fish top is in a straight, in-gauge hole section. Then it may be assumed that the fish top is damaged. The next apparently logical step is to clean the top of the fish by milling. A flat-bottom mill may be run, bypassing the fish top a short distance and cutting through the side of the fish. The upper, excised section falls and lodges beside the fish, leaving two fish in the hole and creating a difficult fishing situation.

The first preventive measure is to KNOW the length of the milling assembly and the correct depth to the top of the fish. Run an induction log if there is any question. Run a skirted mill that does not have any cutting action or hard facing, on the side or the bottom of the skirt. This will not cut the fish if it bypasses the top. If the skirted mill does not go over the top of the fish, then run a bent joint, bent sub, or knuckle joint to help guide it over the fish.

The importance of catching the fish without milling is illustrated in the following example.

Hitting an Obstruction on a Trip

A wildcat well had $10\frac{3}{4}$-in. intermediate casing set in a straight hole at 13,500 ft and was drilling a $9\frac{1}{2}$-in. hole in hard formations at 16,000 ft with 16 lb/gal mud. The hole below the casing was straight with a deviation of less than 2°/100 ft and a dogleg of less than 1°/100 ft. At 15,000 ft inclination began to increase and was corrected with a packed pendulum.

While tripping in the hole on a bit trip, the drilling assembly ran into an obstruction at 14,200 ft, 700 ft below the $10\frac{3}{4}$-in. casing shoe and then dropped free. There was no noticeable change in the pipe weight, but this was not definitive. The pipe hit hard and solid, even though it was being run in the hole at the proper speed, and the driller reacted quickly. After checking the mechanical condition of the equipment, preparations were made to continue running in the hole.

The assembly was lowered 20 ft slowly and stopped. After picking the assembly up several times and touching the obstruction at the same point, it was rotated slowly past the point about 5 ft with increasing torque. The assembly would rotate freely after it was picked up, so it was pulled.

The drill-collar section had parted, leaving the bit, junk sub, angle-measuring tool, shock sub, two drill collars, integral blade stabilizer (IBS), short drill collar and an IBS with a box looking up, and a broken pin in the IBS box. The total was 92 ft of fish. The shock sub was run higher in the

Fishing Procedures

drill collar assembly than normal to obtain the correct spacing on the pendulum.

An overshot dressed to catch the fishing neck on the IBS was run in the hole but would not take hold of the fish. It would torque through the 10-ft section much like the early action of the parted assembly. A slick induction tool located the top of the fish at 14,100 ft and ran about 10 ft past the top of the fish. This and earlier pipe action indicated the top of the fish had laid over in a washed out or out-of-gauge hole section.

The overshot was run on a bent sub in a standard fishing assembly and measured in the hole. The hole was circulated and the assembly lowered until it began to take a slight amount of weight, indicating that it was touching bottom or an obstruction. This occurred at about the same point where the other tools had stopped, 15 ft below the top of the fish by pipe and wireline measurements.

The fishing assembly was picked up 30 ft, and a reference point was marked on the north side of the assembly. It was then lowered slowly to about 14,100 ft, picked up, and rotated 15°, and the procedure was repeated. When the assembly was lowered in the 210° direction, it seemed to touch something and jump off. The lowering procedure was repeated several times in this area, and the overshot went over and grasped the fish. The overshot was seated and the fish jarred about three times before it came free. It was chained out and recovered intact.

The top IBS was marked on the side, indicating that the pipe and fishing assembly had been down beside the fish. The pin Dutchman indicated a new break. The remainder of the fish was normal. The bit was intact, and the cones rolled freely. If the bit had hit a ledge, one or more cones probably would have been sheared.

The cause of the fishing job is unknown. The hole was straight, verified by later open-hole logs, with a minor washout where the fish was recovered. Also, normally the fish would fall to the bottom of the hole, which did not happen. The bit or possibly the IBS probably hit a small ledge. The pin broke, and the top of the fish laid over into a small washed-out section. The pipe then bypassed the top of the fish. The fish was not stuck tightly and came free easily. If a mill had been run to clean off the top of the fish, it would have cut down beside the fish and possibly severed it, increasing the difficulty of the fishing job. Overall, the fishing job was satisfactory.

Inside Catch

The basic inside-catch fishing tools are spears and taper taps. They are used mainly instead of an outside catch tool when the clearance outside the fish is too small for an outside catch tool and when they are strong enough to hold and recover large-diameter fish. Taper taps usually are used to catch smaller, lightweight fish. They are seldom used, except in special cases because of the disadvantages described in Chapter 3.

Smaller spears catch tubing, smaller drillpipe and lightweight fish. They are relatively weak compared to the corresponding outside-catch. The inside diameter is small, so running wireline tools through them is restricted. Larger spears catch large drillpipe and casing. The casing spear is a very strong tool. Most fishing with spears is for casing and large tubulars. Spears overall are less strong compared to overshots and screw-in catches. So, the overall working action is somewhat restricted.

Spears use inverted slips. To release the spear, the body of the tool must be moved down a short distance relative to the slips so they can retract and release the fish. There can be a higher risk of being unable to release tools with inverted slips compared to those with slips run in a normal position that release by picking upward. In some cases, especially with a smaller spear run on larger diameter pipe, it may be run too deeply into the fish so it lacks space to move it further downward for releasing. This generally does not occur with larger spears.

Washing Over

Washing over is the procedure of circulating and drilling or washing over a stuck fish with a larger diameter tubular. It is an important fishing tool but also a very high-risk procedure, often considered a last resort operation. Before washing over, ensure that the fish is worked extensively with jars and a bumper sub located immediately above the fish.

When the fish cannot be worked free, then wash over and release part or all of it. If the entire fish is released, catch and pull it. If part of the fish is released, catch the free section, back off, and recover it. This operation normally is repeated until the entire fish is recovered.

Washing over as used here includes longer strings of wash pipe and using washover spears. It does not include relatively simple procedures such as washing over about 30 ft over the top of a fish to clean it for catching. Washing over is a high-risk operation. Risk increases with closer tolerances, greater depths, longer sections of open hole, and higher mud weights. It is essentially a last resort operation. If the washpipe is left in the hole around the fish as a second fish, it is very difficult to recover. If a small amount of washpipe such as a washover shoe and sub, are left in the hole around the fish, it may be possible to remove these by milling. But, milling is almost an exception because in most cases the lengths and quantity of pipe left in the hole preclude milling as a practical procedure. Generally the only alternative is sidetracking or redrilling. This is not intended to eliminate washing over because in many cases it is justified.

Washing-over may not be considered as risky as noted because the procedure is deceptively simple. However, it warrants repeating that it is a HIGH-RISK procedure, even under the best conditions.

Operations

Wash over operations, like all others, require meticulous attention to detail. This is emphasized by the fact that usually the hole must be sidetracked or plugged and abandoned if it is unsuccessful. Therefore plan and conduct operations carefully.

The washpipe must have an inside diameter large enough to pass freely over the fish and small enough to run safely in the hole. The actual length depends upon hole conditions and relative clearances. A free-pipe log may be a guide of the length of washpipe to run. Other guides are wellbore deviation surveys and free-point data. Normally, the maximum is about 500–800 ft of washpipe in large, straight holes that are in good condition with minimum formation problems. Run shorter lengths in smaller holes and those with hole problems. As an example, a free pipe log may indicate that only the upper 300 ft of a long fish is stuck. Then run about 350 ft of washpipe. Always run the minimum amount necessary.

Connect the washover pipe to the bottom of the fishing string and a washover shoe on the bottom of the washover pipe. Shoes range from a long-tooth design for soft formations to a scallop design for harder formations or complete mill shoe for cutting metal.

Run the washpipe to within 25–50 ft of the top of the fish, and circulate and condition the hole. Then lower it with reduced circulation while rotating slowly. Clean out the annular space around the fish, drilling as necessary until the entire fish is covered or the top of the washpipe is near the top of the fish. Circulate to flush cuttings out of the hole and then pull the washpipe.

Catch the fish with a standard fishing assembly. If the fish was washed over completely, then work it, including bumping and jarring to release it so it can be recovered. If the entire fish was not washed over, then catch and work it, taking free points if applicable. After working extensively, back off and recover the free fish. Then repeat catching and working or washing over as indicated by the condition of the hole and fish.

A fish may be washed over for 300 to 500 ft or more. Then after it is caught with a fishing assembly, it frequently is found to be free for a few hundred feet or less. The additional sticking is attributed to the time delay between washing over and catching with a fishing assembly. Sometimes a fish, stuck off bottom, drops before it is completely washed over. An example of this is an assembly stuck in a keyseat. Then run a fishing assembly, catch the dropped fish, work it free if necessary, and recover it. The dropped assembly usually is damaged, often leaves sheared bit cones on bottom, and may be tightly stuck.

In these situations, run a washover spear to recover part of a fish or prevent it from dropping after washing over. Run the spear inside the bottom of the washpipe in an inverted position. Run the washpipe into the hole in the normal manner. Catch the fish with an overshot connected to the spear. Wash over the fish while the spear remains stationary in the top of the fish. The spear slips engage the washpipe when the fish is washed free and starts to drop. Then pull the assembly and fish out of the hole.

Washing over with a washpipe spear is a high-risk operation. It has all of the risks of a regular washover and the additional risk of problems with the spear. Much of the equipment has close tolerances. The spear procedure is simple in description but complicated in operation. It may be difficult or impossible to release the spear if the washpipe sticks.

Close tubular and hole tolerances in all washover operations increase mechanical friction and resistance to mud flow, causing high drag and torque. If it is severe, the forces may approach or exceed the strength rating of either the washpipe or the fishing assembly and cause a twist-off. Reduce mechanical friction by decreasing the length of washover pipe.

Washing over at a higher penetration rate can overload the mud system or cause a buildup of cuttings, increasing the risk of sticking the washpipe. Prevent this by controlling the penetration rate. Circulate as required to minimize cuttings density in the mud.

Close tolerance between the outside diameter of the washpipe and hole diameter promotes wall sticking. Resistance to flow causes increased pressures leading to possible formation breakdown and lost circulation. Good mud properties are very important in successful washover operations. If a water-base mud system is being used, consider changing to an oil or oil emulsion mud system.

Cased vs. Open Holes

Cased and open-hole washover operations are similar but with some basic differences. Overall, there is less need of washing over in cased holes as compared to open holes. Larger tubulars are used in open holes so they can be worked harder. Generally clearances are slightly larger. Washing over in the open hole is a higher-risk operation because of cutting or caving material in the mud and possible wall sticking. Most washing over in open holes is over tapered-type tool joints or connections. This, combined with the ability to cut the formation, improves washover operations to a limited extent.

Cased holes generally use smaller tubulars with reduced safe working forces. Smaller clearances increase the risk of sticking. Flat-type coupling shoulders such as tubing collars are most common. Also, sand or similar material is the most common cause of sticking. Thus stuck pipe in cased holes usually is firmly fixed in place or sanded up in most cases.

Washing over in cased holes often is extremely difficult, especially with longer sections of tubulars. Washing over sanded-up tubing inside casing is a good example. Tubing lies against the side of the casing because of normal hole inclination. It is tightly wedged in place with sand or other material such as barite. Therefore washing over involves washing or drilling out sand or possibly barite. It also includes cutting the sides of upsets and collars and possibly part of the side of the tubing.

There are many disadvantages in this situation. Milling is extremely difficult. The mill cutting-face is small because of the thin wall of the washpipe. Therefore it cuts slowly, has a short life, requires more trips to run replacement mills, and overall milling efficiency is very poor. It is not uncommon for a mill to cut less than 30 ft before it needs replacing. The new mill may not go to the last milled depth but stops higher in the hole, so part or all of the section must be milled again.

The mill cannot be dressed with cutting material on the outside because it will damage the casing. However, there is still a high risk of casing damage. The bottom of the mill tends to wedge between the casing and the tubing, upsets and collars. This causes casing wear. Metal cuttings and other abrasive particles further increase casing wear. It is relatively common to find holes in casing where washing over and milling occurred.

The wedging effect causes hard running and high torque. This in combination with the additional friction of the close-tolerance tubulars, increases the risk of twisting off.

Excluding milling problems, recovering the fish may be very difficult. The fish may be weakened because of the mill cutting part or all of one side of the upsets and collars. The partially milled and weakened collars can part during fishing. It is not uncommon to recover one joint of fish per fishing run, recovering the fish in multiple pieces. Additionally, a free piece of the fish may fall down beside the remainder of the fish, leaving a second fish in the hole, further complicating milling and recovery procedures.

In summary, milling over tubing sanded-up in casing is a very high-risk washover operation.

Washing Beside a Fish

Washing beside a fish is the procedure of running a washstring down beside a stuck fish. This is seldom done because of limited clearances, collar interference, and pipe entanglement. Also, it may not be efficient because of the material to be drilled or washed out such as junk, hard shale, etc.

There are situations where the normal washover procedure is less efficient, such as washing over sanded-up tubing in a cased hole — a complicated job. In this case there is a distinct advantage of washing out the sand by running the washstring down alongside the fish if conditions are favorable. A case of washing out alongside the fish is illustrated in the following example.

Washing Beside a Fish

A high productivity, high GOR-flowing oil well had $7 5/8$-in. intermediate production casing and a 5-in. liner from 12,200 to 11,100 ft with a commingled completion from several sands in the gross interval from 12,000 to 12,100 ft. Semiflush joint $2 3/8$-in. tubing was seated in a permanent packer set in the liner at 11,900 ft. The packer fluid was the 18 ppg water-base mud used to drill the well. The completion was different because the well was completed during a period of material and pipe shortages. After about 20 years, production declined to the point where pumping was justified. The pump installation had been deferred until necessary because it was recognized that a fishing job would probably be required because of the heavy-weight drilling mud behind the packer.

After starting the workover, the tubing was found stuck as expected. It was perforated above the packer but would not circulate. This was expected but was and should be tried since it can be an easy way to release stuck pipe when it works. A free point showed 100% free tubing in tension and torque to 6,000 ft and 100% stuck at 9,000 ft. A stuck-pipe log indicated the tubing was stuck at random points of varying lengths. The tubing was worked for 4 hours and the free point run again. The 100% free point had moved downhole to 6,800 ft and the 100% stuck point to 10,000 ft, a common occurrence. The tubing was worked for two more hours, the free point was run, and there was no change in the sticking points.

The tubing was cut at 10,000 ft with a chemical cutter. Circulation could not be established, so the tubing was worked but did not release. The deeper cut was made because frequently the partially stuck tubing can be released with a combination of working and circulating. This was not the case, but the free point was worked down to 7,200 ft. The tubing was cut, the hole circulated, and the tubing recovered.

It was decided to wash down the side of the tubing fish because of the clearances and problems when washing over stuck tubing. It appeared reasonably certain that the tubing was stuck with barite that had settled out of the mud. Barite is sometimes very hard to wash out, but washing alongside the fish appeared to be the best action. About 4,000 ft of the recovered fish with a mule shoe on bottom was run on a workstring. The mule shoe guided the workstring past the top of the fish without difficulty. The old mud was washed out, including bridges of settled barite from 7,200 to 10,000 ft. The hole was not washed out past 10,000 ft because the tubing fish had been cut at this point and there was a risk that the free section of fish would fall downhole beside the remainder of the fish and become wedged and stuck. There was also a risk that it would stick the washstring due to a third string, the loose upper fish, falling down by the other two parallel strings. The washstring was pulled, an overshot was run, and 2,800 ft of tubing fish was recovered without difficulty.

The washstring was rerun and washed down beside the tubing fish from 10,000 ft to the top of the liner at 11,100 ft. The fish was caught with a circulating overshot, but circulation could not be established through the perforation above the packer. A free point showed the fish to be free to 11,400 ft. The fish was cut at 11,200 ft and the section of fish was recovered. The remainder of the fish was washed over with washpipe in the conventional manner. The fish was caught with a circulating overshot and found to be plugged. This precluded cutting the fish, and it could not be worked free of the packer. The fish was recovered with a mechanical back-off tool, and the well was restored to production.

This example also illustrates the problems that occur when weighted water-base mud is used as a packer fluid. The report did not mention tapered collars, but they are recommended and were probably used because of the lack of problems. The fish was recovered in this case without the severe problems which are sometimes encountered in this type of fishing operation. In summary, washing alongside the fish expedited the fishing job, minimized costs, reduced the risk associated with the use of conventional washpipe, and permitted the well to be placed on production at an earlier date to start generating income.

PLUG BACK AND SIDETRACK

Plugging back and sidetracking often is the next procedure after fishing unsuccessfully. Except for special situations, any hole from vertical to horizontal can be sidetracked. The procedures are straightforward. It takes between 3 and 6 days to set a plug, dress it off, and sidetrack before resuming drilling. However, a failure at any point in the sequence of operations usually requires drilling out the original plug and repeating the entire process, a costly procedure. Like fishing, plug back and sidetrack failures should not occur but they do. Lack of attention to detail often leads to problems or repeating the operation. So the main preventive measure is to do it right the first time.

Plugging Back

A hard, immobile plug is an important part of sidetracking successfully. Other plugs may be needed, such as isolating plugs to shut off high-pressure formations and lost-circulation zones. They may also be required in accordance with the rules and regulations of the regulatory agencies having jurisdiction. The sidetracking plug serves several purposes. It provides a firm, hard, immobile seat for directional tools, closes the original hole so the tools will not reenter it, and stops problems from lower zones.

Select the kickoff point carefully. It should be as deep as possible to save hole. It should be high enough to allow a normal buildup angle so the fish is bypassed at a safe distance. The deviated hole should be far enough from the original hole to prevent problems of high pressure and lost circulation from the original hole. Sidetrack in a medium or medium-soft formation and avoid very soft or very hard formations when possible. Sidetrack at least 100 ft and preferably 200 ft below a casing shoe.

It should not be necessary to set more than one plug, but there have been cases where over five plugs were required. In one case where this occurred, another operator sidetracked off one plug in a comparable hole in the same area, using oil mud. This emphasizes the importance of setting the plug properly. The most common problems include a plug that is too short; poor displacement practices, including contamination due to failure to use spacers; excess retarder; no sand in the plug slurry; low early strength; and not waiting for the plug to harden. There apparently is a higher risk of failure when setting plugs in oil mud, but there is no valid reason for this except lack of planning and attention to detail.

The plug length is important. DO NOT select a shorter plug to conserve drilled hole or save on cement or time required to drill it out. The dressed-off plug length should be long enough so that the original hole is plugged to a depth below the point where it could interfere with the sidetracked hole. The bottom of the plug should be at a depth where the center lines of the original and new hole are separated by a distance equivalent to at least several bit diameters. Use a longer plug for higher deviation and when the inclination of the original hole is unknown.

Base the plug slurry volume on the calipered or drilled-hole volume with ADEQUATE EXCESS for the total plug length, including the dressed-off section. Design the slurry for an early, high compressive strength of 2,500–3,000 psi or more in 24 hours. Have a slurry density about the same as the mud density or heavier, since intermingling of the slurry and mud due to gravity is negligible. In most cases use class H cement regardless of depth, although class A can be used for shallower depths. Use 20–35% good quality — preferably larger-sized — fracture sand to increase plug hardness. Blasting grade and special flint sands may be considered under special conditions but their use is seldom justified. The sand will not settle appreciably in a good slurry before it hardens. Add accelerators or a very minimum amount of retarder to insure that the slurry remains pumpable during mixing and placement. DO NOT USE EXCESS RETARDERS. Good mixing water is a standard precaution. Test for thickening time and compressive strength with the same water that will be used to mix the plug. Test initially for the proper blend of additives with cement from the same silo storage to be used. A final test verifies thickening time and compressive strength with blended cement from the transport truck that has all additives mixed and blended. Catch wet and dry samples.

Use standard cementing practices with good placement techniques to minimize contamination. Spacers or chemical flushes separate the slurry and mud, reduce contamination and clean the walls of the hole to improve bonding between the cement and formation. Mix the cement slurry (normally batch mixed). Pump the spacers and slurry in the correct order followed by the displacement fluid, usually mud, and displace the slurry to the correct position in the wellbore. Measure fluid volumes accurately and pump at the maximum safe rate, preferably under turbulent flow conditions to minimize contamination.

There are various displacement procedures. Each has advantages and disadvantages, and the unbalanced column method is the safest and most common in most cases. After placement, immediately pick the drillpipe up to a point safely above the top of the slurry to prevent sticking. Then either pull it out of the hole or pull at least 5–10 additional treble stands. Circulate for the equivalent of two thickening times while monitoring pit levels. If high pressure causes a flow, hold 200–500 psi under closed preventers. If lost circulation zones are exposed and there is a risk of loss, displace with a lighter fluid and monitor pit levels without circulating.

Dressing Off

Allow the cement to harden to the required compressive strength. In field practice determine plug hardness by observing the penetration rate while dressing off the excess cement above the kickoff point using Table 4–3 as a guide to plug hardness.

Table 4–3 Approximate plug hardness

Drilling-Milling Rate	Cement Hardness
• 10 ft/hr or 6 min/ft,	eqv – 3,500 psi, very hard
• 20 ft/hr or 3 min/ft,	eqv – 3,000 psi, hard
• 30 ft/hr or 2 min/ft,	eqv – 2,500 psi, firm
• 40 ft/hr or 1.5 min/ft,	eqv – 1,500 psi, soft*
• 50 ft/hr or 1.4 min/ft,	eqv – 1,000 psi, very soft**
• 60 ft/hr or 1 min/ft,	eqv – 500 psi, not set***

*   — Sidetracking questionable
**   — Sidetracking very questionable, try time drilling.
***— Drill or circulate out and reset plug

Table 4–3 is based on dressing-off with about 1,000 lbs of bit weight per inch of bit diameter, with 50 to 60 rotary rpm and 1,000–1,500 psi pump pressure. The information is a rough estimate. Holes have been successfully kicked off of softer plugs and kickoffs have been unsuccessful on harder plugs.

Dress off the plug with a long mill-tooth bit, polycrystalline (PDC) bit or a cement mill run on a limber drill-collar assembly. DO NOT run the assembly into soft (green) cement. This COMMON ERROR causes sticking. Observe standard precautionary procedures. Clean out most of the excess cement while it is soft to save extra time drilling hard cement. Plan to have cement cleaned out to about 150 ft above the estimated kickoff point before the slurry reaches any appreciable compressive strength. Handle cement-contaminated mud in the standard manner by pretreatment, dilution, etc., depending upon the mud system.

Dress off the plug in stages. Drill a short section of cement, and, dependent upon hardness, either drill ahead or wait for cement to gain additional strength. Adjust the periods so the bit will be at the kickoff point when the cement is hard. Plugs characteristically have "hard and soft" sections, especially in the open hole. Stop drilling in a harder section. If the plug has not hardened in 24–30 hours, then additional waiting time is seldom justified. Clean out and set another plug.

Sidetracking

Sidetrack the hole in the conventional manner, normally with a deviation assembly. Sidetrack direction is not selected for blind side-

Fishing Procedures

tracking. In crooked or more highly deviated holes it may be better to deviate toward the low side of the hole so there is less risk of crooked hole.

The hole may deviate off a very soft plug by time drilling at very low rates depending upon the bit, plug hardness and formation. The procedure includes orienting the deviation tools, reducing the circulation rate and holding up on the bit to reduce the penetration rate to about 2 to 4 ft/hr. Drill 10–15 ft at this rate and then increase it slowly until the hole is sidetracked and the bit is drilling in new formation.

Sometimes it is difficult to deviate in very hard formations. An alternate method of deviating is to run a deviating assembly with a maximum kickoff angle sub and reduced (usually minimum) hole size. Sidetrack by drilling a pilot hole with a small bit for a short distance of 10–30 ft of new hole. The longer distance applies to harder formations. Then open the hole to full gauge and repeat as necessary. It is important to drill only a short distance with the small bit and then open the hole to full gauge to prevent a dogleg and potential keyseat problem at a later date.

Cased holes are sidetracked in one of several ways. The most common procedure is to mill out a section of casing, set a plug in the casing, dress off and sidetrack in the conventional manner, drilling the sidetracked hole out through the section where the casing was milled. A less-common method includes milling out a section of casing, setting a plug and then deviating off of a whipstock set on the plug. Another procedure that is seldom used and not normally recommended is to set a plug in the casing. Then run and set a whipstock in the casing above the plug. Mill a hole or window through the casing and then sidetrack.

MILLING CASING AND TUBULARS

Milling casing and tubulars is a common procedure associated with fishing and casing repair. Casing may also be removed for replacement as described in Chapter 7. Mills are also used to remove stuck drill collars, washpipe, and similar tubulars.

There are cases of removing two strings of casing simultaneously. Junk is milled up with regular mills as described in the section on milling junk in Chapter 3. Most casing and large tubulars are milled with an inside casing cutter or section mill. Milling can cause high and often fluctuating torque, so use part of the description in the section "Milling Operations" in Chapter 4 as a guide.

Milling long sections of tubulars in a cased hole, and to a lesser extent in an open hole, is almost a last-resort fishing operation. Most of the milled cuttings are circulated out, but some can settle downhole (Fig. 4–19). During extensive milling, the cuttings pack around the fish and wedge it tighter, practically precluding recovering it with regular fishing tools. Also the mill cuttings must be milled again, further reducing milling efficiency. Junk subs help reduce the problem. Place a wear ring or guide bushing on the section mill to prevent the knives from cutting the regular casing when milling long sections of tubulars.

Figure 4–19 Milling on mill cuttings
Note: Mill is milling on both the object being milled and the mill cuttings. This remilling action decreases the life of mills and increases the time to complete the job. The casing may also be worn excessively, even with the proper type of mill.

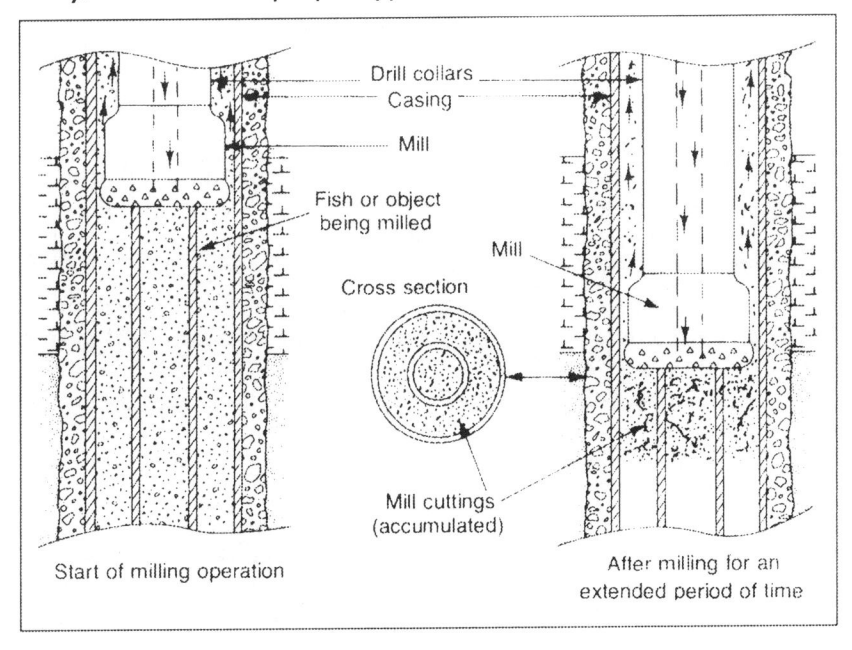

Cuttings removal can be a problem when milling long sections of tubulars. Run junk subs above the mill, as many as three in severe cases. Mud viscosity of 50–80 API funnel sec, sometimes higher, will carry out most cuttings. Higher mud weights may help remove larger cuttings. Remove steel cuttings from the mud system at the surface with fine, 150- to 180-mesh shaker screens and ditch or bar magnets, use 2 or 3 in the flow line (Fig. 4–20). There have been cases where the cuttings stuck to the drill tubulars due to magnetism, causing severe removal problems.

Fishing Procedures

Figure 4-20 Magnet in a flow line
Note: Remove, check, clean, and replace the magnet periodically to determine the approximate amount of metal fragments in the mud. Also check the drill cuttings samples.

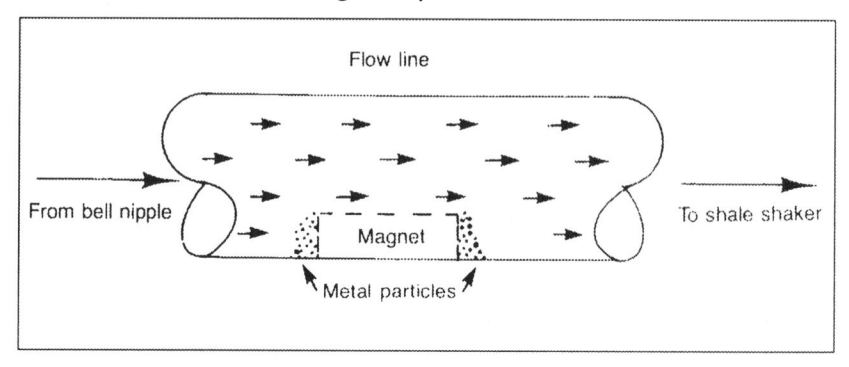

Milling a section of casing is the more common method of deviating in a cased hole. Casing can be milled in both vertical holes and those with high inclination. Run a section mill on a limber assembly with drilling jars and 2 or 3 junk subs or boot baskets above the mill. "Feel" for a casing connection by lowering the mill with expanded arms, pick up, and start cutting about 5 ft above the connection in a well-cemented section of casing. The reason for starting 5 ft above a casing collar is to leave a longer part of a cemented joint above the cut section. This helps prevent backing off the joint while conducting rotating operations below the cut section. Mill and remove at least 50 and preferably 80 ft of casing.

Pick the pipe up at 10–30 minute intervals to ensure that cuttings are not stacking up above the mill. Note that the pipe will continue to rotate through a "bird nest" of stacked cuttings but stick when it is picked up. Use a check valve above the mill if necessary to prevent plugging from mud flowback while making connections and otherwise moving the pipe.

There are various techniques for deviating off a whipstock and milling a hole or window through the wall of the casing. It is faster than milling a section of casing when it works properly. However, there are many problems such as milling the whipstock instead of the casing or rolling off the whipstock while trying to cut the window. Also, tools hang and stick as they are pulled back into the casing, and the whipstock may move and prevent reentry of the whipstocked hole. These problems frequently require fishing and often sidetracking again higher in the hole. Milling a section of casing is more common for these reasons.

REFERENCES AND SUGGESTED READING

Bannister, C.E., and O.G. Benge, "Pipe Flow Rheometer: Rheological Analysis Of A Turbulent Flow System Used For Cement Placement," SPE 10206, Society of Petroleum Engineers of AIME meeting in San Antonio, Oct. 5–7, 1981.

Dees, John M., and William N. Spradlin, Jr., "Successful Deep Openhole Cement Plugs For The Anandarko Basin." SPE 10957, presented at the annual Fall Technical Conference of the Society Of Petroleum Engineers of AIME in New Orleans, Sept. 26–29, 1982.

Dudleson, Bill, Mike Arnold, and Dominic McCann, "Early Detection Of Drillstring Washouts Reduces Fishing Jobs," *World Oil,* Oct. 1990: 43–47.

Haut, Richard Carl, and Ronald James Crook, "Laboratory Investigation of Light-weight, Low Viscosity Cementing Spacer Fluids," SPE No. 10305, 1981.

Kaisi, M.S., J.K. Wang, and U. Chandra, "Transient Dynamics Analysis Of The Drillstring Under Jarring Operations By The FEM." Society Of Petroleum Engineers, Drilling Engineering, March 1987: 47–62.

Sabins, Fred L., and David L. Sutton, "The Relationships Of Thickening Time, Gel Strength And Compressive Strengths of Oilwell Cements," SPE 11205, Fall Technical Conference of the Society Of Petroleum Engineers of AIME in New Orleans, Sept. 26–29, 1982.

Smith, Dwight K., "Cementing, SPE Monograph Vol. 4." H. L. Doherty Series, 1976.

Chapter 5

OPEN-HOLE FISHING

Fishing operations are conducted to recover a fish or restore the hole to a condition in which normal operations can be resumed. The operator must always bear in mind that the overall objective is to drill and complete a producing well. Therefore, alternative procedures such as completing uphole or through the fish and sidetracking should be considered during fishing operations.

The diameter of the fish can range from as small as a few tenths of an inch to the full size of the hole and may vary in diameter over the length of the fish. The fish literally can be as long as the hole is deep. It will usually be circular in cross section, but it can have different shapes, such as bits, reamers, stabilizers, or square drill collars. Thus, there will be a wide variance in fishing operations. However, most fishing operations fall into groups:

- Small objects such as bits, bit cones, tong dies, and hand tools
- Drilling and other downhole assemblies
- Test tools, packers, and plugs
- Wirelines
- Small-diameter tubulars
- Assemblies bent or broken above the rotary
- Miscellaneous

This chapter primarily covers those failures that occur in the open hole.

Fish are milled away, swallowed, worked free and pulled, or washed over and recovered. Other operations may include milling a clean top or free pointing and parting to release and recover a free section. In special circumstances, fish must be released with soaking and dissolving agents. Fishing in directional and horizontal holes is similar, with allowance for deviation. Fishing operations may be restricted by formation problems ranging from a minor disruption to loss of the hole requiring sidetracking, plugging, redrilling, or drilling a kill well.

Common failures and sticking situations are summarized in Table 5–1.

Table 5–1 Sticking indicators

Type Of Failure ———>	Caving Sloughing	Wall Stuck	Key-seat	Fract-ures	Twist Off
Indicators \\|/ \\|/ \\|/ Movement Before Sticking ———>	Free	Free	Free	Free	Free
Pipe Action After Sticking					
Reciprocation —>	Minor to none	None	Down or none	Free or none	Free
Rotation ———>	Minor to none	None	None to free	None to free	Free
Circulation ———>	Minor to none	Full	Full	Full	Full

Sloughing and caving normally are indicated by the lack of circulation, and the assembly usually cannot be rotated or reciprocated, except sometimes downward to a limited extent. The assembly can be stuck at any point, but the most frequent occurrence of sticking is at the bit or the top of the drill collars and in the lower assembly if reamers and stabilizers are used.

Wall sticking is indicated when the assembly cannot be reciprocated or rotated, but circulation generally is not affected. Wall sticking generally occurs after the assembly has been at rest, sometimes for less than 5 minutes. Sticking most commonly is in the drill collars.

Pipe usually keyseats when moving upward. Any obstruction on the assembly may keyseat, but the most common points are at the top of the drill collars, followed by a point in the drill-collar assembly where drill-collar size changes and at the bit. Keyseats usually are indicated by the inability to move the assembly upward. It may be completely stuck or free to move downward or rotate. Circulation is seldom affected. Keyseated assemblies may become wall stuck.

Fractures most commonly cause sticking during drilling when the bit slips into the fracture and sticks. In severe cases all movement is stopped, although circulation normally is not affected.

Miscellaneous causes of sticking vary widely. A metal object such as a tong die or preventer bolt may fall into the hole and stick the assembly at any point. However, it occurs most commonly at the top of the drill collars or at the bit and often is more severe in hard formations and cased holes. Sticking may completely restrict movement of the assembly, but it normally will not affect circulation. When the assembly is sanded up or cemented in, this usually is detected as soon as it happens due to restricted assembly movement and circulation. There are many other miscellaneous causes of sticking.

BIT CONES AND BITS

Fishing for bits and especially roller bit cones is one of the most common fishing operations. Bit companies publish recommended operating conditions for bits based on formation types. However, it may be difficult to predict the type of formation; for this and other reasons, lost bit cones is a common failure.

Prevention includes operating the bits according to manufacturers' specifications, monitoring the bit action carefully, and giving preference to solid-bodied bits. It is not difficult to remove bit cones if proper precautions are observed. Similar procedures are used to remove or recover other small pieces of junk (Fig. 5–1).

Cones

Loose cones can be detected by uneven, erratic torque and generally low-level fluctuations. These are harder to detect in deeper holes and in those with high torque and drag, such as crooked, high-angle directional and horizontal holes.

Bit cones are removed by drilling up or walling off with bits and mills. Otherwise they can be recovered with junk baskets; use magnets less frequently. Recovery methods depend mainly on the size of the junk, formation hardness, depth, hole size, and similar factors. The average size of the bit cones lost in the hole can be estimated from the bit size, how long the cones were drilled on, and the condition of the recovered bit—i.e., did the cones drop off, and did the bit show wear from running on the cones?

Junk shots normally are not recommended for breaking up bit cones since the cones usually are in relatively small pieces. Junk shots may be less effective because fill material or cuttings can cover the cones, causing excess standoff. They can be more effective on larger junk and bigger cones in a cleaner hole. Junk shots usually are not recommended in cased holes because of possible casing damage.

The importance of removing or recovering bit cones properly is sometimes overlooked. This may be due in part to the fact that fish-

Figure 5-1   Recovering or removing small junk

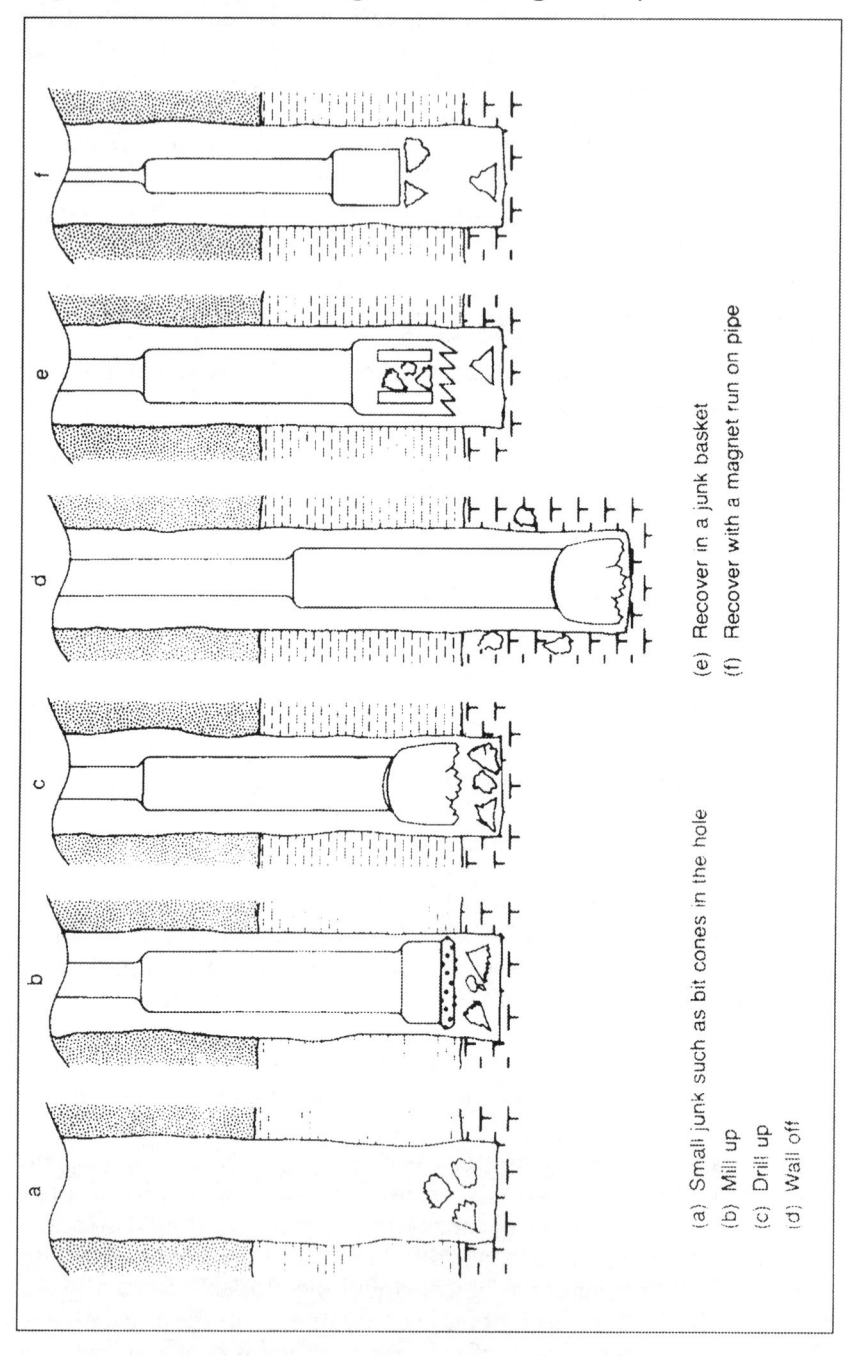

(a) Small junk such as bit cones in the hole
(b) Mill up
(c) Drill up
(d) Wall off
(e) Recover in a junk basket
(f) Recover with a magnet run on pipe

314   Chapter 5

ing for bit cones is a relatively routine fishing job, and more time is probably spent fishing for bits. Various examples illustrate the many situations that may occur when bit cones are lost and the recovery or removal procedures.

Losing and Recovering Bit Cones

- A 6½-in. bit was used to drill a 7⅝-in. retainer, 250 ft of cement and 5 ft of new formation at 14,000 ft. Three bit cones were left in the hole, and were milled up in one mill run. This was followed with a junk basket without any recover. Junk subs were run on both the mill and the bit (Fig. 5–2). A short-tooth, hard-formation bit was run to drill ahead.

The bit was probably run too long, causing the loss of the cones. Overall, the fishing job was successful; but there is some doubt if the junk basket run was justified, especially after following it with a short-tooth bit.

Figure 5–2 Improved junk removal with a junk sub (boat basket)

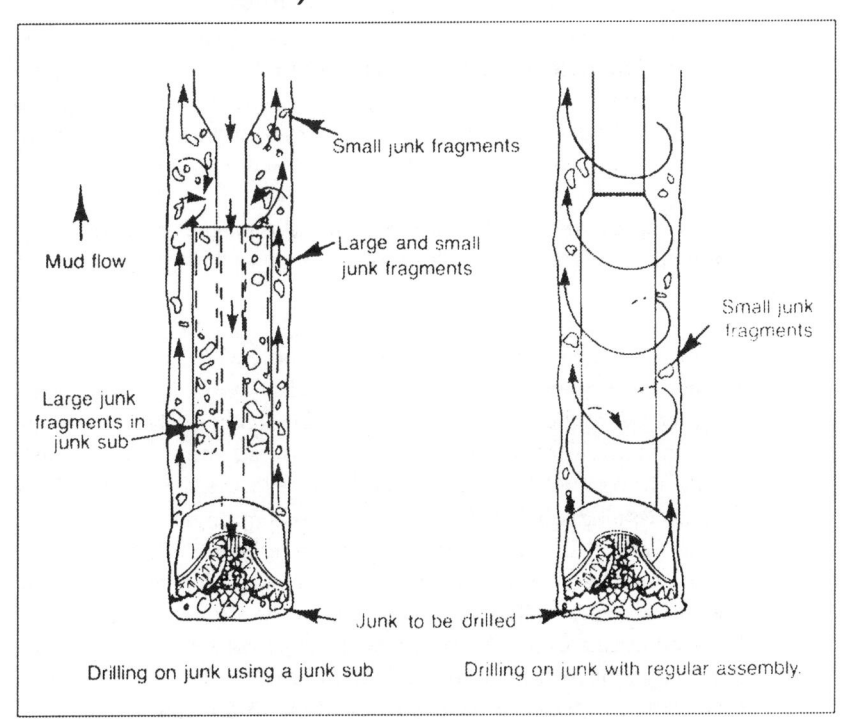

Open-Hole Fishing

- A hard-formation 6-in. bit drilled 43 ft in 10 hours, 4.3 ft/hr with 20,000 lb of bit weight and 54 rpm at 15,000 ft. When the bit was pulled, one cone was missing. A magnet was run and stopped on a bridge at 14,000 ft. A short-tooth, hard-formation bit was run to clean out the bridge and drill on the cone for 30 minutes. Bit bearings were recovered in the junk sub. The cone was then recovered with a magnet. The small bit was probably run too long, causing the lost cone.
- Three 8¾-in. bit cones were lost at 12,000 ft in a medium formation. A junk basket was run and drilled 2 ft with no recovery. The cones were milled for 10 hours. A mill-tooth bit then finished drilling the cones and drilled ahead. Part of the cones were probably walled off. A mill works well in this situation because it can break up the cones into rubble even if it is worn. The junk basket run was a questionable procedure.
- Two 12¼-in. bit cones were left in the hole at 13,000 ft in a medium formation. A magnet was run with no recovery. A mill-tooth bit was run and drilled for 2 hours, but it did not wall off the cones. A junk basket was run with no recovery. An 11¾-in. mill then was run and drilled on the junk for several hours. When it began to run smoothly, it was pulled. It was 60% worn. A bit was run and walled off the remaining junk after 30 minutes of drilling. The hole was drilled ahead. On a later connection, junk apparently fell into the hole and stuck the bit. After working the bit free and pulling it, a mill was run and ground up the remaining junk.

In view of the depth and size of the cones, the magnet run was very questionable. Note this is one of the few cases of walled-off junk falling back in the hole.

- Three 8¾-in. bit cones were left in the hole at 6,000 ft in a hard formation. A junk basket was used to drill 1 ft in 2 hours without recovery. A mill drilled 20 ft in 2 hours, indicating a clean hole. A hard-formation insert bit was run and drilled ahead. As this example shows, junk baskets are ineffective in the hard formation. The first mill probably should have been run longer.
- Three 12¼-in. bit cones were left in the hole at 1,300 ft in a soft formation. They were recovered in two runs with a recessed-face magnet above a mill shoe guide. Normally, a junk basket would be tried at this depth, but the magnet trips were very successful.
- An 8¾-in. bit cone was left downhole at 6,000 ft in a medium formation. Successive trips were made with a magnet, junk basket, then another magnet, all without recovery. A short, mill-tooth bit was run and drilled up or walled off the bit cone in 2½ hours.

- Three 9½-in. bit cones were lost at 9,000 ft in a medium formation. The first cone was recovered with a magnet, the second with a junk basket, and the third was milled up in 3 hours.
- One cone and a cone and shank from an 8¾ by 12½-in. hole opener were left in the hole at 10,000 ft in a medium drillability formation. Both pieces of junk were recovered in one run with a junk basket after drilling 3 ft. The basket had a 1½-ft formation plug in the bottom when it was pulled.
- An 8¾-in. cone was left in the hole at 9,000 ft in a medium formation. It was drilled up or walled off in 2 hours with a hard-formation mill-tooth bit.
- Three 8¾-in. cones were lost at 14,000 ft in a medium-hard formation. Two were recovered in the first run with a junk basket. One-half of a third cone was recovered in a second run. A hard-formation mill-tooth bit drilled up or walled off the remainder in 2 hours.
- Three 8½-in. cones were lost at 12,000 ft in a hard formation. A junk basket was run without recovery. A mill was run and milled on the cones for 3 hours. The hole was conditioned and 7-in. casing was run and cemented. A 6⅛-in. mill-tooth bit was used to drill out of the casing, then drilled on the junk for 3 hours. When the bit was pulled, the cones were left in the hole. Two were re-covered with a magnet. A mill-tooth bit was run and milled on the junk for 2 hours. A mill was run and milled for 5 hours. An insert-type premium bit was run to drill ahead. After 12 hours, it had lost 60% of the inserts. A diamond bit was run and ringed out in 2 hours due to junk. The hole was then cleaned out by making two mill-tooth bit runs. It was then drilled ahead with a diamond bit.

This example shows several obvious mistakes. The cones should have been removed before running 7-in. casing. If this could not have been done, then the cones should have been removed completely after running the casing. The 6⅛-in. bit was run too long on the junk. The insert and diamond bits should not have been run until the hole was clean.

Procedures for recovering bit cones in soft and hard formations are slightly different. Recovery in medium formations can fall in either class. Select the procedures that fit the specific conditions.

In Soft Formations

The method of recovering bit cones in soft formations depends to some extent upon the number of bit cones lost. If only one cone is left in the hole, a hard-formation mill-tooth bit often will break it up and wall it off in very soft formations. Usually when the cone or pieces of the cone are pushed into the wall of the hole, they will not

Open-Hole Fishing

fall back into the hole. The action of the bit during drilling indicates if the cone is walled off.

A junk basket is commonly run if there are two or three cones. The junk basket usually is most effective when it can cut a core or plug of formation from 1½ to 3 ft long. This serves as a plug, retaining the junk inside the basket. Usually running one junk basket is sufficient, but sometimes a second run is needed.

A mill is sometimes used to break up or wall off the cones, especially if there is more than one cone. Compared to flat-bottomed mills, a concave mill or one with a ribbed cutting face is slightly better at breaking up junk but slightly less efficient at walling it off. Run single or double junk subs or boot baskets above the mill or bit.

Magnets run on wire lines normally are less efficient in soft formations since the bit cones often are covered with cuttings and stuck in the bottom of the hole due to the weight of the drilling assembly. However, if the magnet and running line are available and the hole depth and conditions are satisfactory, one or two wireline magnet runs may be justified, especially since they usually can be made in a short time.

A recessed-face magnet with circulating port and a soft-formation-type mill shoe run on the drillpipe has been successful in some cases. The shoe can be rotated and circulated to release the cones so the magnet can catch them. However, it is usually more effective to make another run with a junk basket.

In Hard Formations

The most effective way to remove bit cones and other small pieces of junk in hard formations is to mill them up with concave mills (Fig. 5–3). The milling action usually will indicate when the cones have been completely ground up. A hard-formation mill-tooth bit may then be run as a final cleanup tool. Run the bit only for a short time to prevent leaving teeth, and possibly more cones, in the hole. Run single or double junk subs on both mills and bits. If a diamond or PDC bit is to be used later, make another cleanout bit run with an inset bit to ensure the hole is clean.

Junk baskets are less effective in hard formations. Longer drilling time is required, and the retaining fingers often tear or break off while drilling the junk. The mill shoes frequently wear out before drilling a core long enough to hold the junk in the basket, and it often is difficult to break the core off if it can be drilled.

If the depth and condition of the hole permit and the equipment is available, use a wireline magnet and make several runs. Generally, running a magnet on drillpipe is not justified unless special information and experience in the area indicates it should be run.

The following example describes a special bit problem that occurred in a deep hole.

Figure 5-3   Milling with a concave mill

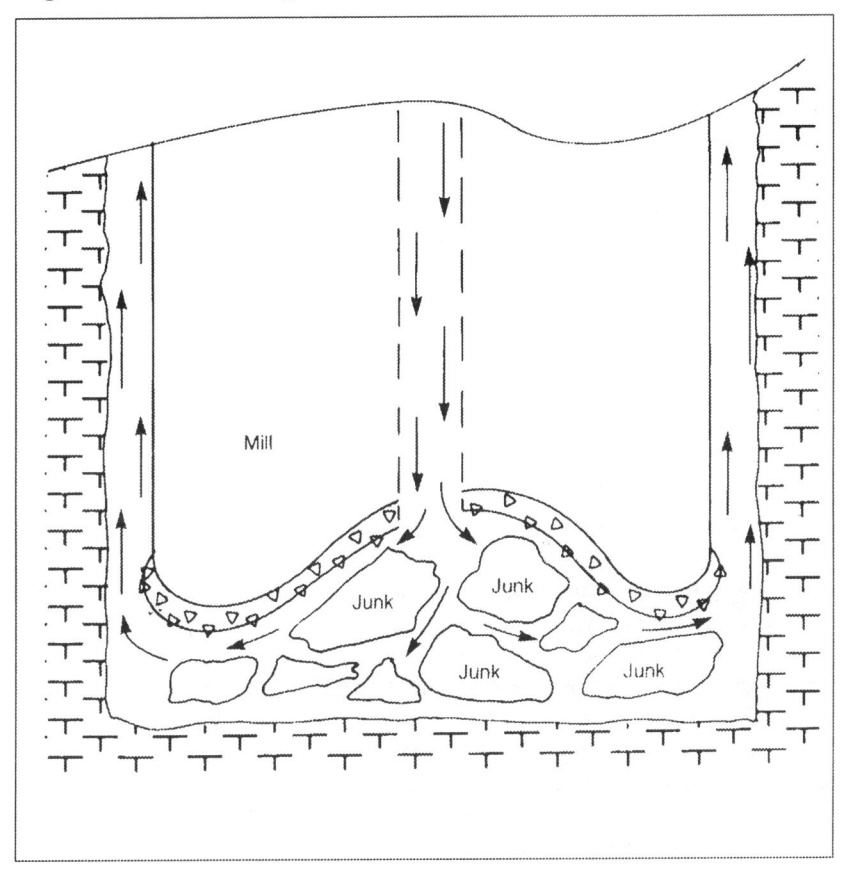

Recovering Bit Cones in a Deep Well

A wildcat well had 13⅜-in. intermediate casing set at 13,000 ft and drilled out with a 9½-in. premium roller bit because of hard and abrasive formations. Chert inclusions precluded using PDC or regular diamond bits. An insert roller bit drilled 580 ft from 15,600 to 16,180 ft in 165 hours, averaging 3.5 ft/hr at 45 rpm and 50,000 lb of bit weight. It graded T-4, B-2, G-1/32-in., a normal bit run for the area and depth.

A similar bit was run, and the bottom 90 ft of hole was reamed as a standard safety procedure since there was no evidence of undergauge hole (except possibly the last 20 to 30 ft, which could have been fill material). It drilled 170 ft to 16,350 ft in 56 hours, averaging 3.0 ft/hr at 45 rpm and 50,000 lb of bit weight. Torque increased slightly in the second hour before the bit was pulled. The drilling rate declined and torque increased during the final hour. A normal bit run would have been at least 100 hours,

Open-Hole Fishing

but the torque and drilling rate indicated a bearing failure. The cones were left in the hole. Marks on the bit shanks indicated running on the cones.

A 9¼-in. concave mill was run, and the last 60 ft of hole required reaming. The mill is obviously an inefficient tool for reaming, but the undergauge hole was unexpected. Otherwise, a short-tooth bit would have been run to break up the cones, even with the higher risk of leaving additional cones in the hole. The mill wore out in 15 hours. This was a long run, but it is a good practice, especially in deeper holes, to run a mill as long as it continues cutting. The mill was heavily dressed and there was little risk of leaving junk in the hole. A second mill finished milling the cones in about 8 hours. It was run an additional 2 hours and drilled smoothly at about 0.5 ft/hr, ensuring the cones were removed. About 10 lb of miscellaneous bit junk was recovered from the junk subs or boot baskets after each mill run. Another premium-grade bit was run and drilling operations were resumed.

The bit cones were recovered efficiently. The mills were run slightly longer than normal, but running the mills longer probably allowed removing the bit cones with two mills; otherwise three might have been required.

The reason for losing the bit cones is not well established. This occurred in a long, sandstone section where similar bits averaged 2.5–4 ft/hr. Except for the last hour, the torque variation was within normal limits. While there is a possibility that the bit was defective, experience indicates this is unlikely. The best conclusion is that the bit failed due to the increased amount of very sharp and abrasive sandstone, and the bit was probably run too long, as indicated by its condition when it was pulled.

Later during a bit trip at about 18,000 ft, fluctuating tool weights indicated a developing keyseat. There had been similar but less definite indications on the prior trip. An insert bit was run with a 9-in. three-point roller reamer on top of the fourth drill collar and reamed from 16,000–18,000 ft at about 50 rpm in about 18 hr. After the bit reached bottom, it was used to drill ahead at about 3 ft/hr with 50 rpm and 50,000 lb of bit weight. During drilling, the drill-collar assembly acted like an inefficient, packed pendulum. This was satisfactory since the formations exhibited an angle-building tendency.

After drilling for about 25 hours, the drilling rate declined and the torque increased. The bit was pulled and the cones were left in the hole. A short-tooth bit was run to ensure the hole was full gauge to total depth and to partially break up the cones, the formations being too hard to wall them off. The tool joint on the junk sub failed, causing an extensive fishing job. This was not directly related to the loss of the bit cones except possibly due to rough running in prior milling operations. Also, if the bit had not failed, the junk sub would not have been run and would not have failed, thus eliminating the fishing job. The bit cones were subsequently cleaned out with two 12-hour mill runs. On each mill run, 5–10 lb of miscellaneous junk was recovered in the junk sub.

At first it was assumed that the bit had become worn while reaming. However, the reaming operation should have been easier on the bit than normal drilling. The total combined reaming and drilling time of about 43 hours was relatively short compared to the average minimum expected bit life of over 100 hours. The drilling recorder showed the bit had drilled 1 foot in the last 2½ hours of drilling. The torque began to increase at the start of the period and continued until it was abnormally high before the bit was pulled. The bit probably failed prematurely. It probably had some wear due to reaming, but this should have been insignificant in a hole with less than 8° deviation. The bit was definitely run too long after it started to fail, which resulted in the lost cones. That this running was excessive was further verified by inspecting the bit. It was obvious from the torque and drilling rate that the bit had been run on after the cones had locked.

The bits that caused the cones to be lost should not have failed based on running time and operating conditions. They were from the same manufacturer and the same batch. The manufacturer did not have any suggestions. Therefore it was assumed that at least part of the failure was due to faulty bits. The remainder of the well was drilled without problems using bits from other manufacturers.

Bits

Individual bits are seldom lost in the hole. Dropped bits occur in a few cases. They are seldom stuck, but they are difficult to catch and generally must be milled. Removing a dropped bit by milling is described in the example "Milling Up a Bit and Junk" in Chapter 4. Bits connected to the drilling tools can stick, but they generally are not too difficult to release. Often a relatively simple fishing job, such as fishing for a bit, can be complicated by other problems. A relatively common example of recovering a bit and bit sub and the associated problems are summarized in the following example.

Recovering a Bit and Sub

The connector on top of the bit sub washed out, leaving the bit and the bit sub in the hole. An overshot run was unsuccessful. A joint of washpipe was run to wash over the fish. The bumper jars failed, leaving the lower half of the jars and a joint of washpipe in the hole as a second fish. The washpipe was recovered with an overshot, the bit and bit sub were recovered with a second overshot, and the cones were removed by milling.

Stuck Bits

Stuck bits are relatively uncommon but do occur. The most common sticking situations are bits stuck in caving, fractures, or a tapered hole, and less commonly, keyseated. In all cases, begin working the pipe in the correct direction immediately. Increase the working force rapidly to the maximum safe working level. The bit normally is released without difficulty, especially if drilling jars and a bumper sub are run. Use maximum circulation to help erode the formation and prevent sticking due to caving higher in the hole.

Sticking in a tapered hole is most common in harder formations and normally occurs near bottom on a trip in the hole with a new bit. The most common cause of a tapered hole is drilling with an out-of-gauge bit—and possibly, failure to run a near-bit reamer. Unless the bit is very lightly stuck and can be worked free easily, always pull it and check for a damaged, pinched bit.

This type of sticking is almost always preventable. It is seldom a problem in softer formations and is indicated by smooth-acting, increased torque. It is more common in harder formations and at deeper depths. Run a stabilizer or roller reamer above the bit to either open the hole to gauge or to cause excess torque to indicate if the bit is out of gauge. When tripping in the hole, run the last few stands in the hole carefully, observing the weight indicator. If tapered hole is suspected, the assembly should be picked up frequently. A good practice is to ream the last 50–100 ft.

The bit may stick in a fracture while drilling. Circulation is not affected except for possible loss in the fracture. Sticking in a fracture is difficult to prevent, but fortunately it can almost always be released with the proper working procedures. Sticking only the bit in caving is relatively uncommon. Normally it can be worked or pumped out of the hole without undue difficulty.

Parted Bits and Large Junk

Bits and other large pieces of junk are almost always removed or walled off by milling. Normally a heavily dressed concave mill should be used. At least two junk subs should be run above the mill. Sometimes the junk rolls or spins, increasing the difficulty of milling it. It normally takes at least three mills, and possibly four, to mill away a 7-in. bit. Hard-formation roller bits are seldom used because they are inefficient, have a relatively short life, and create the risk of leaving cones in the hole.

Junk shots are more commonly used to break larger fish, like bits and similar-sized junk, into rubble. Observe the restrictions for using junk shots on bit cones as described in the section "Bit Cones and Bits." If the top of the fish is clean, then a junk shot run on a wire line may help break up the bit or other large junk. The junk shot may also be run on pipe with a skirt to clean out fill material on top of the junk; however, this is more questionable and the time may be better spent milling.

If the junk is small enough, run a junk basket. However, in most cases, junk (like a bit) is too large to fit into a junk basket. If there is a way to catch the fish, such as with a pin on top of the bit, then the fish should be screwed into and recovered. Normally if the bit is full gauge, it will not turn over in the cased hole. In the open hole, the bit may or may not turn over, depending on the size of the hole. The problem is determining the position of the bit.

Sometimes an impression block can be used to give an imprint of the top of the fish. However, it is less successful in the open hole due to cuttings above the fish obscuring the detail needed to determine the position of the fish. Generally the best procedure in this case is to run a tool joint that fits the bit. The fish top can be circulated clean before attempting to screw into the fish. In a few cases a taper tap may be used, but these cases are relatively uncommon; the taper tap is used more commonly in cased holes.

Consider using a circulating sub, because the bit jets are probably plugged with cuttings, especially if high-pressure zones are exposed in the well bore. Bit jets may be removed with a string shot, but it is not advisable in this case since it is not known how securely the fishing assembly is screwed into the bit. If several runs with the tool-joint box are unsuccessful, then additional runs probably are not justified.

A magnet normally will not be effective except on very small bits. A short grab overshot usually is not recommended because if the bit pin is looking up, there is probably not enough distance to make a catch. There is some question whether even a short grab overshot can take hold of the pin. If such a situation occurs, test the short grab overshot on a similar bit on the surface.

Other than with the methods noted, large pieces of junk normally must be milled up completely or broken up until they are small enough to catch with the procedures described for bit cones. These same general procedures apply to other moderately large pieces of junk, such as small subs, slip segments, and tong links. If the fish has a threaded connection, try to screw it in and recover. Junk shots should be used with discretion; however, do not hesitate to use them under favorable conditions.

Generally, the most positive approach is to mill the junk, including an upside-down bit. Sometimes this is difficult, especially when the junk spins with the mill. Considerable ingenuity is required to mill junk successfully.

DRILL-COLLAR ASSEMBLIES

Stuck drill-collar assemblies are the second most common cause of fishing. There is a high incidence of sticking in drill-collar assemblies, most frequently at the top or bottom. They can stick for most of the reasons listed as causing fishing. A stretch-type free point usually determines where the pipe is stuck down to the top of the drill collars. If it indicates sticking at or below the top of the drill collars then the drill collars are stuck, but it cannot distinguish which part of them are stuck. Wireline free-point tools are needed for this.

The basic procedures for releasing and recovering drilling assemblies are deceptively simple. However, the actual operations can be complex, depending upon specific conditions. When the drilling assembly sticks, either with or without drilling jars and a bumper sub (drilling jar-bumper sub), begin working it immediately using the general procedures described in the section "Working Stuck Tools" in Chapter 4. Use releasing agents if applicable. If the assembly is not released, then back off the free collars and recover them. Catch the remainder of the fish, work it free, and recover it; or back off and recover the free collars. The drill collars may be perforated to establish circulation, but this is less common. The drill collars cannot normally be cut with a jet cutter because of the thick wall.

If the fish is not recovered in this manner, wash over the fish to release it, then catch and recover it. It may be necessary to use different catch tools or to mill a fishing neck. Otherwise, plug back and sidetrack, plug and redrill, or abandon the hole.

Without Drilling Jars and Bumper Sub

If the assembly sticks and drilling jars and a bumper sub were not in the string, then work to free the assembly. Initially, working the stuck assembly, even without jars and a bumper sub, can be effective and is highly recommended. However, it is generally less efficient because the drillpipe acts as a spring, absorbing much of the working force before it reaches the sticking point.

For example, consider a fish stuck at 8,000 ft and the jar-bumper sub at 5,000 ft. The 3,000 ft of pipe between the jar-bumper sub and the sticking point acts as a spring, absorbing most of the force and preventing it from being applied at the point where the fish is stuck, and where the working action is needed most.

If the stuck assembly cannot be worked free, then back off at or shortly above the electric-log free point and recover the free pipe, including free drill collars. Run a fishing assembly with jars and a bumper sub, screw it into the fish, and work it, following the general procedures in the following section.

With Drilling Jars and Bumper Sub

Different companies have different policies for running drilling jars and bumper subs. However, except for very special situations, they are highly recommended, especially in higher-risk holes. Drilling jars and bumper subs can be used immediately if the assembly sticks below them. This greatly increases the chance of releasing the fish shortly after it sticks and eliminates the need of backing off to run the jar-bumper sub.

Jars and a bumper sub are highly effective tools, especially with additional drill-collar weight concentrated above the tools. However, they should be positioned near the sticking point for maximum effect—preferably within one, and not over two drill-collar lengths when possible. Drill collars also cushion the jarring and bumping action, although to a lesser extent than drillpipe because of their rigidity. Since they are an inert mass, drill collars require additional force to initiate movement and overcome inertial effects. Therefore, if several drill collars are free and working seems ineffective, it is best to back off, recover the collars, and reposition the jar-bumper sub closer to the sticking point.

One general rule in fishing is to select the easiest, most efficient method of recovering the fish, as illustrated in the following example.

Recovering Dropped Stand of Drill Collars

An on-structure wildcat well had $7^5/_8$-in. intermediate casing at 14,800 ft and was drilling a $6^1/_2$-in. hole at 15,200 ft with 13.2 lb/gal mud. On a trip to change bits, the last stand of drill collars were picked up through the rotary with a lift nipple. The lifting line, or cat line, was being used to pull the split rotary bushings so the pipe wiper could be removed. The line tangled in the elevator latch and opened it, dropping the stand of collars. The 115-ft-long fish was part of a packed-hole, hard-formation assembly. It included, from the bottom up, a bit, six-point reamer, replaceable blade stabilizer (RBS), short drill collar, RBS, shock sub, RBS, two drill collars, and a lift nipple.

A fishing assembly was made up using a crossover sub with a pin looking down to screw into the lift nipple. It was run and screwed into the fish, and the fish circulated and pulled without difficulty. All of the bit cones were intact.

The relatively simple fishing job was preventable. The fish was recovered in the simplest, most obvious manner with a minimum amount of lost time. It is also interesting that the cones were recovered; in most similar cases they would have sheared off.

Sticking the Drill-Collar Assembly

Sticking the drill collar assembly in sloughing, caving, and drill-cuttings accumulations is a common method of sticking. Keyseating and wall sticking are discussed individually in the respectively named sections later in the chapter.

The most common points of sticking are around the top of bit or drill collars, the top of the larger diameter collars, and at reamers and stabilizers. Sticking often occurs during connections and opposite formations known for sloughing and caving and in out-of-gauge hole sections. The most common indicators are restricted movement of stuck pipe and no circulation with increasing pump pressure. There are various procedures for releasing and recovering a drill-collar assembly stuck by caving and sloughing, as described in the following sections.

Working, Back Off, and Clean Out

After the fish sticks, begin working immediately as noted earlier. In this case, assume the drill-collar assembly, equipped with a jar-bumper sub, is completely stuck and circulation is minimal. Work the stuck assembly for at least 4–6 hours, increasing working force levels rapidly and jarring with the maximum force. Assume that the working is unsuccessful and that the circulation rate is decreasing, which indicates additional sloughing.

Run an electric free point and back off with a string shot at the deepest point, normally leaving at least one drill collar free. If there is no reason to expect problems catching the fish, then recover all the free collars, possibly leaving half a collar free.

In this case, assume that four stuck drill collars have been left in the hole. The choices here are either to go directly after the fish with a fishing assembly or to make a cleanout run first. Usually in a sloughing or caving situation a cleanout trip should be made, which also helps stabilize the hole. Run a cleanout assembly and clean out to the top of the fish. Do not damage the top of the fish; there is little risk in this case since it is a drill collar. Also take precautions to prevent plugging it. Circulate and condition the hole and pull the assembly.

Run the fishing assembly with a circulating sub and a safety joint. Screw into the fish and begin working, concentrating on working upward. In this case, the fish cannot be circulated, but circulation through the circulating sub helps ensure that caving and sloughing material does not settle around and stick the fishing assembly. Work the fish free and pull it out of the hole. If the fish cannot be worked free, back off and recover in multiple sections or by washing over, as described in the following sections.

Recovery in Multiple Sections

Assume the drill collars have stuck, but without a jar-bumper sub, and have been worked for 4–6 hours as described in the preceding section. In this case, the assembly is stuck at the top of the drill collars.

Run an electric free-point log. The best place to back off is at the top of the drill collars as indicated by the free point, assuming the hole is straight and no problem is anticipated in screwing back into the fish. If a problem is expected, such as a highly deviated or crooked hole, leave one or two joints of drillpipe on top of the collars. If the hole is out of gauge, back off in a straight, in-gauge section. In this case, back off at the top of the drill collars and pull the drillpipe.

Make a cleanout trip if necessary. Then run a fishing assembly with a circulating sub and screw into the drill collars. Work the fish with a jarring and bumping action, concentrating on working up. After several hours, run an electric free point. Assume the free point has moved downhole, indicating that part of the drill collar assembly is free. In this case, 4 of the 30 drill collars have been found free.

Continue working and check wireline free point periodically (at about 2–3 hour intervals). Remember to be patient. It is not uncommon to work more drill collars free, though generally a decreasing amount with time. At the same time, the condition of the fishing string may be deteriorating due to heavy working and jarring so it may need servicing. Assume that 12 drill collars have been worked free in about 8 hours. Back off these collars and pull them out of the hole. Be sure to service the fishing assembly.

Next, either make a cleanout trip or catch the fish again and work loose and recover another section. If the fish cannot be worked free, then back off and recover by washing over, as described in the following section.

Fishing and Washing Over

Assume the preceding case, with 14 stuck drill collars remaining in the hole. If they cannot be worked free, usually the next reasonable alternative is to wash. The fish is about 420 ft long, and includes 14 drill collars and a bit. The amount of washpipe to run depends upon various conditions, as described in the section "Washing Over" (Chapter 4). In this case assume 7-in. collars in an 8¾-in. hole with caving that caused the fish to stick. The clearance is very limited. Run about 250 ft of washpipe. This should allow recovery of some collars and others may be partially free. Do not run the inside spear because, after being washed free, the collars will not drop an appreciable distance, if any, and caving would also cushion the falling fish. The internal spear is a higher-risk tool, especially in caving conditions.

Open-Hole Fishing

The procedure for washing over, covered in detail in Chapter 4, is summarized here to clarify this specific example. A scalloped shoe and the washpipe should be run on a limber assembly. Circulate above the fish and then begin washing over it. Wash slowly, taking care not to overload the washpipe annular space with cuttings. Continue washing over for the length of the washpipe or until limited by excessive torque or by the washpipe beginning to stick.

Next, pull out of the hole and run a fishing assembly. Catch the fish and work it, while monitoring the free point and using it as a guide. Recover the entire fish or back off and recover a section of it. Then, depending upon specific conditions, catch and work the remaining fish or wash over.

Pumping Out Pipe

Pumping the pipe from the hole is a specialized "last resort" procedure. It is most commonly used on stuck drill-collar assemblies at shallow depths under favorable conditions and when other efforts to release the stuck tools (including strong working) are unsuccessful. It normally is applied to an assembly that cannot be circulated due to a bridge around it, such as caving. If the next step were to back off and wash over the stuck assembly, an attempt to pump the assembly out of the hole may be justified.

An assembly stuck with a cuttings bridge is like a piston, where the piston area is approximately equal to the cross-sectional area of the assembly at the point where it is stuck. If pressure is applied to the closed hole below the assembly, the force acting upward on the assembly will be equal to the pressure multiplied by the effective cross-sectional area of the assembly at the point where it is stuck. In practice, the pump-out force of the piston in combination with a high overpull applies a MAXIMUM upward force to the stuck drilling assembly. Stroking the jars further increases the force.

The mud weight normally will be the same on both sides of the caving bridge. The additional upward piston force is the amount of pressure that can be placed on the system multiplied by the piston area. The maximum pressure usually is equivalent to the breakdown pressure of the formation with a safety factor and allowing for mud weight.

The procedure can best be explained by considering a specific example. A typical drilling assembly has 4½-in., 16.60 lb/ft grade E drillpipe with 7-in. drill collars in an 8¾-in. hole with 10 lb/gal mud. The formation breakdown pressure gradient is 0.75 psi/ft, equivalent to 14.4 lb/gal mud, so the equivalent of 4.4 lb/gal mud can be applied as pressure.

A typical assembly weighs 190,000 lb in 10 lb/gal mud and is stuck at 10,000 ft. The maximum pull is 298,000 lb, or an overpull of 108,000 lb, assuming a 90% safety factor. An assembly packed off around 7-in. drill collars has an equivalent piston area of about 38 square inches. The 4.4 lb/gal mud equivalent at 10,000 ft is 2,300 psi which, applied to the piston area, represents an upward force of 88,000 lb. Therefore the total upward force of 196,000 lb (88,000 lb + 108,000 lb) can be applied against the formation at the stuck point. This can be compared to the normal overpull force of 108,000 lb at a total weight indicator reading of 298,000 lb as long as the assembly is stuck.

If the assembly was stuck due to a balled bit, the equivalent piston area would be 47 square inches. A pressure of 2,300 psi gives an upward force of 108,000 lb at the sticking point or, in combination with the overpull, a total upward force of 216,000 lb.

High pulling forces such as these normally are not recommended. While they have been successful in some cases, they usually are ineffective. However, if the assembly is stuck and the next alternative is to back off and wash over, attempting to pump the pipe out of the hole may be justified. It should not increase the difficulty of washing over, is easy to apply, and may release the stuck assembly. The higher mud pressure also may be effective in leaking through, cutting out, and removing the bridge, especially if the assembly can be moved a short distance (Fig. 5–4).

Pumping Out Stuck Drill Collars

An assembly was stuck at 8,600 ft. It did not move or release after being worked extensively for 12 hours, including pulling the maximum overpull of 120,000 lb. Based on free-point data, the assembly was stuck near the top of the drill collars. The next step would be to back off and wash over.

A maximum pull was taken and the drillpipe was pressured to about 2,100 psi. There appeared to be a slight movement, but it was hardly detectable. The pressure was reduced to 1,800 psi and the assembly was jarred. After each jarring stroke, the assembly was then pulled to the maximum overpull. It began to move slowly and was moved upward about 10 ft in one hour. A single joint of drillpipe was removed after working the assembly for another hour. A total of six stands were worked out of the hole in 4 hours by a combination of pulling, jarring, and pumping. At that point the assembly pulled free. The hole was then circulated, the assembly was pulled and checked, and a new bit was run. The hole was cleaned out, recovering a large volume of caving and slough material.

Open-Hole Fishing

Figure 5-4 Pumping pipe out of the hole

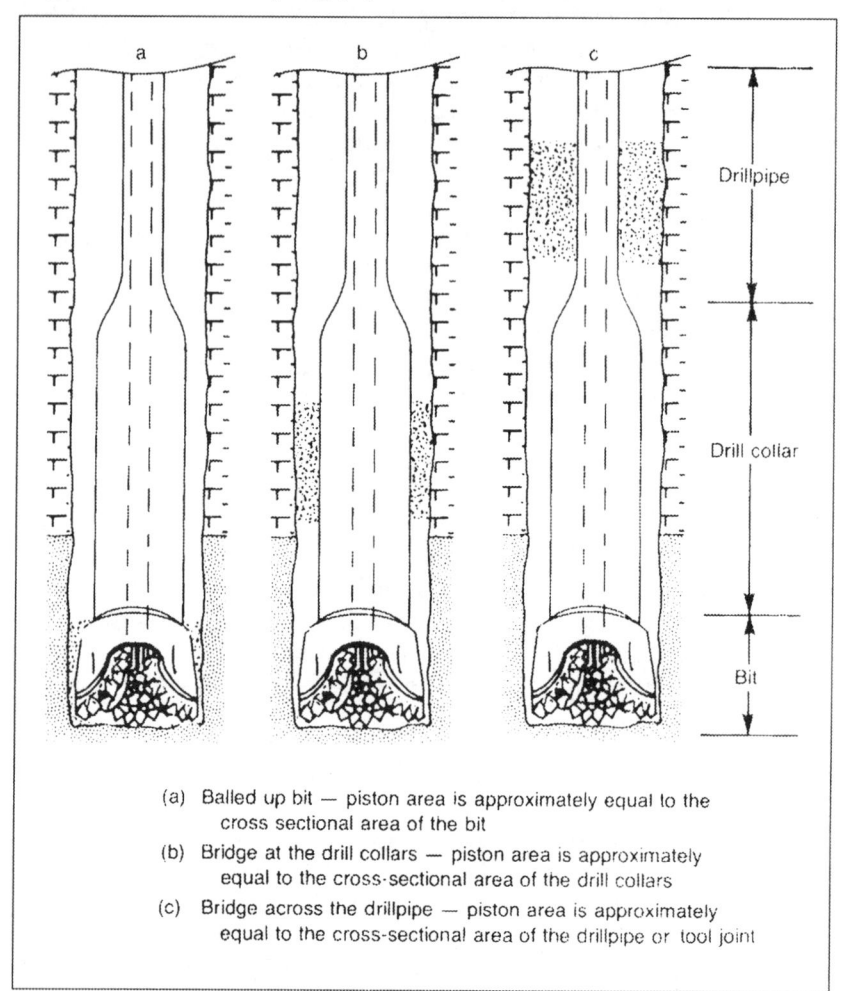

(a) Balled up bit — piston area is approximately equal to the cross sectional area of the bit
(b) Bridge at the drill collars — piston area is approximately equal to the cross-sectional area of the drill collars
(c) Bridge across the drillpipe — piston area is approximately equal to the cross-sectional area of the drillpipe or tool joint

    This procedure is seldom used and the following is offered without comment. The concept of applying a maximum force is akin to having excess overpull available. Excess force may not be the best solution, but sometimes it is the best available.

Inside Catch

    An inside catch may be used when the fish has a relatively large diameter that prevents it being caught with an overshot. This situation occurs most commonly in the drill-collar section and often is an alternative to milling a fishing neck. Fishing with a pipe spear is less common because it is not as strong as an overshot; however, a pipe spear is relatively strong and may be the best procedure under some fishing conditions. Taper taps are seldom considered for catching drill collars except in special cases.

Large drill collars such as a stand of 8-in. collars may be run above the bit in an 8¾-in. hole as a stiff drill-collar assembly to help drill straight. These collars are more likely to stick because of their relatively large diameter compared to the hole size and the annular restriction at the top of the collars and below the smaller collars above them. The top connection on the 8-in. collars can fail, leaving a pin "Dutchman."

The small annular clearance eliminates an outside catch such as an overshot. Therefore the alternatives are an inside catch such as a taper tap or spear to mill a fishing neck before sidetracking. The taper tap is seldom considered in this case. However, the pipe spear is a relatively rugged tool. Depending upon the amount of fish in the hole, its use might be justified. In the worst case, the spear can break off in the top of the fish. If the broken spear is not retrieved, then it can be milled off prior to milling a fishing neck.

The strength of the pipe (casing) spear is illustrated in the following example.

Fishing with a Fishing Spear

A well had 9⅝-in. surface casing at 1,500 ft and an 8¾-in. hole drilled to 10,000 ft. The hole had been logged and 7-in. casing was being run. When about 5,000 ft of 7-in. casing had been run, the pipe parted in the first thread below the collar, dropping the casing. The hole was cleaned out with a bit to the top of the fish at 4,000 ft, but the fish did not fall all the way to the bottom. An overshot could not be used to grab the fish because of the hole size and the lack of clearance over the fish. A casing spear was run on a fishing assembly and took hold of the inside of the fish. The fish was worked free in about 8 hours and pulled.

Fishing Neck

A fishing neck may be milled on drill collars and other large-size fish in order to catch them, usually with an overshot. The fish must have enough metal stock, and usually a smaller inside diameter, so after cutting the fishing neck it is strong enough to recover the fish. A common situation is milling a fishing neck on stuck drill collars (Fig. 5–5).

Figure 5–5 Milling a fishing neck

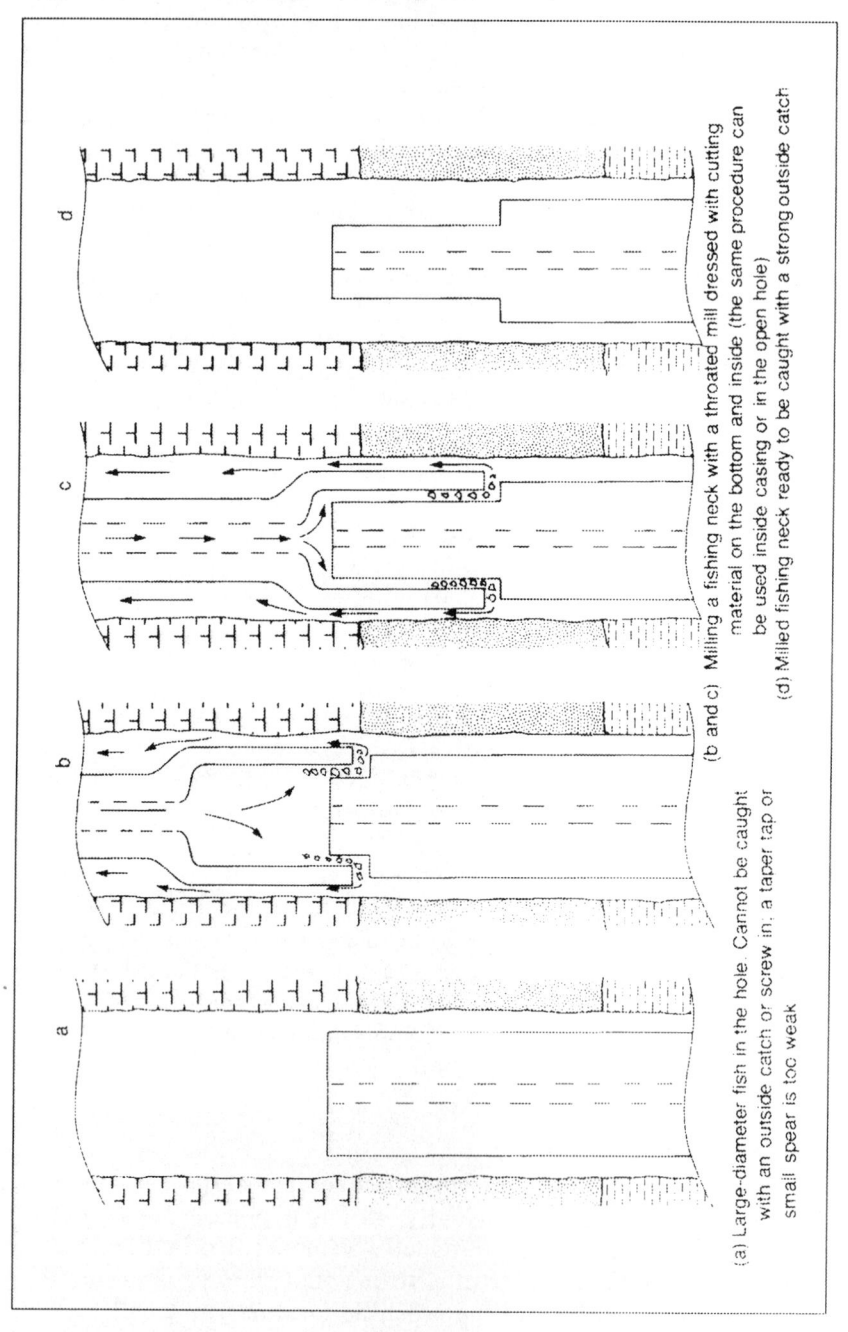

(a) Large-diameter fish in the hole. Cannot be caught with an outside catch or screw in; a taper tap or small spear is too weak

(b and c) Milling a fishing neck with a throated mill dressed with cutting material on the bottom and inside (the same procedure can be used inside casing or in the open hole)

(d) Milled fishing neck ready to be caught with a strong outside catch

Cut the fishing neck using a mill shoe with a flat or slightly scalloped bottom. If there is sufficient clearance, a thin skirt should be run as a guide to hold the mill centered over the fish. It also improves the removal of metal cuttings. Avoid a very thin skirt because of the risk of breaking it off and leaving more junk in the hole.

Mill over the fish conventionally, removing part of the metal on the outside of the fish. Normally the fishing neck should be cut about 2 ft long using one or two mills. Cuttings removal usually is not a problem. Sometimes the fishing neck is slightly out of alignment due to a crooked hole. The cut surface may also be uneven, increasing the risk of breaking the slips when catching them with an overshot. For these reasons it may be necessary to mill a clean top on the fish and cut the fishing neck again.

After cutting the fishing neck, catch it with an overshot and work and recover the fish in a conventional manner. Usually the inside of the fish is open, so a free point and string shot can be run for a partial recovery if necessary. If the fish is not open, try opening it by conventional procedures.

In a few cases, the inside diameter of the fish may be opened by milling to accept a pipe spear.

STUCK DRILLPIPE

Failures in the drillpipe section primarily include sticking, parted tool joints, and twisting off. Sticking is commonly caused by caving, sloughing, and similar actions in the bottomhole assembly, and may also occur during fishing for drill collars or deeper drillpipe. Sticking is a severe condition without circulation, which is an important reason for establishing and maintaining circulation. As with most sticking situations, work the pipe to establish movement and circulation. If this is unsuccessful, the two reasonable alternatives are to back off as deeply as possible and fish for the remainder or, less commonly, to perforate and try to establish circulation, possibly working more pipe free before backing off. The latter course is recommended when applicable and before backing off.

Parted drillpipe usually is due to a tool joint failure. Twisted off drillpipe is caused by over-torquing. Fish for both of these in a conventional manner.

Perforating for Circulation

Perforating to establish circulation is a method of releasing a stuck assembly. It is most common in the drillpipe section, but may be used in the drill collars.

First ensure that the annulus is plugged, such as with a bridge around the drillpipe, and not the pipe itself. The rate of pressure increase and the volume of mud required to pressure up often indicates where the plugging has occurred. A sharp pressure increase and low volume of mud indicates a plugged pipe. A leak-off test may also be used. The formation should take fluid at a maximum pressure equivalent to about 1.0–1.3 psi/ft, calculated at the bit and allowing for mud weight. If the formation does not take fluid at this pressure, then the pipe probably is plugged.

Bits commonly plug. Open them by blowing out the jets with a concentrated string-shot charge. They can be drilled out or walled off later without difficulty in most cases. Plugged bits have been opened by an impact from dropping a joint of one-inch pipe inside the drillpipe.

Sometimes surging may be effective in unplugging bits and removing other obstructions. Surging involves pressuring up the system to a moderately high pressure and releasing it rapidly. Start at relatively low pressures of 200–500 psi and gradually increase to the maximum allowable pressure. Care should be taken to prevent collapsing tubulars with relatively low collapse strength under moderately high surge pressure conditions. For example, when surging large casing with relatively heavy mud, it is possible to collapse the casing under high-pressure surging conditions. Usually there is less risk of collapsing stronger drillpipe.

If the stuck drilling assembly cannot be circulated, it is more difficult to release and recover. However, if part of the assembly can be circulated, that section usually can be released, backed off, and recovered. Normally use all other procedures to establish circulation (as described in the section "Working to Establish Movement and Circulation," Chapter 4) before perforating. Circulation may be established by perforating the drillpipe or, less often, the drill collars, but perforating is seldom used and often only as a last resort procedure.

Two of the more common causes of sticking and lack of circulation in the drillstring are the slumping and caving of accumulated drill cuttings and, to a lesser extent, barite sag, both of which create bridges around the assembly. Although these can occur anywhere, there is a higher frequency of occurrence around the drill-collar assembly. The condition often can be detected during working and fishing by a free-pipe log indicating a gradual, increasing restriction of pipe movement. This may occur over an interval as long as 500 ft. Pipe stuck in this manner often will exhibit vertical movement at a deeper depth than movement in rotation. Both actions are distinct from wall sticking and keyseating, in which a sharp reduction of pipe movement indicates the sticking point.

In an attempt to establish circulation, the pipe should be perforated in one or two places. The point to be perforated depends on various factors. It should be near and slightly above the point where the last pipe movement was detected. Use the stuck-pipe log as a guide. One hole should be made at the point of minimal movement where the pipe is about 10–20% free, if there is a short interval from 100% stuck to 100% free. If circulation is not established, make another perforation higher and try to circulate. If there is a long interval from the point where the pipe first began sticking to the point where it is 100% stuck, usually the lower part of the interval should be perforated. Often it is preferable to perforate too deep, then if circulation is not established perforate again at a shallower depth (Fig. 5–6).

Sometimes after establishing circulation, it may be possible to circulate deeper by using perforation ball sealers. The volume through an individual perforation and the overall circulation pressure may be limited when circulating through a sequence of perforations. This could prevent circulation from being established deeper in the assembly. A few perforation ball sealers added to the flowstream, by making a rotary connection, should preferentially plug the upper perforations, increase pressure, and improve the circulation rate lower in the hole. After pumping, the balls either float or drop, depending on the relative density of the balls and the mud. They generally will plug the jets of a jet bit. Before using them, be sure that ball sealers will not adversely affect future operations such as running a free point or a string shot inside the stuck assembly.

Circulating with a crippled pump may help release additional pipe in a stuck drilling assembly. Removing one or more suction or discharge valves and circulating at a moderately high rate often bypasses the pulsation dampener and creates a pulsating action that is transmitted to the stuck assembly and may help release it.

If the entire assembly is circulated, there is a better chance of releasing it by working. If circulation is established and part of the assembly is released, then back off and recover the released section. In most cases, a cleanout trip should then be made. Be sure to use the correct cleanout assembly. Continue fishing using standard procedures until all the drillpipe is recovered, then fish for and recover the drill collars.

If circulation is not established, back off and recover the free fish. Then either catch and work the remaining fish or wash over, depending upon the situation.

Parted or Twisted Off

Twisted-off or parted drillpipe causes fishing. The cause and prevention of twisted-off and parted drillpipe is covered in various earlier sections of the text.

Figure 5–6   Perforating to establish circulation

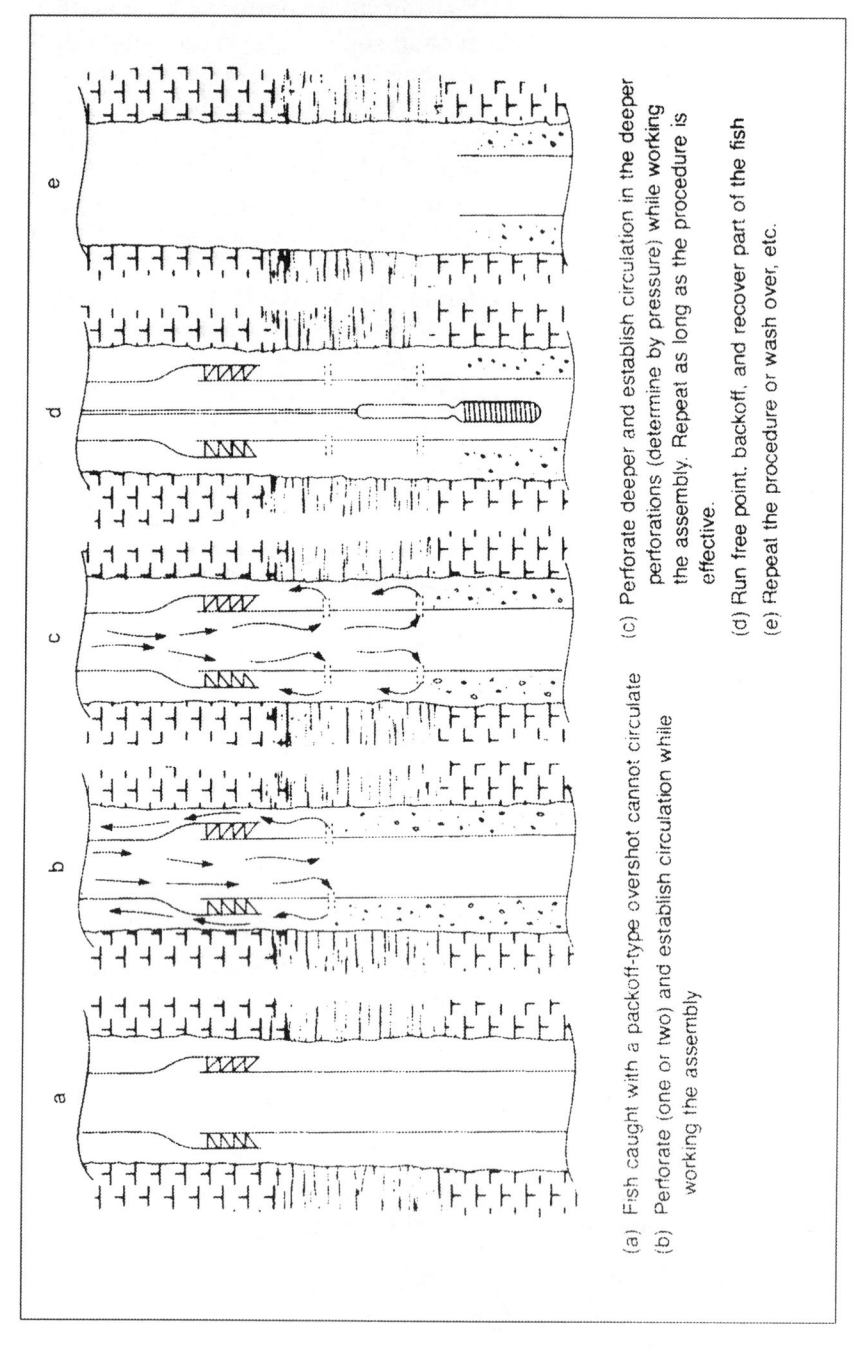

(a) Fish caught with a packoff-type overshot cannot circulate
(b) Perforate (one or two) and establish circulation while working the assembly
(c) Perforate deeper and establish circulation in the deeper perforations (determine by pressure) while working the assembly. Repeat as long as the procedure is effective.
(d) Run free point, backoff, and recover part of the fish
(e) Repeat the procedure or wash over, etc.

336   Chapter 5

The pipe should be pulled out of the hole after it parts. The depth to the point of failure is approximated from the amount of pipe weight remaining, and the correct depth is based upon the amount of pipe pulled. The appearance of the recovered piece of pipe will indicate the type of tool to be run. If the pipe unscrewed, run a mating connection on the fishing assembly and screw into and recover the fish, working as needed. If the pipe parted cleanly in the body without excess distortion, catch it with a regular overshot. If the part is near and above a tool joint, run a double-bowl overshot to catch either the pipe body or the tool joint. If the parted section is necked down, an extended bowl overshot may catch the undamaged pipe below the necked-down section. Dress off or cut a new top if the top of the fish is twisted, bent, or otherwise distorted. Then fish for it in the normal manner.

Out-of-Gauge Hole

A fish top laid over in an out-of-gauge hole section can be very difficult to catch. Out-of-gauge hole sections can occur anywhere in the hole, but they are more common in the upper part of the open hole where there has been more time for out-of-gauge hole to develop. Knuckle and bent joints, bent subs or pins, and extension joints can be used above the overshot, and wall hooks and mule shoes can be used below the overshot or other catch tool. This is described in the following example.

Catching a Fish in an Out-of-Gauge Hole

A hole was being drilled with 4½-in. drillpipe. About 400 psi of pump pressure was lost, and the weight indicator showed a reduced weight. When the assembly was pulled, the pin of the top drill collar had failed, leaving the drill-collar assembly minus one collar in the hole. Overshots with wall hooks were run, but they passed down beside the fish. Several runs were made with a knuckle joint without success.

A joint of drillpipe was bent smoothly to a 7° angle. The overshot was run on this bent joint, but it would not engage the fish. The depth measurement to the top of the fish was verified with the induction log. The angle of the bent joint was increased to 15°, and a wall hook was run on the bottom of the overshot. The bend was so severe that the assembly had to be started into the hole by pulling it down with a lifting line. The fish was caught and recovered. When the hole was subsequently logged, the section where the top of the fish was located was out of gauge past the limit of the logging tool. The change in deviation at this point was about 2°/100 ft.

Accidental Sidetracking

Accidental sidetracking while fishing is relatively uncommon if proper precautions are taken. It can be prevented in most cases. If the hole is sidetracked, it often is very difficult to reenter the old hole. Even if the old hole is reentered, there is a risk that the casing will hang or enter the sidetracked hole instead of the original hole. It may be necessary to intentionally sidetrack higher in the hole to ensure entering the correct hole. If sidetracking occurs while fishing for the drilling assembly, it often can be considerably higher in the hole, resulting in the loss of a longer section of drilled hole and possibly part or all of the fish.

Unintentional sidetracking usually occurs in soft formations and where the top of the fish is in a washed out or crooked section of hole where it is difficult to catch. In this case, it is important to know where the top of the fish is. Then tools such as wall hooks, bent joints, and kick subs can be used to catch the fish.

In some cases the original hole and fish may be abandoned and drilling continued in the unintentional sidetrack. This decision can be made only after evaluating all the circumstances of the specific fishing job. There may be a risk that the sidetracked fish will fall into the hole later and stick the drilling assembly or cause another sidetrack. It may be possible to cement the fish in place by filling as much of the original hole as possible with cement slurry. Applying pressure helps force part of the cement out into the formation and around the fish, thus helping ensure that the fish remains immobile. However, there is a question as to where the cement goes. In a few limited cases, this may be solved by cementing through a stinger run down beside the fish.

The most positive method of ensuring a good hole is to cement the original and sidetracked holes. Then the hole can be sidetracked higher in the normal manner. Magnetic interference from the fish may affect orientation instruments if the direction of the sidetracked hole is controlled.

KEYSEATS

Keyseating can be a major problem. The description, prevention, development, detection and removal of keyseats is described in the section "Keyseats" in Chapter 2. The methods for releasing keyseated assemblies are covered here. These methods depend upon where and how the assembly is keyseated, if it can be rotated, the direction and amount of reciprocal movement, and general hole conditions. When faced with a keyseating problem, begin working the assembly immediately, but work it correctly based on where and how it is stuck.

The similarity of keyseating to wall sticking or the possibility of the keyseated assembly wall sticking should not be overlooked. One of the more common problems that occurs with keyseating is not the keyseat itself, which can be sufficiently severe, but the higher risk of wall sticking while the assembly is stuck in the keyseat. This complicates the problems of releasing the stuck assembly because other releasing methods must be considered, and it increases overall risk. Therefore, if the keyseated assembly is free to rotate, turn it periodically while working to help prevent wall sticking. Otherwise, use torque during working to help release the assembly.

Most keyseated assemblies can be circulated. In most cases the operator should circulate the hole while working the pipe. This serves several purposes. It allows more direct and positive control of the well—a potentially important safety factor, especially if high pore pressure formations are exposed in the wellbore. The movement of the pipe while it is being worked can cause caving. Circulating removes the caving material and helps prevent further sticking. To a limited extent, the erosional effect of the circulating mud helps prevent wall sticking and may help erode away part of the keyseat.

If any procedure or combination of procedures tends to allow upward movement, then repeat those actions. If the drilling assembly moves upward during early working but the rate of upward movement decreases with time, then most likely the keyseated part of the drilling assembly is digging deeper into the formation. When this happens, stop the current operation and try another procedure.

The typical first reaction after sticking in a keyseat is to pull the pipe loose. However, this should not be done, especially early in the working procedure. If the pipe is free to move downward, be sure it can move downward while working since this may provide one of the best methods of recovering it.

Where and How the Assembly Is Stuck

Knowing where and how a keyseated assembly is stuck helps determine the method of working and releasing it. Any point in a downhole assembly that has a larger diameter than that in the adjacent upper section is subject to sticking in a keyseat. In standard drilling assemblies these points include the drillpipe tool joints, the top of the drill collars, the points where drill-collar size changes from smaller to larger collars, the reamers, the stabilizers, and the bit.

The most common point is at the top of the collars and, less frequently, at the bit. Tool joints seldom keyseat because they form most keyseats. Drill collars have stuck when the diametrical difference in collar sizes was in the order of ½ in., but this seldom occurs. Collar diameter variance is not considered to be significant until it approaches 1 in.; this also may indicate an unbalanced connector, especially in smaller collars, and a crossover sub generally is used. Reamers and stabilizers seldom keyseat because they generally have tapered ends that help lift the tool away from the wall so that it rides up out of the keyseat.

The frequency and severity of keyseating at the drilling bit is second only to keyseating at the top of the drill collars. As a general rule, keyseating at the bit is not as severe as keyseating at the top of the drill collars. Usually, the bit can be released if it has not been pulled into the keyseat with an excessive force, especially if drilling jars or a bumper sub are run in the upper drill-collar assembly.

It may be more difficult to release an assembly stuck below the top of the collars in a low-side keyseat. The bit may be forced into the wall of the hole. The amount of force depends upon the relative stiffness of the drill collars and the angle of the dogleg. If these are sufficiently great, the drilling assembly cannot be forced downward because this motion pushes the bit into the wall, increasing the difficulty of releasing the stuck drilling assembly. If working up or down is unsuccessful, this situation generally requires washing over.

There is a better chance of releasing a keyseated assembly by working if drilling jars and a bumper sub are run and the assembly sticks below them. Also, if the correct type of keyseat wiper is run on top of the collars, the assembly generally can be released by working if it sticks above the jar bumper sub.

Pipe Free to Move Downward

There is a better chance of releasing any assembly that cannot move up through a keyseat but is free to move downward. Therefore, do not pull the assembly into a keyseat so hard that it sticks and cannot be moved downward. As long as the drilling assembly is free to move, there is a reasonable chance of working it up though the keyseat or wiping the keyseat out; there also is less risk of wall sticking. However, if the drilling assembly is completely stuck, it may be necessary to wash over, a higher-risk job.

Assuming the drilling assembly is free to move down, connect the kelly and circulate. It is much easier to rotate pipe suspended from the kelly than from the elevators. This is important because rotational movement frequently will release the drilling assembly from a keyseat by rolling it out and upward, or will permit reaming out the keyseat enough to release the pipe. Keep the pipe moving and rotating, even at a slow rate, to prevent or reduce the possibility of wall sticking and to ensure free downward movement. Do not take a stretch-type free point reading at this time. The extra pull required to obtain stretch data may be sufficient to cause sticking or to lose downward movement. Also, stretch data is not accurate enough to determine which part of the drill collar assembly is stuck.

An assembly stuck below the jar-bumper sub is indicated by some free travel in these tools. If this occurs, then the sticking situation probably is not too severe. The drilling assembly often can be worked though the keyseat and pulled out of the hole. Otherwise, if circulation and light working does not release the keyseated assembly, try one or more of the following procedures.

Working and Rolling Out

One releasing procedure is to rotate the pipe out of the keyseat. Pull the drilling assembly up into the keyseat with an overpull of 3,000–6,000 lb. Then rotate it from about one to three turns to the right, depending upon the depth. A reduction in overpull indicates the drilling assembly is rolling out of the keyseat or cutting it deeper. Cutting is undesirable at this point, but generally it will not be severe with only a limited amount of overpull.

The pipe should then be lowered to ensure that it is still free to move downward. Repeat the procedure several times, gradually increasing the torque to the maximum safe level if necessary. In order to ensure downward movement, do not pull excessively. Mark the pipe opposite a fixed reference point to determine if the pipe is being worked upward or the keyseat is wearing deeper. If the assembly is released, then pull it out of the hole and remove the keyseat by reaming.

This method can be very effective if left-hand spiral reamers or stabilizers are keyseated. It is somewhat effective when the drilling assembly is stuck at the bit. Normally the pipe is worked with rotary and kelly as noted. If slips are set around the pipe in the rotary, take the usual precaution of tying the slip handles together. If they jump out of the rotary or break, this may prevent a piece of the slips from falling down the hole.

Wearing Out the Keyseat

The next releasing procedure is to attempt wearing out or wiping out the keyseat. If the correct type of keyseat wiper is on the assembly, then use the procedures described in the section "Keyseat Wiper" in Chapter 3.

Otherwise, the operator should alternately pull up into and drop out of the keyseat, with and without rotation. Mark the pipe opposite a fixed reference point to measure upward movement. Initially the assembly should be pulled up with about 3,000–5,000 lb overpull

Open-Hole Fishing

and then dropped back out of the keyseat. The assembly should then be lowered after every 5–10 pulls to pass tool joints and to ensure that it is still free to move downward. If the assembly keyseats below the top of the collars, do not lower it or else the top of the collars will move below the keyseat.

Repeat the procedure at successively higher increments of overpull. When the pulls are high enough, take a stretch free-point reading to locate the point of keyseating or ensure that the point is at or below the top of the collars. Continue increasing the overpulls to the maximum safe level, always dropping back down to ensure free downward movement.

The maximum safe level of pull in this case is based upon the pipe weight above the keyseat that is available for pushing the pipe back down out of the keyseat. Note that when the assembly is pulled into the keyseat and lightly stuck, the drill-collar weight below the sticking point is temporarily ineffective for pushing the assembly out of the keyseat. Therefore the maximum safe level of pull is about equal to the weight of the pipe above the keyseat, reduced by a safety factor. This limitation is necessary in order to have sufficient pipe weight available to push the drilling assembly down and out of the keyseat, thus ensuring free downward travel. The maximum safe level of pull may also be increased if the bumper sub is free and working, since it increases the efficiency of bumping down.

If the foregoing method is unsuccessful, try bouncing the drilling assembly out of the keyseat. This may be effective, especially with lighter drilling assemblies at lesser depths. Pick the assembly up until the neutral point is at the keyseat and mark the pipe opposite a reference point, similar to the procedure for determining movement. Take an overpull equivalent to about 50% of the assembly weight above the keyseat. Then, drop the pipe rapidly (using good judgment) and stop it abruptly with the free-point mark on the drillpipe slightly above its original position. In this case, the drillpipe acts as a spring. The sudden stop causes the pipe to elongate or stretch, then snap back to its original length. At the point of the keyseat, the assembly will strike the formation with a sharp blow. The operation requires dexterity but can be very effective.

An alternative method is to take an overpull as noted, then lower the assembly in a series of sharp drops to slightly above the original reference mark (again using good judgment). This imparts a whipping motion to the assembly and may help throw it out of the keyseat. Repeat these procedures without allowing the drilling assembly to move below the point where it was marked originally. If the free point appears to move upward as the pipe is picked up, it indicates the drilling assembly is wearing down the keyseat or is riding up and out of it.

These procedures either will release the drilling assembly and allow it to ride over the keyseat or will cause the keyseating part of the drilling assembly to become more deeply imbedded in the formation. Make sure that the pipe is still free to move downward.

In practice, when a keyseat wiper has been installed, the pipe can be worked harder, both by pulling into and bumping out of the keyseat and by rotating, with less risk of sticking in the keyseat. This additional margin of safety justifies the use of a keyseat wiper any time there is a risk of keyseating the drilling assembly. Also, the advantage of having drilling jars or a bumper sub in the upper drilling assembly when the drill collars are keyseated below them is obvious. The tools can be used to help jar the drilling assembly into and bump it out of the keyseat. It is important to ensure that the drilling assembly remains free to move downward. If the drilling assembly is keyseated above the tools located in the upper drill-collar assembly, then these tools are ineffective for helping to release the assembly.

Reaming Out the Keyseat

Do not overlook the possibility of reaming the keyseat with a string reamer if the pipe is free to move downward. This is a special procedure that may be used if the distance from the keyseat to TD is more than the depth to the keyseat. The procedure is described in the section "Reaming" in Chapter 2.

Set Pipe on Bottom

If the recovery procedures discussed earlier are unsuccessful, the next step is to run the assembly back in the hole, set it on bottom, and back off or cut the pipe above the drill collars. The drillpipe usually can be pulled out of the hole past the keyseat without difficulty. Then the keyseat can be reamed out and the fish recovered. This assumes that the stuck tools are free to move downward as emphasized when discussing releasing procedures (Fig. 5–7).

This has several advantages. It is a positive approach to removing the keyseat and recovering the fish. The drilling assembly has been or should have been in the hole for 24–36 hours and has been subjected to relatively heavy working stresses. Therefore, it is time to inspect and repair it and tighten the tool joints before a failure occurs. Doing this allows most of the drillstring to be recovered in case the fishing job is unsuccessful and saves a longer section of hole.

There is always a strong hesitancy to set any portion of the pipe on bottom because of the possibility of wedging or packing off caused by settling drill cuttings or cavings, or less likely, by wall sticking.

Open-Hole Fishing

Figure 5-7 Setting drill collars on bottom and removing keyseat

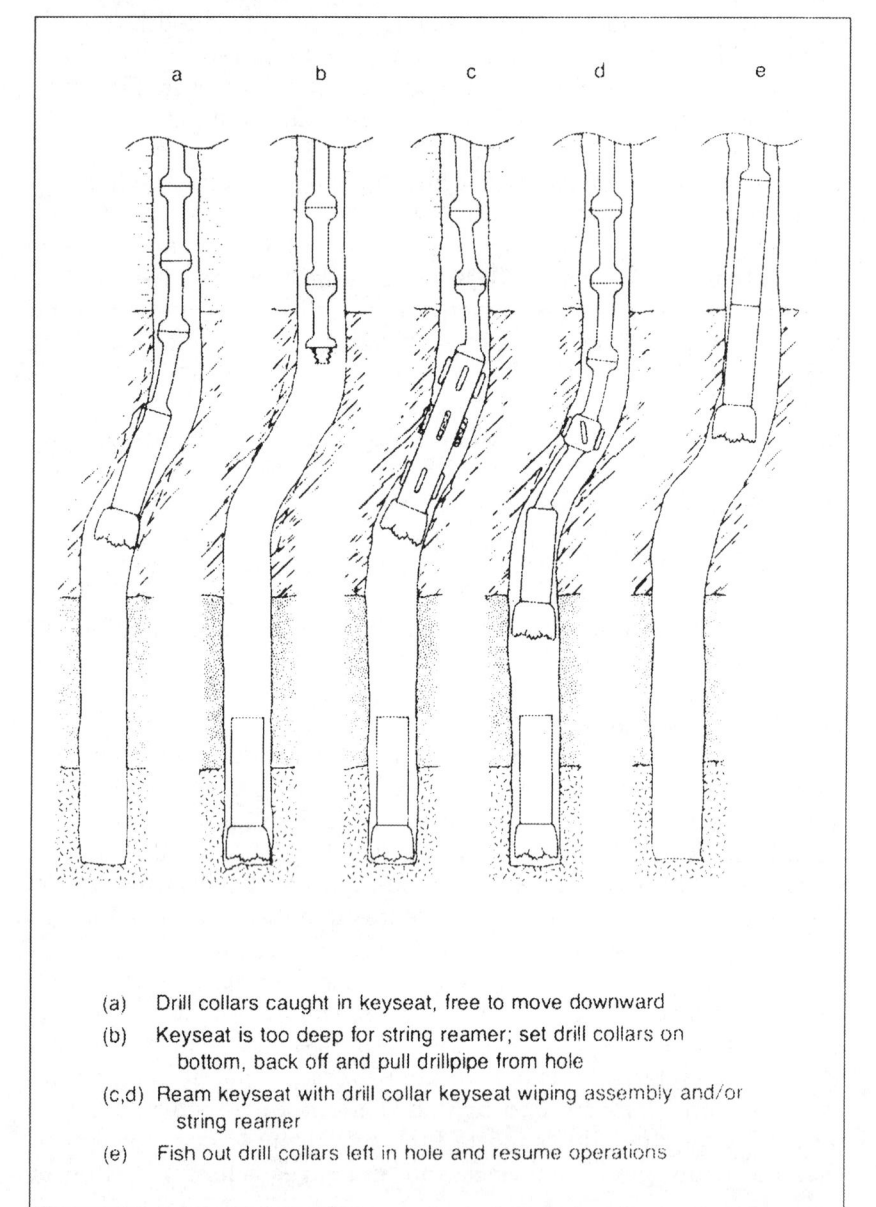

(a) Drill collars caught in keyseat, free to move downward
(b) Keyseat is too deep for string reamer; set drill collars on bottom, back off and pull drillpipe from hole
(c,d) Ream keyseat with drill collar keyseat wiping assembly and/or string reamer
(e) Fish out drill collars left in hole and resume operations

However, this is strongly preferred over continuing to work on the keyseat until the pipe firmly sticks. It is easier to fish for a drill-collar assembly at the bottom of the hole than a full string of drill tools. Also, in hindsight, the keyseat should have been detected and removed before it caused the problem.

The pipe should be run back to the bottom. To help prevent the bottomhole assembly from sticking, mix a pill of higher viscosity mud with about 8% diesel oil and 20 lb/bbl of coarse-grade walnut hulls. Circulate the hole clean and displace the pill around, and about 300 ft over, the part of the assembly to be left in the hole—in most cases the drill collars.

Assuming that the top of the drill collars has keyseated, either back off at the top of the drill collars or leave one joint of pipe on them. If it is necessary to cut the drillpipe, leave about a 10-ft pipe stub. If the collars were keyseating in the drill collar section, consider recovering part of the collars. Generally, however, it is best to leave all the collars in the hole. If the top of the collars does stick in the keyseat while pulling the pipe—and there is a good chance of this—then the overall complexity of the fishing job may be increased considerably. Sometimes the assembly will turn on bottom so torque cannot be worked down for a back-off. Try bumping down slightly to partially stick the collars. Otherwise, cut the pipe with a jet or chemical cutter. If the pipe does not cut off cleanly, cut several times at the same place and break the pipe at the cut by torquing with weight on the pipe.

Recover the workstring and inspect and repair it as needed. Then wipe out and remove the keyseat by reaming, as described in the section "Reaming" in Chapter 2. Clean out to the top of the fish and recover it by the procedures described in the section "Stuck Drill-Collar Assemblies" in this chapter. In many cases it will be surprisingly easy to recover the assembly. Do not overlook the importance of working it, and increase the working force levels rapidly.

Depending on hole conditions and related factors, this is a moderate-risk procedure. It may prevent dropping pipe with conventional back-off. Otherwise, in most cases, this procedure saves more of the tubulars and more of the hole if the fish must be sidetracked.

If there is some good reason for NOT setting the pipe on bottom, then pull the drill collars firmly into the keyseat until they stick. Then recover them using one of the procedures in the following section.

Assembly Firmly Stuck

The drilling assembly may stick firmly in the keyseat, so that it is unable to move up or down. The assembly usually cannot be rotated, but it almost always can be circulated. If the assembly can be rotated, the situation is slightly less severe, since the assembly is not wall stuck and the rotation may lead to freeing the pipe, at least in the downward direction. Overall, however, this sticking condition is more severe and there are less alternatives for releasing the pipe than there are for pipe that is free to move downward. This is an important reason for running the correct keyseat wiper on top of the drill collars and not pulling too hard into the keyseat.

The general working procedure is slightly different for keyseated pipe as compared to other stuck pipe. The assembly commonly sticks at the top of the drill collars, but it also may stick in the lower drill-collar assembly or at the bit. The working procedures are similar. Always continue circulating, sometimes at a reduced rate.

If the drilling assembly is stuck at the top of the drill collars and with the correct, sleeve-type keyseat wiper in the string, then bump and rotate the stuck pipe down and out of the keyseat and treat it as described in the previous section "Pipe Free to Move Downward." There is a better chance of releasing the assembly when this type of keyseat wiper is run above the drill collars. If the assembly cannot be released, then employ the procedures described in the following sections.

Keyseated at the Bit

If the pipe is keyseated at the drillbit without any movement, then there is a better chance of releasing it, especially if the drilling jars and a bumper sub are in the drillstring. This is illustrated in the following example.

Releasing a Keyseated Bit

A well had 9⅝-in. casing set at 2,500 ft and an 8¾-in. hole drilled to 11,000 ft using 13 lb/gal mud, with a 7-in. drill-collar assembly weighing 73,000 lb in mud, including drilling jars and a bumper sub, and 4½-in. drillpipe. The weight of the drilling assembly was about 210,000 lb. The assembly keyseated at about 9,500 ft. The jars and bumper sub were working, which indicated that the keyseating was below them, and most probably at the bit, since all the drill collars were about the same size.

There was a good chance of releasing the assembly keyseated at the bit or in the lower part of the assembly with drilling jars and a bumper sub run and working. The pipe was worked starting by pulling up and stroking the jars at 20,000 lbs over the string weight, then slacking off and bumping down with 20,000 lbs. This was repeated while increasing the set-down weight periodically, but working the pipe down did not release the bit.

The pipe then was worked with higher pulls and greater set-down weights. This procedure was repeated, employing the increasingly higher pulls and correspondingly harder downward jarring. After working for about six hours, the bit came free.

An alternate working procedure that is sometimes used includes bumping down with one weight for a period of time and then jarring upward at a similar weight for about the same period of time. The force levels are increased periodically up to the safe strength limits of the assembly. Depending upon the type of downhole tools being used, right-hand torque may be applied while bumping and jarring.

Stuck at the Top of the Bottomhole Assembly

When the drilling assembly is keyseated at the top of the drill collars, there is little chance of pulling it out of the hole with a straight pulling action. The best approach is to work the pipe down and out of the keyseat. First, the pipe should be worked downward with standard procedures. Working options are limited if jars and a bumper sub were not run or the assembly is stuck above them. The pipe should be worked using the procedures described in the section "Working Stuck Tools" in Chapter 4, concentrating first on working down. Also try applying torque and the snap or bounce procedure. If wall sticking is suspected, use some of the procedures described in the section "Wall-Stuck Assemblies," if applicable.

If the drilling assembly is worked free in the downward direction, release it with the procedures described earlier in this section. Otherwise, work the pipe successively harder until one of the following occurs.

- The pipe is released.
- The overpull approaches the maximum safe tensile strength of the drillpipe.
- The set-down weight approaches the maximum safe working level.
- Hole condition begins to deteriorate, the jars or bumper sub fail, or the assembly needs servicing.

If the drilling assembly cannot be worked free, prepare to back off and recover the fish.

Back Off and Recover

If the assembly is not released by working, then back it off. The last working action should have been to work the stuck pipe upward at increasing force levels and maximum pulls. Even if this was not successful, it helps ensure that the pipe is tightly stuck so there is less risk of it falling when it is backed off.

The fish will almost always drop when it is released by backing off. This can severely damage the dropped string by bending or kinking the drill collars, bending and (less frequently) breaking the drillpipe, and almost invariably shearing the cones off the bit. Sometimes the dropped assembly will break down the lower formations due to the temporarily increased pressure from the piston effect as it falls. The damage to the equipment and the hole and possible multiple fishing jobs are the main reasons for setting the pipe on bottom when it is free to move down.

Ream and wipe out the keyseat if the backed-off assembly drops (Fig. 5–8). Then recover the fish using the procedures described in the section "Sticking the Drill-Collar Assembly" in this chapter.

Figure 5–8 Drill collar keyseat-wiping assemblies

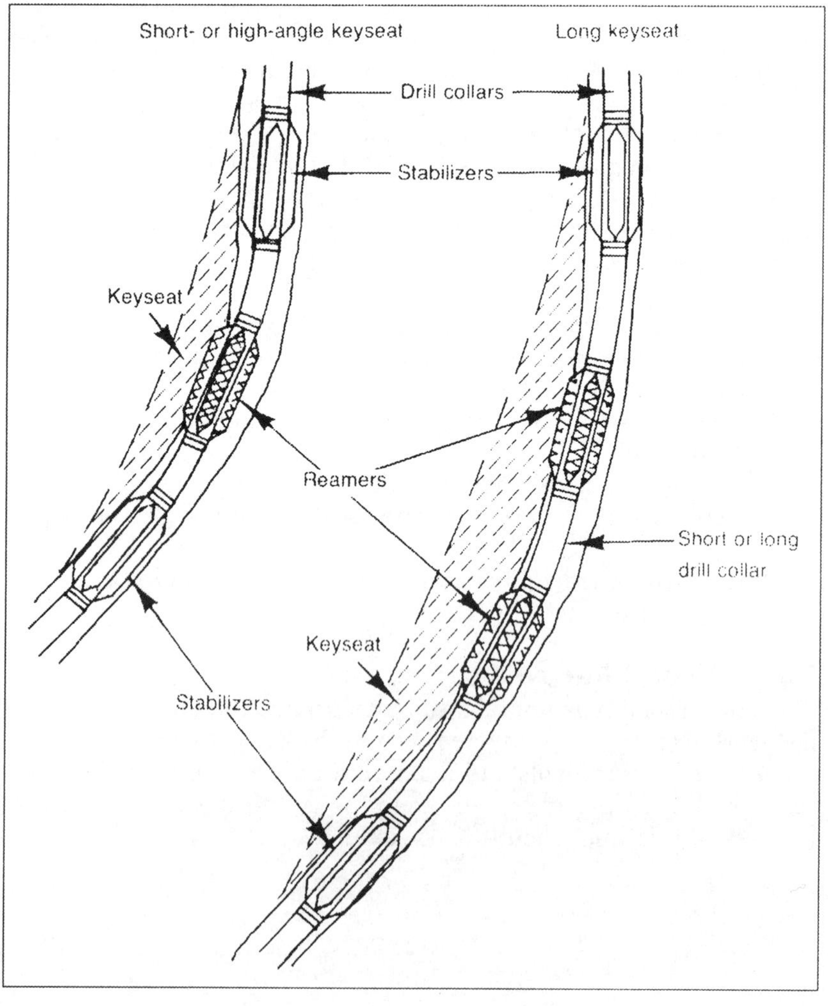

Chapter 5

If the backed-off stuck fish does not drop, there are two reasonable alternatives to try. One is to run a fishing assembly, positioning the jar-bumper sub closer to the stuck assembly in order to work the fish more efficiently. This procedure may be used if the stuck pipe did not have a jar-bumper sub or was stuck above these tools, and depending on how well the stuck pipe was worked before backing off. The working action before backing off may indicate if this fishing procedure is justified. Another consideration is the fact that generally the success of working and releasing assemblies keyseated at the top of the drill collars is only minimal at best. The words "reasonable alternatives" are used above because other alternatives might include sidetracking or redrilling depending upon specific conditions.

If the fish is to be worked, catch it with a fishing assembly and try to work it down first. If successful, set the fish on bottom and continue with the procedures described earlier in this section. Otherwise, the fish should be worked in the standard way. If it cannot be released, prepare to wash over.

The other alternative is to wash over and release the pipe from the keyseat without additional working. Again there are two alternatives. If the fish is washed over without the internal spear, it will drop, causing the possible damage and fishing job(s) as noted. There is a higher risk of sticking the washpipe containing the internal spear. Evaluate this risk against the possibility of severely damaging the fish if it is allowed to fall. This question should be resolved based upon experience and the specific conditions in the hole. As a rough guide, use the internal spear at shallower depths, when there are few formation problems, when clearances are acceptable, and when the keyseat is expected to be relatively short so a minimum amount of washing over is required. The pipe usually is backed off close to the keyseat, so run only 60–90 ft of washpipe to reduce the risk of sticking.

Wash over without the spear in the normal manner, as described in the section "Washing Over" in Chapter 4. Cut the fish loose from the keyseat, allowing it to fall. Ream and wipe out the keyseat and recover the fish using the procedures described in the section "Sticking the Drill-Collar Assembly" in this chapter. Otherwise, wash over the fish with the internal spear and recover the fish. Often this type of washing over is less severe than many other wash-over operations.

Keyseated at the Top of the Drill Collars

An on-structure wildcat had $9\frac{5}{8}$-in. casing set at 2,500 ft and an $8\frac{3}{4}$-in. hole drilled to 11,000 ft with 7-in. drill collars and $4\frac{1}{2}$-in.

drillpipe in 13 lb/gal mud. The weight of the drilling assembly was 209,719 lb ((800 x 114)+(10,200 x 16.6)) x (0.805). The weight indicator showed about 220,000 lb, which included the kelly and traveling block. The top of the drill collars became stuck in a keyseat at 9,500 ft. The weight of the 9,500 ft of free pipe was 137,000 lb in 13 lb/gal mud.

There was no pipe movement so the pipe was worked for about 30 minutes before a connection was made to circulate. The connection was not made immediately because the tool joint was about 15 ft above the rig floor. Normally the kelly should be connected immediately before starting to work the pipe, if it is convenient with the tool joint near floor level. If there is any danger due to high-pressure zones, install the inside blowout preventer to ensure positive control. When working the fish in this manner, it is best not to allow the bottom of the kelly to pass below the rotary.

The pipe was bumped down for about one hour, pulling up to 20,000 lb over the free point and slacking down to 20,000 lb below the free point. The set-down weight was increased to 40,000 lb for the next hour, and to 60,000 lbs for the hour after that. The main objective at this time was to maintain a downward force on the formation at the point of the keyseat with the drill-collar weight. The pipe was worked to alternate this force and further stress the formation in order to release the stuck drill-collar assembly downward.

Since there was no pipe movement, the overpull was increased in 20,000-lb increments at 1-hour intervals while slacking off 60,000 lbs after each pull. The pipe could not be rotated and torque was tried periodically without success. When this still did not release the pipe, it was worked upward at the maximum force level.

Backing off and running a fishing string was considered, but this procedure was not done because the pipe was worked at relatively high force levels without success. It was decided to take the positive actions of backing off and washing over. The job was completed successfully without dropping the drillpipe.

WALL STICKING

Wall sticking, or pressure differential sticking, is a relatively common and often severe problem. Frequently it is not recognized, especially when associated with keyseating. The description, prevention, and detection of wall sticking is described in the section "Wall Sticking" in Chapter 2. The procedures for releasing wall-stuck assemblies are described here.

There are various methods of releasing wall-stuck assemblies. Selecting the correct method depends upon the specific conditions. An operator should be familiar with all procedures and should take action immediately.

Working

Begin working with wall-stuck assemblies immediately, the same as in all sticking situations. Sometimes it is difficult to differentiate between wall sticking and keyseating. It is important to identify the sticking mechanism because, for maximum efficiency, the initial methods of working can be different. Keyseated assemblies should initially be worked downward. If the wall-stuck assembly is not on bottom when it sticks, then it initially should be worked to establish movement and to increase the working force levels rapidly. If there is a question about the type of sticking, then work downward. (It is assumed here that the type of sticking has been ascertained.)

Working a wall-stuck assembly immediately increases the chance of releasing it. It normally sticks in the drill-collar assembly, but a stretch free-point should be run to verify the sticking point. In a few cases the drillpipe wall-sticks, but this is uncommon and the working procedures are similar. Continue full circulation at or near the maximum rate since the erosional action of the mud may help remove the seal around the sticking contact area. Also plan for and obtain the material and equipment needed to release the pipe.

Drilling jars and a bumper sub are a major benefit if the assembly is stuck below them, which is most common. Having them allows hard working to be started immediately. The pipe should be worked even if jars and a bumper sub were not run. Although it is not as effective as other procedures, still it involves a positive action and there is a good chance of releasing the stuck pipe by continued working. Increase the working forces rapidly to the maximum safe level with heavy jarring and bumping. While working, reduce the mud weight to the minimum safe level if possible.

Also, obtaining and mixing releasing fluids normally takes several hours and often longer. Utilize this time to work the pipe, since it is not uncommon to release the pipe by working it in this manner. Working is not as successful at releasing wall-stuck pipe as it is for other types of sticking, but it is still strongly recommended.

Releasing Fluids

Releasing fluids (sometimes called soaking agents or soaking fluids) are fluids that remove or dissolve the sealing filter cake around the perimeter of the contact area and break the pressure seal (Fig. 5–9).

The most frequently used releasing fluids in a water-base mud system are oil mud, invert emulsion mud, and oil (either field crude or diesel with additives). Also included in this category are fluids or additives that increase lubricity or help prevent sticking the assembly. There is less wall sticking in oil-base mud. Diesel oil with additives such as surfactants are used, and there is at least one report of

Figure 5-9 Soaking stuck assemblies

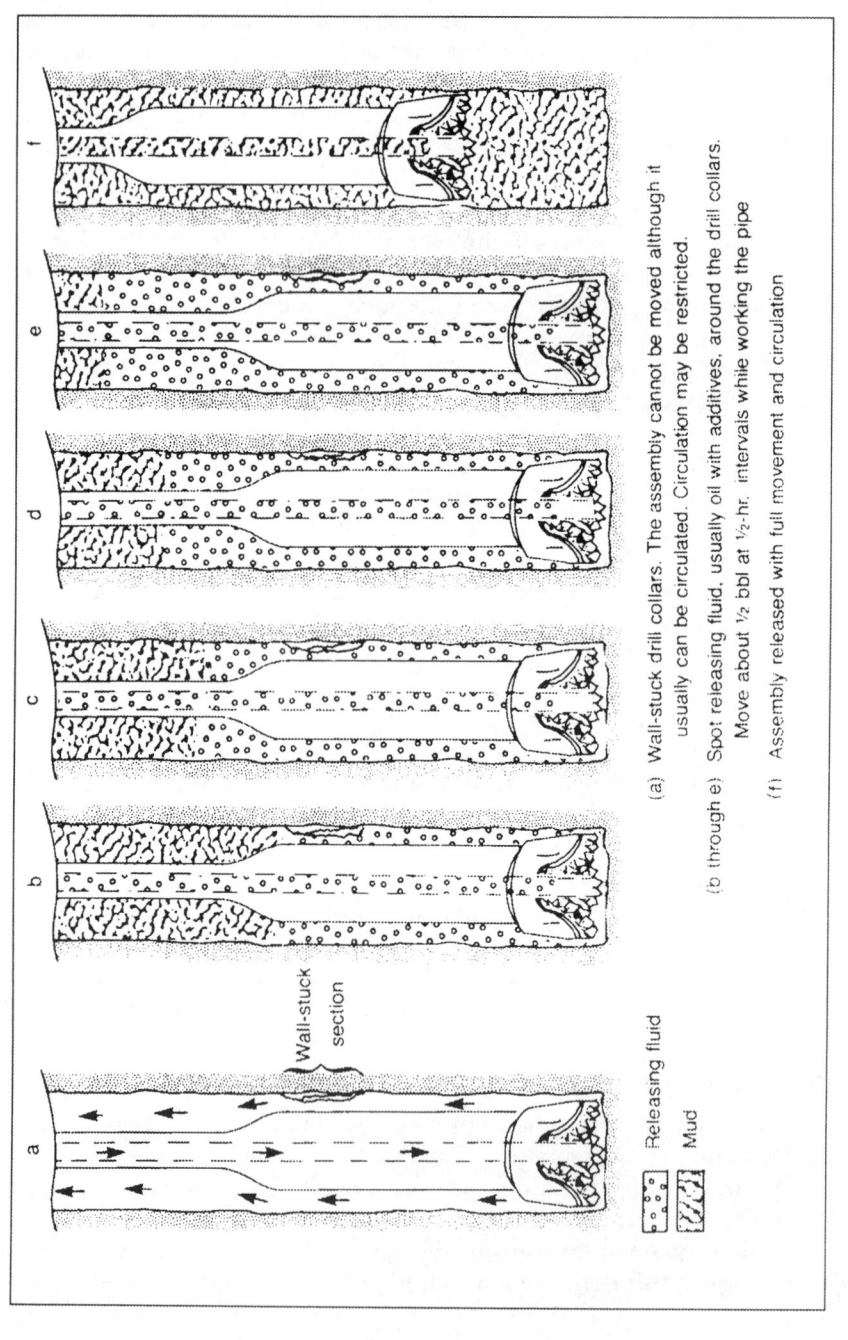

(a) Wall-stuck drill collars. The assembly cannot be moved although it usually can be circulated. Circulation may be restricted.

(b through e) Spot releasing fluid, usually oil with additives, around the drill collars. Move about ½ bbl at ½-hr. intervals while working the pipe

(f) Assembly released with full movement and circulation

352 Chapter 5

using isopropyl alcohol spearheaded with viscosified oil. The wall cake causing the sealing action should soften and deteriorate during the soaking period and release the stuck assembly.

The best method is to continue working the assembly while using releasing fluids, assuming the condition of the hole and the drillstring are satisfactory. If the jars and a bumper sub are not run in the current assembly, then they can be run while soaking as described in the next section. The fluids usually can be added at any stage of the working cycle while continuing to work the stuck assembly.

Usage

Start the releasing procedure by first using pipe stretch data or other means to locate the sticking point. This normally is in the drill-collar section below the drilling jar-bumper sub, but it may be in the drill collars above them. This is a good reason for using heavyweight pipe above the jar-bumper sub since it seldom wall-sticks.

Calculate the volume of soaking fluid needed by first calculating the annular volume of the full length of the bottomhole assembly plus about 200–300 ft of drillpipe annular volume. Then add to this the fluid volume to provide for fluid movement during the soaking period; this is about ½ bbl per 30-minute interval for 12 hours or 24 bbl. Add an excess of about 40–50%. Additional excess may be needed to allow for washouts and other out-of-gauge hole sections and to ensure complete fluid coverage over the stuck zone. It takes about 80 bbl for an average 1,000-ft-long bottomhole assembly in an $8^{3}/_{4}$-in. hole. Never use less than about 50 bbl of releasing fluid for standard-size holes in the range of 6 to $8^{3}/_{4}$-in. Always plan for a total soaking time of at least 12 hours; 18–24 hours is best.

More accurate fluid volumes can be obtained by determining the precise stuck section with a wireline free-point tool or stuck pipe log. However, this procedure seldom is justified, especially considering the time involved to obtain a wireline truck and to record the measurements.

It is best to batch mix the releasing fluid in a tank truck with a pump truck. The rig pump can be used, but the pump truck allows better control and measurements are more accurate. Then pump the fluid into the drillpipe and displace it with mud until the leading edge of the fluid is a short distance past the top of the wall-stuck section. Usually the hole should be filled to about 200 ft above the top of the drill collars so that the entire annular volume, including the entire wall-stuck section, is filled with the soaking agent. Displace at a fast rate to help remove all the mud.

The fluid should then be moved with the rig pump at a rate of about 0.5 bbl at 30-minute intervals for 8–12 hours. The most convenient way to do this is by counting pump strokes and recording these on the regular drilling footage recorder. There have been cases where the pipe was released after soaking as long as 24 hours, but most operators believe that if the assembly is not released within 12 hours, the procedure will not be effective.

Working the pipe during the soaking period is strongly recommended, subject to hole conditions such as caving uphole. The assembly can be worked more efficiently with jars and a bumper sub. If they are not in place, displace the soaking fluid around the drill collars and back off. Back off as deep as possible; often this allows for the recovering of part of the unstuck drill collars. Generally one loose drill collar should be left above the top point of sticking for the convenience of washing over later if it is required. Pull the pipe and run a fishing assembly with jars and a bumper sub. Screw into the fish and begin working it. This allows time for soaking while tripping to run the fishing assembly and the stuck assembly can be worked more efficiently. This approach can be very effective. Sometimes, displace additional soaking fluid to the end of the pipe and then screw in and work the stuck assembly while moving the releasing fluid periodically.

A generally lighter, unweighted soaking fluid column causes relatively low back pressure on the drillpipe, so this is not a problem. Some prefer to add weight, especially in heavier muds, so that the soaking fluid weight is equivalent to the mud weight. This permits balanced mud columns inside the drillpipe and in the annulus. However, it takes additional time to add weight to the soaking fluid. Usually a better method is to get the soaking fluid in place as soon as possible. Delaying may cause the fish to stick tighter due to an increasingly effective sealing process around the perimeter of the contact area, and the contact area may increase in size with time.

Diesel generally is available as rig fuel and unweighted soaking fluids can be mixed and displaced rapidly. Unweighted soaking agents have been used satisfactorily many times, including in high-weight mud systems. These may cause unbalanced mud columns; however, these are easily controlled, and the soaking fluid can be held in place by controlling the pressure in the drillpipe and annulus, even in heavier mud systems. The preventers can be closed to keep the annulus from flowing and opened through an annulus valve when pumping mud down the drillpipe to move the soaking agent. The actual pressure differences involved are relatively small. For example, an 800-ft column of unweighted diesel oil as a soaking agent in the annulus and 15.5 lb/gal mud in the drillpipe will only cause an extra annular pressure of about 375 psi. The operator must evaluate this and select the best option.

There is some concern that the unweighted soaking agents will be contaminated by mud. Field experience indicates this is not a problem, especially with an adequate volume of soaking agent. In one extreme case, unweighted diesel oil was used in a 15.5 lb/gal mud system. The oil was spotted in place and moved periodically for about 10 hours without releasing the stuck assembly. The oil was then circulated out of the hole, through the surface mud system, and back down and around the stuck assembly. The assembly was released during this second soaking period.

Another procedure can be used with unweighted soaking fluids, sometimes with an excess of fluid. Displace the soaking fluid in the normal manner and then over-displace it by about 10 bbl. The drillpipe will still contain a long column of lighter-weight soaking fluid. Allow the extra 10 bbl of fluid to flow back into the drillpipe slowly, and watch for bit plugging. Passing this extra fluid around the stuck pipe section helps ensure complete mud displacement and better fluid contact with the seal around the contact area, and improves the chances of destroying the seal. A longer period of soaking and moving the fluid periodically can then be begun.

Soaking agents soften and release the wall cake. To prepare for this, circulate the hole adequately to ensure it is clean and stabilized after using soaking agents. Monitor the mud properties carefully, making corrections as needed.

Releasing wall-stuck pipe in this manner is effective probably about 50–75% of the time with good working procedures. If the wall-stuck assembly is not released by soaking or other releasing procedures, then recover the stuck assembly using the methods described in the section "Sticking the Drill-Collar Assembly" in this chapter.

Wall Sticking Combined with Keyseating

A wildcat well had 9⅝-in. intermediate casing at 6,000 ft, and an 8¼-in. hole drilled to 14,000 ft. Hole deviation was less than 2° from the surface to 7,000 ft, increased gradually to 8° at 11,000 ft, then began increasing at a faster rate. The hole was drilled with a stiff assembly from below the intermediate casing and a packed-hole pendulum was run at 11,500 ft in a 10° hole. Hole inclination increased to 13° at 12,000 ft and was held at 12–14° to 14,000 ft. The drilling assembly included 4½-in. 16.60 lb/ft drillpipe, 750 ft of 7¼-in. x 2¼-in. drill collars, 117 lb/ft, with a six-point roller reamer as a fulcrum for the packed-hole pendulum. The formations were classified as medium-hard to hard. Various low permeability sands were encountered, and the mud weight was gradually increased to 14.8 lb/gal, buoyancy factor of 0.778, at total depth.

Open-Hole Fishing

During the later trips, there was increasing evidence that a long keyseat was forming in the section from 10,800–11,500 ft. In some cases, the overpull was as much as 50,000 lb. On one trip, the section was reamed with the packed-hole pendulum, which decreased the overpull to 20,000 lb on the next trip. On subsequent trips, the overpull gradually increased. A keyseat wiper was run on top of the drill collars but did not appear to be effective.

When the assembly was pulled to change bits, the pipe keyseated with the bit at 11,750 ft (with the top of the drill collars at 11,000 ft). The keyseating occurred with a tool joint about 15 ft above the rig floor. The pipe was worked for about 1 hour with increasing weight. This moved the pipe downward about 7 ft, at which point the pipe stopped moving, and there was no apparent travel in the sleeve-type keyseat wiper. The pipestring weighed with 210,000 lbs—[(750)(117)+(11,000)(16.6)](0.778).

Next, the top single was unscrewed and the kelly was connected. This took about 1½ hours to do because the connection was 8 ft above the rig floor. After making the connection, circulation was established without difficulty, which is characteristic of keyseating and wall sticking. The pipe was worked for 6 hours at increasing set-downs and pulls from as low as 120,000 lb (22,000 lb below the free point) to as high as 298,000 lb (156,000 lb over the free point, or 90% of the Grade E drillpipe tensile strength). Pipe-stretch data indicated that the assembly was stuck at or below the top of the drill collars. There was no evidence of movement in the sleeved keyseat wiper while the pipe was being worked.

In analyzing the situation, it appeared that the drill collars were keyseated at the top of the collars but could be keyseated lower. There was a strong possibility of wall sticking because of the lack of movement in the sleeve of the keyseat wiper. One item appeared to be somewhat abnormal. Usually, when the assembly is moved downward a short distance out of a keyseat, it will be free to move down. In this case, the assembly was moved down about 7 ft, but it did not release or become free. Either the assembly became wall-stuck at that time, or there was a possibility that the bit was digging in on the low side of the hole, restricting downward movement.

A 50-bbl batch of diesel was mixed with additives. This volume was calculated to cover the drill collars (assuming a 9-in. hole) and leave 25 bbl inside the drillpipe to be pumped out slowly. One half of the oil was displaced into the annular space around the drill collars and the remainder was displaced at the rate of 3 bbl/hr for 8 hours.

As the oil was pumped down the drillpipe, the pressure increased to about 1,500 psi, and then decreased as the oil was pumped out the bit. It was necessary then to close the bag-type preventer and maintain a pressure of about 500 psi on the annulus to hold the oil in place. Since the pipe had to be worked through the closed bag-type preventer, working was limited to about 20% of the time to prevent excess wear.

The pipe was not released, and the oil was circulated out of the hole. It was circulated through the surface mud system, back downhole, and spotted across the drill collars while waiting for the equipment to run a free point and back off. When travel was obtained in the keyseat wiper, the pipe was worked heavily for about 30 minutes. One joint of pipe was worked out of the hole in the next 30 minutes. The assembly was then worked upward with heavy drag until the bit was up to about 11,000 ft. At this point, drag was reduced to normal, and the assembly was tripped out of the hole without difficulty.

A keyseat-wiping assembly was run back into the hole, and the section from 10,700 to 12,000 ft was reamed. This section was reamed periodically during subsequent drilling, and the well was completed without excessive problems.

Two somewhat controversial factors are how long the stuck assembly should be soaked and whether the assembly should be worked while soaking. If soaking agents are effective, the assembly should be released in 12–16 hours. Barring formation and other problems, always work the stuck tools during the soaking period.

Wall-Sticking Examples

A conventional drilling assembly with 5-in. drillpipe drilling a 12¼-in. hole became wall stuck at the drill collars at 12,000 ft while making a connection. A 50-bbl batch of crude oil with additives was displaced around the drill collars. During the soaking procedure, the oil was moved at a rate of about 1 bbl/hr and the pipe was worked about 50% of the time. The assembly came free with a 120,000-lb overpull in 16 hours.

A 7¾-in. hole was being drilled at 4,500 ft when the pipe stuck with the bit about 10 ft off the bottom while making a connection. Pipe-stretch data indicated the sticking point was at or below the top of the drill collars. Diesel with additives was displaced around the drill collars and soaked while working the assembly. After 3 hours, the pipe was backed off above the drill collars and pulled out. A fishing string was run and stuck immediately above the fish before screwing in. Diesel with additives was displaced around the stuck assembly. The assembly was worked steadily while soaking and was released in 3 hours. After the hole was circulated and conditioned, the fishing assembly was screwed into the fish. Diesel oil with additives was spotted around the stuck fish. The fish was worked while soaking and was released after 4 hours.

In critiquing this example, the quick sticking indicates excessive mud weight. The first stuck assembly apparently should have been soaked and worked longer before backing off. However, it may not have had jars and could not be worked as efficiently. Therefore oil was spotted around the assembly to start the soaking process. The pipe was backed off in order to run the more efficient fishing assembly, and the trip time was utilized to soak the stuck assembly.

A drilling well had 9⅝-in. surface casing set at 2,500 ft. While drilling an 8¾-in. hole at 12,000 ft with 4½-in. drillpipe, the tool joint in the top of the shock sub failed, leaving a bit, a near bit reamer, and a shock sub with a broken pin in the top in the hole. A standard fishing assembly was run and wall-stuck at 8,000 ft. Diesel with additives was displaced around the drillpipe. The pipe was worked while soaking and moving the oil at a rate of ½ bbl at 30-minute intervals. The pipe was released after soaking and working for 4 hours, and the jars above the overshot began tripping and working. The hole was conditioned and the original fish was caught, worked free, and recovered.

These examples show the efficiency of releasing wall-stuck assemblies by soaking combined with working. The success ratio indicated by these examples is somewhat higher than actually occurs in field operations, probably due to lack of knowledge of the correct soaking procedures.

If the assembly is not released by soaking, then the operator should consider some of the other methods, such as reducing the annular hydrostatic head, if applicable. Otherwise, run a free point and back off preparatory to washing over.

For Oil Mud

Most soaking-type fluids are either oil based (usually diesel) or are mixed with diesel. Therefore they generally are not considered effective when wall sticking occurs while drilling with oil muds. However, releasing agents have been mixed with diesel and used to release wall-stuck pipe in oil muds many times.

Other fluids have also been proposed for use in oil muds. In one reported case, a well became wall-stuck at 17,900 ft with 12.2 lb/gal oil invert mud while changing pumps. A 5 bbl batch of isopropyl alcohol was spotted, preceded by a 20 bbl batch of viscosified oil spacer, the pipe was soaked and worked free, and operations resumed. This description raises some questions, such as why 12.2 lb/gal mud was used before sticking and 10 lb/gal after sticking. The pill volume was very small compared to standard-size pills used to release pipe that is stuck in water-base mud. However, the principle is the same—to dissolve the wall cake—and the overall procedure is of interest. Handle isopropyl alcohol with CAUTION; it should not present a problem with careful operating procedures.

Reducing Hydrostatic Head

Reducing the hydrostatic head to reduce the differential pressure across the contact area is another method of releasing wall-stuck pipe. This also is the reason for emphasizing the use of the lightest weight mud that can safely be used as a preventive measure. A temporary reduction of very short duration often is sufficient to release the stuck pipe. The procedures described in this section are not standard and there is a higher risk than with other methods. These procedures also require favorable hole, formation, and equipment conditions. However, they have been used successfully to a limited extent.

Lightweight Fluid

This releasing procedure for wall sticking is based on reducing the hydrostatic pressure on the formation temporarily by using a lightweight fluid in the annulus. It is a moderately high-risk procedure, and it may cause a problem with well control and formation caving. Caving and damage to the hole can be reduced by minimizing the contact time with the lightweight fluid. Get the fluid out of the hole as fast as is reasonably possible. Also, ensure that hole conditions permit its usage.

Calculate or estimate the overbalance that is causing the wall sticking—i.e., the difference between the hydrostatic head and the pore pressure. The overbalance should not be more than 0.03–0.05 psi/ft for using this procedure; otherwise, it might be better to consider one of the other methods of releasing wall-stuck pipe.

In selecting a lightweight fluid, use the most convenient, compatible, rapidly obtainable fluid. Formation damage is an important consideration. Diesel is compatible with most water-base and oil-mud systems. Use freshwater for non-fluid-sensitive formations and add sodium chloride (NaCl) or potassium chloride (KCl) for fluid-sensitive formations. Water with 20 lb gel/bbl may be best for some formations.

Calculate the length of the mud column and lightweight fluid column to arrive at the annular hydrostatic head required to underbalance the pressure differential. Normally this should not be over about 600–900 psi pressure reduction at the sticking point. Ignore the hydrostatic pressure inside the drillpipe and drill collars.

As an example, consider wall-stuck collars at 10,000 ft in 12 lb/gal mud (0.624 psi/ft). An 8,000-ft mud column and 2,000-ft saltwater (0.46 psi/ft) column would give a total hydrostatic head of 5,912 psi. This is a reduction of 338 psi (0.0338 psi/ft) from the column of 12 lb/gal mud. Replacing the water column with diesel (0.3471 psi/ft) would give a total hydrostatic head of 4,686 psi. This is a reduction of 554 psi (0.0554 psi/ft) from the column of 12 ppg mud.

Open-Hole Fishing

Pump the volume of lightweight fluid necessary to flow the desired fluid column length into the drillpipe and displace it into the annulus with regular mud. If necessary, the flow rate can be controlled by holding back pressure on the annulus, commonly with the annular preventer. Displacing fluid down the annulus is not recommended. The flow is too restricted, it may not reduce the bottomhole pressure sufficiently, and there is a strong possibility of plugging the bit.

After the lightweight fluid is in the annulus, open the annulus and flow it at the maximum rate. Usually it is best to flow through a maximum opening in the choke manifold or through other controllable openings or outlets, if available. Continue pumping at a moderate rate down the drillpipe. There will be a strong, surging flow from the annulus.

The lighter annular fluid column reduces the hydrostatic head, which often releases the wall-stuck assembly. There also may be some releasing effect due to fluid erosion acting upon the sealing wall cake around the contact area. Work the assembly at low to moderate force levels throughout the displacement.

Nitrogen

There has apparently been a very limited use of nitrogen, even though it can reduce the bottomhole pressure substantially on a temporary basis. The amount of nitrogen needed is calculated similarly to that for lightweight fluid, allowing for nitrogen compression at the existing temperature and pressure.

Nitrogen is used in the same manner as lightweight fluid. Pump it down the drillpipe and out into the annulus. Use the preventers (usually the annular preventer) to control pressures. Then flow the annulus at the maximum rate while continuing to pump down the casing at a moderate rate, thus creating a high surging flow. Work the assembly at low to moderate force levels throughout the displacement.

There have been reports of pumping nitrogen down the annulus, but the procedure is very questionable.

Compression Packer

Releasing a wall-stuck assembly with a compression packer is an older method that is seldom used now. However, it may still be applicable under certain conditions. Hydrostatic head is reduced temporarily with the compression packer. This is a moderately high-risk procedure, so ensure that hole conditions permit its usage.

In operation, back off above the sticking point. Run a compression-type open-hole packer above a screw-in connection on a fishing assembly. Run the drillpipe dry or partially dry, but with sufficient fluid to prevent collapse. Screw into the fish. Set down weight on the fish to expand the packer element and seal the annular space. Set down additional weight to ensure a strong downward force. Open the test tool and allow a maximum flow rate out the drillpipe. This procedure reduces the hydrostatic head and may release the stuck assembly. Release is indicated by loss of the set-down weight and downward packer movement if the pipe was not stuck on bottom. There have been reports of releasing wall-stuck tools when the packer was seated and before opening it.

FISH ON A FISH

A very difficult fishing situation may be created when there is a fish on top of a fish. For example, washpipe stuck while washing over a fish often leads to sidetracking. Metal junk dropped on top of a fish can greatly increase the difficulty of recovering the fish. Other examples include breaking or twisting off an overshot in the bowl section, breaking and dropping the overshot slips, twisting off a mill skirt, and breaking off a wall hook. Milling is often the only reasonable action before sidetracking.

Always work from the top down, handling each problem as it occurs. Recover the top fish and successive lower fish with standard procedures. Try to prevent actions that will restrict the recovery of lower fish. One example is excess milling on the top fish. Steel cuttings can settle downward, sticking the lower fish tighter or increasing the difficulty of releasing it.

Metal junk dropped on top of the drilling assembly causes a similar problem. Pull out of the hole carefully if smaller junk is dropped. It often will bypass the drilling assembly, especially in the open hole. Junk dropped on top of small junk already in the hole should be handled similarly then removed as described in the section on removing small junk. Junk dropped on top of a fish or tools in a cased hole can be a very severe problem. Be sure the hole is cleared of small junk before running a diamond or polycrystalline diamond bit.

Try to treat larger junk in a similar manner. Pull out of the hole very carefully. The junk may ride out of the hole on top of the drill collars or bit. Take special precautions when the drill collars and bit enter any casing shoe or the blowout preventers. Do not rotate the assembly while pulling unless it sticks. Then work and rotate it as necessary.

Several summary examples illustrate the problems and procedures involved when there is a second fish on top of the first fish.

Open-Hole Fishing

- A fish was left in the hole when the tool joint in the top of the bumper sub failed (because of a cracked drill collar pin). The damaged fishing string was laid down, and a new fishing string was picked up and run. In the attempt to take hold of the fish, the skirt of the overshot broke off and was left in the hole. A spear was run without success. Fishing operations continued until the second fish was recovered.
- A cleanout trip was made to clean out to the top of a drill-collar fish. Washover pipe was run, but it became wall-stuck after washing over the fish a short distance. It was soaked free and recovered. Several runs were made with an overshot, but the fish was not recovered. Fishing operations were suspended, and the hole was sidetracked.
- A tool joint washed out on top of a stabilizer located 60 ft above the bit. The bit locked up, probably due to lack of circulation. When the drilling assembly was pulled, it was parted at the stabilizer, leaving 60 ft of fish in the hole. While going in the hole with a fishing assembly, the drillpipe parted in the slip area 2 ½ ft below the tool joint, dropping the 14,000-ft fishing assembly. The fish was caught with an overshot, worked free, and pulled from the hole.

  The overshot on the bottom of the fishing assembly broke, leaving the lower part of the overshot in the hole. It was recovered with a spear. Then the second fish was washed over and recovered with an overshot.
- The tool joint on the bottom of the top drill collar failed when the pin broke, leaving the drill-collar assembly in the hole. It could not be recovered with an overshot, so washpipe was run. The washpipe stuck, and the hole was sidetracked after several more days of fishing.

FORMATION PROBLEMS

Formations can have a strong, often adverse effect on fishing operations. Any formation problem, such as caving and sloughing, lost circulation, or high pressures and flows of saltwater, oil, and gas, can be a major obstacle to fishing. Some of these may cause an unstable wellbore, further increasing the difficulty of fishing and often leading to stuck fishing tools. If formation problems become severe enough to restrict the fishing operation, efforts should be redirected to correct the problem before recovering the fish. Most formation problems are eliminated in cased holes.

Often the best way to solve a formation problem above the fish is to set a plug or bridge the hole over the fish. Use a sand bridge, a barite pill, or a cement plug depending upon hole conditions. In some cases use a polymer or bentonite carried in diesel oil. Correct the formation problem, clean out the plug, and resume fishing. This may take additional time, but often it is the best approach to the problem.

For example, a restricted flow below the fish with a lost circulation zone higher in the hole restricts efficient fishing. The lost circulation zone prevents using heavier weight mud. The operator should then set a plug over the fish, heal the lost circulation zone so it will hold sufficient mud weight to control the high-pressure zone, clean out, and resume fishing.

A formation problem at or below the fish can be more complicated. The general procedure is somewhat similar. Concentrate on correcting the formation problem when it is most severe; otherwise continue fishing. Corrective procedures depend upon the specific situation.

For example, the earlier case of a high-pressure zone below the fish and a lost circulation zone above the fish may be handled in other ways. One possibility is to allow the well to flow slowly into the mud system in a controlled manner until it is semi-depleted locally. Then concentrate on releasing and recovering the fish by repeating the flow and work procedures. An alternate procedure would be to circulate and add lost circulation material to the mud while increasing the mud weight slowly. This may stabilize the hole and allow continued fishing.

The primary objective is to recover the fish. However, it is not uncommon to spend 25–50% or more of the time working on formation problems, as illustrated in the following example.

Fishing with Formation Problems

An on-structure wildcat was designed for drilling to about 15,000 ft. After it was spudded, another well drilling on the structure found production from a deeper zone. A revised plan included extending casing depths on the subject well to test the lower zone.

Surface casing of 13⅜-in. was set at 2,500 ft, 10¾-in. intermediate casing at 8,000 ft, and a 9½-in. hole drilled to 15,000 ft. Lost circulation and gas kicks caused severe problems while drilling the last 1,000 ft of hole. Most of the lost circulation apparently occurred above 14,000 ft as the mud weight was increased to contain the gas zones below 14,000 ft. The original program had provided for casing off the lost circulation zone.

The drillpipe twisted off in the first joint above the drill collars while drilling at 15,000 ft. The free pipe was pulled and, while cleaning out above the fish, the well began kicking. The mud weight was increased to contain the pressure and caused lost circulation uphole. The hole was gradually stabilized and fishing resumed. Circulation could not be established through the drill collars and they could not be released by working and jarring, apparently because of bridging. About 300 ft of the fish was washed over and three drill collars were recovered. The fish was washed over another 200 ft and two more drill collars were recovered. Cavings were sticking the drill collars again after the washpipe was pulled. Short washover runs were necessary because of high torque and drag.

A spear assembly was run inside the washpipe. The washpipe began torquing severely after washing about 100 ft over the fish. The top of the collars had plugged so a string shot could not enter the fish for a back-off. The spear was backed off and the washpipe was pulled. A fishing assembly was run and screwed in but could not release the stuck collars. The washpipe was run again but stuck after washing over about 100 ft of the fish. The washpipe was backed off and the hole was sidetracked.

The sidetrack hole was drilled to 15,500 ft with continuing problems caused by gas kicks and lost circulation. The drilling assembly was stuck and released several times. Because of continuing hole problems, $7^5/_8$-in. FJ casing was run, but circulation was lost while cementing.

The casing was drilled out with 13.9 lb/gal mud. Sand stringers with gas shows were first encountered about 50 ft below the casing shoe and continued at random depths to TD. The sand stringers usually were indicated by a drilling break and very frequently caused a small gas kick. It often was necessary to bypass the shale shaker because of the gas-cut mud, and at times to shut the well in on a choke and circulate the gas out. The mud weight was increased gradually to control the gas.

Lost circulation occurred while drilling at 16,000 ft with 14.5 lb/gal mud. The problem was first controlled with lost circulation material in the mud. As it became more severe, pills of lost circulation material were spotted, followed later by cement plugs. If the mud weight had been reduced, the well would have immediately begun to flow.

All circulation was lost while drilling at about 17,500 ft. While mixing a 75 bbl lost circulation pill, the well began to flow and the pit gained 15 bbl. The well was shut in with 0 psi on the drillpipe and 500 psi on the casing. The lost circulation pill was spotted without returns. A second pill was spotted and the drillpipe pressure increased slightly.

A third lost circulation pill was mixed while waiting for the hole to heal from the last pill. It was spotted while circulating through the choke with 1,500 psi on the drillpipe and 250 psi on the casing. While the pill was being spotted, the casing pressure dropped to zero. After the pill was in place, the pipe rams were opened to work the pipe and the pipe was found to be stuck at the drill collars. It was worked for 3 hours, building up to an overpull of 100,000 lb and setting down about 130,000 lb. The assembly could not be rotated but circulated freely. After the pipe was worked for 4 hours, the well began to flow. The rams were closed and the well was circulated through the choke with a maximum of 1,500 psi on the drillpipe and 1,100 psi on the casing with full returns.

The assembly was apparently wall-stuck in the drill collars, since the hole was straight with no prior indication of keyseats. The pipe had not been moved while spotting the earlier lost-circulation pills. The pipe could have been moved slowly through the closed rams, but this was a higher-risk procedure. The preventer stack included a double set of pipe rams so the worn pipe rams could have been replaced. Also, since the casing pressures were relatively low, they could have been controlled with the inflatable-bag-type preventer. This would have permitted the pipe to be worked or moved safely if the casing pressure were carefully monitored and controlled, but working pipe while circulating out a kick is a high-risk procedure.

Wall-sticking normally is not expected while a well is kicking. Nevertheless, there was a possibility that the gas kicks were coming from the upper section of the hole and there was a sharp pore-pressure reversal, so the lower hole section had a slightly lower pore pressure. This in combination with the fresh-mud lost-circulation pills, which had a relatively high water loss, and the large volume of lost circulation material, which is a good bridging agent, probably accounted for the wall-stuck pipe.

The next four days were spent circulating and trying to kill the well and regain full circulation, building mud and balancing the mud system, and repairing the circulating system and the choke lines. The mud weight was increased to about 15.1 lb/gal, and over 2,000 bbl of mud were lost during the four-day period. A high-pressure pump truck was connected to the rig system as a standby unit and was used to circulate when high drillpipe pressures were encountered.

After the well was stabilized, a 30-bbl diesel pill mixed with a releasing agent was spotted and the pipe was worked for several hours. This included jarring followed by a 100,000-lb overpull and bumping followed by a 120,000-lb setdown and right hand torque, but the fish did not move. The next 10 days were spent trying to free the drill-collar assembly and back off. Most of the time actually was spent circulating and controlling the well, including mixing mud. There was also a number of mechanical and electrical failures in the service company's equipment.

At one point the bit was plugged, apparently due to lost circulation material. Circulation was reestablished by running a junk shot to knock the jets out of the bit. The tool misfired on the first run but worked satisfactorily on the second run.

While the tool was being pulled from the hole, the collar locator stuck, and the line was pulled out of the rope socket and was recovered by fishing with a wireline. During the time when the drillpipe was plugged, the well was controlled by periodically pumping mud down the annulus.

After circulation was established again, the well was circulated until it was more stable. The fish was worked without any noticeable movement. Several times the drillpipe became stuck 500–900 ft above the collars. After the pipe was worked, it would become free and the free point would move down into the drill collars. During the latter part of this period, hole conditions became more stable, although problems were still encountered with lost circulation and gas kicks. The mud weight was increased and maintained at 15.0 lb/gal. After hole conditions stabilized, more time was devoted to working the drillpipe and attempting to back off.

Diesel oil mixed with a releasing agent was spotted across the drill collars and 500 ft of drillpipe. The oil was moved at about 1 bbl/hr while working the pipe. Periodic free points were run. The combination of working the pipe for 24 hours and using the oil freed the pipe down into the drill-collar assembly. The top three collars were worked free and backed off successfully. More collars may have been freed by continued working, but there was some concern about the condition of the overall assembly. The pipe was pulled and the entire assembly was checked.

A fishing string was run and screwed into the fish. Circulation was established with about 25% returns. The fish was worked without any movement. The circulation rate was reduced to prevent losing mud, and returns increased to about 50%. A diesel oil pill mixed with a releasing agent was spotted, and the pipe was worked for about 8 hours while moving the pill about 2 bbl/hr. The fish was not released.

A free point indicated the top two drill collars were about 75% free and the third drill collar was 50% free. A back-off was attempted below the third drill collar, but this was unsuccessful, and a piece of the sinker bar from the back-off shot was left in the hole. The sinker bar effectively plugged the inside of the drill collars and prevented additional free pointing or backing off below that point.

Washpipe was then run, a questionable decision. Obviously, considerable time had been spent on the well, but because of hole conditions very little of that time involved effective fishing. Conditions had recently stabilized somewhat so that fishing operations could be conducted. Other than continuing to jar and work the fish, the next alternative would have been to wash over the fish. Attempting to recover the sinker bar or push it deeper inside the fish was a questionable procedure at this depth and with 15 lb/gal mud. If the drill collar assembly were washed over past this point, it could not be backed off because the inside was plugged. The drill collar assembly in the hole was about 500 ft long, and it would be difficult to wash over this length under the existing conditions.

Washpipe was run, and the fish was washed over for 3 ft. When the torque became excessive, the washpipe was pulled. The washover shoe and the lower part of the washover pipe were split, and part of the washover shoe was left in the hole. A fishing string was run, but it could not be screwed into the fish. A short joint of heavy washpipe with a heavily dressed mill shoe would not wash over the fish. An overshot run on a fishing assembly would not take hold of the fish. Other tools were run to clean out over the fish, recover the junk, or take hold of the fish, but all were unsuccessful. An impression block was run and showed that half of the box on top of the fish had been cut off.

The next several days were spent making washover runs and ultimately washing 200 ft over the fish. The junk in the hole was either milled up or pushed downhole. One drill collar was recovered with an overshot. Another overshot with a screw-in sub would not take hold of the fish. A second overshot caught the fish. It was worked for about 5 hours but did not release or move. Spud bars would not enter the fish. The overshot was released and the string was pulled after 4 hours.

Fishing operations were suspended, and a 5-in. liner was run and cemented.

A flow of oil, gas, or saltwater while fishing probably is one of the most severe situations. If the hole can be circulated, the first approach is to increase the mud weight. This is more effective when circulating near the zone of flow. Add lost circulation material (LCM) if there is any risk of lost circulation. Use the minimum amount of LCM to prevent it from interfering with fishing tools or plugging the bit. If necessary, blow the jet nozzles out of the bit. Consider perforating the drillpipe or drill collars if the pipe is plugged.

Flows above the fish can be corrected by increasing the mud weight if there is full circulation. Otherwise, try to balance the mud weight and lost circulation; this can be very difficult, as demonstrated by the above example. Consider backing off and running casing, even though this may prevent deeper drilling. Another consideration is completing through the pipe.

A flow at or below the fish can be very severe, especially if the fish is relatively high in the hole. It may be possible to shut off the flow by pumping bridging material, such as a barite pill or some type of polymer, through a bridged fish if it is open on the inside. Cementing by pumping through the fish is a high-risk procedure. Pumping into the formation at a sufficiently high rate to block or plug the formation is very unlikely. If the fish is shallow, completing through it is not a reasonable alternative. Evaluate and correct the problem based on existing conditions, but do not overlook the possibility of casing or completing through the pipe and drilling another well later that is properly designed.

An annular flow and split casing or other loss of circulation at a very shallow depth is uncommon, but it can be a similar situation. This is effectively an underground blowout that can develop into an open blowout. It is very difficult—often almost impossible—to pump any type of LCM through the drillpipe that will flow back and bridge and plug the hole in this situation because of the lower fracture pressure of the shallow formations. There have been reported cases where large bridging material such as golf balls and sponges were ineffective.

One alternative is to load the annulus with a heavy fluid that has a hydrostatic head sufficient to stop the flow. This may require a very high flow rate if there is lost circulation or split casing above the fish. It may be necessary to part the drillpipe to achieve the necessary flow rate into the annulus. This is one reason why trying to kill the well by pumping through small tubing is seldom successful.

As a last resort, if the pipe can be tripped, set an open-hole retrievable packer with a long stinger and pump cement through and around the fish. Then either set casing or sidetrack. An alternate method might be to sidetrack a short distance and then run casing. Probably the last alternative is to drill a kill well.

MISCELLANEOUS

A number of operations and procedures that do not fit into any special category can cause a fishing job or may occur during fishing.

Miscellaneous Fishing Situations

- A 5½-in. production packer was stuck at 6,000 ft. While the tubing assembly was being worked, the collar was jarred off the packer mandrel. A short catch overshot caught the mandrel and the packer was worked free and recovered.
- An open-hole drill stem test was run at 8,000 ft in an 8¾-in. hole. After the test was completed, the test tool stuck. It was jarred and worked for about 3 hours while moving it uphole about 100 ft. At that point, the test tool backed off in the body connection above the top packer element. The test tool was caught with an overshot and worked and jarred. It began moving slowly and became relatively free, but it swabbed as it was pulled. At 1,000-ft intervals, it was necessary to work the tool up and down about 30 ft and fill the hole to ensure that the well would not be swabbed in. The dragging rubber was pulled off the packer when it was worked into the surface casing at 2,000 ft. The hole was filled and checked carefully at this point before the packer was pulled from the hole.
- A slip segment was dropped in the hole and lodged in the bradenhead. The drillpipe could be moved, but a tool joint would not pass the slip segment. The preventers were disconnected and

lifted and the slip segment was recovered. Care was taken during recovery to prevent dropping the slip segment deeper into the hole and causing a more serious fishing problem.
- An 8¾-in. hole was drilling at 9,000 ft with a standard drilling assembly. The drill-collar assembly stuck while making a connection. It was worked for 6 hours, and the drillpipe was backed off above the collars. An open-hole drill stem test tool was run and screwed into the drill-collar assembly. The test tool was seated twice but did not release the wall-stuck assembly. The assembly was then washed over and recovered. This method of releasing a wall-stuck assembly is described in more detail in the section "Wall Sticking" in this chapter.
- Pump pressure declined while drilling. The pump and surface equipment were checked and found satisfactory. The declining pump pressure indicated a leak downhole, and the assembly was pulled. The tool joint on top of the shock sub had washed out, leaving the bit, the near bit stabilizer, and the shock sub in the hole. An overshot was run but did not catch the fish. When the overshot was pulled from the hole, the appearance of the grapple indicated it was not going down over the fish. A ⅛-in. larger grapple was run and caught the fish, which was recovered. The inner seals on the shock sub had failed, and the body of the near bit stabilizer was cracked.
- A deep well was being pumped inside 7-in. casing with a submersible electric pump run on 2⅞-in. EUE tubing with the power cable clamped to the side of the tubing. When the pump failed, the tubing was perforated above the pump, and the well was circulated dead. When the assembly was started out of the hole, the tubing moved slightly but there was a heavy drag, indicating collapsed casing. The tubing was chemically cut above the pump and was worked out of the hole. The casing was rolled out to nearly full gauge. The pump was recovered by fishing, a new pump was run, and the well was placed back in production.
- Pipe was pulled to change bits on a drilling well. After the last stand of drill collars was pulled, the blind rams were closed. Later, a pipe cheater (2.385 in OD piece of pipe 2 ft long) was dropped in the hole and lodged on top of the blind rams. The pipe cheater could not be fished out from the surface. The preventer assembly from the bottom up included double drillpipe rams, blind rams, drillpipe rams, a bag-type preventer, a rotating head, and a bell nipple. It was necessary to open the pipe-ram door above the rams to recover the pipe cheater.
- A deep well was being circulated at 19,000 ft preparatory to a trip to change bits. The circulation pressure dropped from 2,550

psi to 2,350 psi. When the assembly was pulled from the hole, the tool joint on the top of the junk sub had washed out. The pin broke, leaving the junk sub and bit in the hole. The junk sub had a 6⅝-in. regular box with a 7¼-in. OD by 2½-in. ID. The section to be caught was about 8 in. long. A short grab overshot dressed with a 7¼-in. grapple and a toothed mill shoe with hard facing material on the teeth was run in the hole. About 20 ft of cuttings and fill material were washed off the top of the fish. It took about 50,000 lb of weight to clean out the cuttings

After cleaning out, the pressure increased to 2,400 psi, indicating the overshot was over the fish. The weight of the fish was too small compared to the weight of the drilling assembly to determine if it was caught. When the assembly was picked up, it lost about 200 psi of circulating pressure. This loss of pressure indicated the assembly had not caught the fish. The hole was cleaned out for another 6 ft using up to 75,000 lb of weight. At this point the assembly could not be rotated with four rounds of torque. While the last 6 ft were being cleaned out, the pump pressure increased to 2,500 psi. After picking up, the mud pressure remained constant at 2,500 psi, indicating the fish was caught. The assembly was pulled and the fish was recovered.

- A conventional 8¾-in. hole was being drilled at 11,000 ft with a standard drilling assembly in an area where caving and sloughing had occurred. The stuck drilling assembly was worked for several hours. A free point found the assembly to be 70% free at 10,000 ft. The assembly was backed off and five joints of pipe were worked out of the hole in 4 hours by a combination of working and pumping the fish out of the hole, which would not circulate. Then the pipe stuck solidly. A free point was run, and the pipe backed off at 7,300 ft. This section was recovered. Then the second fish was caught, worked free in 6 hours, and recovered. The hole was cleaned out to the top of the first fish and conditioned. The fishing assembly was run and screwed into the fish. Limited circulation was established and a releasing fluid (diesel oil with additives) was spotted around the drill collars. After about 8 hours of working and soaking, they were worked free and recovered.
- On a trip to change bits, the first stand of drill collars above the drilling jars was connected, lowered through the rotary, and set into the slips without using a drill-collar clamp. After the elevators were unlatched, the drilling jars opened with a small jarring action. This set up a vibration that let the drill-collar assembly slip through the rotary slips and drop. The tools were recovered after fishing for three days.

These examples are all too common in the industry. They illustrate the severity of the losses from fishing jobs and how fishing often leads to additional fishing. Fishing can place stress on equipment and increases the risk of failure. This, in turn, increases the risk of another fishing job.

Dissolving Agents

Dissolving agents dissolve and remove some formations that cause stuck tools. When the downhole assembly is stuck in a formation that is soluble in a relatively common fluid, then it may be possible to circulate the fluid to dissolve the material that is sticking the pipe.

Massive salt formations usually are drilled with salt-saturated fluids or inert muds. If these formations stick the pipe, slugs of freshwater can be circulated by the sticking point and may release the assembly by dissolving the salt. Similarly, mixtures of hydrochloric and acedic acid will remove carbonates.

Dissolving Carbonates to Release a Fish

A deep wildcat well had 13 3/8-in. surface casing at 5,000 ft, 9 5/8-in. intermediate casing at 14,000 ft, a 7-in. drilling liner from 19,000 ft back to 13,500 ft, and was drilling in a 6 1/8-in. hole at 21,000 ft in a very hard, fractured, interbedded shale and carbonate section. A tapered, 5 1/2-in. by 3 1/2-in. drillstring was being used with a packed-hole assembly, including a diamond bit and a near bit stabilizer.

After a trip to change bits, the last 90 ft of hole was reamed as a standard procedure. After drilling about 10 ft of new hole, the drilling assembly began to torque and the pump pressure increased. The action was similar to a severe heaving, caving condition. An automatic driller was being used, and there was a slight delay disconnecting this before the assembly could be picked up. The drill tools were stuck. The pump pressure was reduced, but circulation was still slightly restricted and the pipe could not be reciprocated or rotated.

The drilling assembly stuck with the bit on bottom. It was worked to move it upward. Pulling, torquing, and jarring actions were used at increasing force levels. After 4 hours, it was worked to a maximum overpull of 125,000 lb. Since the jars were working, the assembly had to be stuck below the top of the drill collars. This was verified by stretch data.

The way the pipe stuck indicated caving. The formations were fractured and heavy mud was being used. Since there was a possibility of wall sticking, the operator decided to spot diesel oil around the drill collars in case they were wall-stuck. If not, the lubricating effect of the diesel might help release the assembly. Thirty bbl of diesel oil with additives was spotted and moved slowly at 1 bbl/10 minutes while the assembly continued to be worked. In 6 hours the pipe was jarred about 5 ft uphole. When it stopped moving, it was jarred back downhole.

The formations contained a high carbonate content so it was decided to try and dissolve them with acid in order to free the assembly. Fifty bbl of acid, 15% hydrochloric and 10% acidic, with additives was displaced to the bottom of the assembly. The acid was spearheaded with 50 bbl of water to help clean the mud off the formation ahead of the acid and to separate the acid from the mud to prevent mud contamination. A 50 bbl spacer of water was pumped behind the acid also to prevent mud contamination. The acid was displaced to the bit and then into the open hole at ½ bbl/minute while the assembly continued to be jarred and worked. When 30 bbl of acid had been displaced through the bit and after one hour's pumping time, the pipe came free.

When the pipe was pulled, the diamond bit was found to be cracked halfway around just above the gauge section. The sides of the bit were heavily marked and scratched. The near bit stabilizer did not show any evidence of being stuck. The drilling assembly was probably stuck at the bit due to the caving of fractured carbonate fragments, which were removed with the acid.

It is difficult to understand how a bit can be stuck to the point where it can withstand the amount of pulling and jarring that occurred in the example. However, this does happen and has been observed in other cases.

Provisional Plugs

During fishing and other operations, it is sometimes necessary to set temporary plugs in the hole. Cement, barite, or sand may be used, depending on the reason for the plug and how it is displaced.

Mechanical plugs are most commonly used in casing. Sand often is spotted on top of bridge plugs. If a shallow casing leak develops during a fishing operation, a bridge plug can be set in the lower casing, the casing repaired, the bridge plug recovered, and fishing operations resumed. Cement plugs can be set in the open hole above the fish to work on the upper section. The cement plug can then be removed by drilling and fishing operations resumed.

Barite can be an effective plugging agent under certain conditions. Mix a high concentration of barite in freshwater and displace it to the point where the plug or bridge will be formed. The mixture must be moved constantly to prevent settling. After motion has stopped, the barite will settle out of the water rapidly, forming a bridge or plug. The solid barite material forms an effective seal that may be almost as hard to drill out as cement. In one case a plug was set in a hole with a slow, high-pressure flow, and the flow was shut off. This is probably an exceptional occurrence, but it demonstrates the effectiveness of barite plugs.

Fishing under Pressure

Few fishing jobs are conducted under pressurized conditions. It is extremely difficult to pull the assembly through a stripping head of a snubbing unit and simultaneously provide internal plugs or pack-offs inside the fishing assembly and any recovered fish. If the well is pressured, a slick assembly can be run or lubricated into the hole to circulate and kill the well so operations can be conducted.

Sometimes stripping units are used through which minor fishing jobs may be conducted. Generally, however, these units are used to run an assembly under pressure to clean out and circulate the hole with weighted fluid so the hole can be reworked under more normal and safer conditions.

Fishing under Pressure

A shallow well was being drilled with air through a rotating head. A short liner of about 300 ft was run and hung in the casing, uncemented. Air drilling continued below the uncemented liner. After drilling a short distance, the liner became unseated and was blown or moved up the hole so the top lodged against the bottom of the rotating head. Lubricating-type equipment was installed to make the repair.

Bridging

Except for a few cases, bridging occurs as an unintentional result of operations or downhole conditions. In a few cases it is an intentional action, such as setting a sand or cement bridge above a plug like a bridge plug, or setting a barite plug above a lost-circulation or high-pressure zone.

Drill cuttings and formation sloughing are the most common causes of bridging, as described in the section "Hole Cleaning" in Chapter 1. Other causes of bridging include high-pressure flows such as saltwater flows and blowouts, and flowing or plastic formations, such as salt and some clays, which reduce hole stability. Clear water fluids used as spacers may cause bridging by accumulating formation particles loosened from the wall of the hole. This can be prevented by using about 20 lb/bbl of bentonite in the water.

Open-Hole Fishing

Observe precautions when cleaning out sand. Do not leave cleanout tools on the bottom without circulating at a rate sufficient to lift the sand. The best cleanout fluid is mud. Water is often used and is satisfactory in most cases. However, there is a risk of sand settling and sticking the tools if circulation is stopped before the sand is cleaned out. In severe cases, use gel water with 20 lb of bentonite/bbl to replace plain water.

Both formation and fracture sands have a high capacity for bridging. A very small volume of sand can bridge around and stick a string of tubing or other downhole equipment. The following examples illustrate the severity of bridging.

Effect and Severity of Bridging

- As an extreme example of bridging, a shallow gas well had dry, gas filled 4½-in. casing. A wireline bridge plug was set over depleted perforations at 4,000 ft before perforating and testing an upper zone.

    Two sacks of 20–40 mesh fracture sand were to be dumped on the bridge plug as a precaution. Since the hole was dry, the sand was dumped by pouring it directly down the casing and placement was checked using the swab bars with the swab cups removed. One sack of sand was dumped downhole rapidly. The swab bars were run and found a bridge at 1,600 ft. Several hours were required to spud it out with the bars. The swab bars were left in the hole and moved slowly. This prevented the sand from bridging when the second sack was poured into the hole slowly at a rate of about 100 lb/10 minutes.

    In a similar situation, 7-in. casing was set on top of a 50-ft sand section. The sand was to be stimulated by loading the openhole section with a pelletized explosive that would be detonated later. The hole was dry. The pelletized explosive bridged continually as it was poured downhole. To prevent this, a swab line was run in the hole with blank bars and was worked slowly while the pelletized explosive was being dumped. This procedure eliminated the bridging problem.

- Casing (5½-in.) was set at 7,000 ft. Two oil zones at 6,500 and 6,800 ft had been perforated, stimulated, and placed on production. After a period of time, the upper zone began making water and some sand. The zone was isolated by setting a wireline packer immediately below the zone and stinging into the packer with a seal assembly to produce the lower zone through the tubing.

Later the well was reworked. A wireline blanking plug was set in a receiving nipple below the packer mill out-extension, a sliding sleeve in the tubing opened about 25 ft above the packer, and the well circulated, recovering some sand. The tubing could not be released from the packer. It was perforated 12 ft above the packer, the sliding sleeve closed, and circulation was established through the perforations. After the tubing was worked for about 3 hours, the seals were gradually worked out of the packer. The tubing was worked up about 500 ft in 8 hours using various procedures before it came free.

The tubing was apparently held by a sand bridge that moved uphole with the tubing as it was worked. The recovered tubing appeared normal, and there was no indication of junk or other foreign material that could have stuck it. A bit was run and cleaned out 15 ft of sand above the packer, and the remaining remedial work was performed.

The tubing was worked up to the minimum tensile strength, and it could circulate through the perforations most of the time. This indicates the tremendous holding force of a 15-ft sand bridge.

REFERENCES AND SUGGESTED READING

Allen, Floyd, "Alcohol Pill Frees Stuck Pipe in Mud," *Oil Patch*, Vol. 10, No. 6, Aug. 1985: 27.

Burton, Jim, "Evaluating Mineral Oils for Low-Toxicity Muds," *Oil and Gas Journal*, July 29, 1985: 129–157.

Brouse, Mile, "How To Handle Stuck Pipe And Fishing Problems," *World Oil;* Part 1 (November 1982): 103–110; Part 2 (December 1982): 75–81; Part 3, Jan. 1983: 123–126.

Dupre, Ron, and Ron Esbaugh, "Spotting Oil Base Fluids As an Alternative to Mechanical Fishing Tools," *Drilling-DCW*, April 1981: 56–61.

Eresman, Dale, "Underbalanced Drilling Guidelines Improve Safety, Efficiency," *Oil and Gas Journal*, Feb. 24, 1994: 39–44.

"Fishing Tool Retrieves MWD Nuclear Source From Deep Well," *Oil and Gas Journal*, Jan. 17, 1994: 64.

"Flow Charts Help Prevent Stuck Pipe Problems," *Oil and Gas Journal:* Part 1 (April 1, 1991): 61–64; Part 2, April 15, 1991: 59–61.

Hinds, Aston A., and William R. Clements, "New Oil Mud Passes Environmental Tests," SPE 11113, Presented at the Annual Fall Technical Conference of the Society of Petroleum Engineers of AIME in New Orleans, LA, Sept. 26–29, 1982.

Love, Tom, "Stickiness Factor: A New Way of Looking at Stuck Pipe," *Oil and Gas Journal*, Oct. 3, 1983: 87–91.

Simpson, Jay P., "A New Approach to Oil-Base Muds for Lower-Cost Drilling," *Journal of Petroleum Technology*, May 1979: 643–650.

Varcoe, Brian E., "Shear Ram Use Affected by Accumulator Size," *Oil and Gas Journal*, Aug. 5, 1991: 34–37.

Chapter 6

CASED HOLE, WIRELINE, AND SURFACE FAILURES

Fishing in cased holes is similar to that in open holes, and many of the same tools are used. Reduced hole diameter may limit tool selection and resulting strength, so more tool failures occur. Formations are less of a problem. Generally, increase working forces faster than those used for fishing in open holes. Minimize milling and take precautions to prevent casing damage.

Stuck packers and retrievable plugs are worked free or milled over and recovered with an overshot. Plugs are milled. Long, small-diameter fish are caught with overshots or may be swallowed in sections with a full-opening overshot. Coil tubing is recovered in this manner, but with limited success. Tubulars stuck in sand are washed and milled over. Tubulars fail in or above the rotary by bending or parting and commonly are held by the slips or elevators. Severely bent pipe will normally break and drop when it straightens during lifting. If this occurs, a catch can prevent a severe fishing job, damaged pipe, and possibly loss of the hole. Each failure has a safe method of repair.

Wireline fish are recovered with tubular or wireline fishing tools. Wirelines are used to fish for small, lightweight fish.

TEST TOOLS, PACKERS, AND PLUGS

Test tools, packers, and plugs are widely used. They normally are very reliable but may fail for various reasons. The most common failures are caused by premature setting, followed by mechanical malfunctions and common sticking.

The manufacturers of test tools, packers, and plugs normally have instructions and recommendations for fishing or drilling out tools that stick or malfunction. Most of these recommendations are based on extensive operating experience and are reliable guides. The manufacturer or a representative should always be consulted when these tools must be fished out of the hole, especially in a difficult fishing situation. Many of these tools are fitted with fishing necks to facilitate fishing.

Malfunctions and Prevention

Test tools, packers, and plugs malfunction in various ways. They may set prematurely while running or setting. Prevent most of these problems by screening the mud before running close-tolerance tools and by having a clean mud system. Otherwise, install a screen on the bottom of the tools before running. Tool bodies may back off or part due to a connection failure. Small, weaker parts may fail. Failures include sheared J slot pin or button and damaged slip mechanisms, bow springs, drag blocks, or hydraulic holddowns. Failures often are caused by hitting a casing collar or other obstruction while running into a hole. Prevent this by first running a gauge ring and then running the tool carefully.

Damaged tools may stick. Remove a tool by working and jarring it out of the hole. Always start working the tool in the correct direction. Hard working and jarring can cause casing damage. It may force slips and other parts outward so they puncture and/or tear the casing, causing leaks or other damage. This problem may be temporarily repaired by squeeze cementing. Otherwise, mill and washover carefully to prevent casing damage.

At times it is necessary to repair a casing failure that is not caused directly by the casing. This is illustrated in the following example.

Recovering a Plug by Removing Casing

Surface casing, 9⅝ in., was set at 1,200 ft and 5½-in. 17-lb/ft, ID 4.892-in. and 23-lb/ft, ID 4.670-in. production casing was set at 9,500 ft. The heavier casing was set opposite thick salt sections. The higher collapse strength was needed due to the flowing tendency of the salt section. Two joints of the 23 lb/ft pipe were run in the top of the production string.

Three zones were perforated in the gross interval from 8,500–9,500 ft and isolated with a retrievable bridge plug and packer for testing and stimulation. The procedure included setting the bridge plug below the zone, moving up a few feet, setting the packer, pressure testing the bridge plug, releasing the packer, resetting it above the zone, testing, stimulating, and testing the zone again. The zones were swab tested, stimulated with acid, carbonate formations, and flowed or swab tested. Surface treating pressure was 4,000–6000 psi, with pressure differentials across the retrievable bridge plug and packer of about 2,000–4,000 psi.

The zones were tested from the bottom up. After testing a zone, the packer was unseated and the hole was reverse circulated with saltwater. The retrievable packer was then lowered to the top of the bridge plug, and the hole was reverse circulated again to clean off the top of the plug. The hole was reverse circulated immediately above the plug to ensure recovery of the perforation ball sealers used in stimulation. The bridge plug was retrieved and the pipe picked up to repeat the test procedure on the next section.

After testing the deepest zone, the retrievable bridge plug was caught, released, and picked up, dragging about 40–50 ft before it moved freely. The plug was reset below the next upper set of perforations, which tested satisfactorily. After taking hold of the bridge plug, it took about 2 hours to release it, including additional reverse circulating for cleanup before the plug would move.

The bridge plug was pulled up slowly but continued to drag and hang up. It was moved uphole about 400 ft without any reduction in drag. At that time, the bridge plug was released because of the excess drag and the difficulty of working it with the packer in the string. Then the packer was pulled from the hole.

A fishing string, including 2⅜-in. drill tubing and 3½-in. drill collars with a jar and bumper sub, was run with the retrieving tool to catch the bridge plug. The bridge plug was caught and jarred down to move it beneath the lowest set of perforations so that it could be left in the hole. The plug stopped at about 9,000 ft, above the lower producing zones, and would not move.

After working, the plug began to move, hanging on each casing collar. This dragging action gradually decreased as the plug moved higher, but the plug was never entirely free. The plug was worked up to 7,000 ft in 12 hours and from 7,000 ft to 70 ft in another 3 hours (all measurements relative to KB). The bottom of the upper two joints of 5⅓-in. 23-lb/ft casing was at about 70 ft, where the plug stuck. The plug had passed through the heavier 23-lb/ft casing sections at about 8,000 ft and 6,000 ft with no discernible difference between movement through the heavier casing as compared to the lighter weight casing. When the plug was pulled up to the bottom of the 23-lb/ft casing at about 70 ft KB, it hung up solidly and could not be moved.

There were four apparent possible actions, including:

1. Run bigger collars so the plug could be worked harder and possibly be released.
2. Mill the plug out.
3. Jack the plug out of the hole.
4. Back off the casing below the plug, pull it to recover the plug, and rerun the casing.

Abandoning the hole was not a reasonable consideration at this point.

The plug-retrieving tool was relatively weak. Jacking the plug out of the hole would probably have broken the tool or pulled the fishing neck out of the plug body, thus requiring milling. The plug could not be milled out easily at the shallow depth because there was not enough distance above it to run a suitable number of large drill collars for adequate milling weight. Also, milling could cause an accumulation of mill cuttings above the plug, sticking it tighter.

The well was effectively dead. The 9.5-lb/gal saltwater overbalanced the bottomhole pressure. The well would take fluid on a vacuum. After several hours, it flowed low GOR oil slowly, but could again be killed by loading with 9.5-lb/gal saltwater.

It was decided to try to jar the plug out. From the bottom up, the fishing assembly extended into the mast and included a plug-retrieving tool, two 4¼-in. drill collars, a bumper sub, an accelerator jar, and an 8-in. drill collar. The plug was jarred and bumped for 8 hours without movement. The strength of the plug-retrieving tool limited the working forces.

Preparations were made to pull the casing. The well had been controlled by loading it with 12 bbl of 9.5-lb/gal saltwater at 3-hour intervals. The loading volume indicated the fluid level dropped about 500 ft between each filling. The water did not affect the carbonate reservoir.

A cement plug was to be displaced down to about 2,500 ft to ensure the well would be under control while removing the casing. About 50 bbl of water were to be pumped ahead of the cement to establish a good, reliable circulating rate and pressure and to ensure the cement would pass freely through the stuck bridge plug. After pumping 10 bbl of water, the well pressured up to 4,000 psi and took fluid at 1 bbl/minute.

The low-permeability formations should have taken fluid at a higher rate and lower pressure, as indicated by prior treating pressures. Therefore, the plug was partially restricting fluid movement through the casing. Since the well had been controlled satisfactorily by dumping water downhole at 3-hour intervals, it was decided to use this same procedure to control the well while removing the casing.

The plug was released, the fishing assembly pulled, and the blowout preventers (BOPs) nippled down and removed. A casing spear was set in the top of the casing about 2 ft below the bradenhead. The entire area was washed down to remove all oil. A 15-ft piece of 5½-in. casing was welded on top of the casing stub and also strap welded. The casing was jacked out of the bradenhead slips with 180,000 lb. About 5,000 ft/lb of right-hand torque was applied to the casing to ensure that it was tight. The torque was then released. This was worked out by lowering and raising with the jacks, and 3,000 ft/lb of left-hand torque was put into the casing. A 300-grain string shot was run into the 5½ by 9⅝-in. annulus and detonated at the bottom of the third joint of casing in the hole, one joint below the plug. The casing did not back off. A second string shot detonated at the same depth released the casing, which was then backed off.

The casing was pulled and the bridge plug removed. Most of the rubber had been torn off the bridge plug. The upper slips were in the 23-lb/ft casing and the lower slips in the 17-lb casing. The slip retainer either broke or was broken while the joint of casing was being unscrewed, and the exact cause of the bridge plug hanging up could not be determined.

New casing was run, screwed into the lower casing, landed in the bradenhead, and cut off, and the BOPs were installed. A cleanout trip was made to ensure the hole was clean. Tubing and rods were run, and the well was placed on pump.

Using two joints of 23-lb/ft casing on the top of the casing string is a conventional procedure. It ensures that any tool that passes through the first two joints of casing will pass through the lower hole sections with the same weight casing. This also prevents running a larger gauge tool that could stick in the lower hole. This is one of the few cases where the use of heavier casing at the top of the casing string has caused a problem. It may have contributed to the sticking in this case, but it still is highly recommended.

The perforation ball sealers lodging on top of the bridge plug slip retainer probably caused the initial problem. This and the dragging plug probably damaged the slips and prevented them from retracting normally. Sometimes it is difficult to circulate all of the balls out, especially if the plug has a long fishing neck. The plug was pulled carefully through the heavier casing. It could not be worked down after sticking in the upper casing because of the lower slips.

Wireline Packers and Plugs

Many packers and plugs are made of drillable material, such as magnesium or cast iron, and are drilled out without excessive difficulty. A few plugs are made of lead, which can be very hard to drill and circulate out of the hole. Drilling two or more plugs will sometimes create a problem. The first plug usually is drilled without difficulty. However, once it releases, the lower part of the plug is pushed downhole. When the lower part lands on the second plug, it may rotate so that the second plug cannot be drilled efficiently. One procedure in this case is to spud and try to break the remnants of the first plug. There is less chance of spinning if the plug can be broken into several pieces. The pieces usually mill or drill rapidly.

Most wireline or permanent packers are drilled with a washover mill shoe. The shoe cuts over the outer part of the packer, allowing it to drop or be removed with a packer retriever tool. This reduces the amount of metal that must be cut and the amount of milling time. Ensure that about 1/16 in. of the cutting material on the bottom periphery of the mill is removed to prevent damaging the casing.

Usually, the packer is free to move upward after milling is completed on the upper slips. These packers are commonly drilled with a spear (sometimes called a packer plucker) run below the mill. The spear passes through the packer and engages the mill-out extension below the packer. After the upper slips have been milled, the packer can be pulled from the hole. This eliminates pushing the packer to the bottom and trying to drill it completely, with the corresponding possibility of it spinning. Using the spear facilitates milling and removing the packer.

The procedure is illustrated in Fig. 6–1. In case 1, the entire packer must be drilled. When part of the bottom slips have been drilled, the packer generally can be pushed to the bottom. If the packer must be completely drilled, such as on top of sand above lower perforations, then it will tend to spin and be difficult to drill. Junk is left in the hole. In case 2, the top slips on the packer are milled, allowing it to be pulled from the hole with the retrieving tool, and thus leaving the hole clean. In comparison to conventional milling procedures, the retrieving tool saves time.

Figure 6–1 Milling out a packer

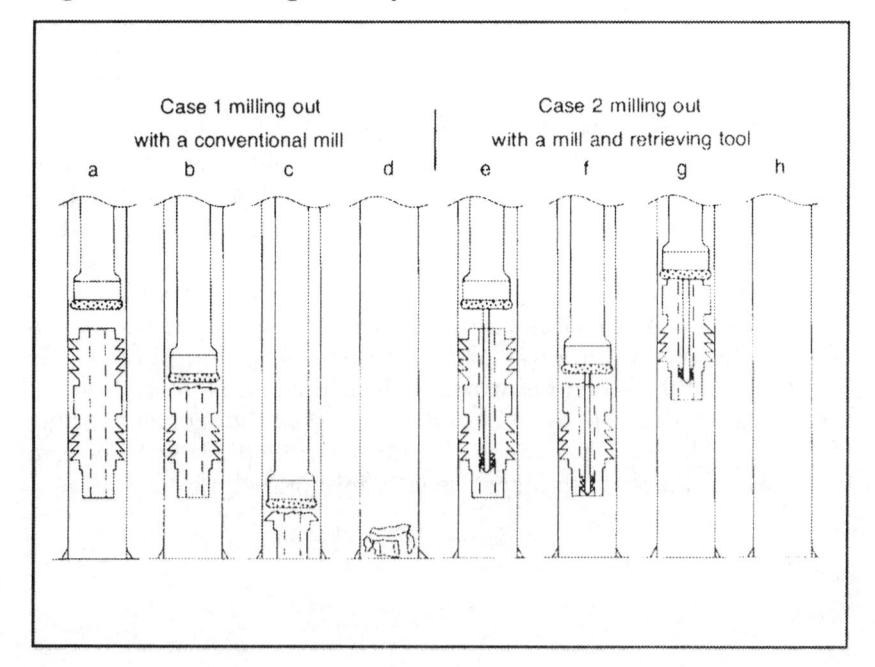

Milling a Packer While Using a Packer Retriever

Cement retainers usually can be drilled with one mill. Packers often are drilled through the upper set of slips with one mill, although as many as three mills may be needed to remove the entire packer. In one case, while a wireline packer was being run, it hung up and set prematurely. The setting tool malfunctioned or hung up in the packer. The setting tool was recovered with an overshot. The packer was milled out using a packer retrieving tool below the mill. The first mill was run for 6 hours, and the second for 5 hours. After milling on the packer for 1 hour with the third mill, the packer released and was pulled.

Test Tools

Test tools for open holes usually are compression- or inflatable-packer types. Cased-hole tools more commonly are hook-wall or hydraulic holddown types. These tools are relatively reliable but can malfunction. Many tools have a strong, heavy-duty fishing neck. If the tool joint on the top of the tool fails, the fishing neck can be caught with an overshot.

Open-hole tools most likely stick in the packer elements, sometimes due to a plugged equalizer valve and line. They may stick in the tail pipe section because of caving. Cased-hole tools may stick in the slips when hydraulic holddowns fail to retract. Packer rubbers may stick on any tool, especially retrievable production packers that have been in service for a long time.

Open-hole test tools are seldom damaged on the trip in the hole, except possibly when they are run too fast and damage the packer elements. Most are run on a fishing-type assembly. They often are set on bottom and may stick during the test, so concentrate on initially working up. If the tools are stuck above the test tool, use "open-hole fishing procedures." If the tools are stuck in the tail pipe, it may be necessary to shear loose and recover the packer, and then either catch the tail pipe and work it loose or wash over and recover it.

Inflatable packers and hook-wall-type tools, including retrievable production and cementing packers, normally are not set on bottom. Concentrate on working for movement. If the fish has not been worked with jars and a bumper sub, then back off at the safety joint if available. Back off at the top of tools fitted with fishing necks. Either back off leaving one joint of pipe or cut 6–10 ft above the top of other tools. Then run a fishing assembly to catch and work the fish. Work open-hole tools harder and for a longer period. If the tool is not recovered, release it and then mill over it with a mill shoe as described earlier.

When working open-hole tools, take precautions in case of caving. Before running the tool, circulate the hole clean as a preventive measure. If the tool is not fitted with a built-in circulating sub, then run one in the fishing assembly. This helps keep the upper hole clean while working the fish.

Cased-hole tools such as a regular hook-wall packer may stick. Concentrate on working the assembly upward as described in the section "Working Stuck Tools" in Chapter 4. The slip segments prevent downward movement until they are retracted. If the packer has hydraulic holddowns, there is a possibility that these are stuck in the extended position. Pressurizing the annulus and surging the drillpipe or tubing may help retract them and release the packer. If there is circulation and the annulus cannot be pressured for some reason, displace a heavier fluid into the annulus while allowing the pipe side to flow back. Torque may help break the grip of the hydraulic holddowns. A leaking test can be run to determine whether the packer rubbers are damaged. Sometimes the rubbers tend to stick to the casing. Work the packer in an upward direction, using torque to help release it.

Stuck tools with inverted slips should be worked downward first. Use the various procedures already mentioned, such as torquing, to release them. Most retrievable test tools are run under jars and a bumper sub. This permits immediate, effective jarring and bumping if the tool sticks. Sometimes the tool has a floating mandrel, so the mandrel travel may be used when working. Also, be aware of possible sand and caving from upper, open perforations when working inside casing.

Another cause of sticking, most commonly in cased holes, is dropping junk such as preventer bolts and nuts, hand tools, or tong dies on top of the test tool. The tools normally are full-gauge, so there is very little room to let the junk bypass. Sometimes it is necessary to mill out a stuck tool, as illustrated in the following example.

Milling Out a Stuck Squeeze Packer

An on-structure wildcat had 13 3/8-in. conductor at 1,500 ft, 9 5/8-in. surface casing at 6,500 ft, a 7-in. drilling liner from 13,500 ft back to 6,100 ft, and a 4 1/2-in. production liner from 15,000 ft back to 13,000 ft. The 7-in. drilling liner was tied back to the surface as production casing.

A retrievable squeeze packer was run to 14,000 ft to squeeze off lower perforations. Afterward, the packer could not be released to reverse out the cement. The fish was worked for 4 hours without obtaining movement, at which time the jars stopped working. The fishing string was backed off, pulled, and checked, and the jars were replaced. The tool was run, and the fish was caught and jarred for 3 hours without any movement. Several additional fishing trips were made, and one joint of the fish was recovered.

A hydraulic pulling-tool assembly was run above 36 joints of tubing so the tool could be set in the 7-in. casing. After being screwed into the fish, the assembly was picked up to set the hydraulic tool and the pipe parted. After the pipe was pulled from the hole, it was found that the lowest joint had failed, leaving a tubing pin in the connection on top of the hydraulic pulling tool. A taper tap was run to screw into the fish, but the tap broke off while being screwed in, leaving the broken thread (called a Dutchman) in the top of the fish.

A 20-in.-long fishing neck was cut over the top of the hydraulic fishing tool with three mills, averaging 8 hours of milling time per mill. A tapered box was used to screw onto the fish, and the hydraulic pulling tool and one joint of tubing were recovered.

The fishing string was run and screwed into the fish. A gauge ring was run on a wireline to see if the fish was open. It stuck in the top of the fish and the line pulled out of the rope socket, leaving the wireline jars and sinker bars. A wireline overshot would not catch the rope socket. A reversing tool and tap were run, and two joints of the fish, including the wireline tools, were recovered. A fishing string was screwed into the fish, and wireline tools again were run and lost. An attempt to back off part of the tubing with a reversing tap was unsuccessful.

It was not known if the wireline tools were sticking up out of the fish. A skirted, flat-bottomed mill was run to dress off the top of the fish and milled 2 ft in 13 hours. The fish was caught with an overshot and jarred and worked for 8 hours. It released and all of the tubing was recovered, leaving a tubing pin in the collar on top of the retrievable cementing tool.

An overshot was used to catch the tool, and it was jarred and worked for 12 hours without being released. The cementing tool was milled out. A total of 5 ft was milled with 5 mills, 1 ft/mill, in 53 hours, 10.6 hours/mill. Normal operations were resumed.

There were several questionable fishing operations in this job, such as using the taper tap and not running a combination fishing string with larger pipe inside the 7-in. casing. A retrievable packer for squeezing at this depth and in the small hole is questionable. Normally, a drillable-type packer should be used. The mills were relatively efficient, considering the depth and hole size. The difficulty of running small wireline tools at this depth is well illustrated by the lost tools.

Pressure Below Packers and Plugs

Pressure may accumulate below plugs and, in some situations, below packers. If the plug or packer must be moved or drilled, this pressure can create a dangerous situation, especially if it exceeds the hydrostatic head. A kick or blowout situation may occur with tubular tools. Wireline tools can be blown up the hole, usually causing a fishing job. A similar situation may occur when drilling out sand, barite, and other types of bridges with drilling tools or a chipping bailer. In all of these cases, be prepared for the pressure. The situation can be more severe at shallow depths, especially with gas pressure below the plug or packer.

One preventive measure is to set packers and plugs as deep as possible to increase the effective hydrostatic head. Another is to run a mill-out extension with a plug receptacle below most packers, especially in high-pressure wells. Set a plug in the receptacle to shut off the pressure. Normally, run a sliding sleeve in the lower tubing in order to circulate the well dead. Unseating the tubing without shutting off the pressure can expose the casing to dangerously high pressures. The well can be killed by loading the tubing, but there is some risk of damaging the producing formation. Sometimes, set a tubing blanking plug in a lower tubing collar with a wireline, perforate the tubing above it, circulate the well with mud of sufficient weight to control the pressure, and then pull the plug and unseat the tubing.

A better procedure in some situations is to set a plug and pressure-relief-pin assembly in the plug-seating nipple of the mill-out extension. Pick up the tubing and circulate weighted fluid above the packer. Then reset the tubing and equalize pressures by pulling the equalizing pin with a wireline. Pull the plug in the mill-out extension. If prior work was uphole, then set the tubing seal assembly in the packer and equalize the pressures as described. In all these cases, the well is under control.

Depending on conditions, several procedures may apply when an equalizing pin is not used. Packers without equalizing-pin assemblies, bridge plugs, and bridges present similar situations. The most common solution is to increase the mud weight if pressure is expected. Then fish or drill out the packer or drill out the bridge plugs and bridges. The main point is to be aware of potential high-pressure situations and be prepared.

It may be possible to close the preventers and drill the plug or sand bridge under pressure. Another alternative is to drill the obstruction under pressure with a snubbing unit. Drilling a small pilot hole through the obstruction may be applicable in very special situations; however, there is not only a risk of breaking the small bit or mill but also the problem of controlling the pressure. Do not overlook the use of a chipping bailer in special situations where it might be appropriate. These tools can be very efficient, such as for drilling some bridge-type plugs in 4–6 hours. Hydrostatic bailers may drill or clean out sand bridges efficiently. It is easy and common to run these types of tools under pressure through a pressure lubricator.

A problem may occur with low-pressure formations open higher in the hole. If these are not worth saving, then close them by squeezing. If they are worth saving, then run a full-opening packer on tubing, set it below the open perforations, and work through the tubing with small-diameter tubulars or wireline tools.

Another solution is close the open perforations by puddle cementing. This procedure is preferable to squeezing in order to temporarily shut off open perforations. These can be reopened easily by reperforating, possibly followed by a light stimulation to restore productivity. It may be possible to use lost-circulation material instead of puddle cementing if the formations have low permeability and a low-pressure differential is expected.

A similar problem can occur when a check valve, usually a flapper, is run below a packer. The flapper may be held closed because of the higher pressure below it. The flapper normally opens when the packer's stinger-seal assembly is inserted in the packer bore. Under high-pressure conditions, it may be impossible to insert and seat the assembly in the packer bore because of pressure below the back pressure flapper valve. This can occur, even with pressure above the packer, in an attempt to equalize the pressure across the flapper. Sometimes, the limited strength of the small, relatively weak, sealing stinger limits the amount of weight that can be set on the flapper valve without breaking the stinger.

Broken Packer Stinger and Seal Assembly

A packer was fitted with a flapper valve. The stinger-seal assembly was run on 2⅞-in. tubing and could not be seated and latched into the packer. The casing was pressurized without success. When the tubing was pulled, the stinger-seal assembly had broken off and remained in the hole, requiring a minor fishing job.

Cased Hole, Wireline, and Surface Failures

When this condition occurs, it may be necessary to displace a heavier fluid in the hole above the packer so that pressure sufficient to open the flapper valve can be applied. Trying to open the flapper with any other tool is a questionable practice. If the tool is small enough to run through the packer bore, it probably is too weak and may break, causing a fishing job. The flapper may be opened by perforating with a vertically oriented jet charge, but this could damage the packer bore.

If this condition may occur, the best prevention is to run a packer with a mill-out extension. Then use a plug-receptacle nipple fitted with a plug that contains a pressure-relief pin, as described earlier in this section (Fig. 6–2).

Fishing with a Taper Tap

A taper tap is most commonly used inside casing where an inside catch is needed because of limited clearance between the top of the fish and the casing. This prevents having to use the stronger outside catch with an overshot. Normally, a spear is the preferred inside catch tool. However, the condition of the top of the fish or the type of fish may prevent the use of a spear. Some situations in which the taper tap should be used instead of the spear include when the inside of the fish is out of round, when the hole inside the fish is too small for the spear, and when the hole is too short and the spear cannot be run deep enough inside the fish to catch it.

A taper tap can be used to recover some types of smaller junk, such as subs and other fish, as described in the section "Taper Tap" in Chapter 3. It can be an efficient tool in special situations, as illustrated in the following examples.

Fishing with a Taper Tap

- A casing scraper and mill were used to clean out cement in 5½-in. casing after a squeeze job. The downhole tools failed, indicated by a loss of torque and lack of penetration rate. When the assembly was pulled out of the hole, the crossover sub between the mill and the casing scraper had failed at the top of the crossover sub, leaving the pin from the sub in the bottom of the recovered casing scraper. The fish was caught with a taper tap and recovered.

  In a similar case, a crossover sub on top of a casing scraper failed, leaving a broken pin (Dutchman) in the top of the scraper. It was recovered with a taper tap.

Figure 6–2   Flapper check vs. retrievable plug

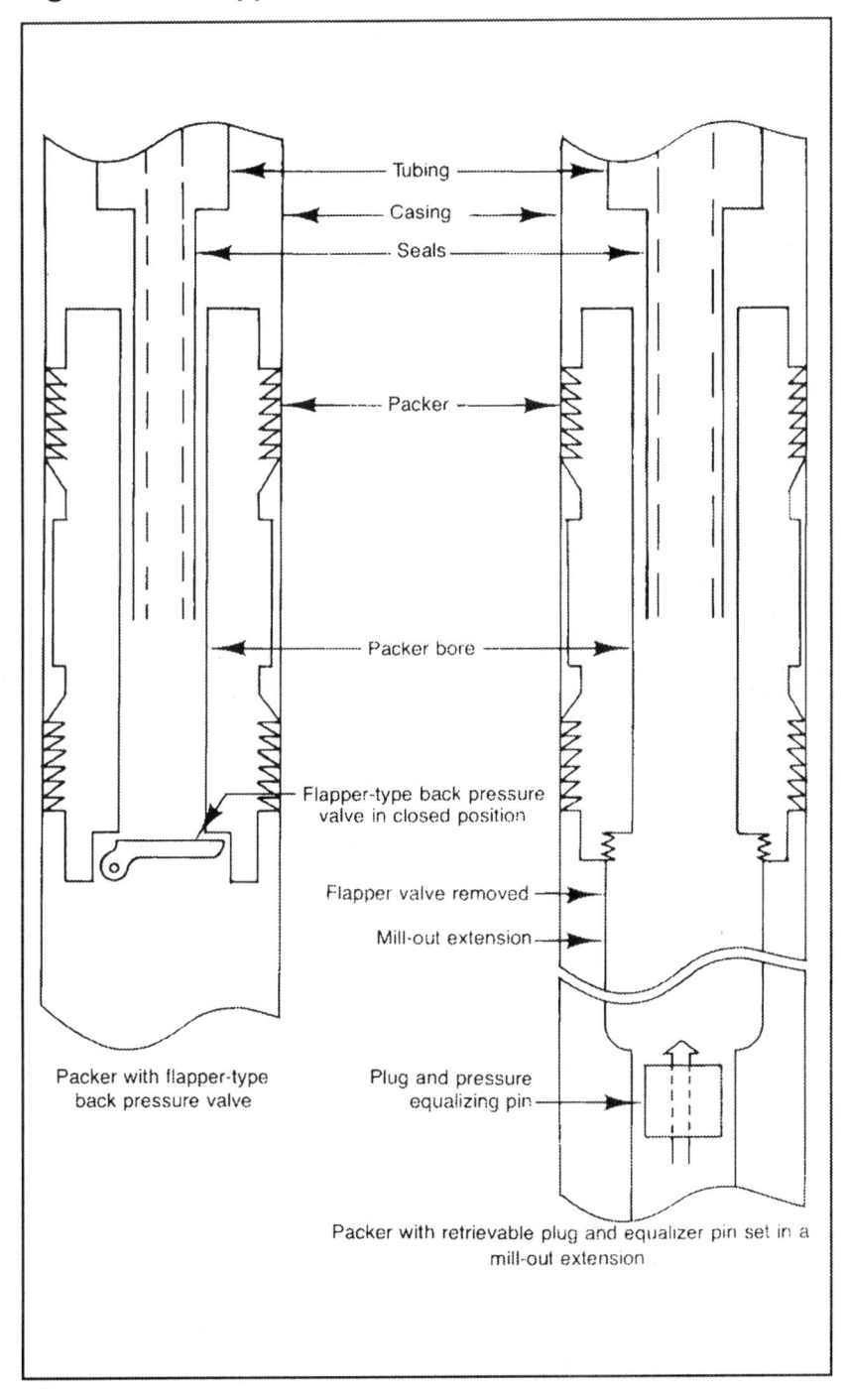

Cased Hole, Wireline, and Surface Failures   389

- A modified whipstock was used to deviate inside casing that was in a very hard section of formation. While drilling off the whipstock, the crossover sub above the bit failed, leaving the bit, crossover sub, and whipstock in the hole. A taper tap was run through the carrying ring of the whipstock so that it could be screwed into the crossover sub, but it would not take hold. A larger tap with a shorter taper was run to screw into the carrying ring. It would take hold, but stripped off each time tension was applied. A cement plug was set over the whipstock and the hole was sidetracked at a shallower depth.
- A well was to be deepened below 7-in. casing. A permanent wireline production packer was milled through the top slips and chased to the bottom of the hole. The remainder of the packer could not be milled efficiently because it rotated under the mill. It was caught with a taper tap and recovered.

SMALL-DIAMETER TUBULARS

Small-diameter tubulars are broadly classified as smaller tubulars most frequently run in cased holes. They are arbitrarily defined as being the following sizes:

- Tubing — 3½-in. outside diameter or less
- Drill Tubing — 2⅞-in. outside diameter or less
- Very Small Tubulars — 1½-in. outside diameter or less
- Coil Tubing — various sizes, normally 1¼ in.

There are intangible problems with small tubulars. Since they are not handled as often as large tubulars operating personnel are less familiar with them, which causes problems. Tubing normally is used in completion and production, a time when the well is generating income and a major part of the expenditure has already been made in drilling and completing the well. If the well is lost at this point due to a fishing job, the entire investment is lost. In other words, the monetary loss is much greater than the loss caused by the unsuccessful completion of a fishing job early in the drilling operation. This emphasizes the importance of successfully completing fishing operations with small-diameter tubulars.

Fishing Operations

Fishing operations with and for small-diameter tubulars are broadly similar to regular operations. The major differences are allowances for the smaller diameters, higher flexibility, reduced strength characteristics, and closer tolerances. The holes generally are smaller, so fishing tools are smaller with correspondingly reduced strength. This also increases the difficulty of washing over, especially for tubing stuck in sand. Limited strength increases the risk of failure in tension sometimes caused by inadvertently overstressing. All these factors increase the difficulty of fishing.

High flexibility also requires careful handling at the surface, including picking up, laying down, and using an extra support in the mast, such as a belly board. The small threaded connections are easy to damage, especially in the smaller sizes. Most small-diameter tubulars have "V" type threads in the range of 8–10 threads/in. These tubulars have less strength and tend to continue screwing together under conditions of relatively high torque, often with resulting damage caused by an overtight connection. Some tubing has special couplings that further increase the risk of connection damage. Connect small-diameter tubulars carefully, using the correct torque. It is best to use butt-shouldered work strings for heavy-duty fishing and related activities, such as milling and cleaning out.

Tubingless Completions

Tubingless completions use 2⅞-in. tubing and, less commonly, 2⅜-in. and 3½-in. tubing as casing. Their main advantage is the relatively low initial cost, but there can be serious disadvantages. Most applications are in shallow gas wells and often cause severe problems in the later life of the well. However, a smaller diameter tubing may be run in a few cases.

There are many examples of failures, but all are monotonously similar. They include stuck and broken coil tubing, stuck and sanded-up tubing, stuck wireline tools, metal junk dropped in the hole, split or burst tubing, and similar problems. Fishing is limited because of small-size tools, their low strength, and the small size of the cased hole.

Holes in tubing may be patched, but these patches further decrease the working inside diameter. Sand and other washable debris may be cleaned out with coiled or small-diameter macaroni tubing.

However, other conventional fishing operations, such as washing over, milling, and drilling metal, mostly are high-risk procedures. Because of these problems, tubingless completions should generally be avoided, except for special applications.

Similar but less severe problems are encountered in slim-hole completions with 2⅜-in. NUE tubing in 4-in. casing and 2⅞-in. NUE tubing in 4½-in. casing and 2⅞-in. EUE tubing in 5-in. casing, especially at deeper depths. Small heavy-duty work strings, such as small-diameter drill tubing, may be used in some of these latter cases, but operations are generally very limited. These types of completions are more common in shallow oil and gas wells and often cause severe problems in the later life of the well, especially for cleanouts, recompletions, etc.

Tubingless completions can occur unintentionally. For example, the tubing might be inside smaller casing and a failure might occur for which the tubing must be treated as a tubingless completion. This is described in the following example.

Collapsed Casing in a Slim-Hole Completion

A slim-hole completion had 5½-in. casing set at 6,000 ft, 2⅞-in. tubing at 5,800 ft, perforations in a marginal gas zone at 5,750–5,800 ft, and a marginal gas zone, not perforated, at 5,000 ft. The well began loading up with water and went off production. A wireline tool indicated a restriction above the perforations. A small string of tubing was run but only cleaned out a few feet. It was pulled, revealing that the cleanout tool was marked as if it had run on metal—an indication of collapsed tubing.

The tubing was found to be stuck; apparently the casing had collapsed over the perforated interval, thus sticking and collapsing the tubing. Economics did not justify cutting the tubing and sidetracking the hole. A tubing swage was run, but stuck while swaging. In the attempt to work it free, the wireline broke, leaving a section of wireline that covered the upper marginal zone at 5,000 ft. The wireline fish could not be recovered, and the well was temporarily abandoned.

There are a large number of examples of tubingless completions lost because the fish could not be recovered. This is probably one of the main reasons that tubingless completions are less frequently employed.

Concentric and Parallel Configurations

Small tubulars can be run in the hole as parallel or concentric tubing strings, the concentric configuration being the most common. Also, a fishing job in this case often is less complicated, depending upon the sizes of the tubing and casing. Sometimes the smaller tubing string can be recovered inside the larger tubing string by pulling the larger tubing. A special case in which this procedure was used under difficult conditions is described in the following example.

Recovering Macaroni Tubing Inside Larger Tubing

A high-pressure, low-volume gas well had 7-in. casing set at 11,000 ft and a 5-in. FJ liner at 15,000 ft. A 2⅞-in. tubing string was latched into a permanent packer set at 14,700 ft. The well was producing from multiple, low-permeability, high-pressure zones below the packer. During completion, the well was fractured. While flowing back to clean up, the tubing plugged with frac sand. A sinker bar encountered a bridge at 3,000 ft. This was too shallow to allow perforating and circulating to kill the well so that the tubing could be pulled. A coiled tubing unit was rigged up and used to clean out to about 3,500 ft, where it stuck. The coiled tubing was worked free and pulled.

The operator decided to clean out the tubing with small slim-hole tubing. A snubbing unit was rigged up, and a 1½-in. mill was run on 1¼-in. tubing. The hole was circulated at about 1.5 bbl/min, and several bridges were cleaned out to 9,000 ft. The lower tubing was still plugged. The 1¼-in. tubing was pulled out of the hole, and 1-in. tubing picked up and run below the 1¼-in. tubing. Several bridges were cleaned out to 12,000 ft, when the well began to flow, making a larger head of fluid and sand. The 1¼-in. by 1-in. tubing string stuck before the annulus between the tubing strings could be shut in.

The 1¼-in. by 1-in. tubing could not be circulated, but it was open so fluid could be pumped into the formation. Special bridging material was mixed and pumped down through the 1¼-in. tubing string and out into the 2⅞-in. tubing. This plugged the tubing and prevented the small string from accidentally falling out of the larger tubing. The annulus between the two strings was dead. The 2⅞-in. tubing string was snubbed out of the hole by stripping out the 1¼-in. by 1-in. tubing. When all of the 1¼-in. by 1-in. tubing had been recovered, the well was flowed on the casing side until the pressure declined. The annulus was loaded with heavy mud, which was bullheaded into the well to kill it.

The remainder of the 2⅞-in. tubing was pulled from the hole. Some problems were encountered with the bridged portion of the tubing. Tubing was snubbed back into the hole to the top of the packer. The well was circulated with heavy mud, the tubing was stung into the packer, and the well was flowed and cleaned up. A wireline was run to ensure the lower perforations were not sanded up.

Parallel tubing strings sometimes are run as two or more tubingless completions. Small tubulars are less commonly run as parallel tubing strings inside casing. Taper the collars on one of the tubing strings, usually the shorter string, to prevent them from hanging up on each other during tripping. If a fishing job occurs with both strings in the hole, pull the undamaged string if possible. Then fish in a conventional manner applicable to small tubulars. Otherwise, if it is necessary to fish with both strings in the hole, fishing is most likely to be very difficult and often unsuccessful. Normal fishing operations in these cases are almost impossible, except for some milling and possibly catching part of the fish with a taper tap or spear. The chances of success increase with larger tubing and casing sizes. This probably accounts, to some extent, for the decline in parallel tubing string completions.

Very Small Tubulars

Very small tubulars, sometimes called macaroni tubing, cause special problems and a high risk of failure when fishing. In most cases, they are used when no other tubular will fit the operation. These tubulars are ineffective for conventional fishing operations and only moderately effective at washing out to the top of a fish. They may be used for cleaning out a bridge inside the fish, but should always be handled carefully.

Very small tubing often is extremely difficult to fish out of the hole, especially in large casing. Small tubing is very flexible with low strength. It generally is crushed by the slips in a regular overshot. It also breaks easily so only a very short section is recovered, if any. If the slips do hold, the low-strength small tubing may break, frequently immediately below the overshot slips, so there is little recovery per fishing run. Overshots with a large slip contact area may help reduce this problem, but the tubing still breaks easily.

Some overshots have specially designed slips and openings to catch small tubulars. One type, a trapdoor socket, usually has one fixed-slip element and one element that slides open on a slanted guide. This tool presents a relatively large opening to the fish and will close or trap the fish inside the assembly. It will catch any reasonable configuration of pipe body.

Run the overshot with a thin wall guide and a full-opening fishing string. While rotating slowly, lower it gently over the fish. The fish passes upward into the fishing string. The amount of tubing that can be recovered per fishing run depends upon the size of the tubing and the friction between the small tubing and the wall of the fishing string. When the friction load exceeds the buckling strength of the small tubing, its walls will buckle below the slips of the overshot. Normally this is the maximum recovery. Catching 500 feet or more of a fish normally is a very good recovery. The biggest risk in this case is pushing the small tubing downhole where the lip of the overshot sets on the fish but does not go over it, and the fish collapses, balls up, bends, and breaks. The only other fishing alternative in this case is milling, as described for milling coil tubing, a procedure which generally is unsuccessful on a long fish of this type.

When possible, one of the best preventive measures is to run the small tubular inside larger tubing fitted with a no go on bottom. If the small tubing fails, causing a fishing job, it usually can be recovered by pulling the larger tubing.

Coiled Tubing

Coiled tubing probably presents one of the most difficult types of fishing for small tubulars. It is not used for the fishing string except in a few limited cases. Normally, it is best to use larger tubing with more strength and rigidity.

As with very small tubing, run the coiled type inside recoverable tubing or the drillpipe string if possible. Then recover it by pulling the tubing or drillpipe if fishing occurs. Take precautions as necessary to prevent the fish from dropping out the end of the pipe.

If the coiled-tubing fish is inside casing, the job is more difficult. It may not be completed successfully, even under relatively favorable conditions, depending upon the depth and length of coiled-tubing fish in the hole. Fishing for coiled tubing is somewhat similar to fishing for very small tubulars, but is often more difficult. Use the same general procedures. A short section may be recovered with a conventional overshot fishing assembly. The guide on the bottom of the overshot should be as thin as can safely be run safely and with an inside taper on the bottom to guide the coiled tubing up into the overshot. The assembly should be as light as possible and should employ a sensitive weight indicator.

Running the fishing tool over coiled tubing and how much tubing can pass into the fishing string are major problems. The situation is similar to fishing for macaroni tubing but is more difficult because of the risk of buckling and breaking below the fishing tool. The overshot slips often crush the fish or may not catch if the fish is distorted, bent, or flattened.

A full-opening assembly above a full-opening trapdoor overshot is probably best, but recovery often is very limited (Fig. 6–3). Set down only a minimum weight on the coiled-tubing fish to prevent buckling. Sometimes 1,000 lb can buckle the coiled tubing. Detecting weights this small is extremely difficult and, for practical purposes, almost impossible. This is another reason for the low success ratio in recovering coiled-tubing fish.

In operation, run a full-opening trap-door socket fitted with a full-gauge, thin-lip guide shoe on a full-gauge ID fishing assembly (Fig. 6–4). Lower it over the fish with slow rotation. The coiled tubing slides up inside the fishing assembly similar to the method described for very small tubulars. Under favorable conditions, as much as 500 ft of fish can be recovered in this manner, depending on the size and stiffness of the coiled tubing fish. The average recovery per run is considerably less. It is important to have a sensitive weight indicator and to watch it carefully. Set the minimum amount of weight on the fish to prevent it from buckling. As the fish slides up through the fishing assembly, a drag force is exerted downward on the fish. This can cause the fish to buckle and is the main reason for limited recovery.

Figure 6–3 Full opening overshot (courtesy Fishing Tools, a Petrolane Co.)

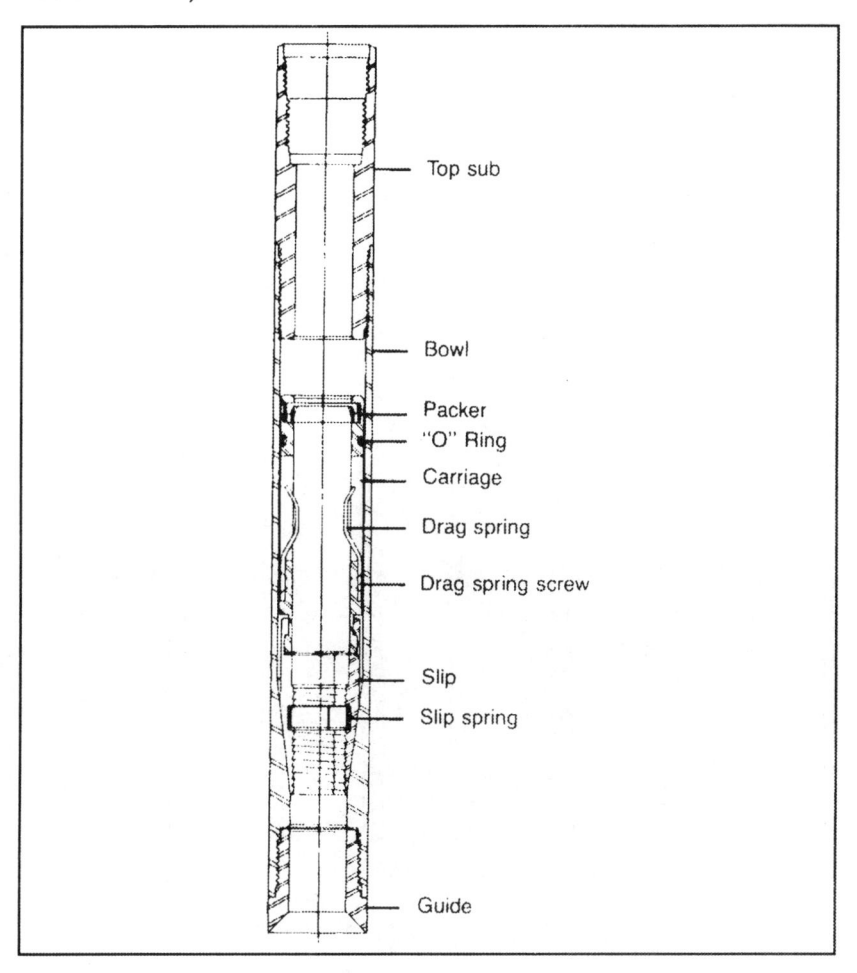

Cased Hole, Wireline, and Surface Failures

Figure 6-4 Recovering coil tubing

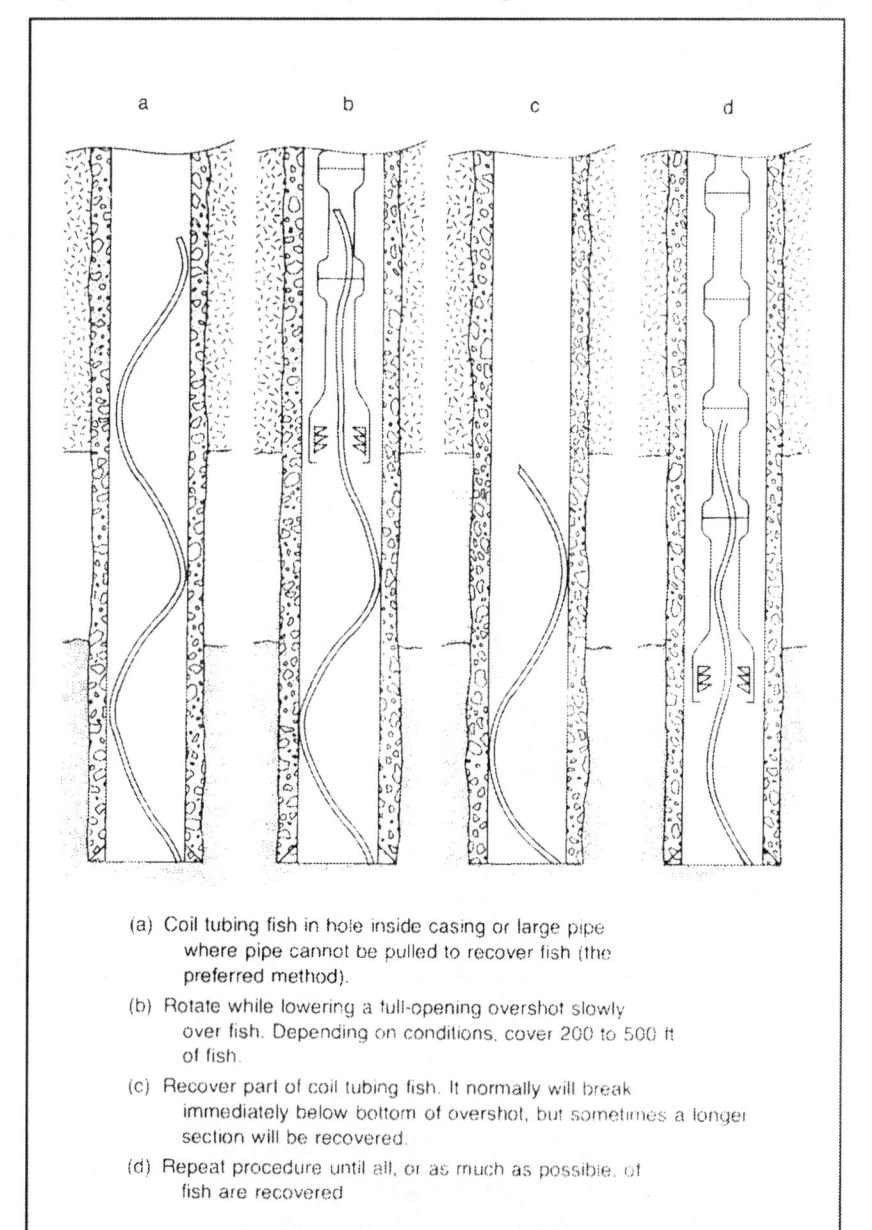

(a) Coil tubing fish in hole inside casing or large pipe where pipe cannot be pulled to recover fish (the preferred method).

(b) Rotate while lowering a full-opening overshot slowly over fish. Depending on conditions, cover 200 to 500 ft of fish.

(c) Recover part of coil tubing fish. It normally will break immediately below bottom of overshot, but sometimes a longer section will be recovered.

(d) Repeat procedure until all, or as much as possible, of fish are recovered.

When coiled tubing buckles or crumples below the fishing tool, the only way to remove it is by milling. As the tubing is milled, it becomes more compact. There is a tendency for some of the cuttings to settle down around the tubing until the hole is effectively filled with solid steel. Mills can be highly efficient, but milling any appreciable amount of steel in these situations is an expensive, high-risk procedure.

When it is necessary to mill coiled tubing in such a case, use a washover mill shoe and a short section of washpipe. Dress the shoe on the bottom and inside with hard facing material. If the amount of material to be milled is not excessive, it can be milled with this assembly and recovered. Short sections of coiled tubing may be milled up or pushed downhole with a flat-bottom mill.

SANDED-UP TUBULARS

Sanded-up small tubulars can be one of the most severe fishing problems, especially if a long section is sanded up in a cased hole. This is fairly common on many production wells. Sand is an extremely effective bridging agent. One of the best preventive actions is to produce through a packer with tail pipe at the bottom of a short perforated interval. The tail pipe may sand up, but recovering this short section is moderately easy. Longer perforated intervals may require some modification.

In most cases, the sanded-up tubing cannot be circulated. Trying to wash on top of it with a bit is ineffective because the circulating fluid only washes a short distance below the fluid nozzle. The main, initial recovery procedure is accomplished by working and jarring, with an emphasis on establishing movement and circulation. Maintain a minimum fluid pressure. There is a possibility of fluid seeping through the sand and establishing circulation, especially with working. There is a reasonable chance of releasing the stuck pipe if circulation can be established. Perforating to establish circulation is seldom successful.

Washing over is almost a last-resort operation in cased holes, especially for longer sanded-up sections. Tubing invariably sands up when laying on the low side of the hole with the collars touching the casing wall. In the open hole, the washover shoe will slip over the collar and cut the formation. Inside casing, the washover shoe most often cuts part of the upset and tubing collar. This severely reduces mill efficiency. Thicker-walled washover shoes cut slower, thereby cutting more metal and running longer. Thinner-walled washover shoes cut faster, wear out quicker, and present a risk of splitting the mill shoe.

Cutting upsets and collars reduces the integrity of the fish. The fish often breaks or is already parted so only short sections are recovered on each overshot run. There is a high risk of pieces of fish dropping alongside the remainder of the fish. Overall, washing out long sections of sanded-up tubing inside casing is seldom successful.

When fishing jobs occur with standard-size tubing, conduct operations using conventional fishing tools and procedures. Stuck tubing due to sanding up is a relatively common problem (Fig. 6–5). The general fishing procedure is described in the following example and indicates the severity of sanded-up tubing.

Figure 6–5 Washing over sanded-up pipe
Note: Figures a, b, and c are the ideal case. In figure a, the sanded-up centralized pipe is in the hole. It is washed over without difficulty in figure b and recovered with an overshot in figure c. Figures d, e, f, and g are the more normal case encountered in operations. The pipe is sanded up while lying near the edge of the hole with collars touching as in figure d. When pipe is washed over as in figure c, part of the collars and pipe body may be cut (see Figure 6–6). This involves extra time, more mills, and possible problems with the washpipe. An overshot is used to catch the fish in figure f. When picked up, the fish will frequently part, dropping joints, parts of joints, collars, etc., in the hole to further complicate the fishing job in figure g.

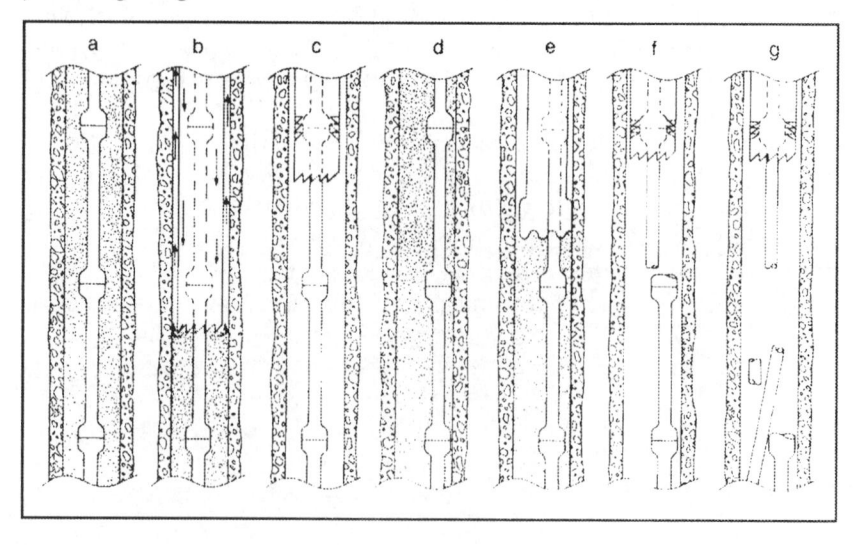

Recovering Tubing Sanded Up Inside Casing

A production well had 5½-in. 17-lb/ft, C-75, LT&C casing set at 9,500 ft (MD). During drilling, it was deviated at 4,000 ft with an angle buildup to 20° at 6,500 ft, dropping off to 5° at 8,500 ft, and maintained at 5° to TD. The 2⅞-in. EUE, J 55, 6.5-lb/ft producing tubing was landed open ended at 9,000 ft. The bottom was fitted with an API top-lock pump holddown as a no go for swabbing and to fit a rod pump at a later date if needed. The well was completed from multiple zones in the gross interval from 9,000 to 9,500 ft. Production from the well had declined, and the well finally stopped producing. The tubing was plugged when an attempt was made to circulate the well with a portable pump. Sanding up was suspected since it was a common problem in the area.

A workover rig was moved in and the tubing was found to be stuck. It could not be circulated and was worked unsuccessfully with low pressure for 5 hours to establish circulation. Working-force levels were gradually increased to 85,000 lb, 35,000 lb overpull over the pipe weight of 50,000 lb in 10-lb/gal saltwater. A free point found the tubing open to 8,000 ft and stuck at the same depth. Backing off at 7,500 ft was unsuccessful. The tubing unscrewed at about 4,500 ft during the placing of left-hand torque for a second back-off and was reconnected.

The tubing was perforated at 8,000 ft with a small hole to establish circulation, but the well would not circulate. It was perforated at 7,800 ft and circulation was established. The tubing was worked while circulating, and circulation was established through the perforation at 8,000 ft, as indicated by the circulating rate and pressure. The tubing was perforated at 8,200 ft and again worked while circulating. Circulation apparently was not established through the lower perforation.

A string shot was run to back off the tubing at 8,000 ft. When the string shot was detonated, the tubing did not back off and the string-shot tool stuck in the hole. It could not be worked loose, so the line was cut and recovered. The string-shot tool could not be recovered with a wireline fishing assembly. The tubing was cut with a chemical cutter at 7,900 ft. A fishing assembly was run, and the fish was jarred and worked for several hours. The fish parted at 8,000 ft, where the back-off shot had been detonated, and 100 ft of tubing fish was recovered. The free-point tool was not recovered.

An impression block was run and indicated that there was a tubing collar on top of the fish. This was expected, but the impression block was run to ensure that the collar was on the fish and not loose in the hole. An overshot did not catch the fish. A pin was run and screwed into the fish, and it was worked but did not release. Blind sinker bars found the top of the fish plugged, apparently with the string shot tool. The fishing assembly was released, and a washover shoe run below two joints of washpipe. The milling action indicated that the shoe was milling on the edge of the tub-

Cased Hole, Wireline, and Surface Failures

ing. After washing over for 2 ft, the washover pipe was pulled. An impression block showed the washover shoe had milled one side of the tubing. An overshot was run but would not take hold of the tubing. The fish was sidetracked and the well completed in a 3½-in. FJ liner.

A very short section of sanded-up tubing may be released by washing down beside it with a very small tubing stinger if size considerations permit. However, there is a high risk of breaking the stinger and leaving a second fish on top of the first.

If the tubing is plugged on the inside for any appreciable length, there is little chance of opening it up except by using coil tubing. This is a high-risk procedure for long sections of plugged tubing. It is moderately successful at washing sand out of the tubing under favorable conditions. Small positive-displacement motors developed for use on coil tubing may be run if sizes permit.

Milling is the last fishing alternative and seldom used on long sections of sanded-up tubing. In many cases, the best option is abandonment or sidetracking higher in the hole (Fig. 6–6).

Figure 6–6   Sanded-up pipe — cross section

Note: The shaded area is the minimum part of the 2⅞-in. tubing collar milled out with the 4½-in. by 3⅞-in. mill shoe.

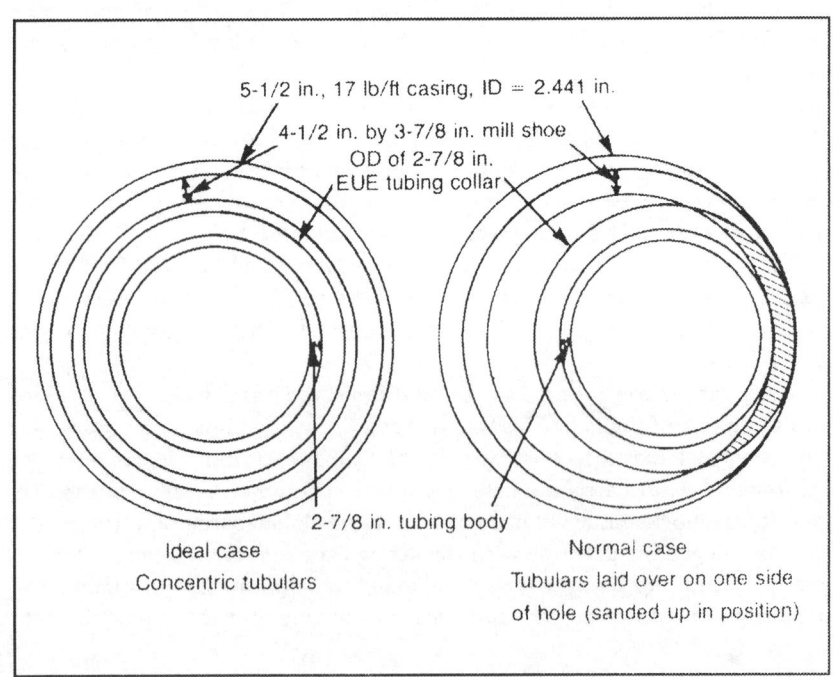

5-1/2 in., 17 lb/ft casing, ID = 2.441 in.

4-1/2 in. by 3-7/8 in. mill shoe

OD of 2-7/8 in. EUE tubing collar

2-7/8 in. tubing body

Ideal case
Concentric tubulars

Normal case
Tubulars laid over on one side of hole (sanded up in position)

TUBULAR FAILURES AT THE SURFACE

Surface failures are classified as tubular failures that occur in or above the blowout preventers. These range from bent pipe to broken tool joints, pipe, and kellys. They generally occur in small drill collars of about 6-in. diameter or less, drillpipe, and tubing. Surface failures create a severe situation that must be handled carefully to prevent further damage and loss. These failures are almost always preventable. Prevention obviously includes avoiding the original problem. However, it also includes recognizing the problem when it occurs and handling it properly.

Tubulars may bend—often severely—but do not immediately break and drop associated with fatiguing. This may occur with or without the slips in the rotary and with the pipe held by the elevators or by a friction hold in the rotary due to the bend—usually very light loads in this latter case. Other tubulars break or part at the surface and do not drop but cannot be caught or picked up conventionally. An example of this is breaking or parting the drillpipe a few inches above the seated slips.

Failures at the surface can cause danger to personnel and the risk of damaging equipment. Provide for these cases in the best manner possible by evaluating each situation. The risk increases with heavier pipe weights. Some of the remedial procedures require taking a risk of parting a bent assembly at the surface. This can also be dangerous, but less so since it generally involves very light weights.

Causes of Failures

Tubulars are bent or parted at the surface by a number of actions, most of which are preventable. In most cases, the description of the failure also includes the obvious preventive action. Some of the more common causes of failure are listed below.

- Setting the slips unevenly or otherwise improperly.
- Setting descending elevators on improperly set slips or on a slip segment.
- Slips set on fast descending pipe or top of the slips are caught on a tool joint.
- Incorrect use of double elevators, such as failure to secure both elevator bails correctly.
- Hitting a joint setting in the rotary with a heavy object, such as a swinging drill collar, pulling one or more joints into the "V" door, a stand of drillpipe, or a dropped traveling block.
- Tubulars dropped at the surface that do not fall into the hole.

- Slip and tong-die cutting.
- Tool-joint failure.
- Breaking a tool joint against a locked rotary.
- Tool joint set too high above the rotary.
- Excessive or improper pulling with pipe tongs.
- Inadvertently closing blind rams on pipe.

Dropped traveling blocks create a severe problem. Excluding danger to personnel, they can create situations ranging from bent to dropped pipe. Dropped blocks most commonly are caused by hitting the crown and cutting a drilling line. Other less common causes include brake failure and human error. Many of these actions may cause the pipe to break or part, dropping a section in the hole. In this case, begin fishing by using standard procedures.

Wellhead Security

Control and security of the well is a primary concern when a surface failure occurs. Corrective procedures depend upon the nature of the problem, well pressures, and similar concerns. Keep the hole full, and monitor hole and pit fluid levels.

Always have an inside blow-out preventer on the rig floor and set in the open position. Ensure that all personnel know where this tool and the opening/closing tool are kept since the possibility of an uncontrolled flow through the parted pipe is always present if high-pressure zones are exposed in the open hole.

If necessary, the inside blow-out preventer should be used. Connect it and bump it tight with a sledgehammer. Do not use drillpipe tongs if the pipe is bent except in an emergency, since there is a risk of breaking it. Install a check valve or plug the well with an equalizing pin inside the bent or broken pipe if this is applicable and the equipment is available. As an additional safety precaution, load or slug bent tubulars with heavy mud, either by pumping it through a connection or with a fill-up hose. A pack-off overshot or drillpipe spear may be used for circulation.

The casing side can be controlled with preventers in the regular manner. The pipe rams normally should not be closed immediately after the pipe is bent since the objective is to keep the assembly still until a way to catch it is installed. However, if the well is kicking, it is better to move the assembly and close the rams, even at the risk of dropping the pipe, since fishing is obviously a lower risk than a blow-out. In extreme cases, such as a blowout, close the shear rams, drop the assembly, and close the blind rams.

Tubular Bent, Not Parted

Tubulars may bend at the surface. If they break and drop, then recover them by standard fishing procedures. In some cases tubulars may bend, sometimes severely, but not drop. Using the correct equipment and operating techniques may prevent dropping them and causing a fishing job.

When the pipe is bent, there is a high probability that the metal in the pipe is overstressed in tension on the outside of the bend and in compression on the inside of the bend (Fig. 6–7). When the pipe is bent and does not drop the assembly, the metal still has sufficient strength to support the weight of the assembly below the bend. At this point, the common reaction is to pick up the assembly and replace the bent section. This often is a mistake. If the bend is severe, the pipe almost invariably will break and part when picked up. Therefore, before picking it up, make arrangements to catch or hold the pipe at a point below the bent section in order to prevent dropping it.

Figure 6–7 Tubulars bent or parted above the rotary
Note: The bent tool joint is caused by dropping blocks, etc. Install a drill-collar clamp. Pick up slowly while rotary slips slide in the rotary (hold slips down). Continue until the next tool joint is above the rotary and in a position to replace the damaged joint.

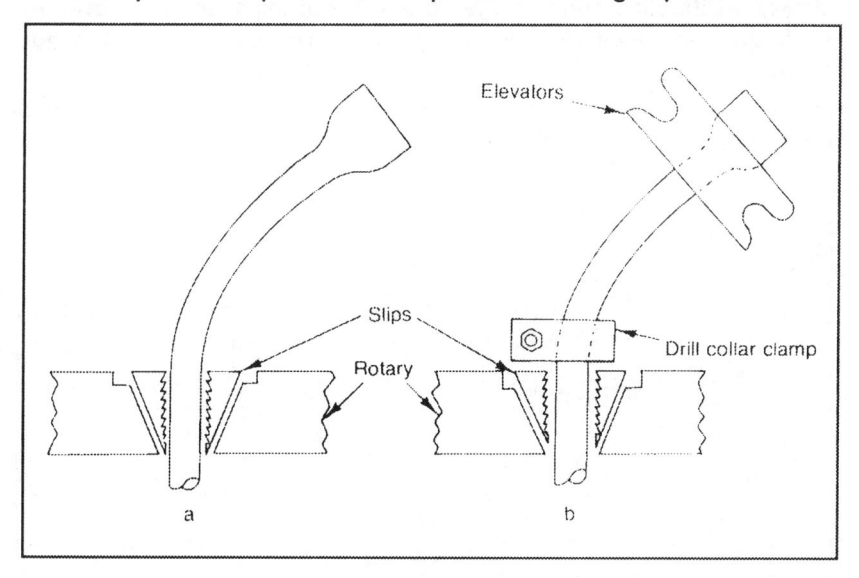

Cased Hole, Wireline, and Surface Failures

One problem is determining how much the pipe can be bent and still picked up in the conventional manner without breaking and dropping the assembly. Picking up the assembly reverses the stresses at the bend. The action is similar to severe, intense fatiguing. These stresses can be high, depending on the degree of bending and the assembly weight below the bend. Any bend of over 5–10° is strongly suspect, especially if it occurs with a relatively heavy assembly over a distance of a few feet or less. The amount of allowable bending without failure when the pipe is picked up depends upon various factors. These include the amount of bending, the diameter of the tubular, and the load on the tubular when it is picked up.

The important factor with bent pipe is that these cannot be predicted accurately. If there is any question, arrange to catch the pipe. If the bend is very small—i.e., less than 5°—it may be possible to pick up the assembly without special precautions (Fig. 6–8). However, if there is an appreciable bend and heavy assembly weight, arrange to hold and catch the pipe while picking it up in case it breaks.

Figure 6–8   Tubulars parted above the rotary
Note: The pipe has broken off leaving a short stub above the slips. Catch with slip-type elevators, a short grab overshot, or a pipe spear, depending on the size of spear and weight of the fish used. Pick up slowly while slips slide in the rotary (hold slips down). Continue until the next tool joint is in a position to replace the damaged joint.

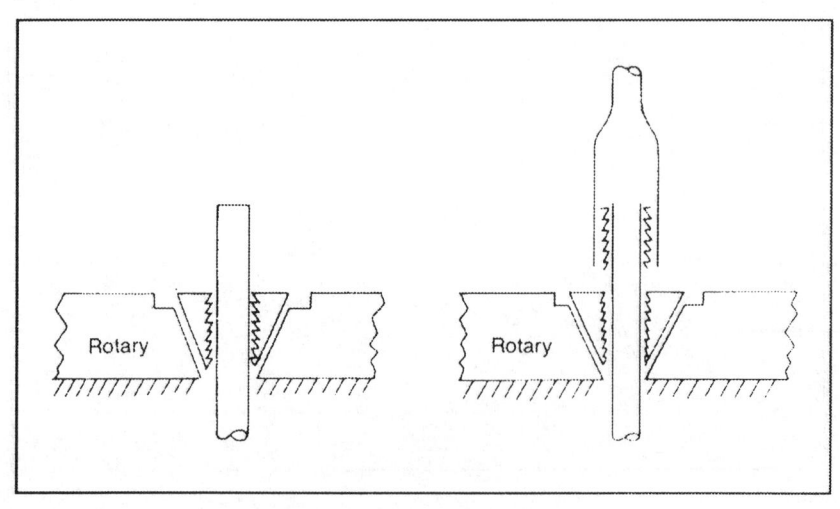

Handling Bent Tubulars

Handling bent assemblies is important, not only to prevent dropping the pipe but also because there is a higher incidence of bent than broken assemblies at the surface, at least before the bent assembly is picked up. The method of catching the assembly to prevent it from breaking and falling depends on the location of the bend relative to the rotary, whether the slips are in the rotary, BOP installation, clearance between the bottom of the rotary (or rotary beams) and the top of the bell nipple, and the general location of the equipment relative to the bend.

The method or device used to catch the assembly must be installed at a point far enough below the bend to ensure that it is opposite an undamaged section of pipe. It also must be placed firmly against the tubular so that it will catch the tubular almost immediately, before the tubular can fall any appreciable distance. The momentum of a falling tubular increases with the distance it falls. This increases the difficulty of catching it, so it is important to limit the falling distance and catch the tubular as fast as possible. Other methods of handling bent tubulars are described with the methods of handling them with various degrees of bending.

Moderate Bend

A tubular with a slight to moderate bend usually is bent from a few degrees up to about 5–10° over a relatively long section. Normally it can be picked up without breaking, subject to the limitations discussed earlier. However, there is always a risk of breaking and dropping, so take precautions accordingly.

The most common and easiest way to prepare for catching the assembly is to use the slips in the rotary, assuming that the pipe is bent well above the rotary. Seat the slips securely, tamping them in place. Tie the slip handles together and attach a drill-collar clamp securely around the pipe immediately above the slips. If the rotary slips cannot be used, then arrange to catch the pipe with a slip and spider or slip-type elevator set on the preventers, as described in the next section.

Close and lock the drillpipe rams. Do not jam them too tightly and pour a little oil on the pipe so it will slip easily through the rams. The pipe rams may not catch the pipe if it breaks, parting below a tool joint. However, they will prevent any large metal fragment from falling in the hole. Then pick the pipe up slowly until the bent section is supporting the full weight of the assembly. If the assembly does not part, open the pipe rams and continue pulling the pipe upward slowly. Let it slide through the dragging slips until the next tool joint is in a position so the bent joint can be replaced. Hold the slips firmly in the rotary. Hold down hydraulically actuated slips in a similar manner, but use the power source. If the assembly parts, it will catch in the slips and not fall through. Close the pipe rams again as soon as the lower tool joint passes.

Leave the slips on the pipe above the tool joint as it passes through the rotary. Place another set of slips around the tool joint and hold them in place as the tool joint passes up through the rotary. If another set of slips are not available or the bend is minor, then hold the first set of slips firmly in the rotary as the tool joint passes through it. Then replace the damaged tool joint and resume operations.

If the pipe breaks and is caught by the slips, treat it the same as pipe parted above the rotary. If it breaks and is not caught by the slips, then fish for it in the normal manner.

Severe Bend

A severe bend in the tubular normally is one that is more than 5–10°, depending upon the diameter of the tubular and the length of the bend. There is a high risk that the pipe will break and drop if it is picked up without taking proper precautions. Generally, the same precautions should be taken as described above for a tubular with a moderate bend. However, there are two alternate actions.

One alternative is to take proper precautions to catch the bent tubular and then pick it up. If it breaks, then catch it and pick it up by one of the methods described in the following section, "Types Of Catches." An example of this procedure is a short bend near the top of the set slips. Picking up the pipe might partially straighten the bend before the pipe can part, making it easier to catch the pipe stub. This procedure, with some variation, is described in the following example.

Picking Up Drillpipe Bent above the Rotary

A development well was drilling in an 8¾-in. hole at 10,000 ft with a standard drilling assembly and 4½-in. drillpipe. A connection was made at about 4:30 A.M. The kelly was made up on the joint in the mousehole and picked up. The kelly and joint were lowered to connect the single to the pipe in rotary. As the drillpipe pin was set in the box, it immediately was spun tight and bumped. When the kelly and joint were first lowered, the driller apparently did not stop the downward movement of the traveling block. The block continued moving downward and the kelly and single swung out through the V-door and laid down on the ramp and catwalk. The pipe extending out of the slips in the rotary formed about a 90° bend with a 6-ft radius.

The drilling lines were straightened out, the traveling block was picked up, and the equipment was checked. The single was disconnected from below the kelly with heating and tongs, and the joint was caught with the elevators. The slip handles were tied and the drill-collar clamp was installed on the drillpipe above the slips.

The assembly was picked up slowly. When the blocks began picking up the weight of the assembly off the slips, the pipe parted about 6 in. below the tool joint, leaving a stub about 2½ ft above the rotary. The assembly was caught by the slips in the rotary. A set of slip-type elevators were then used to catch the stub immediately above the rotary. The entire assembly was picked up slowly, allowing the rotary slips to drag. After the assembly had been moved upward about 6 ft, it was set in the slips and the elevators were moved down to a straight section of drillpipe. The assembly was picked up, still allowing the slips to drag until the next tool joint was above the rotary. The damaged joint was laid down and the pipe string was pulled so that the bits could be changed while waiting on another kelly. This was installed, and drilling operations resumed.

An overshot could have been used to pick up the assembly after it broke, but the elevators were readily available. Using a drillpipe spear would have been questionable because of strength limitations and the relatively sharp bend. Dragging the pipe through the slips prevented a fishing job.

The second alternative is to cut the pipe with a cutting torch and then catch the stub. This may be applicable when the pipe has a very sharp bend near the rotary so that the stub can be caught with a spear. If the bend is a considerable distance above the slips and the pipe near the slips is undamaged, cut the pipe at a higher point to prevent the risk of damaging the pipe stub, which can then be caught with a slip-type elevator or overshot. When cutting the tubular, take proper precautions to prevent fire.

Tubular Bent, Not Caught in Slips

A bent tubular not caught in slips can be a difficult problem, depending upon the circumstances and the specific point where the bend occurs. Bent pipe not caught in the slips is either held by the elevators or kelly or is partially supported by the bend in the rotary and, possibly, by some weight set on bottom.

Tubulars bent in this manner are handled similarly to those bent and caught in the slips, except that a method of supporting the pipe below the bend must be devised. Sometimes this is not the easiest job, but it is better than fishing the assembly out of the hole (Fig. 6–9). First, install an inside blowout preventer if appropriate. The pipe rams in the blowout preventer normally should not be closed since this may cause unwanted pipe movement.

When selecting a rig and planning the rig-up design, always consider the possible need of installing slip-type catching tools on top of or below the blowout preventers. Different procedures are used depending upon the situation, and most are site specific.

Normally, the easiest procedure in dealing with bent tubulars not caught in slips is to set a slip-type elevator or slip and spider on top of the preventers as a catching tool. In some cases, these may be used as a lifting tool when the rotary slips cannot be used. Installing the catching tool on top of the preventers may cause a space problem when clearance is limited between the top of the bell nipple or rotating head, if used, and the bottom of the rotary or rotary beams. The top of the bell nipple frequently will pass partway up the rotary, providing more clearance. For very close clearances, it may be necessary to remove the bell nipple by cutting it free.

If the pipe is held by the elevators, then use the slip and spider or side-door slip-type elevator as a safety catch and let the pipe drag through these as it is picked up. Pick it up until a tool joint is above the catch tool and then replace the damaged joint in the normal manner. Otherwise, remove the rotary, catch the side-door slip-type elevators with the bails, and pick up and replace the bent joint. As a safety device install another slip and spider on top of the preventer as soon as there is clearance. In either case, if the pipe breaks and is caught, treat it as pipe parted above the rotary. If it breaks and is not caught, then fish for it in the normal manner.

In many cases, installing a slip-type elevator or slip and spider below a rotary is difficult, time consuming, and involves a considerable amount of extra work. However, it is a secure backup system and well justified if it prevents dropping the pipe. There have been a number of cases where bent assemblies were picked up and the pipe parted, dropping the assembly as noted. The resulting fishing job is also a considerable amount of extra work, generally with a much higher risk. Therefore, take all necessary precautions to prevent dropping the assembly.

Figure 6–9 Tubulars bent in or below the rotary
Note: When a failure or accident similar to that shown occurs and the pipe is not dropped immediately, the first step is to install a slip and spider or slip-type elevator on top of the preventers. This ensures that the pipe is not dropped in ensuing operations. The method of repairing the failure will depend on specific circumstances.

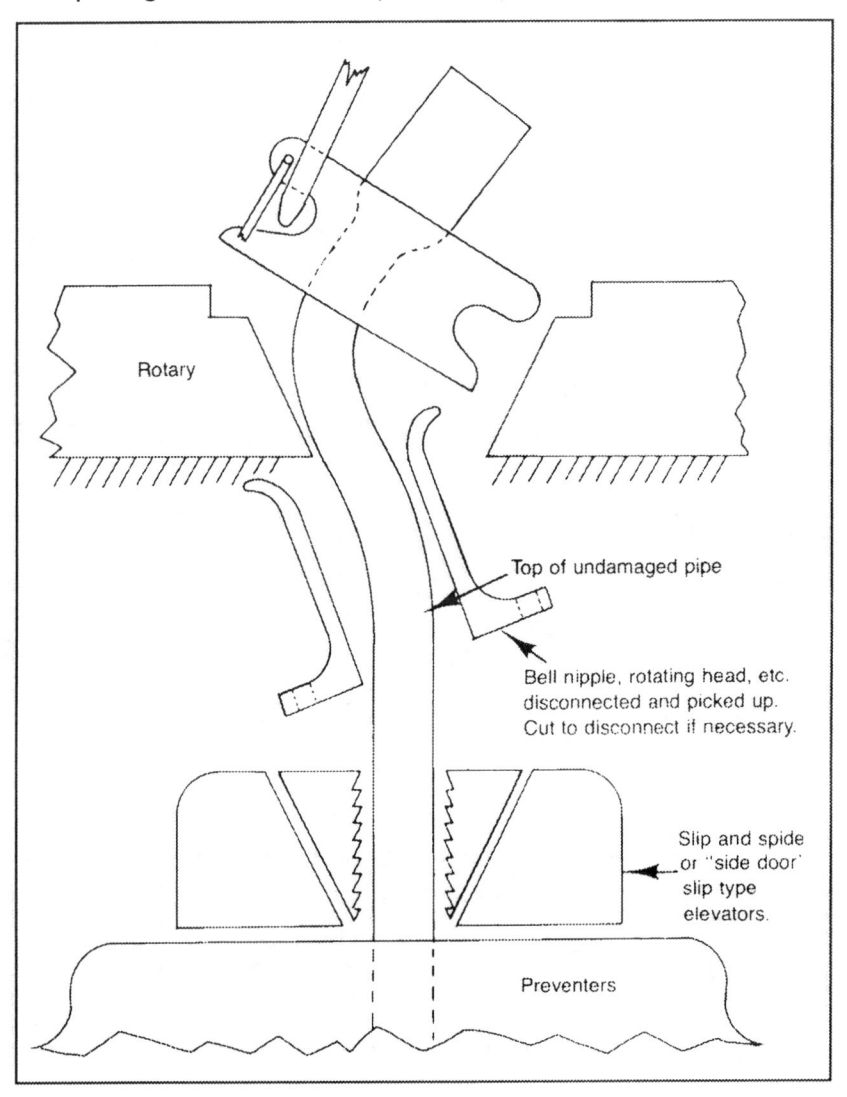

Cased Hole, Wireline, and Surface Failures

Picking Up Bent Pipe Incorrectly

A development well was drilled in an area where lost circulation, water flow, and caving formation problems were relatively common. Surface casing, 9⅝-in., was set at 1,500 ft, and an 8¾-in. hole was drilled to 8,000 ft. Operations were severely hampered by formation problems while drilling the last 1,000 ft.

A trip was made to change bits. While going in the hole and about 1,000 ft off bottom, the slips were set incorrectly. The bottom of one slip segment set in the recess for the kelly drive bushing, and the top of the segment caught under the descending elevators. The elevators tilted toward the unsupported side and bent the drillpipe severely at about a 25° angle almost 18 in. below the bottom of the tool joint (Fig. 6–10). Initially, the pipe bent but did not break. The driller elected to pick up the drillpipe and set the slips correctly. When the pipe was picked up, it broke at the bend and dropped the drilling assembly.

Figure 6–10   Drillpipe bent in the rotary

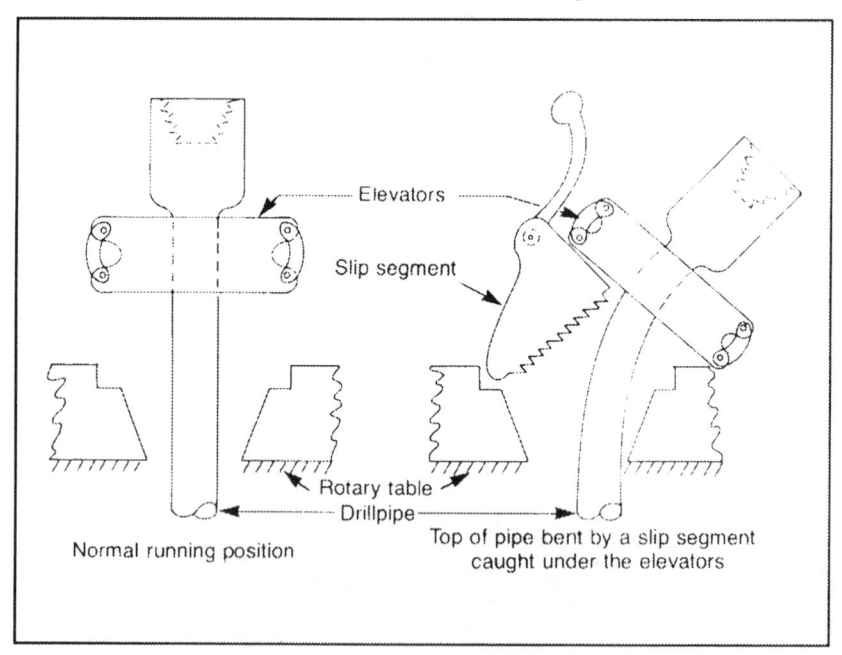

The resulting fishing job was only partially successful. Most of the fish was recovered after extended fishing. In view of the time spent and the deteriorating condition of the open hole, it was decided to set casing as deep as possible. The fish could have been sidetracked, but it was decided that this was not justified because of the worsening hole condition. Casing was run and cemented near the top of the fish. This allowed a completion in the zones behind the casing, and lower zones were lost.

The original failure was caused by setting the slips improperly. It may have been worsened by running the pipe in the hole too fast; regardless, the repair was not handled correctly. The metal fibers were overstressed when the pipe bent, but the pipe was still intact. When the pipe was picked up, it straightened at the bend and caused a failure, resulting in the dropped assembly. The WRONG ACTION caused an extensive fishing job and, finally, reduced well productivity, all of which could have been prevented.

If the overall pipe load is low, an alternate procedure may be used in cases similar to the preceding. It may be possible to pick up the rotary and pipe with lifting cables and hang them off. Check the capacity of the lifting cables first. If satisfactory, disconnect the bell nipple or other point in the blowout preventers, and chain this to the bottom of the rotary. With lifting cables, lift the rotary with part of the preventer section attached and the pipe. After lifting the required distance, chain the preventer section to the rig floor beams, lower the assembly a short distance, and remove the chains from the preventer section. Set the catching tool so that the pipe can be caught, and then lower the rotary and pipe until the pipe is set in the catching tool. This releases the elevators and pipe in the rotary. Set the rotary aside, catch the pipe with another slip-type elevator, and treat it as a regular bent-pipe problem.

In most cases, it will not be possible to lift the pipe with lifting cables as described in the preceding case, so use an alternate procedure. The bent-pipe section probably is short in this case since the elevators caused the bending. First, verify if there is sufficient clearance for the catching tool by lifting as a single unit the bell nipple and part of the other preventer equipment as needed. If so, lift these with lifting cables, set the catch tool on a preventer flange, and make the repairs in the manner described earlier.

There may not be enough clearance to install the catching tool after lifting the equipment. In this case, see if cutting off the bell nipple will provide sufficient clearance. If so, remove it, lift the equipment, and make the repairs. Normally it is not difficult to replace a bell nipple.

Another alternative is to lift the rotary and pipe with casing jacks. Block the rotary, lift the necessary part of the blowout preventer, and make the repairs by one of the methods described earlier. Jack placement may be difficult in this case.

Another case is when pipe is bent below the rotary and not held by the elevators or slips. This might be a situation where part of pipe weight is set on bottom. First, pick up the rotary and set it aside. Remove the bell nipple and as much of the preventer equipment as required to provide room for setting a slip-type elevator. Strip the equipment over the bent pipe or pipe stub. Pick the pipe up with the slip-type elevator or use another one higher on the pipe, depending upon the situation. Replace the damaged joint, replace the other equipment, and resume operations.

If the pipe is stuck in the situation described above, then back off at a shallow depth, several joints below the rotary. Replace the blowout preventer equipment, rotary, etc., and then run good pipe, screw into the fish, and begin standard fishing procedures. Do not back off deeper above the sticking point in the normal manner. There is a risk of the damaged joint(s) parting at or near the surface. Restore the surface equipment so the well is under control and then begin fishing.

Types of Catches

A pipe stub extending above the rotary is a common type of catch. Slip-type elevators are among the strongest catching tools. They usually catch a pipe stub extending 12–18 in. above the slips. Once the tubular stub is caught with the slip-type elevator, pick it up slowly, allowing the pipe to slide upward through the rotary slips or other catch tool. Hold the rotary slips firmly in the rotary until the damaged joint is in a position to be replaced. Do not close the blowout preventer rams until the tool joint is above the rotary and the slips set. Closing the preventers may move the pipe laterally a small amount, causing it to break and drop.

Overshots also provide a strong catch. A regular overshot without a skirt and fitted with spiral-type grapples will catch an 8–12 in. stub. An unskirted, short catch overshot fitted with a basket grapple will catch a 6–8 in. stub. Catch the tubular stub with the overshot and replace the damaged joint using procedures similar to those for catching the stub with a slip-type elevator.

The overshot may crush the end of a tubular, depending upon the condition of the stub and the weight of the pipe string. In this case consider using a pipe spear. The pipe spear can also catch shorter stubs that are too short to catch with slip-type elevators or an overshot. In most cases the pipe spear is strong enough to lift the pipe string. Catch the tubular stub with the pipe spear and replace the damaged joint using procedures similar to those for catching the stub with a slip-type elevator or overshot.

The use of a pipe spear may be limited by several conditions. The tubular stub may be damaged so that the pipe spear will not enter it. In this case, consider using a taper tap or die collar, although they are seldom used because of the limited strength of the catch. The spear may be too weak to lift the pipe string. In these cases consider other options, such as installing a slip-type elevator on top of the blowout preventers or welding on a lift nipple.

Welding procedures may be used in some conditions. Normally welding is not recommended except when special heat-treatment procedures can be used. However, sometimes welding is justified, depending on the assembly weight and other factors. Use deep "V" welds with double or triple passes and/or strapped securely with narrow vertical strips. Take all precautions to prevent a fire when welding.

A lift nipple may be welded on top of the tubular stub if there is sufficient space—normally at least 4–10 in. of stub length. Use a regular lift nipple with the same diameter as the tubular stub, and cut off the bottom tool joint. Alternately, cut off the top 5–10 ft of a joint of pipe with the same diameter as the tubular stub. Weld this piece to the tubular stub by using standard welding techniques. A ring or shoulder that fits regular elevators may be welded onto a longer tubular stub. This allows the operator to use regular elevators. It may be possible to cut holes in the stub and to use a lifting bar in the elevator or elevator bails for lighter fish.

After welding a lifting connection, then lift the pipe string, replace the damaged joint by one of the applicable procedures, and resume operations. In all cases, especially after welding, when the pipe is stuck downhole, it should be worked carefully for a short period of time. If it is not released, then back off a few joints below the rotary using the procedures described in the section "Tubular Bent, Not Parted" in this chapter. An example of using welding techniques is illustrated in the following example.

Welding a Lift Nipple on Drillpipe Parted at the Surface

A drilling well had 9⅝-in. surface casing set at 1,100 ft and was drilling at 8,000 ft in an 8¾-in. hole with a conventional drilling assembly and 4½-in. drillpipe. On a trip for a new bit, a tool joint twisted off about 6 in. above the slips in the slip area while a stand of drillpipe was being disconnected with the bit at 7,000 ft. It was decided that a spear could not be used because of the condition of the stub (which may have been a questionable conclusion) and there was a question about whether a short-grab overshot would catch or crush the stub. A tool joint was cut from the top of a joint of drillpipe and welded onto the drillpipe stub. The pipe was picked up, but it was stuck. The assembly was worked for a short time at low force levels, but then the weld failed, dropping the drilling assembly.

The pipe fish fell to the bottom of the hole. A cleanout trip was made with a bit. An overshot was then run, and three joints of drillpipe, including the parted joint, were backed off and recovered. The fishing assembly was screwed into the fish and, after working the fish for 4 hours, the fish parted, leaving a 2-ft stub of drillpipe on top of the drill collars. The lower part of the recovered fish was badly bent due to dropping, and the point where the fish parted was kinked. This is relatively common with a dropped fish; most of the drillpipe can be recovered by working the fish until the pipe parts, if it is not already parted. A double-bowl overshot was run and caught the top of the collars. After working them for 4 hours, the collars were released and all the fish were recovered, except the bit cones, which had sheared off. The bit cones were recovered and removed with one junk basket run and two mill runs. Drilling operations were resumed 12 days after the tool joint parted.

The overall fishing job was a good operation and probably better than average considering the formation problems in the area. There is no reason why some of the procedures discussed earlier were not used. They may have saved rig time, extra fishing costs, and replacement and repair charges. When the pipe was found to be stuck, the welded assembly should have been set down and the fish caught below the rotary.

Transfer Pipe String Weight

It may be necessary to transfer the weight of the pipe string from one catching tool to another. One case would be when the pipe parts in or above the rotary slips and is caught by them and the stub cannot be caught by one of the common methods. The catching procedure here would be to install a slip-type elevator or slip and spider on the blowout preventers. The assembly weight must be transferred from the upper slips in the rotary to the lower elevator or slip and spider set on top of the preventers. This is necessary in order to remove the rotary and lift the pipe with the slip-type elevators so that the damaged joint can be replaced.

Transfer the pipe-string weight by one of several methods. One is to cut the drillpipe immediately below the upper slips with a cutting torch. Depending on the assembly weight, this could create a safety hazard. Some of the risk could be alleviated by using a drillpipe spear, if possible, to engage the pipe below the point to be cut, and support part of the assembly weight.

Another method is to cut the drillpipe below the slips with an inside mechanical cutter actuated by a power swivel or similar means. It may also be possible to use casing jacks to move the rotary up a short distance in order to transfer the weight and set the lower slips.

Special Situations

There are various special situations that may occur. Some are described below.

A kelly broken in the square or hexagon normally protrudes above the rotary, so a pipe spear may be used. In order to lift the kelly, first install a drill-collar clamp firmly. Then catch the clamp above the elevators and lift the pipe string. Since the kelly may slip through the clamps, it might be wise to weld a catching device and place it above the elevators. Cutting a hole and using a lifting bar may be applicable in some cases. Don't overlook setting a tool joint on top of the closed drillpipe rams.

Slips may stick on the pipe. Pick up and hold the pipe with another set of slips and vibrate the stuck slips loose with a sledgehammer. In extreme cases, bump down on slips with a joint of drillpipe suspended from a lifting line.

Slips may stick in the rotary. Lift against the rotary with the pipe several times. If the slips are not released, hold an upward force on the rotary with the pipe and bump down on the rotary with a joint of drillpipe suspended from a lifting line. In an extreme case, bump down on the rotary with a drill collar. If the pipe breaks, treat it in the manner described in the section "Tubulars Bent Not Parted" in this chapter.

WIRELINE OPERATIONS

Wireline tools are used for many operations, such as logging, setting packers and plugs, opening and closing or shifting sleeves, running free points, swabbing, some fishing, and many other operations. In some cases, these operations are almost indispensable. However, they also can cause problems, many of which result in fishing. Like other causes of fishing, most if not all of these can be prevented by conducting wireline operations in a careful manner.

Fishing with Tubulars vs. Wirelines

In some cases, there may be an option of running wireline fishing tools on either tubulars or a wireline. A wireline is faster, but it is weaker and can only be reciprocated. Pipe is stronger, can be rotated, and has additional weight to push the fishing tool if needed, but may be slower. Heavier pipe weight can be a disadvantage in some cases. It is less sensitive to small weight changes, such as when setting the fishing tool on the wireline fish. There may be a higher risk of bypassing the line and sticking the fishing assembly, especially in the open hole. However, the pipe assembly is stronger, and if it sticks, there is a better chance of working it free.

The selection of tubulars or a wireline also depends on the fish. If the fish is a wireline that is smaller than the line being used in the fishing assembly, then the wireline fishing assembly generally should be used. It is faster, and the larger wire in the fishing assembly has sufficient strength to break the wireline in the fish, especially if the rope socket is crippled. Otherwise, select pipe in most cases.

If the fish is a wireline tool that normally has a fishing neck, tubulars often are the best choice. The overshot can be handled more efficiently with the pipe since the pipe can both apply weight and rotation. Rotation will frequently help, especially when a few short wires are still on the rope socket, which commonly occurs when the wireline is pulled out of the rope socket. The overshot on pipe has sufficient weight and rotating action so that it can be pushed down to take hold of the fishing neck, bumping if necessary. It may be more difficult to push the wireline overshot down over extended short wires, unless a shear pin release is omitted or a jar-up type is used. This in turn may limit jarring up to release the fish.

Types of Wirelines

Wirelines are available in various sizes and configurations for different uses. A good understanding of their construction helps in using them correctly and may prevent fishing or help in recovering them if fishing occurs.

Single-strand wirelines or piano wire are made of high-strength steel. Sizes range from 0.06 in. (less common) to slick lines of 0.072–0.12 in. diameter. Slick lines have a minimal frictional resistance when run through a pressure packoff. Uses include taking depth measurements and running small, lightweight tools such as pressure recorders, pull pressure equalizing plugs, and open and close sliding sleeves. They are not used for common wireline fishing and are difficult to fish for because they are relatively stiff and hard to catch in a small-diameter hole.

General purpose wirelines are used for swabbing, pulling cores, fishing, and similar operations. Sizes range from ¼ in. for very light service at shallow depths to the more common 7/16–5/8 in. sizes for heavier service and at greater depths. Smaller-size lines have individual strands of wire twisted and laid around a core strand. Larger size lines have multi-twisted lines. A group of 5–12 single strands are twisted around a single core strand, usually steel, to form a twisted group. About 5–12 twisted groups are then twisted or "laid" around a core strand, which itself may be a twisted group. Core strands can be made of fiber, plastic, and single- or multi-strand steel.

Both right and left twist, or lay, are available. Under tension, with both ends fixed, the line will torque in the direction opposite to the lay. Short lay lines have more twists per unit length and are more flexible than long lay lines. Heavy-duty lines have woven strands around a woven core. For improved wear and torque, the twist of the individual wire in each strand may be opposite to the twist of the strands.

Wirelines can be spliced. A properly spliced line will have a strength rating approaching the strength of the original line that was spliced, not of a new line. Rotary drilling lines generally are NOT spliced.

Logging and perforating lines or insulated conductor cables have carbon-steel strands laid over a central core comprised of one or more insulated conductor wires. Smaller lines in the size range of 5/16 in. have single strands laid over a single insulated conductor. Larger-size cable in the range of ½ in. and over have multiple twisted strands laid over a central core of multi-strand insulated conductor wires. These lines are damaged for electrical conductivity purposes at pulls in tension considerably below their actual maximum safe tensile failure rating because the conductor wires mash together and short out.

Operating Precautions

Problems frequently occur when running wireline tools. The success ratio of fishing with or for wirelines is only fair at best. Additionally, wirelines are used most frequently near the end of drilling, such as for logging and in completion operations. A major part of the well's costs have been expended at this point. Any action that may cause a failure at this time is very significant. Therefore take all precautions to ensure wirelines are in good condition and operated properly.

Running Close-Tolerance Tools

Wireline packers and other close-tolerance tools that are run on a wireline may stick. Always run a gauge ring first in order to detect flat places in the casing, mud cuttings accumulations on casing walls, and other obstructions that can cause close-tolerance tools to stick or packers to preset. These small obstructions can hold up wireline tools that would never be noticed while running tubing and other heavier downhole assemblies.

Use a gauge ring with a diameter larger than the tool diameter to be run in the hole. It should approach the drift diameter of the tubular. In a few cases, the gauge ring is run on a tool, such as a packer. Normally, it should be run separately because of the danger of sticking.

Run a tubing or casing scraper before making the wireline gauge run if the inside walls of the tubular contain small pieces of cement or metal cuttings, which are often magnetically stuck. This also helps prevent sticking the wireline gauge ring tool. Often, a scraper can be run to clean the tubular wall if the gauge ring fails to pass freely. If the scraper fails to pass, the most probable cause is flattening, so run a tapered swage in tubing or a casing roller in casing. Depending upon use and severity, run a tubing inspection log; then pull and replace the failed tubing. Replace, repair, or pack off the damaged casing.

Run a junk catcher to remove floating debris that may cause close-tolerance tools to stick or packers to set prematurely. As a preventive measure, either run the junk catcher one or more times on a wireline, run it below tolerance tools, or run both to remove any suspended solids that may cause sticking. Normally, remove preset wireline packers or plugs by drilling or milling.

Drilling a Preset Packer

A drillable packer was run on tubing and preset in 5½-in. production casing about 5 ft below the bradenhead. Since this was a drillable packer, it was elected to drill the packer instead of replacing the upper joint of casing. This occurred on a workover rig with a low substructure. Because of the close spacing, only a 6-ft drill collar could be run above the bit and still allow the kelly drive bushing to enter the rotary. The packer was drilled out in 8 hours.

When running a gauge ring, it is usually best to cripple the rope socket a short distance inside of and below the top so that there is a full wireline or cable through the top of the rope socket. This will help prevent flexure failures. Running large diameter gauge rings is a calculated risk. However, it is much easier to fish for and recover a gauge ring tool than a packer or bridge plug. There is always a risk of sticking these tools while running them, but less risk if a gauge ring was run first.

Jumping the Sheave

One of the more common problems with wirelines is jumping the sheave. Depending on the weight of the line and the line size, the line may be cut and the free section with tools attached may drop down into the hole. As a preventive measure, have the sheave guarded with a shield or hood so that the line cannot jump out. Position the hood about ½ to ⅔ of the wireline diameter from the sheave.

Lines jump the sheave when they are run in the hole too fast and there is excess slack at the surface. Always run the line under tension and operate it at controllable speeds. Wirelines also may jump the sheave when the wireline tool runs into a fluid surface or other obstruction that whips the line. Do not enter fluid at an excessive speed. Another common cause of broken lines is pulling out too fast and running into an obstruction or the packoff at the top of the tubular. The line literally is overstressed and pulled apart before upward movement can be stopped.

Worn Wirelines

Worn lines are detected by examining the wear on the individual outer wires. The line is considered worn and needs replacement when about one-third of the diameter of each of the outer wires has worn away. Always replace lines that have broken wires or strands, or stranded lines. This also applies to lines damaged due to overstressing or exposure to contaminating or corrosive fluids.

Flagging Wirelines

Flagging or marking wirelines helps to determine depths, the locations of tools relative to the surface when pulling, and similar concerns. A main reason is to provide a measurement of the amount of wireline fish left on top of the tool in case the wireline parts. The length of the wireline fish determines the fishing procedure and type of fishing tool. If the wire breaks cleanly at the rope socket and all of the line is recovered, then direct fishing operations at catching the rope socket. Use a blind box to cut off very short pieces of line left on the rope socket. Otherwise, fish for the line.

The amount of line left on the rope socket can be determined relatively easily and accurately by taking several simple precautions. Mark or flag wirelines at a precise distance from the rope socket.

The measurement's accuracy should be in the order of one inch. The mark usually is a flag of nylon fibers, positioned about 100 ft from the rope socket. If the line breaks and the flag is recovered, the amount of line from the flag to the broken point can be measured to determine exactly how much line is left on the rope socket. If the flag is not recovered, then over 100 ft of line is left in the hole. The fishing procedure is approximately the same for any length of line over 100 ft.

Swab lines commonly are marked with other flags at distances of approximately 250 and 500 ft from the rope socket. These are warning flags that allow the pulling speed to be reduced as the swabbing tools approach the surface.

Electric lines cannot be marked with nylon fibers but are marked magnetically. In this case, measure the exact distance from the marker to the rope socket. Lines frequently are reheaded, so repeat the measurement each time this is done.

Swabbing Precautions

Swabbing operations are susceptible to failures that cause fishing. The main failure is broken lines. Causes include pulling into the crown, jumping the sheave, running into fluid levels too fast, and similar actions. These problems are prevented mainly by eliminating the causes.

A major procedure for facilitating fishing is to equip the bottom of the tubing with an obstruction so broken swab lines and tools cannot drop out of the tubing and into the casing. A common obstruction is a bar or pinned-collar or pump-seating nipple. Recover the wireline fish from inside the tubing by fishing or pulling the tubing to recover the fish. Either method involves less risk than fishing in the casing.

When swabbing, it often is necessary to run the bare swab tools out the end of the tubing to check for sand in the well and to clean it out if there is sand entry. This can be accomplished while still helping to prevent the loss of the swab tools. Run a top lock pump seating nipple or similar restriction near the bottom of the tubing and a mill tooth or saw tooth collar on the bottom of the tubing. Run a no-go locking nut on the bottom of the swab assembly that is large enough not to pass through the seating nipple. If the swab line breaks, the line and tools will then be held inside the tubing.

When checking for sand fill-up, remove the large nut from the bottom of the swab bars so they can pass through the tubing restriction and into the lower hole. If sand occurs and must be washed out, the mill tooth collar normally provides sufficient digging action to remove the sand while circulating and rotating as the tubing is lowered. These are important preventive procedures and failure to observe them may cause fishing (Fig. 6–11).

Figure 6–11   Tubing restriction and no-go on swab bars

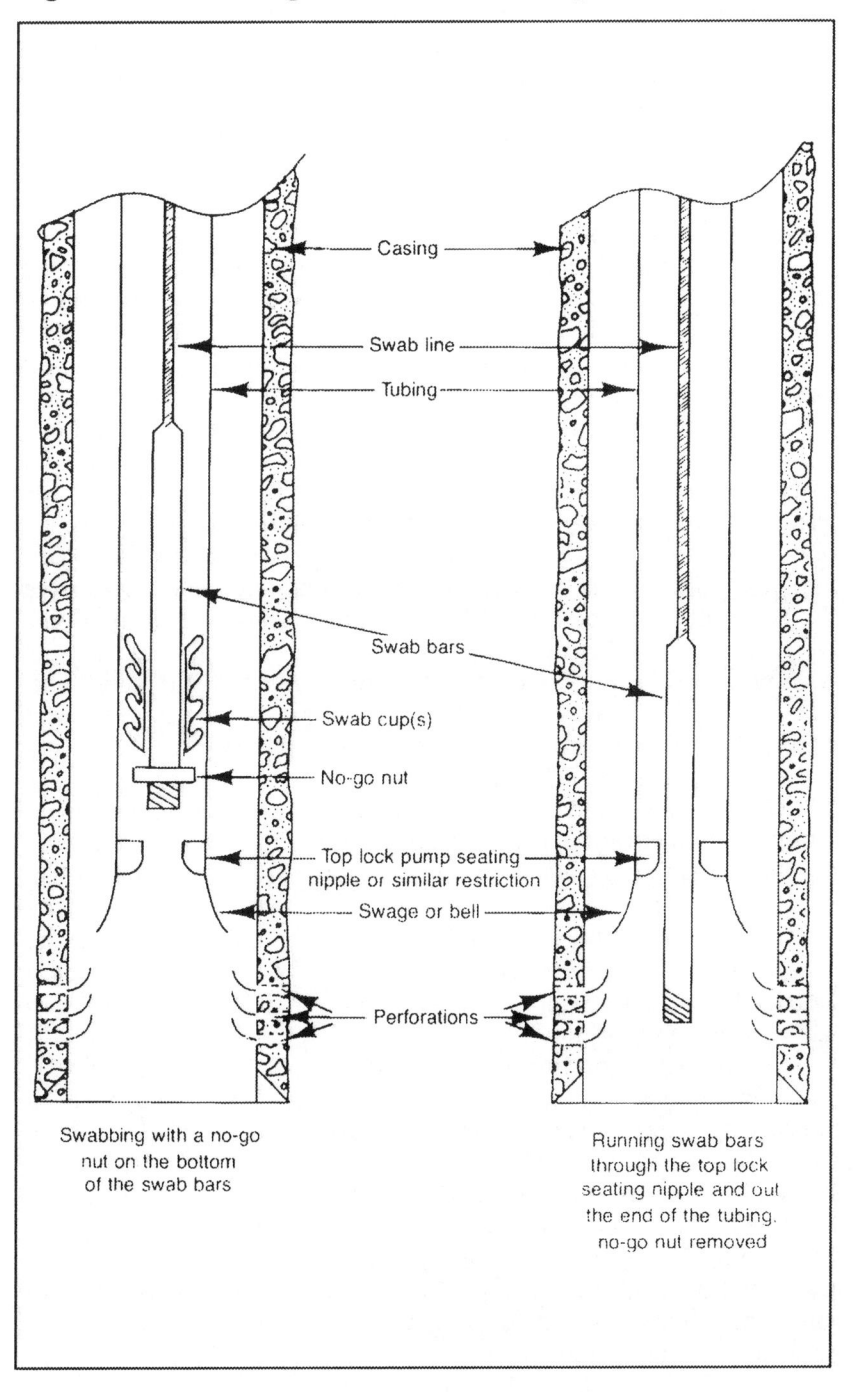

Swabbing with a no-go nut on the bottom of the swab bars

Running swab bars through the top lock seating nipple and out the end of the tubing, no-go nut removed

Cased Hole, Wireline, and Surface Failures

When wirelines are run out the bottom of tubing or other downhole tubulars, take measurements to ensure that the tools can be pulled back into the tubing without hanging up and causing a fishing job. Connect a bell nipple, swage, or similar opening on the bottom of the tubing. Make sure that the flat on top of the wireline rope socket is narrow, not over 1/8 in. wide, and that it is ground to a round, tapered shape.

Crippled Rope Socket

A crippled rope socket has part of the wirelines cut off inside the cable head. This reduces line strength at the top of the rope socket. Assuming line weight and breaking strength are favorable, crippling the rope socket provides a method of breaking the line at the lowest point when necessary (Fig. 6–12).

Crippling the rope socket has various advantages. The wireline will break at the top of the rope socket if the tools stick and cannot be recovered. Otherwise, the wireline may break higher in the hole and leave a long wireline fish that may be difficult to remove by fishing. Also, the line breaks clean at the top of the rope socket. Normally, this eliminates the necessity of cutting strands off the top of the rope socket. It permits fishing directly for the wireline tool—generally a less difficult fishing job. Most commercial electric wirelines are crippled, sometimes to the point that the cable will pull out of the rope socket before damaging the conductors. Run crippled rope sockets all or most of time, especially on gauge ring and close-tolerance tool runs.

There is some question about crippling the rope socket when fishing with a wireline. It may limit the amount of force that can be pulled on the fish. However, if it is necessary to pull with this force, the force required probably approaches the breaking strength of the fishing line. If the line must be broken, it is better to break it at the rope socket. This permits fishing directly for the wireline tool, which usually has a fishing neck. This is a more positive procedure than fishing for the wireline.

Another situation in which the use of a crippled rope socket is questionable is when the weight of the line in the hole approaches the safe working strength of the line. In this case, the line normally would break near the surface at the point of maximum tension instead of at the crippled rope socket. Crippling the rope socket is still preferred. If the line fails, there is a better chance of breaking the wireline fish off at the crippled rope socket and recovering most or all of the line. Running a crippled rope socket when the fishing line can be cut when tools stick is also subject to question, but the crippled rope socket is preferred.

Figure 6-12 Rope sockets

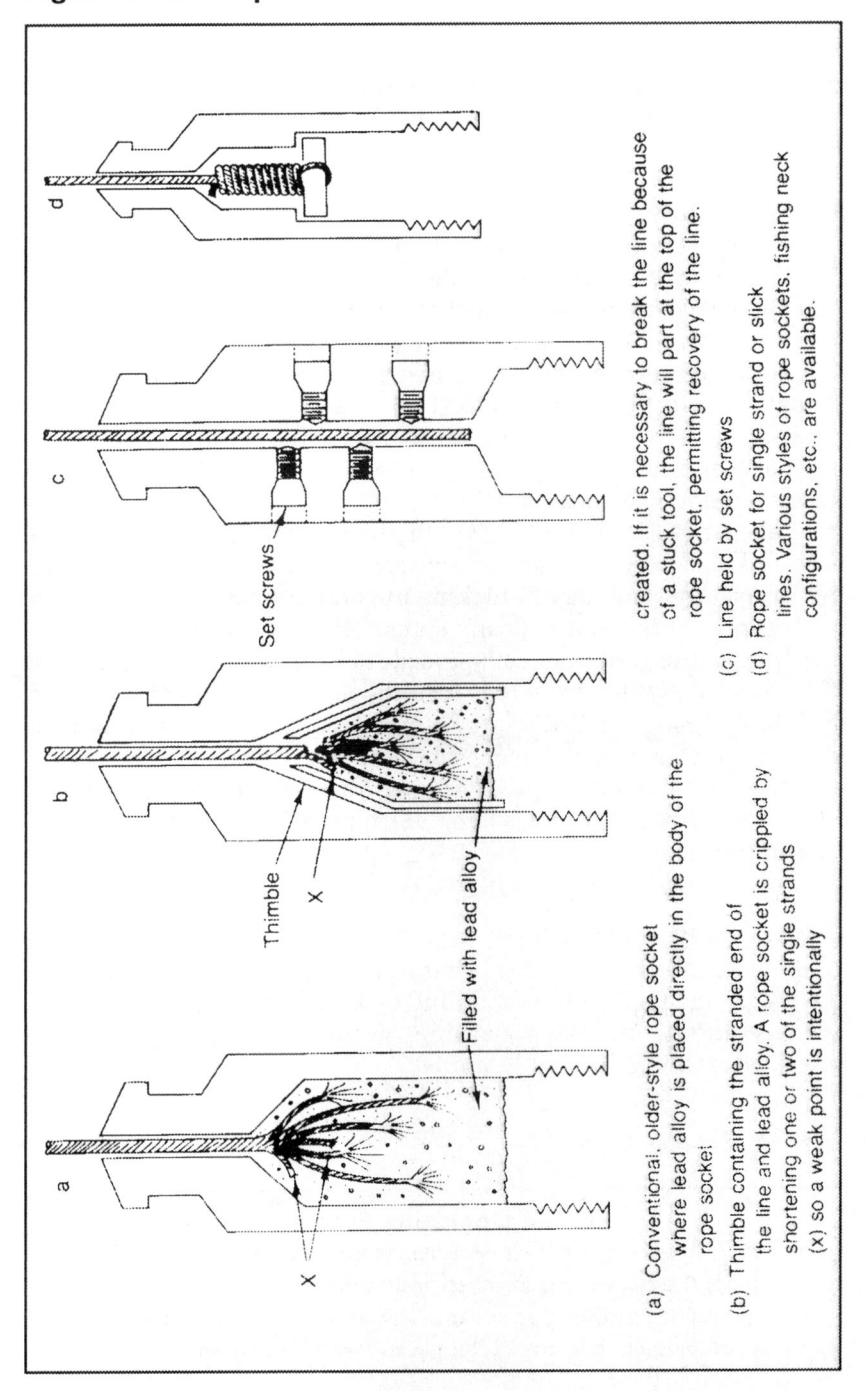

(a) Conventional, older-style rope socket where lead alloy is placed directly in the body of the rope socket.

(b) Thimble containing the stranded end of the line and lead alloy. A rope socket is crippled by shortening one or two of the single strands (x) so a weak point is intentionally created. If it is necessary to break the line because of a stuck tool, the line will part at the top of the rope socket, permitting recovery of the line.

(c) Line held by set screws.

(d) Rope socket for single strand or slick lines. Various styles of rope sockets, fishing neck configurations, etc., are available.

Cased Hole, Wireline, and Surface Failures   425

Another method of parting the wireline near the stuck tools is snapping the line. This flexes the line and fatigues it, so that it generally breaks off at the rope socket. This procedure often is effective, depending on the size of the line, depth in the hole, and pipe size. There is a better chance of parting the line by this method when the rope socket is crippled.

Wireline Keyseats

Wireline keyseats occur when the wireline wears a slot or groove in the side of the open hole. These most commonly occur after a number of logging runs and sometimes while fishing or working stuck wireline tools. They are more common in a crooked or deviated hole. In a few cases, a wireline keyseat may occur in the bottom of the last string of casing run in the hole. The wireline tool sticks when the rope socket is pulled into the keyseat with the wireline. Keyseats, especially multiple keyseats, may be detected by erratic, excess pull. Remove them by a cleanout trip.

Releasing keyseated tools is difficult, especially when they have been pulled forcefully into the keyseat. Try to work the tool out of the keyseat by slacking off, picking up, and bouncing. This seldom is successful if the tool is firmly stuck. The recovery procedure depends upon the situation. One procedure is to cut the line at the top of the keyseat, which usually is near the top of the tool, and recover by fishing. It may be difficult to catch the wireline tool after parting the line, especially in the open hole. Another procedure is stripping over the wireline with pipe. Stripping over is tedious and requires extra time, as noted in the section "Stripping Over Stuck Wireline" later in this chapter.

Operating Problems

Most of the problems that occur in wireline operations involve a few areas, such as general handling, logging, fishing, and completion operations, including swabbing. Some of these problems are described in the following example.

Wireline Operating Problems
- A 7-in. wireline packer was run during completion. While running, the packer hit an obstruction and stuck. The wireline was pulled out of the rope socket. The packer-setting tool was recovered by pulling it out of the packer with an overshot run on tub-

ing. The packer was then removed by milling. This is a relatively common occurrence and emphasizes the need of a gauge-ring run prior to running close-tolerance wireline tools, running a junk catcher or screen on the bottom of the wireline tool, and running the tool slowly and carefully.

- A well was being swabbed in during completion. While running in the hole, the swab hit the top of the fluid, causing the swab line to jump out of the sheave. This cut the swab line, dropping the swabbing tools and 5,000 ft of ⅝-in. swab line. The tubing was open-ended so it could not be pulled to recover the fish. After several days, the fish was recovered with a wireline spear. Also, the well was killed, which may have caused some formation damage. The combination of not knowing the location of the fluid top, running in the hole too fast, and using an unhooded sheave caused the problem.

- A wireline packer was to be set at 12,000 ft to isolate a lower, high-pressure, low-volume zone. While running in the hole, the packer reportedly hit a gas bubble, causing the wireline to jump the sheave. This cut the line, dropping the packer and 6,000 ft of line into the hole. The wireline was recovered with a wireline spear. The wireline broke inside the crippled rope socket. An overshot run on tubing recovered the setting tool. This set the packer, if it was not seated, and it was removed by milling.

  If the packer hit a gas bubble while running, it was being run into the hole too fast for existing conditions. There is a good possibility that the packer hit an obstruction or junk, indicating the need of running a junk catcher below the packer and of possibly making a gauge run before running the packer.

- A series of new zones over a long interval were perforated in a pumping well. A Christmas tree was screwed on the tubing, which was set in the slips and pipe rams closed. A short lubricator to cover the rope socket and collar locator was connected to the top of the tree with a collet-type latch. The well was to be perforated with multiple runs of long lengths of expendable gun. The well began to flow while running the third gun in the hole at 3,000 ft. The flow began blowing the gun out of the hole. The collar lock on the bottom of the lubricator was not seated properly or had failed and could not be locked. The shooting line was pulled out of the hole, outrunning the gun.

  After the gun was flowed out of the hole, the Christmas tree master valve was closed preparatory to circulating the well dead. A faulty casinghead valve failed, causing a high, uncontrolled flow of gas and oil out of the side of the casing. A line was later laid into the well and tied into the Christmas tree. High-pressure pump trucks, located a safe distance from the well, circulated saltwater at a high rate until the well was almost dead. The casing valve was replaced, and operations resumed. The combination of the faulty valve and collar lock caused a severe problem. The overall perforation procedure was also questionable.

- A bottomhole pressure bomb was run on a slick line in a high-pressure, low-volume gas well. The tool stuck due to freezing hydrates. Alcohol was displaced slowly down the tubing for 10 days before the pressure bomb was released and pulled.
- An open-hole logging tool hit an obstruction while running into the hole and the tool and 5,000 ft of line was lost. A spear recovered 2,000 ft of line on the first run but would not catch the fish on a second run. A sprung two-prong grab recovered the rest of the wireline fish, breaking the line off in the crippled rope socket. The logging tool was recovered with an overshot.

Wireline Fishing

Wireline fishing, as discussed here, is the process of fishing for wirelines and wireline tools with wirelines or tubulars. There are many wireline fishing operations. They are conducted by a wide variety of wireline fishing tools that are run on either tubulars or wirelines depending upon specific conditions, as described in the section "Fishing with Tubulars vs. Wirelines" in this chapter. When fishing for wirelines, observe all precautions.

Working with wirelines and using wireline tools can be highly efficient. For example, running a magnet on a wireline takes less time than running it on tubulars. Wirelines operate most efficiently at shallow depths, and efficiency decreases with increasing depth. For example, jarring action is crisp and well defined with wireline tools at 5,000 to 8,000 ft. At 14,000 ft and deeper, the tool action is less distinct and more difficult to determine. This is caused by line stretch due to increased weight and friction against the walls of the hole. This stretch increases the difficulty of determining specific tool action.

Handling wirelines correctly requires experienced personnel. There are many examples, a few of which are listed here. Slacking the line off excessively on a downstroke while bumping causes line flexure at the top of the rope socket and can lead to line failure. When a tool is stuck and the wireline cannot be cut or broken, it can be worked in this manner to break it near the rope socket. When fishing for a wireline with a wireline, it is important to have a full-gauge guide above the wireline fishing tool. If the tool bypasses the top of the fish, there is a higher risk of sticking by entangling it in the wireline fish.

Fishing Assemblies

Wireline fishing assemblies are similar to a standard wireline assembly, but they are designed specifically for fishing. Run wireline jars and a bumper sub immediately above the catch or fishing tool. Link or floating-mandrel types serve as both jars and a bumper sub. Hydraulic jars are more efficient than the link or floating-mandrel types for jarring upward. Weight bars increase both the force of the jarring action and assembly rigidity.

Use shear-pin releasing-type tools when applicable to provide a positive release if the fish cannot be recovered. Sometimes shear pins release too soon, which is a disadvantage. If this type is not used, sometimes the fish cannot be released. The only way to release the overshot is to cut or break the line or to pull it out of the rope socket. Then catch and recover the rope socket with an overshot.

Fishing for Wirelines

Fishing for wirelines includes the procedures used for fishing with either tubulars or wirelines. Common wireline fish include measuring lines, single- and multiconduit cables, swab and coring lines, and wirelines for handling packers, plugs, and similar types of tools. Like all fishing, there are preventive measures to prevent or alleviate fishing.

Probably the most common fish are wirelines. Which of the various methods to use in fishing for these depends upon the type and length of line lost, whether it is an open or cased hole, and related factors. The wireline fish normally has a wireline tool on the end. In many cases this tool sticks, causing a fishing job. The tool may or may not be recovered with the wireline fish. The wireline fish often parts above the rope socket leaving some line in the hole. The fishing job may be less complicated, especially if the line parts in a crippled rope socket. This leaves the stuck wireline tool, which often is less difficult to recover.

Wirelines, including electric cables, are often difficult to recover by fishing. Unlike fishing for tubulars, the option of drilling up and walling off is seldom available. It also is common to recover the wireline in sections of varying lengths.

There is always a definite risk of losing the hole or plugging back and sidetracking.

Wireline Inside Pipe

Frequently, a wireline fish will be inside pipe, such as a parted swab line inside tubing that has a restriction, perhaps a bar collar or no-go device, on the bottom. This leaves the positive action of pulling the tubing to recover the fish when it cannot be recovered conveniently by fishing inside the pipe. Pull the tubing until the top of the wireline is exposed, then recover the wireline. Normally, the well must be circulated dead or otherwise be under control.

If the tubing does not have a restriction on the bottom, the fish may fall through and out the bottom. In this case, do not move the tubing or other pipe and fish for the wireline inside the tubing. This often is easier than fishing for it in the larger casing or open hole.

Sometimes, if it is difficult to catch the fish and the tubing can be circulated, then the bottom of the tubing may be plugged with sand or other bridging material in order to stick the wireline fish inside the tubing. Then the wireline fish can be recovered by pulling the tubing. This is a specialized procedure that can be performed only under certain conditions.

Grab and Spear

The grab and spear, described in the section "Wireline Fishing Tools" in Chapter 3, are the two main tools used to fish for wirelines. In operation, connect the grab or spear to the bottom of a tubular or wireline fishing assembly and lower it to the top of the wireline fish, sometimes bumping down lightly. This creates a loose ball or bird nest of loops of wire that become entangled in the barbs. Normally, do not bump down after catching the wireline fish. Try pulling on the fish with a steady pull to release it. If it is stuck, work it carefully. Limit jarring, or jar with a relatively low force; heavy jarring often breaks the wireline fish at the point where it passes over the barbs.

The wireline spear generally is stronger than the grab. The spear has a larger, stronger prong, and its barbs are stronger for comparably sized tools. This factor is important because one failure relatively common for both tools is when the barbs or prongs break off and are left in the hole, which further complicates the fishing job. The spear can catch "hard to catch" fish more easily than a grab because the spear is stronger and can be bumped down over the fish. It commonly is used in fishing for larger, longer lines because of its superior strength. The spear cannot catch a short fish efficiently and is most frequently used for a slightly longer fish, one longer than about 15 ft.

The grab is weaker than the spear in the same size range. It is more efficient at catching smaller diameter lines and frequently can be used to catch a shorter fish, one of 5–10 ft. There is less risk of bypassing the top of the fish with a grab because of its multipronged configuration. In this respect the tool is considerably more effective in the open hole, where there is a higher risk of bypassing the top of the fish.

The prongs of the grab can be sprung outward (called a sprung grab) and then pulled together and tied with a small wire. This permits running the grab through the blowout preventer section. The small wire wears off as the sprung grab is run through the casing. When the tool reaches the open hole, the prongs spring outward and slide down the edge of the open hole. This increases the efficiency of the tool in the open hole and decreases the risk of bypassing the fish.

Use a full-gauge sub immediately above the prongs in the upper body of the spear or grab. This will help prevent bypassing the top of the wireline fish (Fig. 6–13). Special subs are available, or the body can be built up by welding. Ideally, the clearance between the outside diameter of this full-gauge section and the inside diameter of the casing should be less than or equal to one-half the diameter of the wireline in the fish. This clearance prevents bypassing the top of the wireline fish. Otherwise, if the tool bypasses the top of the fish for an appreciable distance, the section of wireline fish above the tool can become entangled with the fishing assembly, causing it to stick (Fig. 6–14). This is especially important when fishing in cased holes.

Figure 6-13 Recovering a wireline fish
Note: The gauge ring above the spear prevents the fishing tool from bypassing part of the wireline and becoming stuck.

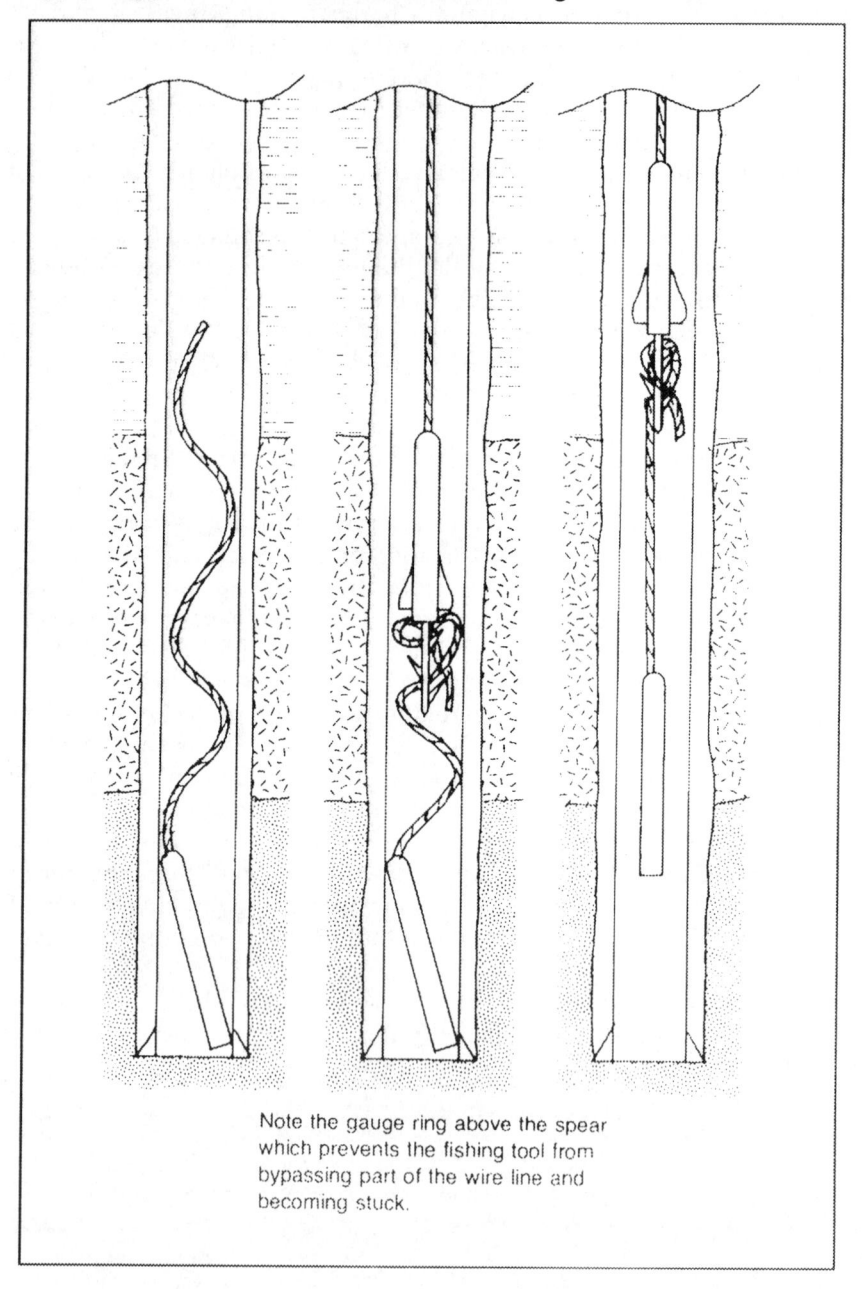

Note the gauge ring above the spear which prevents the fishing tool from bypassing part of the wire line and becoming stuck.

Figure 6–14   Precautions fishing for wirelines
Note: When running an undergauge tool, the tool should be run a short distance below the top of the wireline. Pick the tool up to ensure line is not accumulating on top of it. Then lower it and repeat the procedure until the fish is caught.

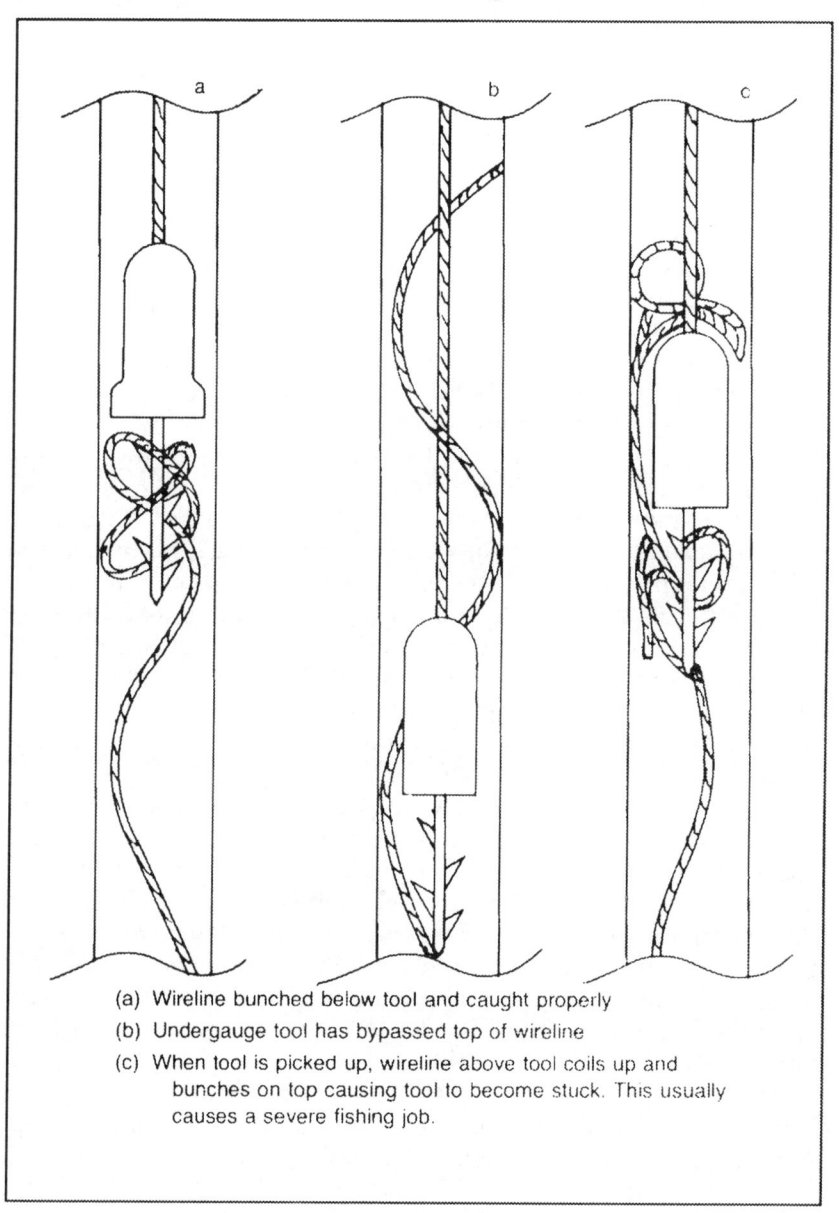

(a) Wireline bunched below tool and caught properly
(b) Undergauge tool has bypassed top of wireline
(c) When tool is picked up, wireline above tool coils up and bunches on top causing tool to become stuck. This usually causes a severe fishing job.

During fishing, be sure the spear or grab is not run too deeply in the vicinity of the fish without picking up and checking to see if the fish is caught or if the wireline fish is bypassing the top of the fishing tool with a possible risk of sticking. While lowering the tool near the top of the fish, pick it up 20–30 ft to ensure it is not stuck. Then lower it slightly and pick up to test again. Repeat the procedure until the top of the fish is located and the fish caught.

There is less risk of bypassing the top of the fish and entangling the fishing tool when fishing with a wireline rather than tubulars. There is also less weight, and this can be detected more easily with a wireline.

When fishing with tubulars, lower the assembly over the fish slowly with a minimum amount of weight to prevent pushing the wireline fish downhole until it becomes a solid mass. If this occurs, it may be difficult to force the prongs into the bundle or "bird nest" of wire. There is also a high risk of breaking the prongs as they are pushed into the bunched up wireline fish. There is a higher risk of this occurring in cased holes, but the action can occur in an open hole, especially in hard formations and in a gauge hole.

Sometimes, if the wireline fish cannot be caught, the operator should run a blind box, pack the wireline lightly in the hole, and possibly unravel a few stands. The line should be packed very lightly. If it is packed too heavily and densely, the prong(s) of the spear or grab cannot be forced into the bunched wireline to catch it.

A wireline fish bundled and packed in the hole is difficult to catch. Sometimes a fish may be opened slightly by drilling it lightly for a short time with a full-gauge, thin-wall, long-toothed washover shoe. This may help untangle and loosen the top part of the balled-up fish. Then run a two-prong sprung grab so that the prongs follow the edges of the hole to help catch and recover the fish.

Take a steady pull after catching the fish. If it does not release, alternately pull and jar lightly. Jarring may break the hold that the spear or grab has on the fish, thus releasing it. Heavy jarring may either break the wire at the point where it is caught on the barbs of the fishing tool or break off the barbs. The fish may break so that it must be recovered in shorter sections. This is probably satisfactory as a last resort, but try to recover the fish in one piece whenever possible. Continue pulling and jarring lightly, increasing the force periodically as necessary until fish is released or parts.

Pull the fish slowly and steadily out of the hole. Do not slack off except when the fish tends to stick. Pull the wireline fish very carefully into a casing shoe. It may stick in the restricted, reduced hole diameter. Work the fish up and down with alternate passes part of the way into the casing. This helps reduce the size of the fish so it will pass into the casing. The fish usually is found in the same place if it is accidentally released.

Stripping over Stuck Wireline

Stripping over stuck wirelines, sometime called the cable guided method, is a useful method of recovering wirelines and tools. It is also a tedious, time-consuming procedure that requires special attention to detail and coordinated action.

High pressures can increase the risk while stripping over, since there is poor well control during this procedure. Normally these pressures are controlled with mud weight, but in some cases high pressures can eliminate stripping over. If the well pressures up while stripping over, try to control it with a bag-type preventer and heavier fluid inside the pipe. Otherwise, cut and drop the line, connect the kelly or inside preventer, and control the well in the normal manner. Then continue pulling the fish out of the hole if it is still caught in the overshot, or fish for it in the normal manner. In severe cases, cut the pipe and lines with shear rams, drop all tools into the hole, and shut the well in.

The general procedure for stripping over is summarized below.

- Pull the stuck line to remove slack and install a line hanger to support the line.
- Cut the line about 4–6 ft above the hanger and attach a quick-connect, slip-type rope socket to the end of each line.
- Use a weight bar above the connection in the line above the rotary (the free line).
- Connect the free line to the stuck line in the hole and test the strength of the connection. It should be strong enough to hold the stuck line.
- Remove the line hanger and support the stuck line with a slotted plate under the rope socket.
- Disconnect and pass the upper free line to the crew member in the derrick.
- Run a standard tubular fishing assembly with an overshot to catch the rope socket on the fish. (This rope socket usually is supplied by the wireline service company.) Connect the overshot to the first stand of the fishing assembly.

- Thread the free line through the overshot and first stand of the fishing assembly and connect it to the stuck line with the quick connector.
- Tighten the free line to pick up the stuck line a short distance with a powered drum mounted on a truck, and remove the slotted plate under the stuck-line rope socket.
- Lower the tubular fishing assembly and the free and stuck lines simultaneously while holding the free line taut with the powered drum.
- Set the fishing assembly in slips and hold the stuck line with the slotted plate.
- Disconnect the lines, and pass the free line to the person in the derrick.
- Catch and pick up a stand of pipe, thread the free line down through it, and connect the free line to the stuck line.
- Pick up the line with the powered drum, remove the slotted plate, and connect the stand of pipe to the pipe held in the rotary slips.
- Lower the stand of pipe while holding the line taut with the powered drum.
- Repeat this procedure until the tubular overshot catches the rope socket.
- Work and pull the fish free while holding the line taut with the powered drum.
- Remove the line by holding the stuck line in the elevators, and then pull it out of the fish rope socket with the elevators. (Note: See the alternate procedures described below.)
- Pull the released stuck line out of the hole, spooling it on the powered drum.
- Pull the pipe and fish out of the hole in the normal manner.

There is a risk that the fish will jump or pop out of the overshot and drop free in the hole when the stuck line is pulled out of the rope socket of the fish. One alternate procedure is to ensure that the fish is set on bottom, which helps retain it in the overshot as the line is pulled out of the rope socket.

Another alternate, more positive recovery procedure includes leaving the fishing line laying slack in the tubular fishing assembly while pulling it out of the hole. Rotate out with pipe tongs or chain out, cutting the fishing line periodically as the pipe is pulled out of the hole.

A summarized procedure of stripping over is illustrated in the following example.

Stripping Over a Stuck Wireline

A completed well had 7⅝-in. production casing set at 10,000 ft with a 5½-in. production liner at 14,000 ft. A permanent packer was set in the liner at 12,000 ft. The packer had a conventional mill-out extension and several 2⅜-in. pup joints and plug receptacle nipples. Another permanent packer was set in the 7⅝-in. casing at 9,000 ft with a conventional mill-out extension. Tubing, 2⅝ in., was latched into the packer with a plug-seating nipple set above the latch in. A plug seated in the nipple was stuck and could not be released with the plug-running tool. If the wireline pulled out of the rope socket, the plug could drop downhole and land on the lower permanent packer. Therefore, the wireline could not be used without fishing out the plug, which would also require drilling the upper packer. Also, the line could break near the surface, leaving the line to be fished out of the hole. To prevent these problems, the tubing was pulled out of the hole while stripping over and maintaining tension on the wireline.

Stripping over is an important fishing procedure when applicable, as illustrated in the following example.

Stripping Over a Keyseated Logging Tool

A development well had 100 ft of 13⅜-in. conductor pipe, 9⅝-in. surface casing at 3,000 ft, and 8¾-in. open hole to 13,000 ft. The formations varied from medium to medium hard and were highly layered with a mixture of sands, shales, and carbonates. The hole was widely varied, ranging from in gauge to highly out of gauge.

After reaching TD, the hole was circulated and the drillpipe was pulled in preparation for logging. An induction-log tool was run, followed by a combination suite, including a neutron tool, both of which ran to bottom. While logging up, the tool stuck at 6,500 ft and could not be moved (Fig. 6–15). The tool was worked by alternately pulling tension, within safe limits, and slacking off for about an hour without moving the logging tool.

Cased Hole, Wireline, and Surface Failures

Figure 6-15 Wireline keyseat
Note: In figure a the wireline has cut into the wall of the hole in a crooked section, causing a wireline keyseat. When the wireline tool is pulled, it can stick in the keyseat, figure b. If the tool is not pulled into the keyseat too tightly, it may be lowered and worked up past the keyseat. If the tool is stuck tightly in the keyseat, alternatives are to cut the wireline near the top of the tool and fish the tool from the hole, or fish for the tool by stripping over. (Vertical scale is distorted for illustrative purposes.)

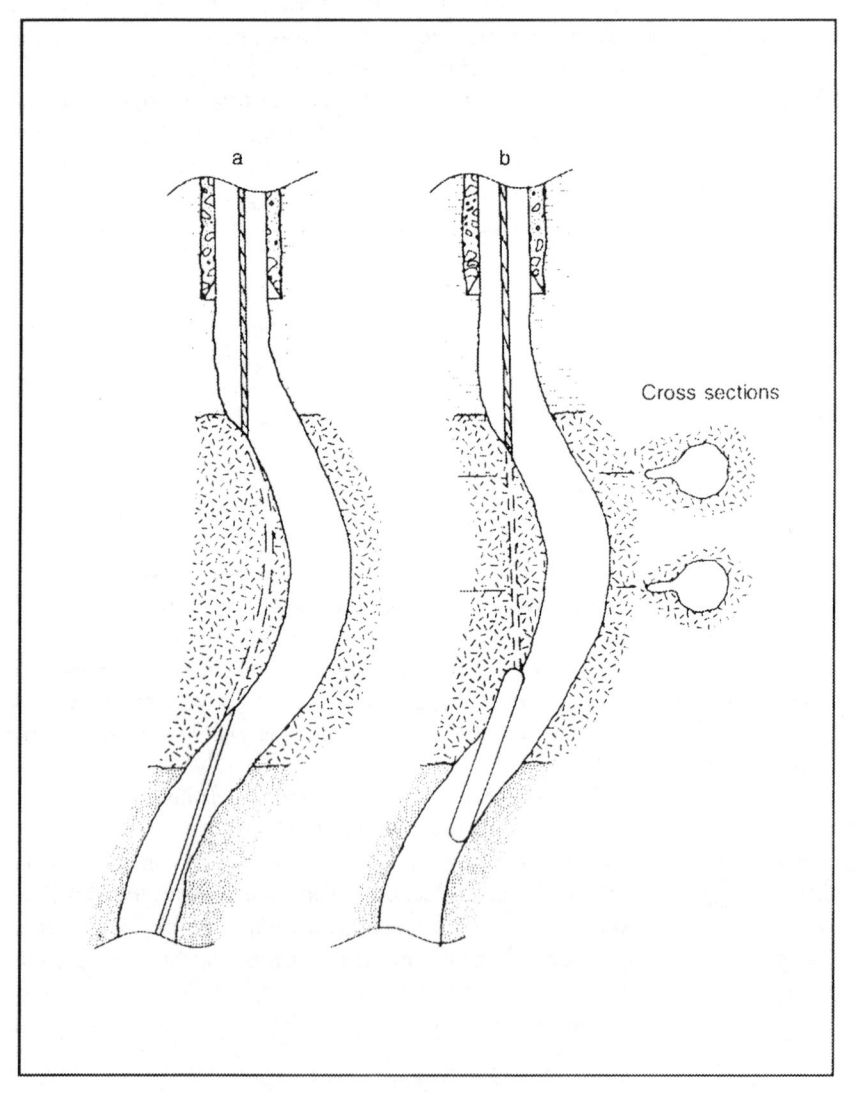

At this time there were two reasonable actions. One was to pull the line until it parted or pulled out of the rope socket and fish in the conventional manner. The tool probably could have been recovered by regular fishing, but there was a risk because of the hole's condition, especially if the wireline was lost in the hole.

The second alternative was to strip over the line with the drillpipe and recover the fish with an overshot. The logging-tool fish contained a neutron source. Regulatory agencies require extensive efforts to recover such a fish. If it could not be recovered, the lower part of the hole would have had to be cemented off. There were also limits on how close another well could be completed to the cemented neutron tool. While stripping over is a slow process, it involves less risk in catching and recovering the fish, so that method was selected.

The service company supplied an overshot to catch the fishing neck on the logging tool. The overshot was connected below two stands of drill collars. The logging line was cut and fitted with special connections to strip the line through the drill collars and drillpipe. It took about 36 hours to strip the drillpipe in the hole. When the overshot approached the logging tool, the logging line was placed under slightly more tension than when running in the hole with the drillpipe. As the overshot set down on the logging tool, the drillpipe appeared to lose about 3,000 lb and then dropped back to normal weight. The weight-change reading was not a strong indicator, but it did seem to show that the fish was caught in the overshot. The drillpipe was then moved up and down a short distance, and the movement was reflected on the logging truck's weight indicator. This showed that the fish was caught in the overshot.

The logging line was pulled to a maximum safe tension to help ensure that the fish was securely seated in the overshot. Pulling the line out of the crippled rope socket on top of the logging tool was considered and rejected. While doing this would have made pulling the drillpipe out of the hole easier, there was a risk that the reactive impact as the line parted would cause the logging tool to drop or be knocked out of the overshot.

The logging line was lowered until it was completely slack, anchored, and cut near the drillpipe box. As the drillpipe was pulled from the hole, the logging line was secured and tied off at each connection. The stands were rotated out with drillpipe tongs.

The logging tool was recovered, securely caught in the overshot. The complete job took about 48 hours. A cleanout trip was made, and the well was then logged without further problems. The caliper log showed that the hole was out of gauge at the point where the logging tool had stuck and in gauge above that point. This reading and the tool action indicated that the logging tool had been stuck in a wireline keyseat.

Parting Wirelines

Wirelines are intentionally parted when wireline tools stick and cannot be released. Fishing lines are parted similarly. Methods of parting wirelines include cutting, pulling out of the (preferably crippled) rope socket, and breaking. Sometimes, the snapping procedure may be used to part the line near the rope socket.

Before parting the line, give full consideration to stripping over it. This procedure has advantages and disadvantages but, depending on the specific situation, it may be justified as an alternative to regular parting.

It is important to be able to part the wireline at the lowest point, preferably at the top of the tools so the wireline can be recovered. The tools used to fish for wirelines are not as efficient as desired, and handling techniques are important. In regular fishing jobs, mills often can be used to mill up the fish, mill over the fish, or cut a fishing neck. A long piece of wireline junk cannot be milled efficiently; therefore, remove as much wireline from the fish as possible.

Various wireline cutting tools are available, as described in the section "Wireline Fishing Tools" in Chapter 3. Select the tool most applicable for the job, whether cutters dropped as a free cutting tool or cutters run on another wireline. Cutters run on another wireline often cause entangled lines. They twist together, so they often must be cut as they are pulled.

Snapping the wirelines is a method of releasing that is seldom used yet often works efficiently, especially if the rope socket is crippled. The success of snapping the wirelines also depends on line size and length and the skill of the operator. The procedure takes time but seldom, if ever, as long as fishing for the wireline, so be patient.

In operation, tighten the wireline, release it rapidly, and stop it abruptly. Stop the line while it still has a small amount of tension at the point where it is stuck or at the top of the rope socket. Mark the stopping point. When the line is stopped abruptly, momentum causes the lower section continue moving downward a short distance. Line elasticity then causes it to move or snap upward. The whipping action of the line causes flexing at the stuck point. Repeated action tends to fatigue the line, causing it to fail and part. In most cases this will not damage the line, except near the broken end.

Sometimes large-diameter lines must be parted with a traveling block. Normally the snapping procedure is preferred to this method, but it is not always successful. The usual procedure is to take all tension out of the line and then either tie it to the elevator bails with a chain or clamp it. Lift the block slowly until the line breaks. Frequently this will be near the surface, although the line may have a weak point deeper in the hole, which can improve line recovery. Sometimes the line parts at the rope socket, and in a few cases the fish is recovered.

Often the line parts at the point where it is tied with the chain. As an alternative, tie the line to the elevator bails and secure the end with line clamps. Another procedure is to tie a larger line to the traveling block. Then clamp the line to be parted to the larger line with the ring of the clamps on the larger line. This may help break the line deeper in the hole.

Ensure that all personnel are at a safe distance when parting lines, especially when pulling them apart.

Fishing for Wireline Tools

Common wireline tools may stick, requiring fishing. These include pressure recorders, sleeve shifting tools, electrical logging equipment, core cutting tools, directional and course measuring instruments, swabbing equipment, packers, and plugs.

If there is a long length of wireline connected to the wireline tool fish, then fish for the wireline with a spear or grab. Recover the wireline and possibly the stuck wireline tool. Otherwise, recover the wireline with the procedures for fishing with a spear or grab. Then fish for the stuck tool.

In most cases an overshot should be run on a tubular or wireline fishing assembly. Lower it over the fish, bumping down lightly if necessary. Work the fish upward to release it, and pull it out of the hole. Run a grab to try to catch a shorter length of line on top of the fish. This usually does not take long with wireline fishing tools. If there is a chance of catching the fish in this manner, the time spent is justified (Fig. 6–16).

Figure 6-16 Recovering a wireline tool fish

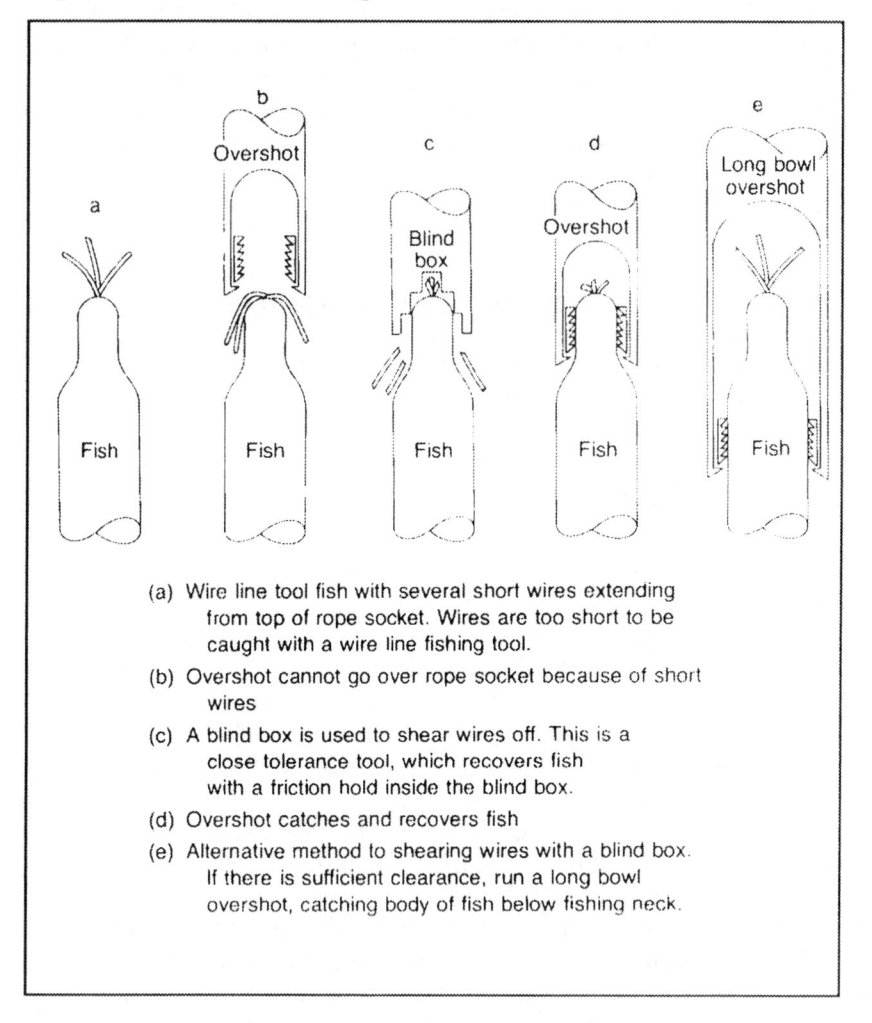

(a) Wire line tool fish with several short wires extending from top of rope socket. Wires are too short to be caught with a wire line fishing tool.

(b) Overshot cannot go over rope socket because of short wires

(c) A blind box is used to shear wires off. This is a close tolerance tool, which recovers fish with a friction hold inside the blind box.

(d) Overshot catches and recovers fish

(e) Alternative method to shearing wires with a blind box. If there is sufficient clearance, run a long bowl overshot, catching body of fish below fishing neck.

A small amount of line on top of the fish, ranging from a few strands to several feet, is usually too short to be caught by either the spear or grab, and may prevent catching the fish with an overshot. Run a long basket to recover the entire tool if clearances are favorable. Otherwise run a blind box, or sometimes a die collar, below a wireline fishing assembly and cut the wires from the top of the rope socket. Then run a wireline overshot, catch the rope socket, and recover the fish. If the cut wires are lying around the neck of the rope socket so that the overshot cannot catch it, run a rasp and clean out the debris. This generally pushes the debris below the fishing neck, but it can stick the fish tighter.

With tubular fishing tools, run an overshot with a mill-tooth skirt and washover the top of the fish for a short distance. Catch the fish and pull it out of the hole. Try a full-opening overshot if there are a few longer wires on top of the fish.

Fishing for Wireline Tools

A wildcat well had 7-in. casing at 15,000 ft and a 5-in. liner at 16,500 ft, with several zones having been perforated and stimulated in the liner. The perforations were found to be producing marginal, high-pressure, low-volume gas, so a wireline packer was run to plug them off. The packer was fitted with a mill-out extension, a plug-seating nipple, a 30-ft joint of $2^3/_8$-in. tubing, a tubing pup joint, another plug-seating nipple, and a 4-ft tubing pup joint so that the zones could be put into production at a later date if desirable.

The packer was seated at 16,000 ft in the liner, and a blanking plug with an equalizer pin was run and seated in the lower nipple. Several zones in the upper liner were completed. The tubing was landed in the PBR on top of the liner, and the zones were tested through a sliding sleeve. After testing the upper zones, the blanking plug was to be removed to commingle all zones.

A wireline tool was run to pull the equalizing pin and equalize the pressures, but it stopped several feet above the plug. An impression block had a very indistinct imprint, which frequently indicates sand. A bailer was run but did not make any recovery. All work was being done under pressure through a pressure lubricator. Working wireline tools at this depth is difficult because of line stretch and the difficulty in determining the tool action.

It was decided to cut the tubing above the obstruction and drop the lower part of the packer tailpipe to allow production of the lower zones. A chemical cutter was run to the top of the obstruction, picked up above the packer to be sure the tool was free, and reset 1 ft above the obstruction. The tubing was pressurized not only to equalize the pressure but also to prevent the cutter from being blown uphole. The cutter was discharged and wellhead pressures increased, indicating the pipe was either cut or partially cut. As the cutter was being pulled out of the hole, it became stuck. The cutter had been worked for about 6 hours when the wireline pulled out of the crippled rope socket.

Cased Hole, Wireline, and Surface Failures

A long lubricator was installed on the Christmas tree to cover the fishing tools and fish. A shear-pin release overshot was run below double jars, since double jars provide a more positive tool action at this depth.

The fish had been caught and jarred for 3 hours when the shear pin sheared, releasing the overshot. When the tool was pulled, the bowl of the overshot was split, indicating that there had been hard jarring action. A second overshot was run but would not take hold of the fish. The overshot was worked for several hours in another attempt to catch the fish. When the overshot was pulled, its guide lip was slightly burred, indicating it may have been working in hard sand. A hydrostatic bailer was run, and a small amount of sand recovered. The bailer was rerun and recovered a very small amount of sand. The overshot was run again and took hold of the fish without difficulty. The fish was jarred for about 2 hours when it began to move slowly. It came free after jarring upward about 6 ft.

A special, high-temperature jet cutter was run and fired but did not cut the tubing. A second cutter was run and did part the tubing. The cutter tagged the bottom to ensure it was cut and dropped. The well was then placed into production and tested.

REFERENCES AND SUGGESTED READING

Littleton, Jeff, "Refined Slimhole Drilling Technology Renews Operator Interest," *Petroleum Engineer International,* June, 1992: 19–24.

Pursell, John C., and Brian E. Moore, "How to Sand Wash Large Tubulars with Coiled Tubing," *Petroleum International,* Aug. 1992: 42–45.

Chapter 7

CASING AND LINER REPAIR

The ideal objective of casing repair is to perform the work in an economical manner so that the inside diameter is not restricted and the casing is as strong as the original. However, casing repairs are seldom ideal. Generally casing should be repaired in the simplest manner possible that is consistent with the original requirements for running the casing.

There are a variety of casing repairs. Always consider the long producing life of a well when planning the repair. Avoid restricting the inside diameter. Often the best repair is to cover the failure with another string of casing, tie-back casing, or a stub liner. If this option is not available, it may be necessary to repair the casing in a less preferable manner as the only alternative to either plugging back and sidetracking, or plugging and abandonment followed by redrilling. Normally, the casing should not be pulled for repairs when it fails during running because of the high risk and the danger of losing the hole. Usually, it is better, and more economical, to end up with a poorly cased hole than no hole at all. A major consideration is that most costs have already been paid, and if the casing failure cannot be repaired then all of this money is wasted. Therefore, the practical objective is to repair the casing so that future operations are cost-effective.

CASING PROGRAM

A good, workable casing program is an important preventive measure. Such a program should be developed with care, making reasonable allowances for predictable future situations. Directional and horizontal wells often cause special design considerations. Competent, experienced personnel should be able to develop efficient programs.

The most common failures occur because casing programs are designed improperly due to error, unforeseeable future operating conditions, or a later change in scope. To reduce or eliminate errors, conduct a detailed review of all available data and have others check the program. Always allow for an extra string of casing on high-risk wells, especially those for which there is less information. Modifying a drilling program during drilling (such as a change in scope) reduces flexibility and often leads to reduced hole sizes and the associated higher risks of drilling problems.

Program Considerations

The casing program should provide sufficient information to enable knowledgeable personnel to drill and complete the well. However, there are especially significant items that may be overlooked or may not be considered important.

Casing Sizes, Weights, and Grades

Programs should be designed for the correct sizes, weights, and grades of casing. Designs often call for a relatively large number of different weights and grades, which increases the risk of running some casing in the wrong position. This is an operational problem, but it could be minimized by restricting the number of different weights and grades. Use the least number of combinations — preferably three or less. Run two joints of the heaviest casing at the surface.

Provide for casing wear, especially on the lower sections of surface and intermediate casing that often are exposed to longer drilling times. Consider using heavier grades in crooked-hole sections. Allow for casing of adequate strength at liner seats to prevent splitting, especially when using heavy liners. Run casing with higher collapse resistance opposite formations that may heave or flow, such as salt sections. Always consider the need for an extra string of casing or an alternate casing program to prevent RUNNING OUT OF HOLE. Situations requiring this precaution include lost circulation and caving formations, excess drag and torque, and pressure transition zones.

It is difficult to change casing sizes after spudding, but some alterations may be possible, such as running casing with closer tolerances. For example, normally run 7-in. casing in an 8¾-in. hole, and drill out with a 6 ¼-in. bit, depending upon the weight of the 7-in. casing. If there is a possibility of "running out of hole" and conditions are favorable, consider running 7 ½-in. flush joint in the 8 ¾-in. hole. Then drill out with a correspondingly larger bit, depending upon the weight of the flush joint casing. The increased hole size may be significant in deeper drilling.

Operation Failures

Some casing failures that occur early in the life of a well are caused by operations. A classic (and unfortunate) example of this occurred in a number of development wells drilled in a large field. Intermediate, 10¾-in. casing of various weights and grades was set at about 12,000 ft. As for most casing strings at this depth, improved grades and higher weights were used in the upper section for tension and in the lower hole for collapse. The completion section was drilled with various-sized bits. An increasing number of wells had problems with bits hanging up in the lower intermediate hole section, and in some cases causing casing leaks. Problems became so severe that a detailed review of records and an inspection of the casing and the used bits were conducted, indicating that some of the bits were marginally oversized for the casing.

Casing program design can contribute to operational failure. For example, the casing should not be subjected to pressure exceeding its design-burst rating. Nevertheless, there are times in the drilling operation when this may occur. Although this may be an operational problem, the designer is obligated to submit a program that can be used to drill, test, and complete the well successfully. A major cause of casing failure is wear, so be sure to provide for it in the program when possible.

Change in Scope

Changing the scope of the well after spudding almost invariably involves drilling deeper. Doing so may cause various drilling problems such as working in an undersized hole at deeper depths and drilling longer sections of open hole; frequently there is also additional exposure to hazardous situations such as caving, fractured, and high-pressure formations, and lost circulation zones. Other problems may include drilling transition zones (normally only one; infrequently two or more) and extending casing lengths, which are often undersized, for drilling deeper objectives.

Casing types and lengths can be exchanged for stronger, heavier casing to provide for deeper drilling, but this can expose the casing to higher stresses, thus increasing the risk of casing failure. For example, a higher-strength production casing split while perforating long sections. This may or may not have been overlooked since the perforating procedure was satisfactory for the original completion casing design. The cased hole size often is reduced, causing additional problems with working and completing in the smaller cased holes. Therefore, careful consideration should be given to the potential increased risk; do not underestimate it, before changing the scope.

Before finalizing the casing program, you can anticipate future needs by talking to the geologist, the geophysicist, and the originator of the prospect. Having studied all the wells in the area, these experts may have information that will affect the drilling program. Inquire about the possibility of drilling deeper and what conditions may be encountered. A detailed study of other wells in the area, talks with other operators in the area, and experience help prevent mistakes.

Designing a program for an unknown future condition is a difficult problem. Always evaluate the casing program so that it will be adequate for any expected drilling depth. If the hole may be drilled substantially deeper, then provide for an extra string of casing if warranted. Overdesigning or allowing for an extra string of casing may be justified, especially for wildcat wells. Plan to use improved, more economical casing designs for subsequent development wells after obtaining information and data from the first exploration well.

Problems may occur when information obtained during drilling indicates the need to drill deeper. For example a drilling program is designed for one depth (about 10,000 ft) and geological information obtained during later operations indicates a high-productive potential objective at a substantially deeper depth (about 14,000 ft). In the process of drilling deeper, the hole becomes too small, a situation commonly called "running out of hole." This can lead to numerous severe problems and a high potential for drastic fishing jobs, possibly resulting in a lost hole.

Obviously the original well plan should have provided for an extra string of casing, but such contingencies cannot always be foreseen. Therefore, at 10,000 ft the operator must choose from among several alternatives. One is to drill the well to the deeper objective with the potential risk of fishing and other drilling problems, including losing the hole. Another is to complete the well to the original objective and drill another well designed to test the lower objective, often at a considerably higher cost. Neither alternative is satisfactory, but the decision must be made.

A similar situation occurs when information obtained during drilling indicates that the objective is more favorable at a slightly different horizontal position; crossing a fault is a typical example. Frequently this is done simply by plugging back and sidetracking. Again, this may involve a longer hole, drilling directionally, and possibly an extra string of casing.

Similar problems and decisions occur with crooked hole, a long open-hole section, and extra transition zones. The correct decision requires not only general experience but specific knowledge of the site and area.

Casing Inspections

All casing should be inspected, since this is an important preventive action. Casing programs are based upon casing with specific physical characteristics according to the manufacturer's specifications. Inspections ensure that forces and stresses can be applied within these specifications. Casing should be inspected visually after every move and before running it.

The first formal inspection usually is made at the mill, so all casing supplied by the mill should be within specifications. The number of inspections required after this depends in part on the reputation of the supplier, which often is indicated by the number of rejects found while inspecting other casing strings from the same source. Inspection requirements also depend on the type of casing string to be run and the work to be performed inside the casing.

If the casing has been fully tested at the mill, then defective tubulars normally should have been rejected there (Fig. 7–1). Subsequent inspections at the storage yard or the wellsite also should detect defective pipe. However, this is not always the case, as illustrated in the following summary examples.

Figure 7–1 Tubular defects
Note: Slag inclusion does not have the same strength as steel in the pipe body. The net effect is reduced strength. For example, the strength of the cross-sectional area (AA') is much less than that of area (BB'), and there is a higher risk of a failure at (AA'). The results are similar with a split or crack. These can be increased in depth due to corrosion and fatigue.

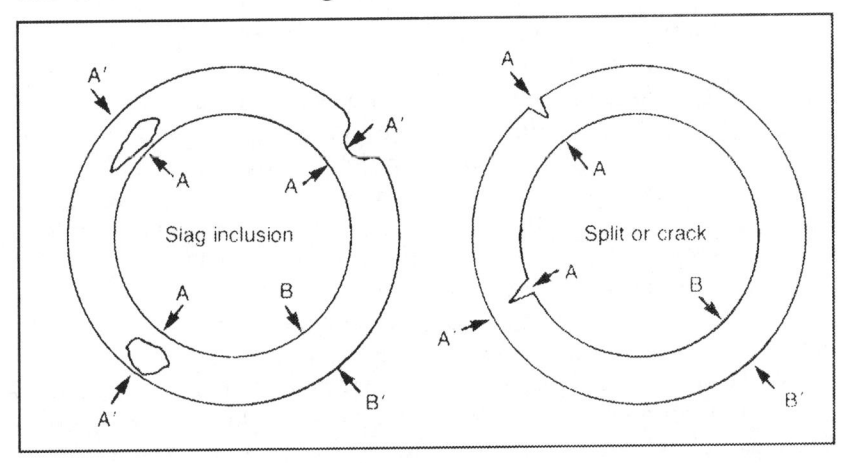

Casing and Liner Repair

Inspection Failures

- Tubing, 3½ in., run to 15,000 ft, was tested and failed at 5,000 psi, well below the burst pressure. The failure resulted in a split 1½ ft long.
- Deep 10¾-in. intermediate casing parted at a very high safety factor. It had been inspected at the mill and again at the wellsite.
- A deep 3½-in. tubing work string had severely damaged upsets. An inspection and subsequent investigation indicated that the upsetting procedure and/or the upsetting tools were faulty.
- A string of 7⅝-in. casing was inspected at the storage yard and again at the wellsite. After running, the casing split near bottom.
- Casing, 5½ in., was run in a 8,500-ft hole. Near bottom, the casing hung up. It was worked and parted. The recovered end of the casing showed it had slipped out of a collar. The threads were cut out of round, with less wall thickness on one side than on the opposite side.

The tubulars were inspected at the mill and again at the wellsite in most, if not all, of the above cases. The inspections were obviously inadequate.

There is always the possibility of damage to casing during transit or while loading or unloading. Therefore some testing at the wellsite is desirable. However, the casing can be tested more conveniently at the mill or the storage yard. Permanently housed testing equipment may be less susceptible to damage and accuracy errors than portable equipment. Each operator must resolve these questions.

One solution, especially for deeper, more critical casing strings, is to do a complete inspection in the yard, including electromagnetic, special-end area, and full-length drift inspections. These types of inspections will eliminate damage to pipe caused by the following: pitted, pulled, smashed, or worn threads; thin walls, internal or external cracks, unequal wall thicknesses, mill marks, slag inclusions; and failure in drift due to mashed, out-of-round, or uneven wall thickness. Third party inspection services are strongly recommended, especially if the operator's inspection personnel do not have much experience. After the casing is moved to the wellsite, run a full-length drift, end-area, and detailed visual inspection.

Completion tubing is often pressure tested on high-pressure wells or where high treating pressures are expected. There are three common testing procedures: above the rotary, below the rotary, and after running the full string into the hole. Using all these has the advantage of pressure testing both the body and the completed connection. Testing above the rotary is safer and slower. Testing below the rotary is slightly faster and more complete because the connection is tested in tension. However, it is also riskier because a failure may cause the tubing to drop, requiring fishing. Testing after running to bottom is the fastest and most complete method, but with a higher risk. Select the procedure which is most applicable for the situation.

Otherwise, burst testing is seldom used. Reportedly, one operator's total inspection program was to move the uninspected pipe to the wellsite and have it burst tested and have an end-area inspection on the pipe racks performed before running. This operator had an enviable record of very few, if any, casing failures.

Casing and tubing in the hole needs inspection at intervals based upon use. The purpose of these inspections is primarily to check the inside diameter for casing wear, corrosion, erosion, pitting, and cracks. Run a base inspection log in deep intermediate casing before drilling out. Then log the casing periodically while drilling deeper to determine the amount of wear. Check the tubing in high-pressure, high-volume gas wells periodically in later life for erosion. Check the casing in older wells before drilling deeper. Consider inspecting the casing where a liner hanger will be seated, especially for casing that has been drilled through for a long period of time, exposing it to excessive wear, and for long, heavy liners. It is a good practice to inspect any casing string for wear and other damage after extensive fishing, milling, etc., especially when these activities are concentrated in a relatively short section (Fig. 7–2).

Running Casing

The time spent casing the hole is normally a very small part of the total drilling and completion time. Running casing is a non-routine operation, however, and subject to increased risk. Failures are less common while running casing but problems may occur. Barite sag can cause problems in high-angle directional and horizontal holes, especially when heavyweight mud is being used, so observe all precautions accordingly.

Is the Hole Ready to Run Casing?

One potential question when running casing, usually in holes with formation problems, is whether the casing will go to the bottom. The risk of running casing may be less because at this point the operator should have a reasonable idea of how long the hole will stay open after cleaning out, the depths where tight hole and bridges occur, and how hard they are to clean out. This information provides a good estimate of whether casing can be run successfully.

Casing and Liner Repair 451

Figure 7–2   Tubular inspections (courtesy Dia-Log)

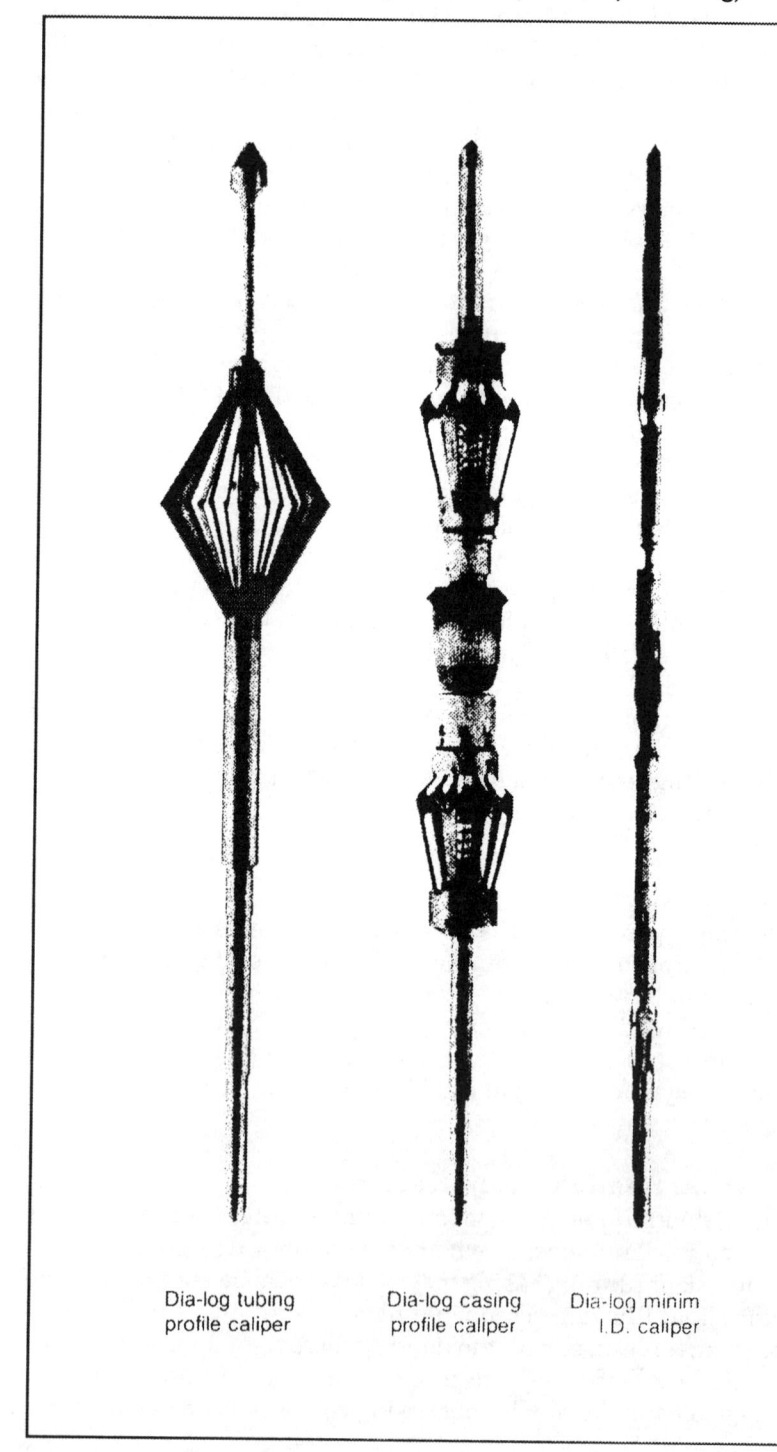

Dia-log tubing profile caliper

Dia-log casing profile caliper

Dia-log minim I.D. caliper

Casing and Liner Repair

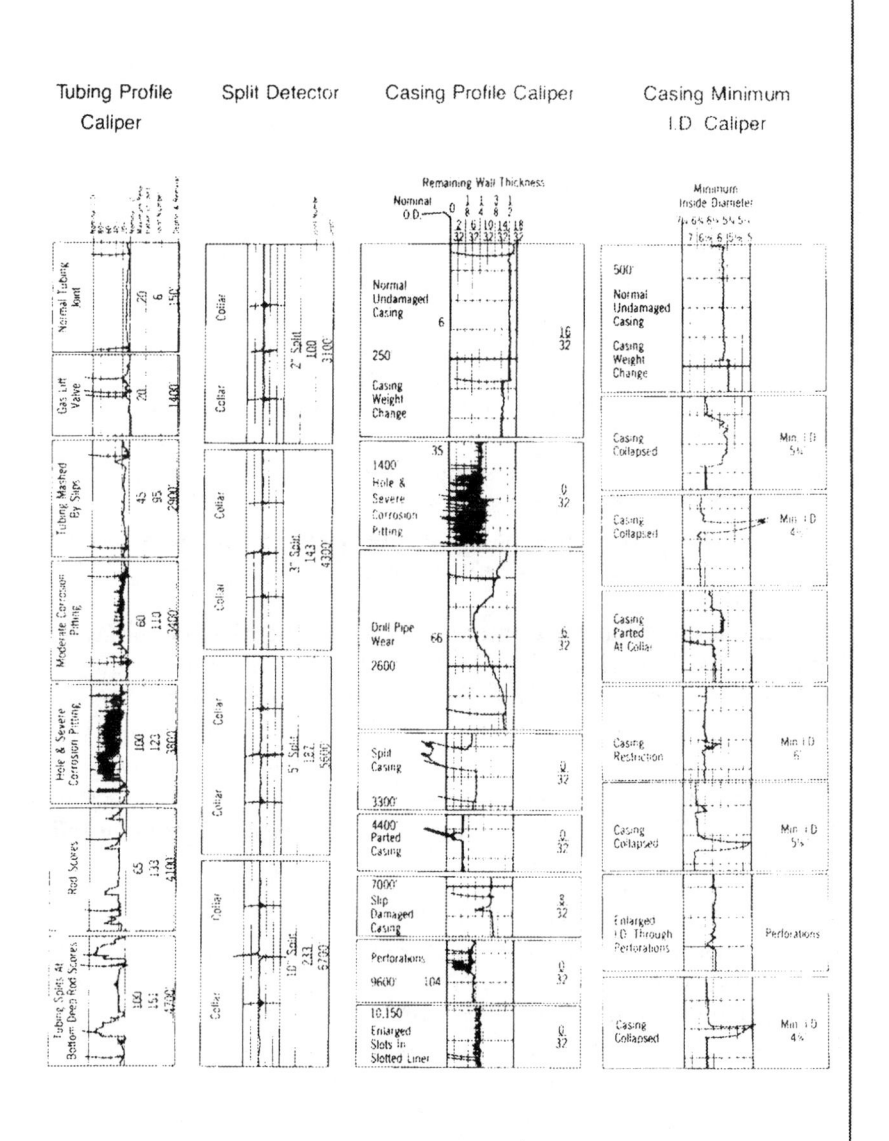

Normally the casing should not be pulled after it has been run below the last string of casing. It is better to have a partially cased hole than to lose the entire hole. Also, if bridging is not excessive, casing will almost always continue running to bottom, even if it cannot be picked up to any reasonable height. This effect, which is due to the relaxed casing conforming to the hole configuration, is not accepted by all but has generally been proven in the field.

Casing normally can be run through any hole that a bit and drill collars will pass through without difficulty. Run the casing as fast as is safely possible, but do not break down the formations and cause lost circulation. Most of the work of rigging up to run casing should be done before starting to pull the bit. Stand two or three doubles or trebles of casing, including the float and shoe, back in the mast to reduce the normally longer time required to connect and run the first few joints. In more severe cases, replace the float shoe with another float collar and run a mill or Texas Pattern shoe below it to help clean out bridges. None of these procedures will guarantee that the casing can be run to bottom, but they do reduce the risk and improve the chances of running the casing successfully. It may be necessary, but less desirable, to omit customary logging before running casing in holes with severe problems and to evaluate with cased hole logs.

Another problem encountered when running casing in holes with exposed high formations is whether the hole will remain quiet and static until the casing is run to bottom. Prior operations will serve as a guide to this. If hole maintenance is very questionable, then when logging, take extra time or run the drillpipe back in the hole to the bottom of the last string of casing and wait for a period that is representative or partially representative of the time needed to run casing. Then go to bottom and circulate out, observing the severity of gas-cut mud. Doing this will give a good indication of whether there is sufficient time to run casing and still keep the well under control. In some cases, a long pill of heavier mud can be spotted above the high-pressure zone as additional security before pulling out to run casing. This procedure requires care, however, because it can cause lost circulation when circulating the heavier mud out.

Other problems may also occur. Losing circulation while running casing is relatively common. Be sure the mud is in good condition, and consider leaving mud with a lower viscosity in the lower hole section. Wall sticking may occur. Add about 3–5 lb/bbl of coarse and medium-size ground walnut hulls to the mud left in the lower hole section.

Precautions

Procedures for running casing are well known and are covered in the literature. Observe good work practices and use the correct tools when casing the hole. The basic criteria include maintaining good inspection practices, having a clean and preferably straight hole, pretesting the cement and mixing water, and running and cementing the casing or liner correctly. Related activities include selecting the correct casing tools, rigging up and starting to run casing, drilling out, and testing and squeezing the casing seat or liner top. Preplanning and allowance for contingencies are standard procedures.

Ensure that all rig equipment and tools to run casing are correctly sized and in good condition. This includes the cementing head and plug container, the wiper plugs, and the liner hanger (when used). Some of this equipment may need inspection when heavy casing loads are being run. Other general preventive measures include the following.

- String up extra lines to handle long, heavy strings of casing as needed.
- Repair or replace cables and ropes with frayed or broken strands as necessary. Tighten all clamps and ensure that all cables have double clamps that are properly installed, with the ring on the free end of the cable.
- Pay particular attention to spring-loaded-type equipment. The elevator latch spring should have a strong, smooth acting tension so that it closes securely and will not jar open. Sometimes lighter tools are used to run the first part of the casing. Be sure that the locking pins or the bolts and nuts that secure the elevator to the elevator bails are tight and properly locked.
- Some running equipment may need to be inspected by sound or by similar means at critical section areas. Check all points that will be subjected to severe stress and repair or replace parts as required. Examples are the hook, elevator shoulders and links, the drawworks shaft, and break bands.
- Connect a crossover sub to the inside preventer that fits the casing, and have the tool ready for use if the well kicks.
- Change one set of preventer rams to fit the casing and pressure-test the preventers.
- Use torque and turn measuring and recording instruments to obtain the correct make-up torque and record it.

Casing and Liner Repair

- Use the drill-collar clamp as an important safety device.
- Pay particular attention to pick-up lines or elevators. Failures in these have caused injury to personnel and equipment damage. Pick-up elevators are recommended. Always catch the casing in the "V door" with the latch up and the hinge toward the rig floor.
- Install the casing shoe and float collar correctly, usually on the pipe rack, and tighten them after they have been picked up. Keep the hole covered until casing is actually run.
- Never lower casing into the open rotary while holding it with a lifting or pick-up line.
- Cross threading during stabbing and makeup usually is caused by incorrect joint alignment and the threads can become crossed and effectively locked. Correct this by unscrewing the joint, picking it up, aligning it correctly, and then lowering it into the casing collar and connecting it.

End of the Tubing

The position of the bottom of the tubing is an important consideration in preventing sticking and fishing and especially in washing over tubing during the producing life of the well. The bottom of the tubing should be set either at the top or bottom of the production interval. The supporting opinions for these two options are as follows.

One opinion maintains that produced sand, the small amount of sand which enters the wellbore over time with the produced fluid, will flow out of the hole with produced fluids, so the tubing should be run to the bottom of the formation. This is the most common procedure.

The other opinion is that if tubing is run to the bottom of the production interval, then produced sand will ultimately collect around the tubing and stick it. Therefore the bottom of the tubing should be left at the top of the formation. In this case there is a risk that the sand will settle down into the lower hole and plug part of the producing formation.

When very long intervals are perforated, the operator must evaluate which procedure has the least risk of sanding up the tubing.

FAILURE AND REPAIR

Failures may occur for many reasons. Casing can part while being run or by excessive pull when stuck. Excess pressure, such as that caused by bumping the cementing plug too hard, may cause a burst failure or, less frequently, a tension failure. Drillpipe wear may cause a failure while drilling below the casing, or the drillpipe can be damaged while fishing inside or below the casing. The movement of plastic formations, or completion operations, can cause collapse failures. Casing may fail due to internal or external corrosion or rod wear during the producing life of the well. Sometimes the failure is more subtle, resulting from an undetected defect. The casing can be landed with excess tension and fail before the normal period of wear has elapsed.

The severity of a casing failure depends on the nature of the failure, the point of occurrence, the investment in the well, the projected future net revenue, and related factors. Severity increases with depth. There are various causes for this. The risk of inadequate casing string design increases. There is an increasingly hostile environment of heat and pressure; higher pressures increase the risk of operations such as squeezing, treating, and stimulation. There is also more risk of not being able to complete the repair successfully due to longer work strings and smaller, lower strength tools. The time needed for operations is extended (for example, longer trip times), which increases exposure to time-dependent problems.

The general approach to casing repairs is similar to that for fishing. Experienced, competent personnel are most effective. Prevention is always the best answer. Be prepared by being familiar with the condition of the open hole, the formations behind the casing, and any previous casing failures and repairs in the area.

Research all details when a casing failure occurs. This often helps determine how to make the repair. If necessary, hold a conference to select a plan of action. At the same time, a contingency plan should be devised that includes approximately how long repair operations should be conducted before changing plans, such as electing to sidetrack.

Some casing failures can be anticipated — for example, casing set in a crooked hole. To reduce the risk of wearing a hole in the casing, an operator can reduce drillstring weight by such means as using a minimum number of drill collars. Double-rubbered pipe should be run through the crooked section along with good mud to improve lubricity.

Casing repair may be hampered by well conditions, such as high pressure or lost circulation zones. It is often necessary or advisable to eliminate these problems, at least temporarily, before continuing the repair. A common example is trying to repair casing at a shallow depth with an exposed high-pressure formation downhole. Since the problem formation can severely hamper repair operations, isolate it with plugs, dump cementing, or other means (Fig. 7–3).

Casing and Liner Repair

Figure 7-3 Dump cementing with wireline tools
Note: Sand may be dumped in a similar manner. Procedure is applicable to open and cased holes subject to hole conditions.

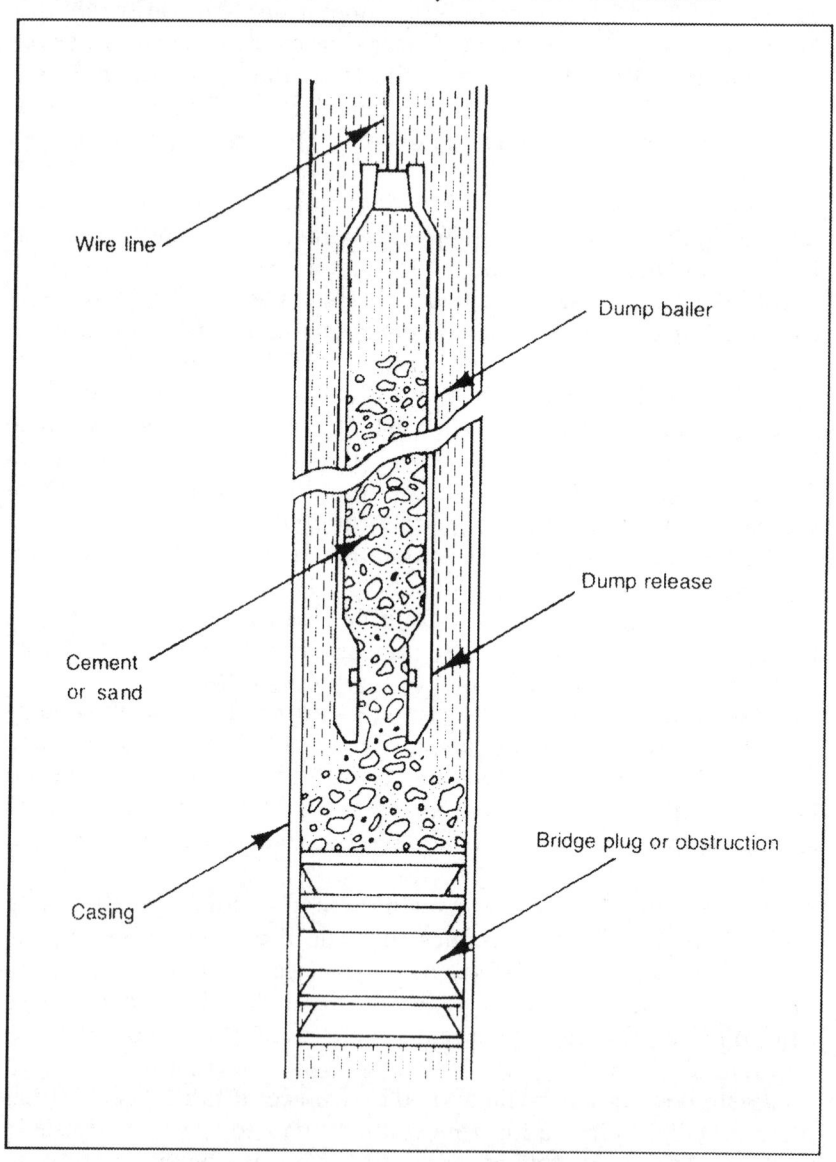

Types of Casing Failures

Casing failures may be broadly classified into two basic categories of design and operations. For purposes of explanation, more definitive classification is based on the type of casing string. Well casings are divided into conductor or drive-pipe and the three basic strings of surface, intermediate, and completion or production casing. Liners are more common in the completion or production type, but may be included with intermediate casing. This is a logical method of classification, although the same failures may occur in several types of casing.

All casing strings generally are subject to all types of failures. However, some failures tend to occur more frequently in one or two types of casing and less in other types. For example, leaks due to drillpipe wear are more common in surface and intermediate casing and less common in production casing. Collapse is more common in production casing.

Casing failures range from simple leaks to long, parted, collapsed strings. Less severe failures include casing leaks, loose shoe joints, or collapsed casing. Repair of these is relatively straightforward; other failures can be more complex, requiring considerable ingenuity to make a satisfactory repair.

Casing failures and their general causes are listed in the following sections. They generally are listed in the order of increasing severity and decreasing likelihood of occurrence.

Leaks

Casing leaks are the most common failure and often occur in association with most other failures. The major causes for leaks are improper makeup during running; drillpipe wear during drilling; wear due to running stabilizers in the cased hole; mechanical wear while fishing and especially milling; corrosion and, to a lesser extent, erosion; and mechanical wear during the producing life of the well.

Casing leaks may be caused by mechanical action during operations (Fig. 7–4). For example, casing can be punctured by jarring a packer loose with a malfunctioning slip segment. The working and jarring action forces the slips upward and outward, driving the slip segment into the wall of the casing and causing a leak and other damage. Other packer and plug parts may cause similar failures.

Casing and Liner Repair

Figure 7–4 Casing damage due to mechanical action

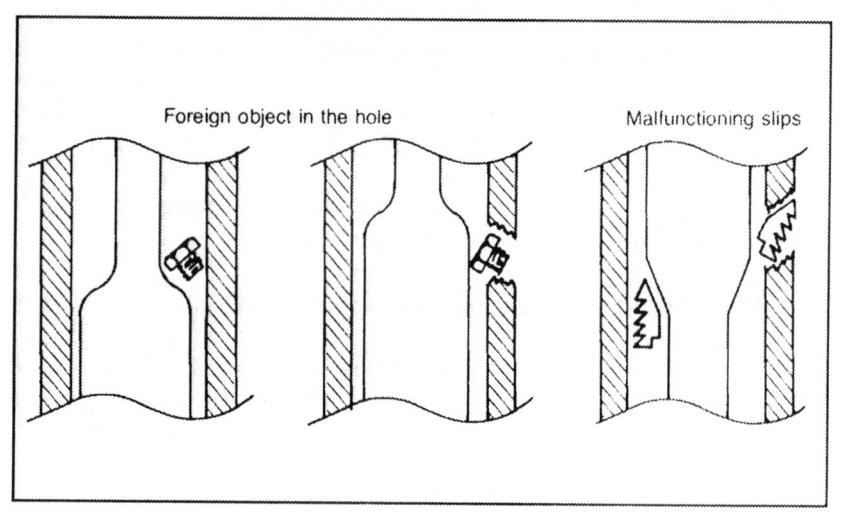

Foreign object in the hole   *Malfunctioning slips*

The same type of action may occur when foreign objects such as preventer bolts and nuts and hand tools are dropped in the hole on top of the drill collars, packers, and other full-gauge tools. As the tools are jarred out of the hole, the object can be forced into the wall of the casing, causing a failure.

Split or Burst

Split or burst casing occurs from many causes, including bumping the plug too hard during cementing or applying excess internal pressure, sometimes in combination with high tensile loading. Excess internal pressures may occur while testing liner tops or testing the casing before drilling out. Some failures that cause casing leaks may also cause split or burst casing.

Test pressures of new casing should normally be limited to 90% of the burst rating and used casing to 80% or less depending upon the amount of wear and type of service. Avoid using excess test pressures, even if they are within the casing specifications. Limit test pressure to 20% over the maximum pressure expected. Be sure to allow for the hydrostatic head if there is an existing differential across the casing wall. Test pressures apply to the burst rating of the casing exposed to the pressure.

Casing may split when hanging long, heavy liners. General causes include inadequate strength due to improper design and equipment selection, incorrect operations, or worn casing at the point where the liner setting-tool slips engage the casing. Use the correct test pressure when testing a liner top. The liner top is tested by setting a packer in the casing above the liner. The liner usually has a higher pressure rating than the upper, larger-diameter casing, and excessive pressure will burst the casing above the liner top.

Casing may be split by high-density perforating, especially in the higher-strength steels. Inadequate design may be an indirect cause if stronger casing with a higher burst rating should have been used. Casing also may split or burst because of structural defects, including slag inclusions in the casing wall and uneven wall thickness (Fig. 7–5). Eliminate these defects with good casing inspection procedures.

Figure 7–5   Split or burst casing
Note: Burst casing is subject to failure in tension. The split does not reduce the cross-sectional area (vertical split) but the structural integrity of the casing is altered and tensile forces may not be evenly distributed. Concentrated tensile forces can initiate a failure at one point. This increases the tensile loading on the remaining body area and may result in complete failure.

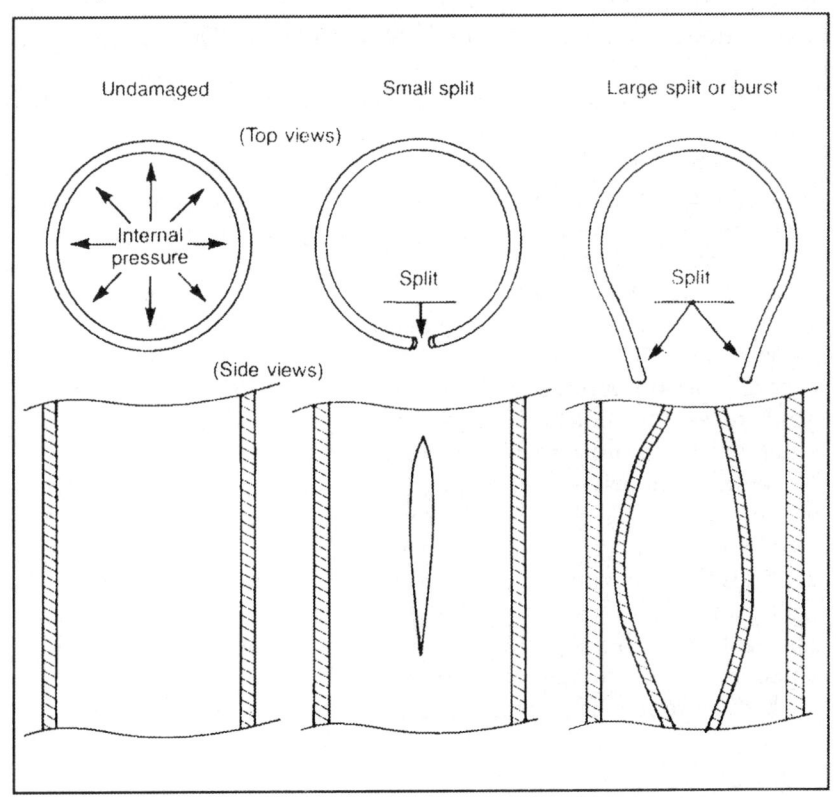

Casing and Liner Repair

Parted

Casing failures due to parting are commonly caused by design, operations, or a mechanical failure due to improper construction (such as a factory defect, which emphasizes the importance of good inspection procedures). Parted casing occurs most frequently at a connection — i.e., it is more common in special, less strong connections than in solid sections with greater body strength. Split or burst casing may part because of the loss of structural integrity.

Casing may part during running due to rough handling, such as being lowered rapidly and stopped abruptly, especially with long, heavy casing strings. Other causes of parting include excess wear and the resulting loss of tensile strength, pulling the casing apart with excess tensile force while working stuck casing, and bumping the plug too hard during cementing. Bumping can exert excess longitudinal tensile force, causing a failure in tension.

Collapsed

Casing collapses for various reasons, some of which are the same as those that cause casing leaks. For example, wear reduces body strength so that external pressures may then cause the casing to collapse. Anything that reduces wall thickness, such as wear or corrosive action, increases susceptibility to collapse. This is more common in older wells. Pressure surging infrequently may be a contributing factor. The operator must ensure that the casing is designed correctly.

Casing may collapse while squeezing or treating below a packer set in the casing (Fig. 7–6). Cement may channel outside the casing. Pressure such as that from squeezing below a packer may be transmitted through these channels and collapse the casing above the packer if the rated collapse pressure is exceeded. Poor cementing is an obvious cause of collapse. Preventive steps include setting the packer higher and pressurizing the casing-tubing annulus. This is generally an operational error, since the condition of the cement behind the casing should be known, and always pressure the annulus when applying pressure below a packer.

Worn or poorly designed production casing may collapse when it is evacuated by reducing the hydrostatic head inside the casing. This may occur in flowing gas wells when the formation pressure declines to a low value. It also occurs in pumping oil wells when the well pumps off, evacuating the casing. Worn and corroded casing can contribute to the problem. As a remedy, produce the well under a packer with a full head of fluid in the annular space above the packer in order to maintain the hydrostatic pressure.

Figure 7–6   Collapsing casing during squeezing
Note: Hydraulic pressure is transmitted through the channel. The casing exterior is exposed to this pressure and is subject to collapse, depending on the pressure differential across the casing wall and the strength of the casing. The risk of collapse is eliminated or greatly reduced by pressurizing the casing-tubing annulus. The casing is exposed to possible collapse any time fluid is pumped through a packer into the annular space outside of the casing without pressurizing the casing-tubing annulus.

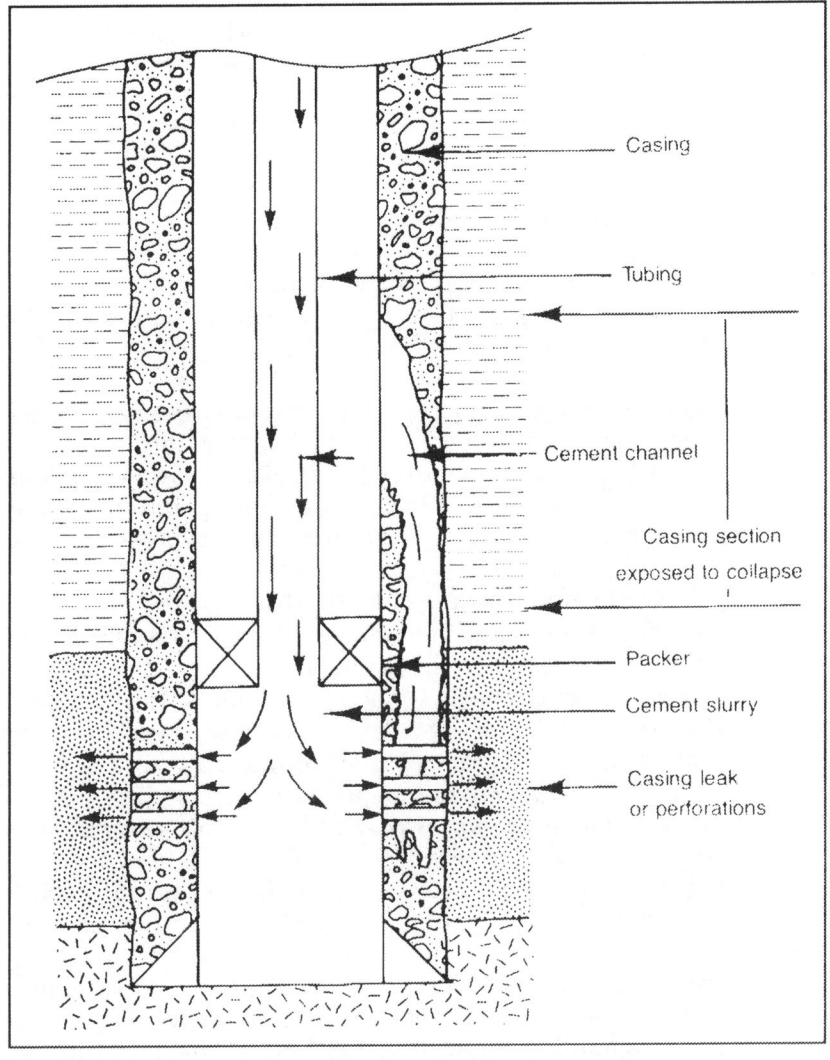

Casing and Liner Repair 463

An infrequent cause of collapsed casing is shifting or flowing formations. Massive salt sections will exhibit a plasticity or tendency to flow. Design the casing that is run through these sections with an adequate safety factor in collapse — normally the equivalent of the earth gradient. Earthquake zones and permafrost sections require special collapse-design factors and operational procedures.

Factors Affecting Casing Repairs

Many factors affect casing repairs, but the primary items include casing type and size, the depth of failure, whether the problem is in cemented or uncemented casing, and whether the option of using an extra string of casing is available. Formations may have a strong effect on casing repairs, such as type, pressure, fluid in the formation, and transition zones. Other factors include the age of the well, its current phase (drilling or production), its productivity, and the severity of the failure.

These factors affect casing repairs in various ways. Casing repairs are best reviewed and evaluated by considering the following questions.

- *How does the failure affect current and future operations?* One of the first things to consider in evaluating a casing repair is whether the failure needs to be repaired. The failure may not have an adverse effect on operations; if not, and assuming no other hazardous conditions, do not repair the casing.

    One example is a small leak in the lower part of the intermediate casing. Frequently the pore pressure at this point is very close to that in the open hole below the casing, so it may not be an immediate problem. Furthermore, the leak probably will be covered with production casing after drilling operations are completed. In this case, repairing the leak may not be justified.

- *Is the inside diameter restricted?* An immediate repair may be needed if the inside diameter is restricted during drilling or completion since this prevents the running of full-gauge tools and other operations. One exception might be when the failure occurs in larger casing and operations are being conducted through a lower liner (for example, a 7-in. liner inside damaged, partially collapsed $9^5/_8$-in. casing). Depending upon the situation, the damaged casing may be covered with another string of casing or a stub liner at a later date.

Casing partially collapsed above a conventional completion with the tubing and packer intact may not need repair, depending upon the situation. It may be possible to stimulate and produce the well through the tubing.

- *Can the failure be repaired by normal future operations?* A casing failure may not present a problem in current operations and can be repaired or effectively eliminated by future operations. For example, the drillpipe can wear a hole in deep-surface or intermediate casing. If this does not create an immediate drilling hazard, drilling operations may continue without making the repair. The hole can be covered later with production casing or a liner that extends over the hole.
- *Can the failure be patched or packed off?* Some casing repairs can be made with a simple patch or by packing off the leak. For example, if production casing develops a leak or becomes damaged, a simple repair is a casing pack-off. Tubing may be patched.
- *Can the casing be plugged off and repaired later?* This question ordinarily applies to production casing. If the casing failed in a lower section, one procedure would be to set a plug and isolate the failure. Uphole operations such as production testing could be performed in a normal manner. If the well is deepened later, consider repairing the failure at that time.
- *Can an extra string of casing be run?* It may be possible to repair the casing failure by running another string of casing, a liner, or a stub liner.
- *Is the failure inside another string of casing?* Generally, the failure must be repaired. If the casing is not cemented, then cement it using primary methods. If it is cemented, then ream it to full gauge and squeeze if necessary.

Summary of Casing Repair Methods

Each casing failure must be evaluated based on the specific conditions of the failure before a method of repair can be determined. All repairs must be economically justified. In general, make the simplest repair possible that will accomplish the desired results.

Most operating personnel and engineers work on casing repairs until they are completed or until excessive costs are incurred. Sometimes they don't know when to quit, which can mean unnecessary expenditures. In severe cases, always consider the alternatives of plugging back and sidetracking, plugging and drilling another well, or plugging and abandoning the hole.

Figure 7-7 Squeezing casing failures

Note: Retrievable equipment normally is used for shallow, low-pressure squeeze jobs and permanent-drillable equipment is used in deeper, higher pressure work. The selection of a squeezing method depends on depth, casing size and strength, formation pressures, and squeeze pressures expected. The follow-up procedure depends on squeeze procedures, types of plug, etc. Common types of plugs include cement plug, retrievable plug, permanent plug, or drillable plug.

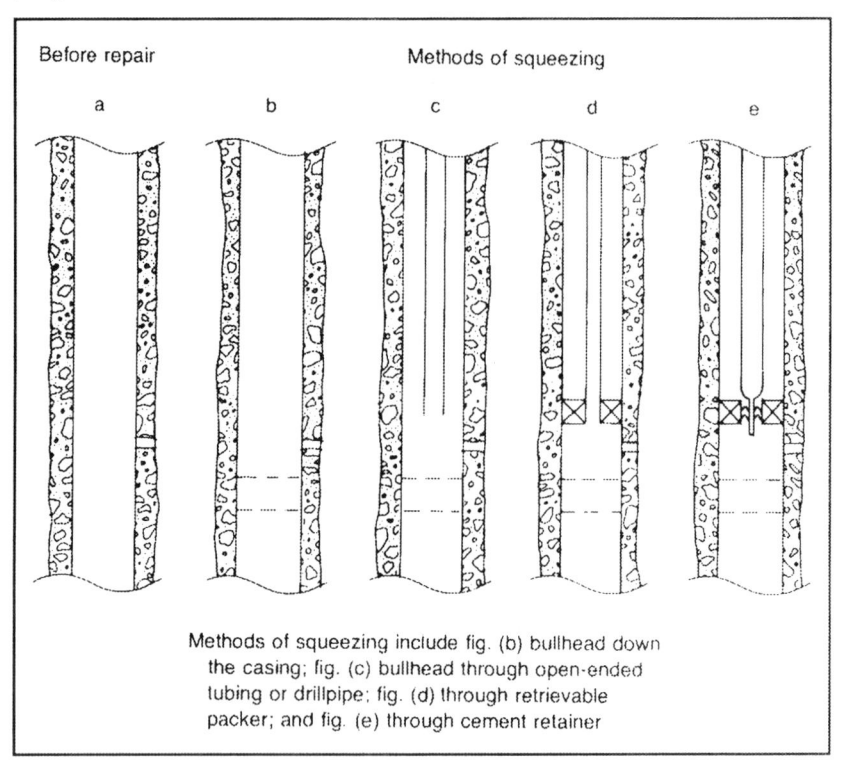

Before repair   Methods of squeezing
a   b   c   d   e

Methods of squeezing include fig. (b) bullhead down the casing; fig. (c) bullhead through open-ended tubing or drillpipe; fig. (d) through retrievable packer; and fig. (e) through cement retainer

The following is a summary of the more common methods of repairing casing failures.

- *Do not repair the casing.* Evaluate the failure in view of expected future operations. If the failure does not adversely affect these, then do not repair it.
- *Loose casing shoe joints.* These are generally repaired by cementing the joints to fix them in place and prevent movement. Then run a full-gauge mill and ream as necessary to ensure a full-gauge hole. In a few cases, if the casing was set near bottom in a gauge hole, it may not be necessary to cement the loose joint and operations can continue.

- *Squeeze and clean out.* Generally, this is the simplest method of repairing a casing failure such as a leak (Fig. 7–7). Squeeze the section and run a full-gauge tool through it to ensure that the hole is full gauge. Ream the section as necessary. The disadvantage of this type of casing repair is that it always leaves a potentially weak section that must be considered during future operations.
- *Pack off the failure.* Pack off the failure by isolating it. A gauge ring normally should be run to ensure a full-gauge hole. Run a hook-wall packer below a compression packer, separated by tubing or, less frequently, by small diameter casing. Adjust the distance below the two packers so that they straddle the failure and isolate it. As an alternative replace the hook-wall packer with a permanent packer. In some cases use two permanent packers.

    The packers can be pulled if needed to permit the running of full-gauge tools. The disadvantage of packing off is that it reduces the working inside diameter of the hole and usually restricts operations below the failure. Nevertheless, it often is the fastest and most economical type repair (Fig. 7–8).

Figure 7–8 Packing off damaged casing
Note: The casing patch is omitted as a method of repair here because it restricts the inside diameter of the casing.

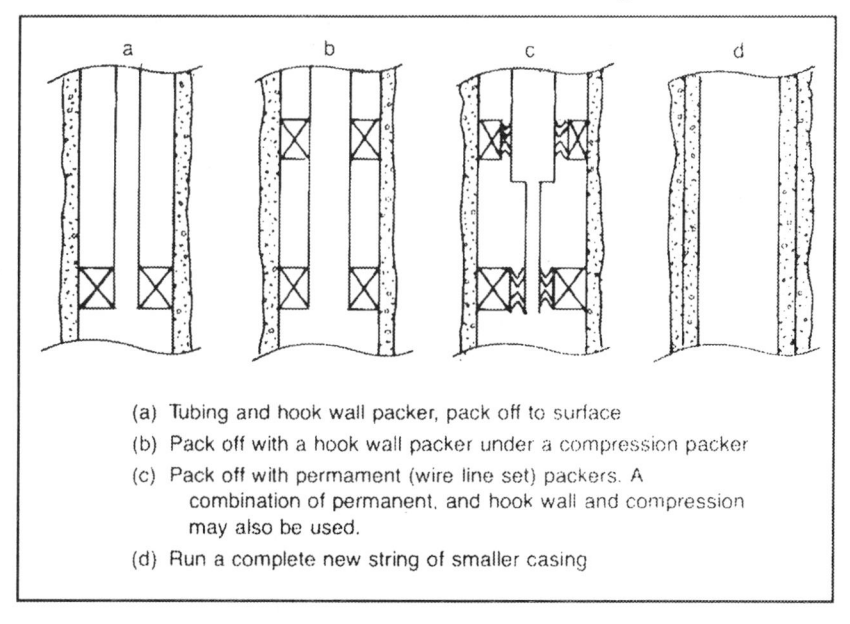

(a) Tubing and hook wall packer, pack off to surface
(b) Pack off with a hook wall packer under a compression packer
(c) Pack off with permament (wire line set) packers. A combination of permanent, and hook wall and compression may also be used.
(d) Run a complete new string of smaller casing

Casing and Liner Repair

Figure 7-9 Internal casing patch (courtesy Homco)
(1) The setting tool is assembled. (2) Tubing is raised to close the circulating valve. (3) Hydraulic pressure is applied to force out buttons on the hydraulic hold-down. (4) Pressure on the underside of the pistons pulls the expander assembly into the bottom of the corrugated patch. (5) The expander assembly is again forced through the corrugated patch, expanding it against the inside wall until the operation is complete.

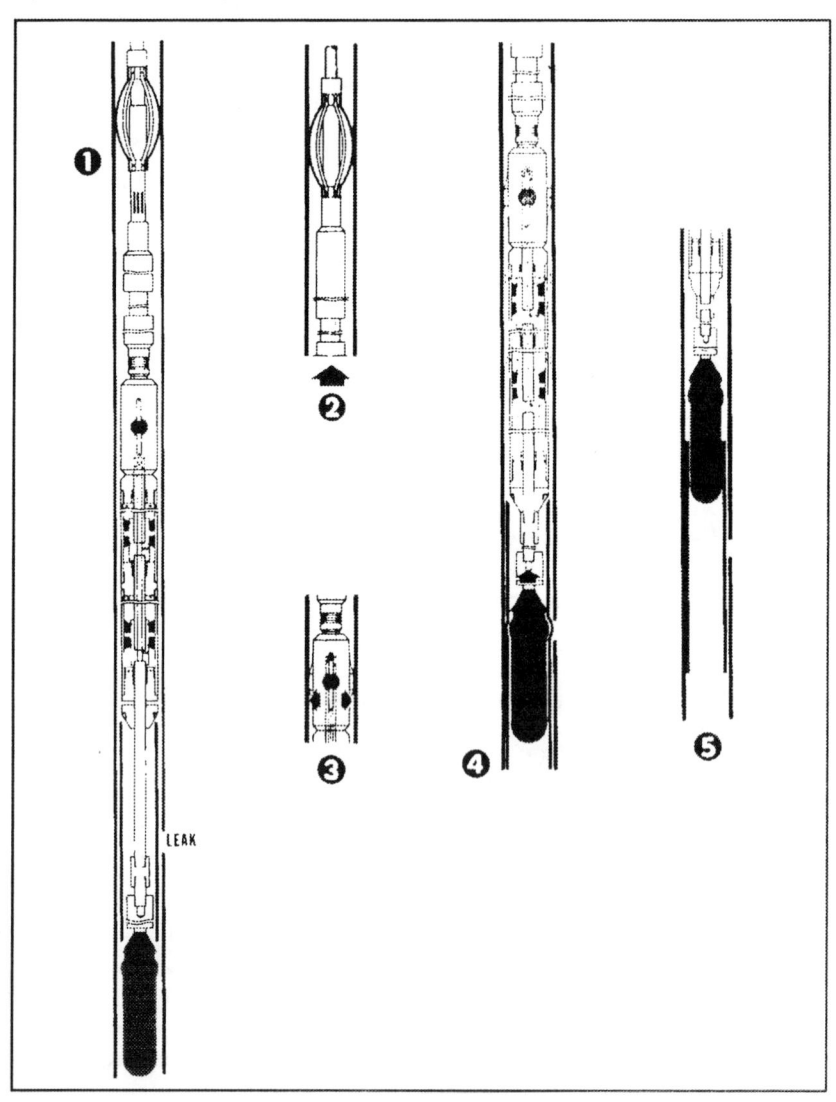

- *Patch the failure.* Various types of casing patches are available (Fig. 7–9). Generally these include a ribbed or corrugated, thin-wall steel cylinder. Run it into the cased hole and position it over the failure (usually a hole). Expand the sleeve by pulling a mandrel through it to form a sheet of metal inside the casing. Patches reduce the inside diameter of the casing a small amount, they cause burst and collapse pressures to be reduced, and they are limited to relatively short lengths.
- *Repair parted casing in place.* One of the best methods of casing repair, when applicable, is to establish circulation through the failed section and perform a primary cement job under a retrievable packer or cement retainer. Alternately, perforate below the failed section and perform a primary cement job in a similar manner. In the latter case, pressurize the casing and pack off the failure or take other steps to ensure that the cement does not flow through the hole into the casing and stick the cementing assembly.

    Another method is to squeeze the section until it will hold the required pressure. Clean out, run a full-gauge mill, and ream the ends of the casing so tools pass unrestricted. An alternate repair method is to run an alignment tool, realign the parted sections, then squeeze to a satisfactory pressure and drill out the alignment tool.
- *Pull, repair, rerun, and reconnect parted casing.* This is one of the best casing repairs, but it is not applicable in many cases. Back off or cut off the casing below the failed section and pull it. Replace the damaged joints, run the casing, and either screw back into the lower section or make a casing bowl connection. Alternatives include using a seal-type bowl, perforating below the bowl and cementing by primary means, squeezing through an unsealed bowl, or employing some combination of these methods depending upon well conditions (Fig. 7–10).

    If the casing is stuck and cannot be pulled, treat it as a failure in uncemented or cemented casing — whichever is applicable.

Figure 7-10 Summarized casing repair flow chart
Note: Plug back, sidetrack, redrill or run a liner, stub liner, and smaller casing were omitted.

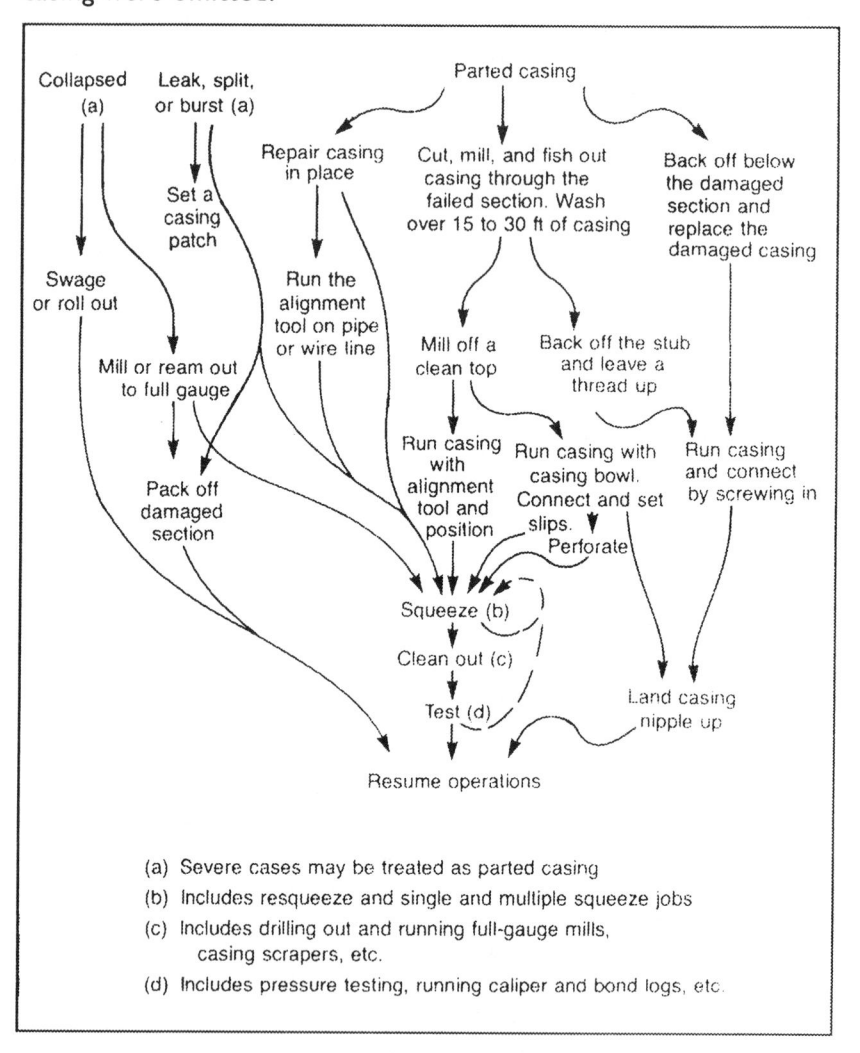

```
Collapsed      Leak, split,              Parted casing
   (a)         or burst (a)
                                 Repair casing   Cut, mill, and fish out    Back off below
                    Set a        in place        casing through the         the damaged
                    casing                       failed section. Wash       section and
                    patch                        over 15 to 30 ft of casing replace the
                                                                            damaged casing
  Swage                          Run the
  or roll out                    alignment
                                 tool on pipe    Mill off a    Back off the stub
                                 or wire line    clean top    and leave a
              Mill or ream out                                 thread up
              to full gauge
                                                 Run casing  Run casing with  Run casing
                                                 with        casing bowl.     and connect
                    Pack off                     alignment   Connect and set  by screwing in
                    damaged                      tool and    slips.
                    section                      position    Perforate

                                        Squeeze (b)

                                        Clean out (c)
                                                             Land casing
                                        Test (d)             nipple up

                                        Resume operations
```

(a) Severe cases may be treated as parted casing
(b) Includes resqueeze and single and multiple squeeze jobs
(c) Includes drilling out and running full-gauge mills, casing scrapers, etc.
(d) Includes pressure testing, running caliper and bond logs, etc.

- *Run another string of casing or a stub or tie-back liner.* If the casing is large enough, run another casing string or a liner. A failure near the bottom of intermediate casing should be covered with production casing. Assuming the casing has an adequate inside diameter to perform expected future operations, this is one of the better repair methods.

Repair a failure near the bottom of the hole above a liner with a stub or tie-back liner. If the failure creates an immediate hazard, it may be squeezed off and covered by casing or a stub liner later.

For example, a 7-in. production casing string may have failed. The string could be repaired by running a 5-in. or 5 ½-in. flush joint liner or as a complete string of casing, depending on depth. The reduced casing size may restrict future operations in deeper holes. A better method in a deeper hole could be to run the 5-in. as a liner and leave the upper section of 7-in. open to allow the running of a tapered work string for remedial work. Depending on clearances and equipment availability, the casing can be set on a collapsible packer for isolation or cemented in with a primary cementing job. Perforating should not be a problem, since there are many tools capable of perforating both sections of casing and also penetrating the formation to an adequate depth (Fig. 7–11).

Figure 7–11 Repairing casing above a liner
Note: The first method of repair to be considered is squeezing. An alternate method is to pack off over the leak.

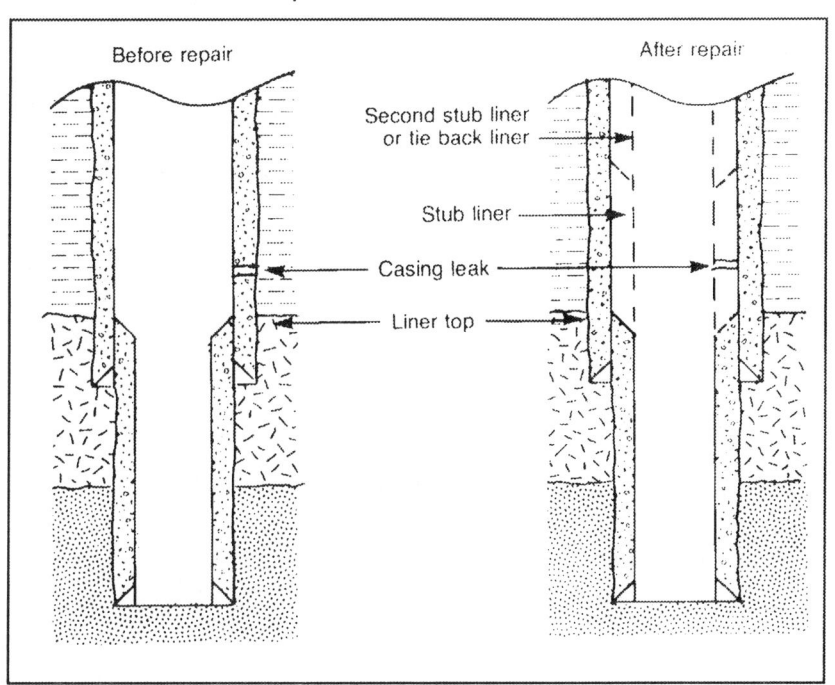

Casing and Liner Repair

- *Failure in casing not cemented.* When possible, pull the casing, replace it, and run and reconnect it in a way similar to parted casing. Otherwise, try to cement the casing with a primary cement job. Running another string of smaller-diameter casing or tie-back casing or a stub liner may be applicable.
- *Failure in cemented casing.* This is one of the more difficult casing repair situations. Generally, cemented casing cannot be repaired by replacing the casing above the failure. It may be possible to ream out all casing above the failure and then replace it, making a casing bowl connection. However, this procedure seldom is used because of the difficulty of milling long sections of casing. If the failure occurs in an inner string of casing inside of the cemented casing, it usually is repaired by reaming the failed section to full gauge with a mill and squeezing if necessary.

  Pulling the damaged casing for a primary cement job is highly unlikely, although covering it with another string of casing or liner may be applicable. Otherwise, squeezing, staging, or making multiple consecutive squeezes is the only alternative.
- *Salvaging casing.* If a well is to be abandoned because of a casing failure, extra effort may be taken to recover the salvageable casing. This may not be justified, especially in view of the risk involved and the questionable amount and value of the salvaged tubular goods. Estimate the value of the salvageable tubulars and the cost of salvage. Generally, in order to attempt salvaging, the value of the salvaged material should be at least 3–5 times the estimated cost to recover it. Any time a well is abandoned, consider selling the well in place or plugging it without salvaging. Although this is only indirectly connected with casing failure, there have been situations where the amount spent salvaging casing from a well was greater than the value of the salvaged casing.

CEMENTING CASING FAILURES

Cementing is an important method of repairing casing failures. Cementing is especially applicable as a preventive action, since incorrect cementing procedures often cause fishing and cementing is also used in many remedial situations. Conventional and squeeze cementing techniques have been covered in the literature. However, less attention has been given to the special applications of cementing related to casing repairs. Some casing failures are best repaired by modifications of standard techniques or by special applications, such as multi-consecutive-squeeze cementing.

Cementing casing repairs and similar remedial jobs are relatively common procedures. If not conducted correctly, these procedures hold a high risk of failure. The result may be as minor as having to squeeze again; in the worst case, however, tools malfunction or operational failures may result in cementing the tools in the hole, requiring a severe fishing job and possibly leading either to plugging back and sidetracking or to abandonment and redrilling.

Cement Applications in Casing Repair

The concept of the change of state from cement slurry, a semi-liquid, to cement, a solid, and the associated hydraulics is important to understanding the use of cement in casing repairs and also the controlling of lost circulation. Conventional primary cementing techniques are used for a covering casing string, such as tie-back casing, liner, or stub liner. All other casing failures that require cementing usually are repaired by squeeze cementing techniques. Because most squeeze cementing involves fracturing, the relationship of squeeze cementing and fracturing is another important concept in casing repairs.

General Review

Cement compressive strength is measured by standard compression tests and the results are given in psi (pounds per square inch). Since hydraulic pressures are measured in the same units, there may be some tendency to equate the two. This may not apply since the two units of measurement are not compatible, especially when cement changes from a liquid to a solid.

Cement slurry acts as a liquid and the flow mechanism is approximated by conventional hydraulic fluid flow concepts. As the slurry changes to a solid, it acts like a liquid with an increasing viscosity. As the viscosity increases, the hydraulic pressure required to move the slurry also increases. When the viscosity reaches very high values, the pressure required to move the cement becomes correspondingly high. This is defined by the thickening time or the pumpability, which are both relative measures.

As cement begins to thicken and harden, it can be moved only as a unit mass and its strength becomes a function of its size and the strength and amount of material reinforcing it. For example, an approximately 4-ft cement plug was inadvertently allowed to harden in a 6-ft, 2⅞-in., N-80, 6.40 lb/ft, 10,570 psi burst tubing nipple. The nipple was connected to a low-volume, high-pressure pump and pressure was applied in stages. The nipple burst at about 12,000 psi without moving the cement. The cement probably had a compressive strength in the range of 3,000 psi. There are a number of cases where sand plugs or bridges in tubing have held pressure differentials in excess of 5,000 psi without moving. Therefore, cement will hold pressure under confining conditions, and formations will act in a similar but slightly less efficient manner.

Squeezing Relation to Fracturing

Fracture gradients and efficient squeeze pressures are related. Fractures are related to the natural earth stresses due to normal sedimentation, uplifts, warping, basin development, bedding planes, formation strength, natural fracturing, and related factors. Fracture initiation pressure induces a fracture in the formation, which is extended by a lesser fracture propagation pressure. All fluids initiate and propagate fractures but, for practical purposes, penetrating fluids such as water and oil generally require less pressure than a less-penetrating fluid such as mud or cement slurry. Propagation pressure is less sensitive to the nature of the fluid than initiation pressure. Therefore, while it is better to fracture or break down the formation with water, cement slurry will extend the fracture in a way similar to water.

Resistance to fracturing increases with time in an open hole. One reason for this is the creation of filter cake — i.e., the movement of drill tools against the wellbore wall tends to pack solids into the wall and these solids invade a short distance into the formation. This is easily verified with leak-off tests in a normal-pressured sequence of formations. Ordinarily, the casing shoe is tested after drilling out a few feet. Other leak-off tests run periodically while drilling deeper will be successively higher. This assumes leaking off at the casing shoe; if they leak off deeper, this still indicates an increasing formation strength at the casing shoe. Another method is to run a leak-off test before plugging a dry hole; it will normally leak at the casing shoe. Compare this leak-off test to the original test of the casing shoe. It usually is greater.

Figure 7-12 Increasing apparent formation strength by multiple squeezing

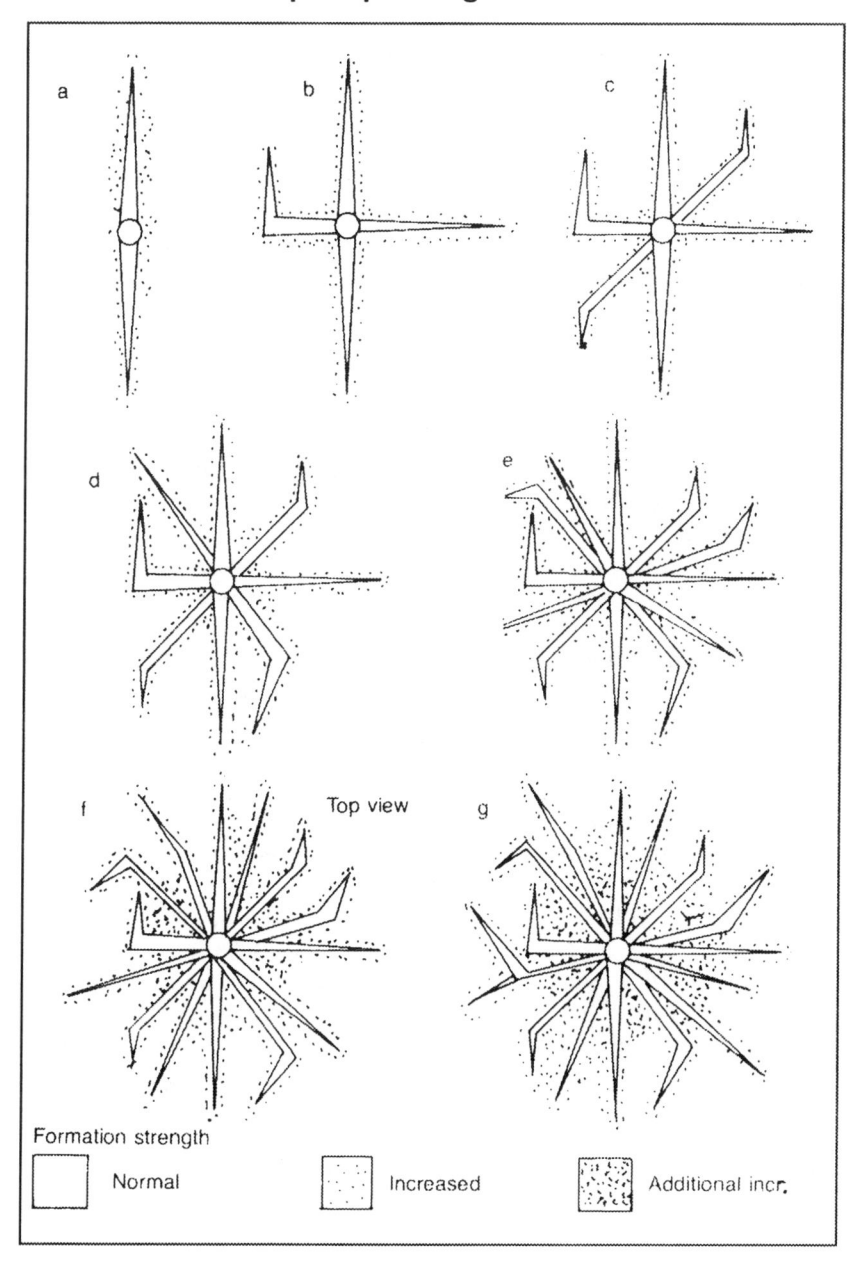

Casing and Liner Repair 475

Fracture orientation normally is horizontal at shallow depths of 1,000–1,500 ft and begins to rotate to the vertical at deeper depths. Orientation is related to the overburden pressure, which averages about 0.70 psi/ft, 12.5 lb/gal. Average fracture initiation pressures are equivalent to the initial breakdown pressure in squeezing. These are about 0.83 psi/ft, 16.0 lb/gal, probably increasing below 10,000 ft. Gradients of 1.1 psi/ft, 21.2 lb/gal, have been observed at 18,000 ft. These are averages, and in "Hydraulic Fracturing" (*SPE,* 1970) Howard and Fast indicate there is wide variation from the average. Propagation pressures are equivalent to pump-in pressure during squeezing. Howard and Fast show the normal fracture propagation gradient is 0.2–0.5 psi/ft less than the fracture initiation gradient.

The main point here is that, when squeezing, the breakdown pressure should approximate the fracture initiation pressure unless the formation has already been broken down. The pump-in pressure should approximate the fracture propagation pressure. Pressures may decrease or, more commonly, increase depending upon the type of fluid, the rate of injection, and the pumpability of the slurry. Otherwise, the cement slurry acts as a displacing medium and is not squeezing.

Multiple squeezes will strengthen the formations adjacent to the wellbore. The exact mechanism by which this occurs is unknown. However, it appears that slurry is squeezed out into the formation into an artificially created fracture, where it hardens. This effectively props the fracture and compacts the adjacent formation. Multiple squeezes increase formation compaction adjacent to the fractures, and ultimately adjacent to the wellbore, so overall they are more compact and resistant to subsequent fracturing and breakdown (Fig. 7–12). The increase in formation strength depends upon formation characteristics, the number of squeezes, and squeeze placement.

It is reasonable to expect that natural earth stresses can be reinforced and strengthened by fracturing the formation and propping the fractures in a localized area. Two questions are: (1) How many factures does it take to increase the strength of the formation by a given amount? and (2) How long can these additional stresses remain localized around the wellbore before they dissipate into the formation and the stresses around the wellbore return to normal, assuming this occurs? These questions cannot be answered directly, but theory and experience provide partial answers.

It appears that a fracture gradient of over 19.2 lb/gal (in one case) can be achieved with about 4 or 5 successive squeeze jobs. In one field application, a formation with a pore pressure of about 8.3 lb/gal and a fracture gradient of about 11.5 lb/gal was treated with successive squeeze jobs, and the apparent formation fracture gradient increased to over 19 lb/gal and was sustained for over two months.

Howard and Fast compare formation-lifting factors. They draw a curve that becomes asymptotic at 19.6 lb/gal after five squeeze jobs. The data could also be interpreted, by extrapolation, to show a gradient of 23.0 lb/gal after 9 or 10 squeeze jobs. If the equalized earth stresses can be increased horizontally, then it stands to reason they could also be increased vertically. Therefore, it should be possible to achieve fracture gradients higher than 1 psi/ft or equivalent to the overburden pressure. The higher gradient of 23.0 lb/ft is hypothetical and has not been proven in the field while squeezing into the open formation. However, it has been approached as described in the section "Multiple Consecutive Squeezes."

Perforations have been squeezed and tested to pressures of this magnitude and slightly higher. However, this may not compare directly with squeezing the open formation through a large hole such as a split or parted casing. However, if the open formation can be strengthened by squeezing through a large casing opening, squeezing could be done through perforations.

Squeeze Cementing

Squeeze cementing is the procedure of displacing cement slurry into the formation. The basic objective is placement of the required amount of slurry while obtaining a predetermined pressure. In a few cases the objective is placement of a fixed amount of cement regardless of pressure (within reason), or obtaining a desired pressure regardless of placement volume. All pressures used here refer to pressure at bottomhole conditions. Surface pressures should be modified based upon the weight of the fluid column.

Squeezing serves various purposes. The main purpose dealt with here is casing repair. These procedures generally are similar to most common squeezing operations with modifications based upon repair requirements. Recognize that the pump-in pressure must exceed the approximate fracture propagation pressure. Otherwise, the squeeze becomes a displacing mechanism, which may or may not be desired.

Conventional

Conventional squeezing is used for casing repairs and various common procedures such as shutting off water and block squeezing.

Normally it is best to use a good quality Class "H" cement and a normal weight slurry, retarded or accelerated based upon the projected pumping time needed. The projected pumping time used should be about 150% of the estimated actual pumping time. DO NOT retard excessively. Use additives for defoaming, viscosity control, etc., as required. Sand is used in some cases, if applicable. In most cases use at least 10 bbls, and preferably over 20 bbls, of batch mixed slurry. Use spacers as needed, normally unweighted. The most common spacer is water. Area squeezing practices may be a guide to the use of spacers.

The best method of placement for common squeezing, especially at higher pressures, is through a drillable cementing tool, especially for higher pressures and deeper depths. Such a tool is safer to operate than a retrievable tool under these conditions; since there is less tool manipulation, it is easier to hold back pressure when required, and there is minimal risk of tool failure and sticking. Using a drillable cementing tool does require drilling, except when squeezing is employed as a plugging action, which can be a disadvantage if the squeeze must be repeated.

The retrievable cementer is a satisfactory tool for less-demanding situations and is commonly used, especially if a second squeeze is anticipated. However, there is a higher risk of tool malfunction, especially sticking the tool or cementing it in place. Using this tool also allows the option of running a retrievable plug below the tool if there are deeper open perforations. Do not overlook pressurizing the annulus to reduce the pressure differential across the squeeze tool and prevent a possible collapse of the casing above the tool. Bullhead squeezing by pumping directly down the casing is seldom used except in shallow holes, low pressures, and similar situations.

A retrievable hook-wall cementing or squeeze tool will be used here for illustration. A drillable cementing tool can be used in a similar manner with some modifications in the procedure.

Run a retrievable hook-wall cementing tool and set it 100 ft to 200 ft above the point to be squeezed. Normally select the tool height above the squeeze point based on the casing volume. The tool height should be sufficient so that the top of the slurry is in the casing below the tool before reaching higher squeeze pressures, in order to reduce the risk of slurry leaking around the tool and sticking it. Also, there is less risk of the cement setting up in the pipe and squeeze tool. Proper tool height also provides additional safe for pumping slower to obtain a higher pressure. These are important actions to prevent fishing and help ensure a good squeeze.

Open the circulating valve, displace water to the tool, and close the circulating valve. Pressurize the annulus and monitor it to detect leaks from the tool. Start injecting water slowly and increase it to a maximum rate and pressure. If necessary, surge to help reduce the pressure. In some cases, spearhead with cleanout acid to reduce pressures and increase volumes. Inject the acid until a stable rate is established that will permit placing at least half of the slurry into the formation within a reasonable time — often about a third of the pumping time or less. The slurry may require a slightly higher injection pressure at the formation, but usually will require less surface pressure after allowing for the heavier weight slurry column.

The most common procedure is to leave the well shut in and use the water in the tubing as spearhead fluid. Otherwise, reverse the water out and circulate slowly while batch mixing the slurry. Modify the slurry volume as necessary based upon the pump-in pressure and volume. If water was left in the work string, pressurize the casing and pump the slurry and tail spacer into the work string. Otherwise, leave the circulating valve open and pump the spearhead spacer, batch-mixed slurry, and tail spacer. Close the circulating valve when about half of the spearhead spacer is through the cementing tool. Pressurize the annulus and continue pumping or displacing the slurry.

Pump about half of the slurry into the formation at a rate determined by the injection test and slurry pumping time. If the rate is too high, reduce the rate when about ¼ of the slurry is in the formation. If the rate is too low, or if pressure starts to increase very rapidly, immediately INCREASE the rate and pressure rapidly for a short time to prevent reaching a premature squeeze pressure. Surge and flow back quickly, follow this by pressuring, and repeat if necessary should the squeeze pressure be reached before ½ of the slurry is in the formation. Then repressure to break down and continue pumping additional slurry into the formation.

Otherwise, continue pumping the slurry into the formation. The pumping rate should be adjusted so that about 80% of the slurry is in the formation at the desired squeeze pressure. Pressure drops of about 6% of the surface pump-in pressure followed by pressure buildups and repeated drops at subsequently higher pressures are a good

Casing and Liner Repair

indication of a successful squeeze. Normally an operator should try to obtain the final squeeze pressure with the top of the slurry in the casing and well below the cementing tool. Hold this pressure for a short time, commonly 10 to 20 minutes, release it, and check for flowback. A good squeeze will have a small amount of flow that stops in a few minutes. If flowback continues, repressurize to 80% of squeeze pressure and repeat as necessary. Excess flowback indicates the formation was not squeezed properly, so it may be necessary to squeeze again. Then reverse out and check for any cement in the returns by opening the circulating valve and reverse-circulating slowly.

There are various alternate procedures at this point. The recommended procedure is to unseat the retrievable cementing tool and reverse out for one circulation while moving the tool slowly. This maintains a moderate pressure on the squeeze. Verify the progress by monitoring pit levels, then pick up slowly to prevent swabbing. Pick up about 700 ft, reset the packer, and apply about 30% of the squeeze pressure against the squeezed section. Hold this pressure for the equivalent of about twice the cement thickening time, then pull out of the hole. Run a bit or mill, often with a casing scraper, and circulate above the squeezed zone until the cement reaches the desired compressive strength. DO NOT RUN INTO GREEN CEMENT. Then clean out the cement and run a test tool to determine if the casing has been squeezed satisfactorily.

If the squeeze pressure is not obtained, inject all the slurry into the formation and overflush with about 10 bbls. Unseat the retrievable cementing tool and reverse-circulate, moving the pipe to ensure that the tool is free and clean. Circulate slowly while waiting for about two thickening times, then squeeze again. Adjust the slurry volume and injection pressure based on the results of the first squeeze.

The squeezing procedure has been detailed here because many casing repairs are squeezed in a similar manner.

Staging

Stage squeezing is a less frequently used procedure for squeezing situations in which it is difficult to obtain a satisfactory squeeze pressure conventionally. Stage squeezing is similar to conventional squeezing except that a slightly larger slurry volume normally is used and it is displaced into the formation in steps or stages. The object is for the slurry to start thickening in the formation (a process often associated with dehydration) so that the higher pressure forces the following slurry into a different sector of the formation at each stage.

Stage squeezing begins with the same procedures as for conventional squeezing, continues until about half of the slurry is in the formation, then is stopped for a short time. Often in the early part of the squeeze, the pressure will drop slightly. Pump for a short period and continue alternately pumping and waiting until the desired squeeze pressure is reached. Try to maintain a relatively constant injection rate while pumping and observe the changes in pressure. Adjust both the injection rate and the length of the waiting period as necessary to achieve the squeeze pressure when about 80% of the slurry is in the formation.

Stage squeezing should start with lower rates and longer waiting periods. Then the rate should be increased and the length of the waiting period decreased as the squeeze pressure is approached. If small pressure breaks occur as described for conventional squeezing, it may be possible to pump continually at a slower rate. Increase the rate substantially if the pressure increases too fast. Pressure up and release rapidly if the squeeze pressure is obtained too early. Take care not to collapse the tubulars while surging. The annular space should be pressurized as noted earlier.

Cement Backed by Metal

A special application of squeezing that often creates a strong repair is to cement failures where the cement is backed by metal. Cementing casing that is run inside of other casing normally is a very strong repair, not only due to the strength of the newly run casing but also because of the cement behind the casing. Primary cement in the open hole is not as reliable. A stronger repair is possible if the failure occurs where the cement is backed by metal. If there is an option, try to leave the casing failure in a position that allows it to be cemented with this procedure. It is a relatively easy method of repair and usually very strong.

Often, a failure in a repair using cement backed by metal results in a relatively thin opening or crack almost surrounded by metal, as listed in the following (Fig. 7–13).

Casing and Liner Repair

Figure 7-13 Cement backed by metal

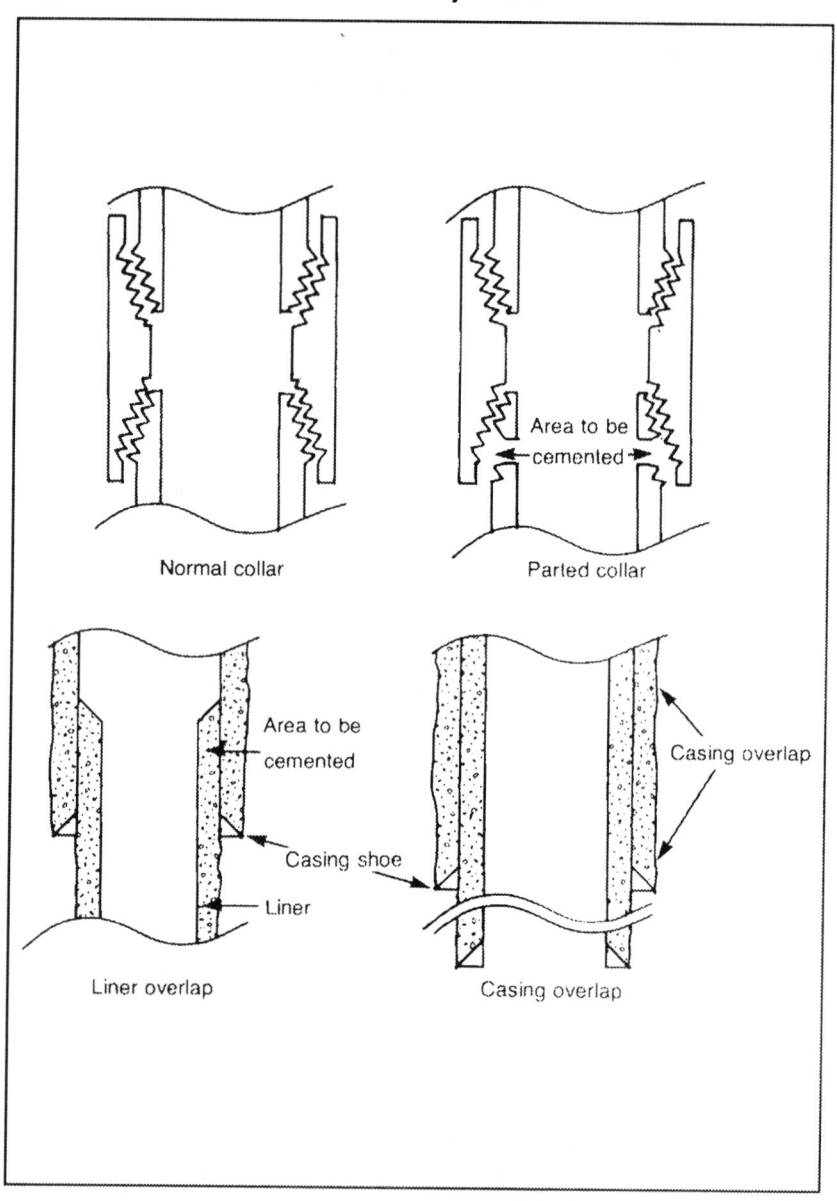

- Casing may part at the root thread. If the two pieces are set back together, there will be a thin crack where the ends of the casing butt against each other and another crack between the edge of the casing wall and the inner lip of the collar.
- A hole may be worn in casing at a point where it is inside another length of casing. The metal backing is the wall of the hole in the inner casing and the inner wall of the outer string.
- A leaking liner top is a similar situation, in which the inner wall of the casing effectively acts as the metal backing for the overlap section.

These types of failures are relatively easy to repair. It is best to have the parted sections in contact or as close together as possible. In some cases it may be necessary to cement behind the backup metal to prevent the metal backing from moving and reopening the point of failure. Failures that are repaired correctly using this procedure have a burst strength approaching the rating of the surrounding casing.

In operation, displace a good quality cement slurry into the opening between the walls of metal backing. No specific pressure requirements apply, except that the pressure must be sufficient to displace the cement, which often includes squeezing. The pumping rate and volume should be sufficient to sweep out any mud or formation debris and replace it with cement. Under-displace to ensure leaving slurry between the walls of metal backing and inside the hole. Frequently, two or more stages are needed to provide adequate backing. After completing the repair, drill out the cement and ream the section out to full gauge as necessary.

Multiple Consecutive Squeezes

Multiple consecutive squeezes are used to seal holes and splits in casing and to plug lost-circulation and fractured zones. It is a temporary repair that requires a later final repair such as running another string of casing. It is faster and more economical than doing conventional squeeze cementing a number of times. The procedure includes applying a series of rapid, consecutive squeezes to increase the apparent natural strength of the formation — in some cases to 18 lb/gal and higher — in a relatively short time at both shallow and deeper depths.

Conventional squeezing is time consuming and costly. A single squeeze job takes at least 36–48 hours of rig time, including drilling out and testing, and often longer in deeper wells. Depending upon the situation, conventional squeezing usually requires 4–6 squeezes to repair the failure. After two or three squeezes and four or more days of rig time, the operator frequently will give up and resume operations with a relatively weak repair and its associated risks. Using the faster, more efficient method of multistage consecutive squeezing, an individual squeeze job can be completed in 6–8 hours.

Design the procedure to create fractures and pack them with cement by using multiple squeezes. Determine a final squeeze pressure. Normally, use the maximum pore pressure expected, the maximum pressure expected with a safety factor, the bursting strength of the casing with a safety factor, or the lowest reasonable value.

Batch-mix the slurry with class G or class A cement at deeper depths. Use the minimum amount of additives, generally only retarders. Select a thickening time that is about 30 minutes longer than the expected actual pumping time. Generally, a slightly increased pumping time should be used for successive squeezes to allow for additional staging. Use water spacers.

A plug should be set several hundred feet below the point of failure, which is often a split or hole. Normally, use drillpipe or tubing with a sawtooth collar, small mill, or core head or similar tool on bottom. When the end of the work string is near the point of failure, connect the cementing line above an inside blowout preventer and adjust the spacing so that the end of the work string is near the point of failure (preferably within about 10 ft), a tool joint is about 3 ft below the pipe rams, and there is about 20 ft of pipe above the rotary.

Close the pipe rams and pump into the formation to determine injection rate and pressure. On subsequent squeezes, perform the same injection procedure but start it as a leak-off test to determine fracture pressure. Open the pipe rams, batch-mix about 100 sacks of cement, and displace the cement to the bottom of the work string. If the cement falls too fast, displace with lighter fluid or close the pipe rams and control the flow rate with the choke manifold.

Close the rams and displace the slurry very slowly for very low pore pressures or lost-circulation zones. If necessary, displace with water to minimize hydrostatic head. Displace the first half of the cement into the formation at a relatively high rate and the remainder by staging. Stage the first squeeze job at a moderately fast rate. Use a longer waiting period between stages on subsequent squeezes. When the last of the slurry is at the failure, over displace about 2–4 bbls.

Displacement volumes are critical, so use the minimum cement to ensure leaving an open channel from the casing into the formation. If cement is left inside the casing, it must be drilled out before proceeding, causing additional delay and expense. Hold the pressure about 15 minutes, then release it. Reverse-circulate for about two work string volumes while lowering the work string slowly at least 10–20 ft below the split to ensure a clean hole. Control the circulation rate to minimize pressure on the formation. Do not stop reversing with the bottom of the pipe below the point of failure. If the hole is taking fluid, pick up without reversing. If a retrievable packer is used (as described later), do not run it below the point of failure.

Pick up the pipe about 500 ft and shut the well in for several hours until the cement reaches an initial set. In some cases, slurry may flow back into the casing, requiring a more extensive cleanout. Help prevent this by pressurizing the casing to about 50% of the last-squeeze or displacement pressure. This may include pumping 1–2 bbl of fluid into the formation after shutting the well in.

Run the work string shoe through the cemented section to ensure the hole is clean. Circulate down the last 2 or 3 stands. DO NOT RUN THE PIPE INTO GREEN CEMENT. Then reposition the pipe and run a leak-off test. Repeat the squeeze procedure, then repeat the entire procedure until the desired squeeze pressure is obtained.

In the last squeeze, leave 10–20 ft of cement inside the casing and wait until the cement reaches the design compressive strength. Then clean out and test in a manner similar to that for a conventional squeeze. Repeat squeezing if necessary.

Each squeeze job should not take over 6–8 hours maximum, including waiting time. A chart plotting of the final squeeze pressure after each squeeze job vs. the number of successive squeeze jobs will help predict the total number needed. A plot using breakdown pressure from the leak-off test gives similar results, so use both.

An alternate placement procedure may apply, depending on the type of casing failure, the casing strength, and pressures. This procedure includes running a hook-wall packer with a 60-ft stinger of aluminum pipe. The squeezing procedure is similar to that already described, with modifications for using the packer. Do not allow cement slurry around the outside of the packer. Reverse slowly to prevent damaging the packer elements. Pull the packer during the waiting period and check it if necessary.

Casing and Liner Repair

Lower the work string to clean out cement through the failure, but do not allow the packer to pass below the point of failure. The aluminum stinger can be pulled off if it sticks and is relatively easy to drill out. DO NOT replace the aluminum stinger with a steel stinger below a shear pin release because it can be very difficult to mill out or wash over the steel stinger.

Multiple squeezes could be performed with a permanent packer with a large bore, although this procedure has not been reported in field practice. Set the packer about 50 ft above the point of failure. Run a small diameter aluminum pipe stinger below the seal assembly, spaced so the end is about 15 ft below the point of failure when the seal assembly is seated. Use the same general multiple squeezing procedure, except squeeze with the end of the stinger below the failure and reverse out carefully to prevent erosion in the packer bore.

The repair of split casing, including problems encountered, is illustrated in the following example.

Multiple Squeeze Split Casing

Intermediate 7-in., S-95, 20 lb/ft, 9,690 psi burst, LT&C casing was set at 15,000 ft in 11.5 lb/gal mud and cemented back to about 9,000 ft. A 5-in., N-80, 18 lb/ft, FJ liner was set at 16,000 ft, TOL at 14,700 ft, in 14 lb/gal mud with several high pressure gas sands behind it. Gas-cut mud was recovered while cleaning out the top of the liner, indicating leaking (Fig. 7–14).

The liner top was tested with a retrievable packer run on 3½-in. tubing and set at 14,500 ft. Water was displaced to the top of the packer for an in-flow test. The well flowed 5 bbl/hr and shut in at 2,500 psi. The 7-in. by 3½-in. annulus was pressured to 2,500 psi and water was pumped into the top of the liner at ¼ bbl/min at 5,600 psi. After releasing the pressure, the packer was unseated, moved, and reseated at 13,800 ft to squeeze the liner top. The annulus was pressured to 3,000 psi while the batch-mixing of the squeeze slurry was started. About 20 minutes after pressure was applied, the annulus lost all pressure almost instantly. It was filled with about 75 bbl of water and the premixed slurry was discarded.

At this time the well was in a potentially dangerous condition. The zones behind the liner had over 10,000 psi BHP, and the intermediate casing would not hold pressure. The retrievable packer was set with the well under control. The obvious next step was to squeeze the liner. This would require high pressures, with the possibility of leaving cement in the tubing or packer and no way to pressurize the casing while squeezing or

Figure 7–14 Repairing split 7-in. casing at 3,500 ft

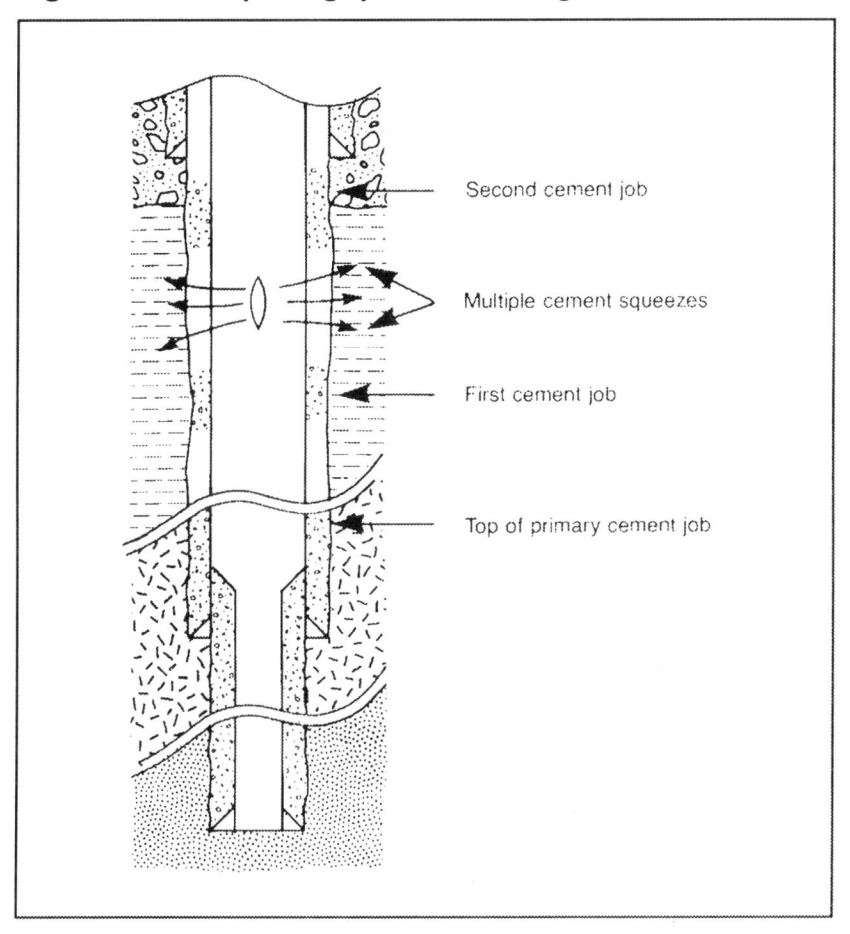

- Second cement job
- Multiple cement squeezes
- First cement job
- Top of primary cement job

to reverse out the slurry. The decision was made to squeeze the liner top and accept the risk.

Fifty sacks of class G cement with 30% silica flour and standard additives were mixed and displaced to the packer, and the circulating valve was closed. Nine bbl of slurry were squeezed into the top of the liner at ¼ bbl/min to a final pressure of 7,000 psi, a high differential across the packer, and 5 bbl of cement was left in the 7-in. casing. Pressure was released, and the packer was worked and released in about 30 minutes. It was raised five stands and reset and a back pressure of 4,500 psi was held on the cement for 4 hours.

The point of failure was located at 3,700 ft with a retrievable packer. When the intermediate casing failed, it was filled with 10.0 lb/gal mud after having been set in 11.5 lb/gal mud. The casing failed at an internal pressure of (10.0)(0.052)(3,700)+3,000 = 4,924 psi. The external hydrostatic pressure was (11.5)(0.052)(3,700) = 2,213 psi. The safety factor in tension at failure was a high (9,690)/(4,924-2,213) = 3.57, excluding tensional effects. The casing was tested prior to running and there was no reasonable explanation for the failure except faulty inspection.

A casing inspection log was run. The preliminary evaluation was a 4-ft casing part at 3,700 ft. A retrievable bridge plug was run and set at 5,000 ft as an added precaution because of the high-pressure lower zones. It was decided to try and pull the casing. BOPs were removed, and a casing spear was used to pull 385,000 lb. The casing was stuck in the bradenhead and could not be moved. The casing inspection log was reviewed again before additional steps were taken to release the casing. It was concluded that the 7-in. casing was split but intact.

Repairing by cutting or backing off below the failure, pulling, replacing the bad joints, or running and reconnecting was eliminated because of problem formations and the difficulty of releasing the casing from the bradenhead. Running 5-in. tie-back casing was not practical because of the reduced hole size and depth. Also, the casing would be protected by covering tubing. Because the deeper zones were questionably commercial due to low permeability, the risk of a sustained blowout was reduced and the easiest, most economical repair was further justified. Therefore the option of repairing by multistage cementing to about 18–20 lb/gal was selected. As an additional safety precaution, the 7-in. by 13⅜-in. annulus would be fitted with a pressure-relief valve set to vent at 1,500 psi in case the repair failed.

The casing was repaired as follows:

- *Day 1* - Displaced fresh water and set a retrievable packer at 3,500 ft. Pumped 50 bbl of water with free circulation up the 7-in. by 13⅜-in. annulus. Closed the annulus and squeezed (Job 1) 150 sacks with 3% CaCl, and overdisplaced with 4 bbl of water. Waited on cement (WOC) for 6 hours with 600 psi on squeeze. This gave a bottom block of cement below the split.
- *Day 2* - Opened the annulus and established circulation behind 13⅜-in. Cemented (Job 2) behind the 7-in. using 150 sacks with 3% CaCl with good circulation and overdisplaced 4 bbl. This gave a top block of cement above the split. WOC for 6 hours and squeezed (Job 3) with 150 sacks with 3% CaCl. Pumped in 100 sacks at 2 bbl/min at 600 psi and staged the last 50 sacks to 2,500

psi in 1¼ hours. Held pressure for 2 hours, released packer, and pulled out of the hole.
- *Days 3 and 4* - Drilled out the cement and pressure-tested the squeeze job. Broke down at 2,400 psi, equivalent to 21.9 lb/gal with 9.5 lb/gal saltwater in the hole, and pumped in 5 bbl at 2½ bbl/min at 1,200 psi. Ran a retrievable packer and set it at 3,500 ft. Pressured the annulus to 1,000 psi and squeezed (Job 4) with 150 sacks with 3% CaCl. Pumped 2 BPM at 900 psi, staged to 1,200 psi, and overdisplaced with 3 bbl of water. WOC for 4 hours.
- *Day 5* - Drilled out cement, ran packer, and squeezed (Job 5) with 150 sacks as above. Pumped 2 bbl/min and the pressure increased from 1,500 psi to 3,200 psi. WOC, ran a bit, drilled out cement, and pressure-tested the squeeze job to 2,000 psi with 9.9 lb/gal in the hole, equivalent to 20.3 lb/gal.

Resumed operations, the liner top tested satisfactorily, and the well was completed without any problem due to the 7-in. casing failure.

The increasing pumping pressures indicated the increased formation stresses. The failure could have been squeezed with open-ended tubing, but as the failure cast some doubt on the integrity of the 7-in. casing it was protected by using the retrievable packer. The repaired section was tested with a leakage test about two weeks later, and the split held the equivalent of 19.0 lb/gal.

DRIVE PIPE AND CONDUCTOR

Normally, there are few failures in conductor casing or drive pipe. These, and sometimes the mud riser, are not classified as regular casing. One exception is conductor run as deep conductor, as described in the section "Surface Casing." These special-purpose tubulars provide a method of returning mud to the circulating system and holding back unconsolidated surface formations such as sand and gravel. They are not designed to hold any appreciable pressure. If some pressure is expected, a rotating head or diverter should be installed.

The most common failures are misalignment and leaking around the base of the pipe. Other failures are relatively uncommon and most are repairable. It may be prudent to consider moving and redrilling for very severe failures, if the option is available.

Casing, Rotary, Mast, and Crown Alignment

Align the conductor casing, rotary, mast, and crown vertically, centering the conductor casing under the rotary, and follow with a good cement job. The drive pipe may be driven, or the conductor pipe set, usually by a rathole digger or in a large hole dug by a backhoe. Ensure that the rig is moved in and centered over the tubulars (Fig. 7–15). If they are not set correctly and aligned, the drilling assembly will hit and rub against them. This may not cause a problem in early drilling operations because the shallow hole is drilled relatively fast and the drilling assembly is fairly lightweight, so lateral forces are lower and cause fewer problems than those in deeper drilling.

Figure 7–15   Mast-conductor hole misalignment

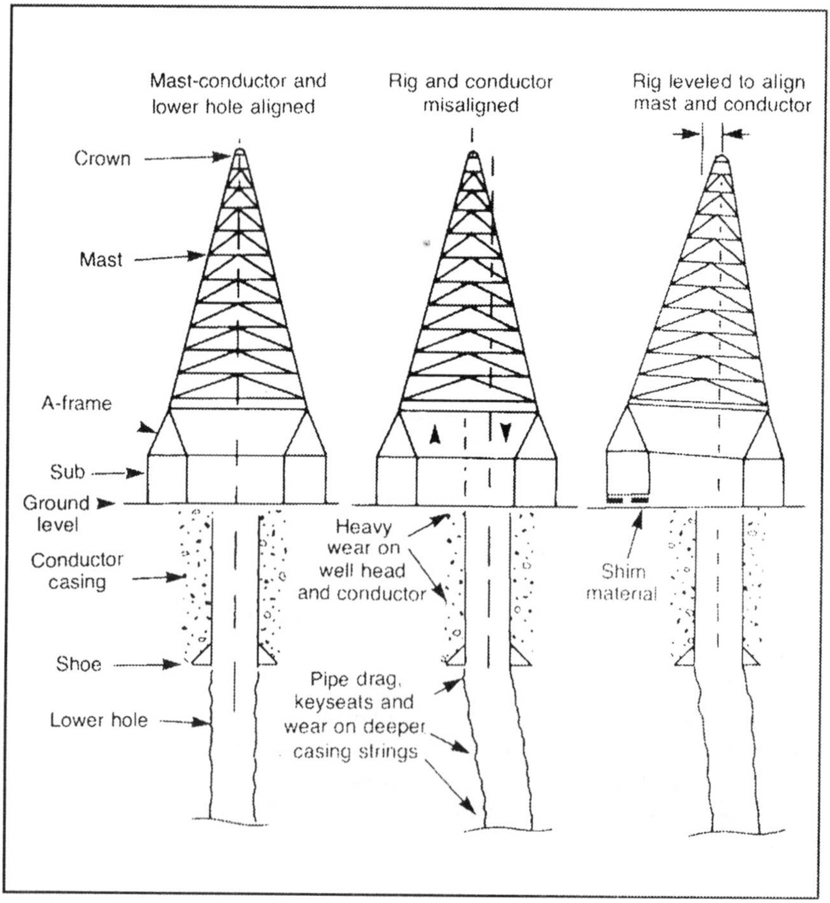

Conductor casing that is off center or out of alignment can cause wear on the casing, the drilling assembly, and even the preventers in severe cases if a wear bushing is not used (Fig. 7–16). It also hinders setting and pulling the slips, seating the kelly drive bushing, dragging the split bushings out of the rotary on trips, and maintaining pipe alignment while making connections in the rotary. The severity of the problem usually increases with depth due to the increased weight and lateral force of the drilling assembly.

Figure 7–16   Inclined conductor casing

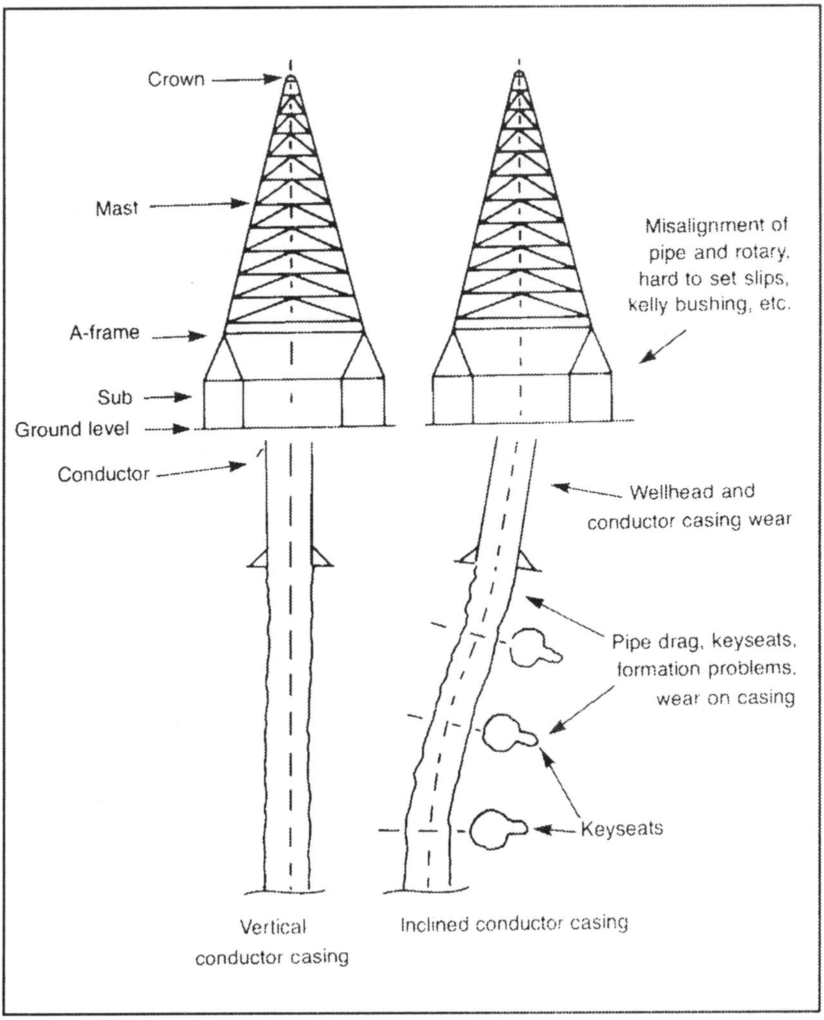

Casing and Liner Repair

A good preventive measure is to drill a larger than normal conductor hole, conditions permitting. This automatically corrects for small misalignments. The conductor tends to hang nearer vertical in a larger hole that has additional clearance (Fig. 7–17). As an example, drill an 18-in. hole instead of a 15-in., or a 20-in. hole instead of an 18-in. — or larger sizes if possible. Drill the hole straight and vertical, especially the first 500 ft. Allow for and correct problems due to dipping formations, boulders, and other conditions.

Conductor casing, rotary, or mast that is off center or out of alignment even a few inches can cause problems, as indicated in the following example.

Misaligned Conductor

A deep well had 20-in. conductor casing set at 100 ft with a rathole digger and cemented. A concrete pad was built to support the heavy rig. The program called for a 12¼-in. hole to 5,000 ft and opened to 17½ in for 13⅜-in. surface casing.

The pipe had continually dragged on one side of the rotary during earlier drilling and hole-opening operations. Difficulties were encountered while making connections due to pipe misalignment. The kelly drive bushing would not slide into the rotary without pulling the pipe over with a spinning chain. It was hard to set the slips because of pipe drag on the side of the rotary. The large-diameter drill collars dragged the split bushings out of the rotary and were excessively hard to handle. The problems became progressively worse with depth. It was estimated that the problems caused a loss of 5–10 minutes per connection and 1–1½ hours per trip.

The condition was evaluated with a stand of pipe in the elevators hanging free and allowed to come to a free rest. The free end of the stand hung 7½ in. south-southwest from the center of the hole, "V" door facing north. The conductor casing was centered directly under the rotary. When the stand was reconnected to the drillstring and allowed to hang, it pushed solidly against the side of the rotary, and one crew member could hardly push the pipe away from the side of the rotary at a string weight of 105,000 lb. After additional checking, the rotary was found to be 2 in. off center toward the east or driller's side, measuring from the center of the rotary to the center line between each A-frame leg.

Figure 7–17 Effect of inclined conductor hole size on hole inclination

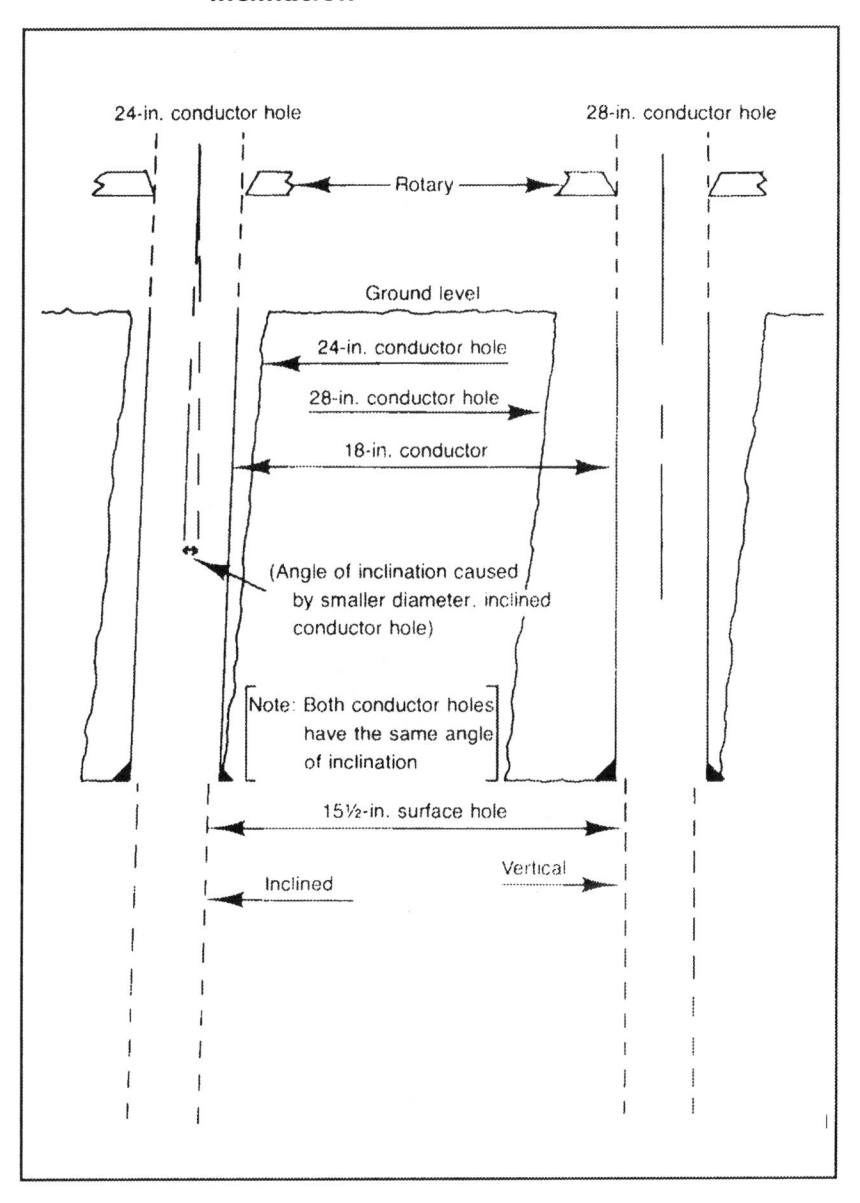

Casing and Liner Repair 493

The free-hanging pipe indicated that the crown was not centered vertically over the rotary. This normally would be caused by an out-of-level concrete pad. The pad elevation was checked at each end and at the middle on the outside and inside of each substructure and found to be within ¼ in. of level. With this particular rig and A-frame leg spacing, if one side of the pad were ¼ in. higher than the opposite side, the crown would move off center a maximum distance of less than 1 in. — which did not account for the free-hanging pipe being 7½ in. off center.

The mast, A-frame, and substructure were jacked up and shimmed until the free-hanging pipe was centered in the rotary. After the substructure was jacked up about 1½ in. on one side, the pipe hung about 2 in. off center to the east when the string weight was suspended from the crown.

The substructure was then jacked up and readjusted so the pipe hung in the center of the rotary with the full string weight suspended from the crown. A free hanging stand would hang with the free end about 3 in. off center to the west. The adjustment was necessary to cause the pipe to hang in the middle of the hole during drilling and tripping. This eliminated the off-center problem. The crown was approximately 6 in. off center after the adjustments; however, the pipe could run freely through the rotary. The well was completed without any further problems caused by the rig misalignment.

When the substructure was jacked up and repositioned, the drillpipe dragged at a point downhole; however, this dragging was not severe. Since the pipe was rubbered, no additional problems were encountered. The intermediate casing was straight and the main problem was the offset rotary, which resulted in a hole that was actually 2 in. off center. The difference in the heights of the substructures on either side of the rig or the slight twisting or deforming of the lower mast and substructure beams further contributed to the problem.

The complete cause of the problem was not determined, but probably it was a combination of the off-center rotary and deformation or unequal substructure and possibly A-frame heights. Leveling the substructure relieved the problem to the point where the well was completed satisfactorily (Fig. 7–18).

When checking the level of the rig base or substructure, it is important, especially on larger rigs, to check the level at the ends and center on each side of both substructures or the approximate equivalent with a unit base. When leveling a rig, use wooden shim material such as ¼-, ½-, and 1-in. thick boards. Place the shim material, under the major part of the base, in some cases up to 80% or more of the base area, and 100% under the main load-bearing areas such as the mast, where the heavier weights occur to keep the substructure base from warping or bending.

Figure 7-18 Shimming a substructure

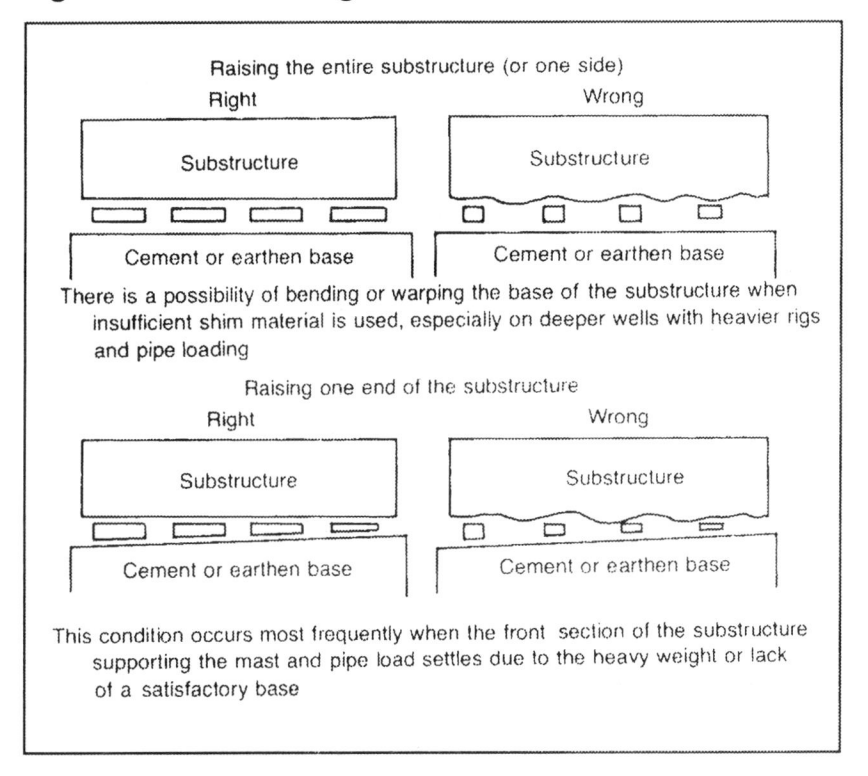

```
Raising the entire substructure (or one side)
         Right                              Wrong
    ┌──────────────────┐          ┌──────────────────┐
    │   Substructure   │          │   Substructure   │
    └──────────────────┘          └──────────────────┘
    ▭   ▭   ▭   ▭                  ▭   ▭   ▭   ▭
    ┌──────────────────┐          ┌──────────────────┐
    │Cement or earthen base│      │Cement or earthen base│
```

There is a possibility of bending or warping the base of the substructure when insufficient shim material is used, especially on deeper wells with heavier rigs and pipe loading

```
Raising one end of the substructure
         Right                              Wrong
    ┌──────────────────┐          ┌──────────────────┐
    │   Substructure   │          │   Substructure   │
    └──────────────────┘          └──────────────────┘
    ▭   ▭   ▭   ▭                  ▭   ▭   ▭   ▭
    ┌──────────────────┐          ┌──────────────────┐
    │Cement or earthen base│      │Cement or earthen base│
```

This condition occurs most frequently when the front section of the substructure supporting the mast and pipe load settles due to the heavy weight or lack of a satisfactory base

---

The rig may be slightly misaligned on purpose during installation to provide for settling. Then when it settles, it is aligned with the hole. Most shallower land rigs are set on matting boards on leveled ground. Sometimes the front or V-door side of the rig is set slightly higher, in the order of 1 in., for pipe racked on both sides of the V-door. This allows for settling with time on the heaviest section of substructure. If the pipe is racked on one side, then it may be set about ¾ in. high to allow for settling. Knowledgeable personnel should be familiar with the procedure.

Parted Conductor Casing

Parted conductor casing usually is different from regular parting because of large sizes. It can be a very difficult fishing job because the large-size pipe is frequently bigger than normal-size fishing tools and the large hole can be washed out to a larger diameter, making it difficult to catch the fish.

The pipe most commonly fails in a connection while running or before cementing. Smaller diameter casing is recovered and repaired similarly to parted surface casing. Normally, reconnect very large diameter casing. Run a new box or pin and screw into the casing with minimum torque. A shop-built welded cut lip guide may be needed to catch the fish in out-of-gauge holes. Lift the casing only enough to straighten it. Then cement with a primary job, using a high percentage of excess cement. After cementing, mill or roll the inside of the parted connection to ensure full gauge and squeeze if necessary.

Parting in the body of the pipe is less frequent and more difficult to repair. One method is a shop-built guided modified casing collar with an outer guide and used as a die collar. Run it, and set down over the fish and rotate it lightly (similar to the die collar). Use a light tension. Because of the limited-strength connection, do not lift the casing. Then cement by bull heading. Later roll and mill the inside as necessary.

A conductor casing failure and repair is illustrated in the following example.

Conductor Casing Failure

Collared 20-in. conductor casing was being picked up and run in a 30-in. diameter hole drilled to 800 ft. The second from the last joint was being lowered slowly when it began to take weight, indicating it was setting on fill material. The casing was picked up to replace the top single with a joint that had the circulating head connected to wash and circulate the casing down. The top of a casing collar hung on a rotary beam and pulled the thread out of the collar. The pipe dropped so that the top was about 40 ft below ground level. There was no reasonable explanation of how the collar could catch on the beam, but the net result was a 20-in. casing fish in the hole with a collar looking up.

Two joints of casing were picked up, connected, and lowered to try to screw back into the casing. The pin would set on top of the casing collar but would not connect. Apparently, the top of the collar had laid over in a washed-out section, because the joint could not be rotated into the collar of the fish and several times it dropped off and went down beside the fish for a distance of about 5 ft. The two joints were pulled out of the hole. A guide skirt was built and welded onto the bottom joint and the joint was run into the hole. The casing pin was screwed into the collar after fishing for several hours to work the shop-built guide skirt over the fish.

The two joints were screwed together with a minimum torque. The casing was then picked up enough to place the connection under a small tension load and ensure the casing string was reconnected and to hold it straight in the hole. A very light tension was used because the threads in the collar were partially stripped and the weakened connection would part at a higher tension. Circulation was established by pressuring the casing slowly and then gradually increasing the pump rate. The casing was cemented by circulating cement to the surface in the normal manner but without moving the casing. A low pressure was applied after bumping the plug and it held satisfactorily, indicating that the casing was intact. After nippling up, the first bit was run slowly to prevent sticking in case the casing was partially collapsed and passed freely. The surface hole was drilled, cased, and cemented without problems due to the repaired conductor casing.

SURFACE CASING

Surface casing or deep conductor is the first basic casing. The casings are similar and the names are sometimes used interchangeably. These casings hold back poorly consolidated formations, contain moderate pressure, and support all subsequent casing strings.

The most common shallow-surface casing failures are loose casing at the surface and leaking either around the casing or through a hole, usually due to wear. These failures are often attributed to incorrect cementing. Always consider placing cement down and outside of the surface casing after primary cementing. The top of the casing may vibrate and work loose if the blowout-preventer assembly is not anchored correctly. Misalignment occurs to a lesser extent. Other common surface casing problems include casing wear at or near the bradenhead and buckling.

Loose at the Surface

Casing may become loose at the surface. This usually is the first string of casing that connects to the bradenhead. The most common cause is not cementing the casing properly to obtain good cement at the surface. In some cases the wellhead may be nippled up too early and not properly secured; therefore, if operations are started early, the movement of the drill tools may cause the casing to loosen.

It is a common practice to cement the casing at the surface separately, even if the primary cement is good. This includes running about 50 ft of 1½-in. or 2-in. pipe down and outside of the casing, then cementing it with good slurry. In many cases this is done as a standard practice. Another procedure is to dump 5 or 10 sacks of dry cement around the casing and tamp it down, adding a small amount of water as necessary.

The casing may become loose after drilling for some time. This generally is caused for the same reasons as the loose casing described earlier. The best procedure is to fill the annular space with graded, rounded pea gravel and add to it periodically. This often is preferable to recommitting down the outside of the casing. Pea gravel appears to work better than sand or crushed rock. Filling with dry cement and tamping normally is not recommended at this time.

Leaking Around the Surface

One type of failure is a leak around the outside of the surface casing. The most common cause is not cementing the surface casing correctly. A leak also may be caused by fractured or unconsolidated shallow formations such as gravel and sand. In this situation, always suspect a hole in the casing.

It may be possible to continue operating with a small leak that does not erode around the base of the rig. Often, as drilling continues, the leak will plug later with bit cuttings. If the leak is larger and continues to flow, add fresh gel and low concentrations of lost-circulation material such as coarse fiber to the mud. These materials often plug the leak.

Another procedure is to spot a lost-circulation pill. Run the pipe with a bit, fitted with large jets or with jets removed, to the casing seat. Spot the lost-circulation pill and bradenhead squeeze with a very low pressure, taking care not to break down the casing seat.

If the leak is more severe, it may be necessary to plug it by cementing. Normally, a second primary cement should not be tried because of the open hole. Bullhead the cement and move it very slowly through the open hole. Some of the cement may seep or flow into the point of leakage. Drill it out after the cement reaches an initial set and resume operations.

Another method is to cement again or squeeze. Set a bridge plug, barite pill, sand bridge, or cement plug in the lower surface casing, usually about 50 ft from bottom. Perforate the casing above the plug and establish circulation. Then cementing can be done with a primary-type cementing job, often by bullheading.

If circulation cannot be established, squeeze the perforations. Use common, fast-setting cement and squeeze, preferably by bullheading for shallower casing, otherwise with a retrievable packer. Squeeze by stages to a moderately low-pressure equivalent to 1–2 lb/gal less than the fracture pressure at the depth of the perforations. Hold pressure on the cement until it reaches an initial set to prevent the cement from flowing back into the casing. With standard precautions, an operator can prevent sticking or cementing the tools into the hole. This is the advantage of bullheading or bradenhead cementing.

Wear at or Near Bradenhead

If casing wear becomes excessive, a hole may be worn through the casing and into the slips in the bradenhead. This creates the risk of a blowout, which is always a severe situation. For example, an offshore well had a shallow casing leak that could not be repaired and had to be controlled by drilling a kill well.

Normal drilling immediately below the surface casing is conducted at moderately high rotary speeds, usually with a relatively light drilling weight. As a result, the lower end of the kelly and the upper drillstring may whip or wobble, causing casing wear. A bent kelly, especially at the lower end, contributes to and accelerates the wear. Other common causes are misalignment, a deviated or crooked hole at shallow depths, and incorrectly hard-banded drillpipe. A wear bushing in the lower part of the preventer can reduce or eliminate the problem, depending on the type of preventers and the casing and drillpipe sizes.

A strong indication of excess wear is when the bit or flat bottom tools hang up while going into the hole, in some cases requiring rotation to drop through. Other indications are excess metal cuttings in the mud or minor to severe lost circulation and/or mud leaks at the surface. If conductor casing is not used or has been cut off below the surface casing, mud may leak and flow around the base of the rig. If the bradenhead is installed on the conductor casing and the surface casing landed in a casing spool, a hole may be indicated by lost circulation. If the surface casing is cemented to the surface, the only evidence of a hole is the bit hanging up. Deeper leaks may be detected by pressurizing above a plug or packer seated at various depths.

A hole may be detected at shallow depths by a visual inspection. One method is to pull the last two stands of drill collars without filling the hole, leaving the mud level below the bradenhead. Lower a light into the hole on a stout piece of string or soft line. If the casing is worn or there is a hole completely through the surface casing at or near the bradenhead, it usually can be detected. Otherwise, it may be necessary to disconnect and lift the preventers to inspect the area.

Casing and Liner Repair

The type of repair depends upon well conditions. Generally, squeezing and/or welding a casing patch is not an option. In a few cases, such as in conductor or shallow surface casing, it may be possible to drill ahead with caution and run the next string of casing early. Otherwise, repair by replacement.

Set a drillable or retrievable plug, depending upon the situation, as deep in the casing as reasonably possible. This reduces the problem of pressure buildup below the plug when it is pulled. Pressurize the casing to verify the hole. Pick up the blowout preventers, and hang them on chains or remove them subject to space. Excavate a work space below and around the bradenhead.

If a regular bradenhead is damaged and an inner casing is not landed on it, cut it off and replace it with a weld-on bradenhead, using a short nipple for spacing if necessary. If an inner uncemented casing, landed on the bradenhead, is damaged, release it from the bradenhead and then use a casing spear to back off a joint. Run and connect a new joint and land it in the bradenhead. If both strings are damaged, back off a joint of the inner casing, replace the bradenhead, and then replace the inner casing. Take all necessary precautions to prevent a fire and use approved welding procedures.

It may be necessary to remove a longer section of cemented casing if it is worn. Mill out the damaged casing with a section mill, then mill off the top with a long washover shoe for a short distance to allow room for a casing bowl. Then replace and reconnect the damaged casing with a casing bowl. If there is an inner string of casing, pick it up with a casing spear to release it from the bradenhead. If the casing is stuck, use the procedures described in the section "Releasing Casing from a Bradenhead or Hanger Spool." Set it down and repair or replace the bradenhead, stripping it over the inner casing if necessary. Then land the inner casing. These procedures are seldom required near the surface.

The repair procedure for badly worn surface casing near the bradenhead is illustrated in the following example.

Repairing a Leak at the Bradenhead

A well had 13 3/8-in. conductor set at 600 ft and 9 5/8-in. surface casing set at 2,500 ft and landed in the bradenhead, which was screwed onto the 13 3/8-in. conductor casing. The surface casing was cemented, but not circulated to the surface. The well was drilled to 7,500 ft, and a fish was lost and sidetracked after an unsuccessful fishing job. Another fish was sidetracked at 9,000 ft and the well was drilled down to about 13,000 ft with numerous problems. After about $2\frac{1}{2}$ months of operation, there was some evidence of constant but low volume lost circulation, and several new bits hung up near the bradenhead. Visual inspections were inconclusive, but there did appear to be a hole through the surface casing into the bradenhead.

A retrievable bridge plug was run on a stand of drillpipe, set, and the rams closed. Mud was pumped into the hole at less than 100 psi. The plug was released, run to 2,000 ft, set, and the drillpipe pulled. The preventers were removed to inspect the surface casing. The 9⅝-in. surface had a hole about 4 in. wide and 10 in. long, worn completely through the surface casing and about halfway through the casing slips.

The surface casing was caught with a casing spear, but it did not release with a pull of 250,000 lb because of the slips stuck in the casing hanger. Hammering on the flange was ineffective. A bell hole was dug about 4 ft below the bradenhead, the 13⅜-in. conductor was cut, and a 2-ft section was removed. The casing slips were released by bumping down on top of the casing hanger flange with a stand of drillpipe while tension was held on the 9⅝-in. casing with a casing spear. The bradenhead was cut off and removed. A ¾-in. pipe was run to about 200 ft in the 9⅝-in. by 13⅜-in. annulus and the annular space was cemented while tension was held on the 9⅝-in. casing.

The damaged 9⅝-in. casing was cut off and a short nipple was welded on it. A nipple was welded on the 13⅜-in. conductor and a slip-on bradenhead was welded on the nipple. The 9⅝-in. casing was landed in the bradenhead. The preventers were nippled up, a wear bushing was installed, and the casing was tested to 1,500 psi. The bridge plug was recovered and operations resumed without further problems.

The retrievable bridge plug was set at 2,000 ft so if the well pressured up during the repair, there would be enough drillpipe in the hole to release the plug and control and kill the well if necessary. The kelly may have had a slight bend that contributed to the wear, but it had a rubber protector on bottom. Normally the centralizing effect of the kelly drive bushing centralizes the kelly. The substructure had settled slightly to the side opposite the failure, but this should not have been a major cause of the failure. There were about 6 million revolutions. The main cause of the failure was the lack of a wear bushing.

Overall Wear

Casing wear is most common in the lower sections of deeper surface and intermediate casing. Wear is difficult to detect before it causes a problem except by running a casing inspection log. Metal particles on flow-line magnets may indicate wear.

Surface casing is especially subject to wear because of the extended drilling time and high weights. Drillpipe movement causes most casing wear and incorrect hard banding accelerates it — especially coarse, large-grain-size hard banding, Use smooth, small, or very fine grain-size in 4-in. or wider bands. Other causes of wear include deviated and crooked holes, bent or crooked drillpipe, abrasive muds, excess milling and incorrect milling practices, and unrubbered drillpipe.

Always run drillpipe rubbers on deeper casing strings, conventionally on top of every second joint. Run double rubbered drillpipe, with a rubber on top of every joint, where severe wear is expected, such as in crooked and deviated hole sections. Do not allow rubbers to run into the open hole while drilling (Fig. 7–19).

Wear leads to many failures such as leaks, splitting, and parting. If liners are set below these casings, ensure that the casing is designed to support the weight of the liner. Often the casing tends to wear near the bottom; therefore a longer liner overlap of 300 ft or more should be used to ensure covering these worn sections and seating the hanger in good casing. Casing wear can cause severe problems, as illustrated in the following example.

Recovering Worn, Parted Surface Casing

Conductor casing, 13 3/8-in. OD, 54.5 lb/ft, ST&C, was set at 100 ft and 8 5/8-in., 24 lb/ft, 8 rd, ST&C, K-55 surface casing was set in a 12 1/4-in. hole at 2,500 ft. The hole was drilled to 5,000 ft and, on a trip to change bits, could not be pulled up past 850 ft. It appeared that the 8 5/8-in. casing had parted or split and might have collapsed. The bit stopped at the same point every time it was raised. If the casing had collapsed, then pulling the bit or the top of the drill collars into the collapsed section should have caused a sticking action, requiring extra weight to release the pipe. However, the bit would go back downhole without any evidence of sticking, regardless of how hard it had been pulled upward.

A heavy barite mud pill was spotted in the open hole to overbalance the underbalanced mud, and the drillpipe was set on bottom and backed off at about 100 ft. The preventers were nippled down, and a casing spear was used to try to pick up the casing. The casing slips were stuck in the bradenhead and cut free with a torch. The preventers were nippled up.

A casing spear recovered a 50-ft piece of casing, including the 10-ft piece cut off in the casing slips and one full joint with split and broken threads on the bottom. The casing spear was rerun and would not enter the casing. An impression block was run, but the impression was indistinct.

Figure 7-19 Double rubbered drillpipe through a dogleg

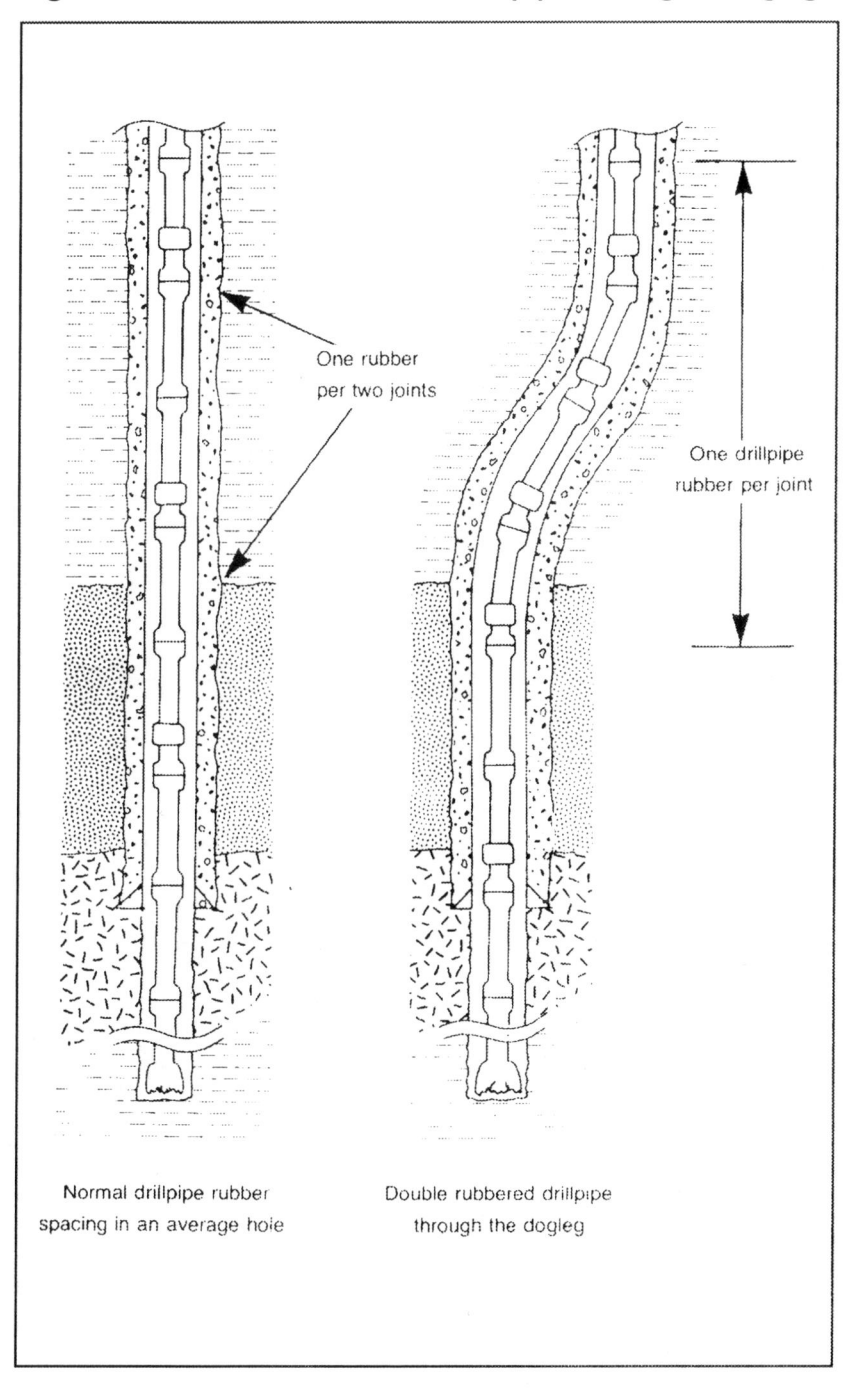

Normal drillpipe rubber spacing in an average hole

Double rubbered drillpipe through the dogleg

One rubber per two joints

One drillpipe rubber per joint

Casing and Liner Repair 503

About 6 in. was milled off the top of the 8⅝-in. casing fish with a 10¾-in. flat-bottom mill. Drillpipe was run and screwed into the drillpipe fish. At this time, the drillpipe had been sitting on bottom for 48 hours. It was worked for about 10 minutes with pressure until circulation was established, and then it moved freely.

The assembly was started out of the hole. It had about 10,000 lb of extra weight, indicating some casing was riding on the drillpipe. It was pulled up until the 8⅝-in. casing hung in the bottom of the preventers. The preventers were picked up to guide the casing through. About 600 ft of casing was stripped out and cut off where it had hung on the drillpipe. The bit was run to 8,300 ft and circulated. When it was pulled, about 280 ft of casing was recovered, hanging on the drillpipe.

The next 11 days were spent running various tools, including flat-bottom mills, taper mills, string mills, overshots, and spears. The hole was cleaned out to the top of the good casing at about 1,100 ft, and apparently to the top of the cement outside the 8⅝ in. casing, as indicated by cuttings. A full-gauge bit stopped at about 1,100 ft, and the drilling action indicated the bit was running on junk. The bit sidetracked and drilled about 20 ft before this was detected.

Most of the prospective productive zones were drilled in the original hole so the decision was made to clean out to the original TD and run 4½-in. completion casing. A 6⅛-in. tapered mill was run on the bottom of one stand of drill collars with a 6½-in. tapered mill on top of the drill collars. It was run to 1,100 ft and then drilled with light weight until it broke through into the original hole. The assembly was then run to TD, the hole was conditioned, and the drillpipe was pulled. Completion casing, 4½-in., was run to TD and cemented. Open-hole logs were not run because of the possibility of sticking. The formations were evaluated with cased-hole logs and the well was completed.

Consideration was given to running a string of 8⅝-in. casing after cleaning the hole out to 1,000 ft. However, the condition of the casing was very questionable, so the well was completed. The casing wear occurred while rotating about 600 hours and 1.5 million cycles. The well was drilled with unrubbered drillpipe that had relatively new smooth-type hard banding. There was no evidence that the uncemented casing contributed to the wear.

If there is a question about casing wear, run a base casing inspection log before drilling out of the casing. Then run the inspection log again periodically and compare it to the base log to evaluate wear.

A major problem can be caused by setting casing in a crooked hole. After the casing is run, the crooked hole section becomes a point of potential failure. The casing will wear excessively with resulting casing leaks and possible parted casing, as illustrated in the following example.

Casing Leak at 9,000 ft

Intermediate 9⅝-in. casing was set at 15,000 ft in a 12¼-in. hole. A deviation survey showed a crooked section at 9,000 ft, which had an absolute dogleg over 5°. Two fish were lost and sidetracked in this interval during drilling. A base casing inspection log was run since this was an obvious point for excess casing wear. Precautions taken to prevent casing wear included running double-rubbered pipe through the section, adding about 6% oil to the water-base mud to improve lubricity, and running the minimum number of drill collars to reduce the string weight.

After the hole was drilled to about 19,000 ft in 3 months, a casing leak was evidenced by minor loss of circulation and some tendency for the bit to hang up at 9,000 ft on trips in the hole. A casing inspection log indicated a hole in the casing. A retrievable bridge plug and packer was run. The bridge plug was set near the bottom of the intermediate casing, and the lower section of casing was tested at 1,000-ft intervals up to the point of failure. The packer was reset above the casing leak and it was confirmed with a leak-off test. The casing was squeezed to the equivalent of 15.5 lb/gal, cleaned out, and tested to the same pressure. The bridge plug was recovered, and drilling operations resumed.

After drilling to 21,000 ft, the casing was tested again and found to hold the equivalent of 15 lb/gal with a small leak-off at the equivalent of 15.5 lb/gal. Since the leak was both small and deep, the casing was not squeezed again. The well was completed satisfactorily without any problem due to the leak in the intermediate casing.

The failure was obviously caused initially by the dogleg in the intermediate casing hole, which in turn was caused by deviating too sharply to sidetrack the fish. This could have been avoided by sidetracking higher, possibly controlling the direction of the second sidetrack to minimize the dogleg, and extensively reaming after sidetracking. However, the casing wear was handled correctly after it was detected as a potential problem.

This failure can be analyzed by considering the approximate forces applied over three months to the failed section while drilling from 15,000 ft to 19,000 ft (Fig. 7–20). The drillstring assembly weight suspended below 9,000 ft was about 139,000 lb in 11.0 lb/gal mud. This figure assumes one-third of the initial weight at 15,000 ft and two-thirds of the final weight at 19,000 ft, reducing by 50,000 lb the weight on the bit for about 70% of the time. This interval was drilled in 3 months with about 5 million cycles. The lateral force exerted by drillpipe on the casing wall over an interval of about 100 ft at an angle of 5.75 deg is about 13,000 lb.

Casing and Liner Repair

Figure 7-20 Force on a dogleg

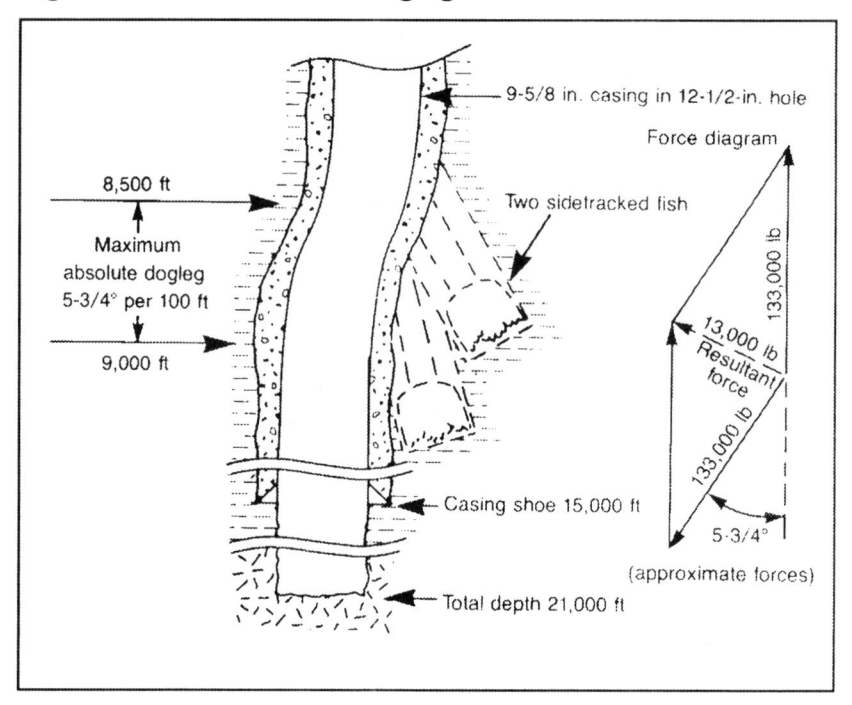

Casing wear causes leaks as illustrated in the following example.

Casing Leak at 13,000 ft

Deep 9⅝-in. surface casing was set at 6,000 ft and an 8¾-in. hole was drilled to 16,000 ft. A fish was lost and sidetracked at 13,000 ft. Intermediate casing, 7-in., was set at 16,000 ft in a normal manner.

A 6¼-in. hole was drilled to 18,000 ft, when a leak was suspected. The leak was located at about 13,000 ft in the 7-in. intermediate casing and was squeezed conventionally. It was squeezed again during deeper drilling. Otherwise, the failed section did not cause any additional problems.

The average weight of the drilling assembly below 13,000 ft was 54,000 lb. Assuming an angle of 4°, the lateral force exerted on the casing by the drillpipe was about 4,000 lb. The section from 16,000 to 18,000 ft was drilled in about 2½ months with 4 million cycles.

These examples indicate the length of time that drilling can be conducted inside casing under certain conditions before a failure occurs. They also indicate the efficiency of repairing by cementing.

Buckling

Buckled casing can be a severe problem and may cause either a lost hole or restricted drilling depth. It may buckle if it does not have sufficient strength when heavy intermediate and/or production casing strings are landed on it. Shock loading also may cause collapsed casing. Install a casing support plate when heavy casing loads are expected, and ensure that it is installed properly, usually sitting on cement. Be sure the surface casing is well cemented and to the surface or ground level when using a conductor.

Casing often buckles near the deepest point of uncemented supporting casing, usually at a point of maximum loading. This point is at or near the surface for surface casing cemented to the surface. Worn surface casing is more susceptible to buckling because of lost strength due to reduced wall thickness. It normally buckles at the point of highest wear. Occasionally, two concentric casing strings collapse, but this is rare.

The best procedure for repairing buckled casing in an uncemented section is to pull the free casing, replace it and run, and connect it to the cemented casing with a casing bowl. Then run a primary-type cement job. The procedure is similar to repairing parted casing as described in the section "Parted Casing." Less severe buckling may be repaired with a casing roller to restore the hole diameter. Severely buckled casing in a cemented section may be milled and opened to full gauge, followed by squeeze cementing. However, this leaves a weak casing section.

When there is an option, cover the buckled section by running either another string of casing, a stub, or a tie-back liner. The potential for buckling in these cases emphasizes the need to use good casing cementing and landing practices, surface casing supporting plates, and precautions to prevent wear.

INTERMEDIATE CASING

Intermediate casing is the second most basic casing. It cases off shallow formations and holds higher pressures that are encountered in the deeper formations, frequently in transition zones. In a few cases two intermediate casing strings may be used. Shallow intermediate casing is subject to the same types of failures as shallow surface casing and are repaired similarly. Deeper intermediate casing failures are more like those in completion casing and are also repaired similarly.

Failures in intermediate and deeper casing strings are more severe than in shallower casing strings. They are harder to repair because of depth, smaller hole size and generally more hostile environment. Abandoning a deeper intermediate hole is more costly compared to shallow holes because of the higher investment. Thus, there is a higher risk in repairing intermediate casing failures, and risk increases with depth.

Most failures common to other casing strings also occur in intermediate casing. Casing wear is common, often due to the longer drilling time. There is a higher incidence of parting in deeper casing strings. Some of the causes are heavier casing loads, longer operating periods, and more difficult operating conditions. Collapsed casing is more common, often caused by flowing formations, improper squeezing or testing procedures, and inadequate casing-string design. Burst casing is less common but may be caused by higher treating pressures due to greater depths. Releasing casing from the bradenhead or casing hanger spool for repairs may be difficult.

Some repairs are straightforward, and others require considerable ingenuity. The difficulty of repairs increases with depth. Always consider not repairing the casing if it does not interfere with operations. It may be possible to repair it later, maybe by covering it with another casing within a short time. However, this is seldom the case.

Parted Casing

Parted casing is one of the more serious casing failures. It is common to all casings but occurs more frequently in intermediate and completion casing. It is a higher risk failure because of the possibility of losing the hole while making repairs.

Parted casing occurs most commonly while running casing or while working stuck casing, and is often due to improper handling. In the latter case, the failure usually is near the surface because of overloading. Parting is more likely in uncemented sections. Other causes of failure include casing wear, landing with excess tension and not down-grading burst, and collapse for tension.

It may be possible to repair the casing in place by cementing. This procedure is seldom used except on conductor or shallow-surface casing where pressure is not a problem and also when the parted casing does not interfere with operations. An example is stabilizing a loose casing shoe joint(s) with cement.

The following summary repairs are subject to modification depending upon the specific well conditions.

While Running

Casing that parts during running is repaired by one of various procedures, depending upon the situation. The parted section frequently drops. It may be possible to lower the casing and reconnect it to the dropped casing. If the casing is free, pull it out of the hole and repair the failed section. Consider cementing if the dropped casing is in good condition and the formations may prevent running it back or may cause serious problems. Cement by primary methods if the casing circulates. If circulation cannot be established, run a stuck-pipe log to locate the top of the stuck pipe. Perforate above this, establish circulation, and cement by primary methods. Then squeeze the casing shoe securely to fix it in place and higher in the stuck section as needed.

Another alternative is to pull the free casing and recover the lower parted section by fishing. In the simplest case, catch the fish with a casing spear, work it free, and recover it. A pack-off spear may be used to establish circulation so the fish is easier to recover. This is seldom successful on a heavy casing fish over about 500 ft long. If the casing fish is recovered, clean out the hole and run a new string, or inspect and rerun the casing. Follow these steps with primary cementing. Casing and hole sizes generally prevent washing-over. Backing off or cutting and recovering part of the fish are seldom options. The casing fish may be sidetracked if it is severely damaged by dropping.

If the lower casing cannot be recovered by fishing, then clean out the hole and run casing to the top of the fish. Connect by screwing-in or with a casing-bowl connection, whichever is applicable. For the screw-in connection, run the casing and screw-in, possibly with a screw-in nipple. Establish circulation, and cement by primary methods. If the casing does not circulate, cement with the procedures described for casing reconnected by screwing-in.

For a casing bowl connection, it may be necessary to mill a new top. Use a flat-bottom mill with an inside guide. Then clean out around the top of the fish with a long-skirt, saw-tooth mill to provide clearance for the casing-bowl catch if necessary. Run the casing-bowl below the casing and connect it to the fish. This includes going over the fish and engaging it with slips. Then pull tension to ensure that the slips are holding and to collapse and close the seals if they were run. Run the seals if prior fishing indicated that the lower fish could be circulated. Establish circulation and cement by primary methods. If circulation was not established during fishing, do not run the seals. Establish circulation through the casing bowl and cement by primary methods. Then squeeze the lower section as needed, usually below the casing bowl, and ensure that the shoe joints are well cemented to prevent backing off.

In a few cases, casing parted during or shortly after may be repaired by using an alignment tool as described in that section.

Casing and Liner Repair

In an Uncemented Section

Repair casing parted in an uncemented section by one of various methods, depending upon the specific conditions. In most cases at least the lower section of casing is cemented in place. If the upper section of casing can be recovered, then set a cement or mechanical plug in the lower casing to secure the hole.

If the upper casing cannot be recovered, then pressurize to establish circulation. If circulation is established, cement by primary methods. Then block-squeeze below the parted section, and follow by squeezing the parted section. If the upper casing cannot be circulated, then repair by squeeze-cementing. Perforate and block-squeeze below and above the part. In either case, squeeze the open formation, sometimes with multiple stages, to obtain a satisfactory squeeze pressure. Ream the parted section with a tapered mill to ensure a full inside diameter if necessary.

Leaving the parted casing open, even after squeezing, may not be acceptable because of conditions of the hole or other conditions. The best repair is to cover the parted section with another string of casing, tie-back casing, or stub liner if the option is available. Another alternative that may be available is to recover all casing that is inside and for some distance below the larger diameter upper casing, and sidetrack. Otherwise plug, abandon, and redrill the well if justified.

Parted, Uncemented Casing

Surface casing, 13 3/8-in., was run and cemented at 100 ft; an 11-in. hole was drilled to 2,800 ft and 8 5/8-in., and 24 lb/ft, K-55, ST&C intermediate casing was run and cemented, without returns during the final displacement. Circulation was lost while drilling at about 1,800 ft. Since no lost circulation occurred after running and cementing casing, it was assumed the casing was satisfactorily cemented.

A 7 7/8-in. hole was drilled to 7,000 ft, and the mud weight gradually increased from 9.3 to 9.6 lb/gal. A slow, steady mud loss began, which was uncommon for the area. A bit had to be worked through a tight place at 120 ft during a trip. A casing inspection log verified 4 ft of casing separation at 120 ft.

A 50-sack, 150-ft-long cement plug was spotted in the casing at 1,800 ft. The preventers were nippled down, and a casing spear was used to pull three joints of casing from the hole. The casing was badly worn, and the bottom joint had a Dutchman in the collar, leaving a broken thread looking up (Fig. 7–21).

Figure 7–21 Repairing 8⅝-in. casing

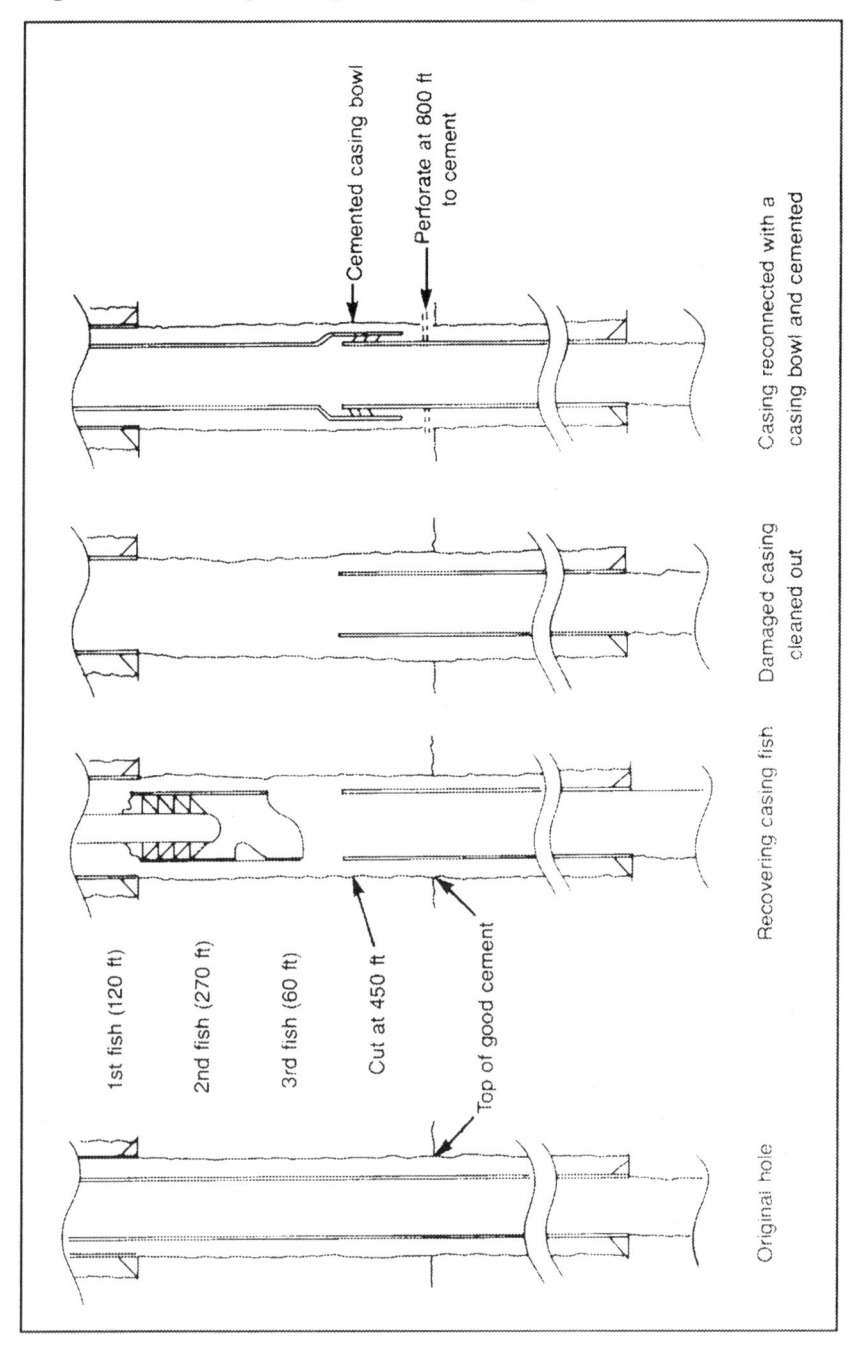

A mechanical cut was attempted at 130 ft. The cutter was pulled with extra string weight, and about 270 ft of worn 8⅝-in. casing was recovered. A casing spear was run to test the strength of the casing, and it held a 40,000 lb pull. A log located the top of the cement at 1,100 ft. The casing was cut at 450 ft, and the cut piece was recovered with a casing spear.

A seal-type casing bowl was run on 8⅝-in. casing. It was latched to the 8⅝-in. casing in the hole, and the seals collapsed with a pull of about 20,000 lb. Four holes were perforated at 800 ft, and the casing was circulated and cemented with a primary-type cement job. The casing was landed, BOPs were nippled up, and operations were resumed.

The failure occurred after drilling through the casing for 330 hr with 1.6 million cycles. The 8⅝-in. casing was set in a straight, vertical hole with less than 1° of deviation and 1°/100 ft dogleg. The drillpipe was hardbanded with material that reportedly would not damage the casing. Drillpipe rubbers were not used because the smooth, hard facing was not supposed to damage the casing.

The failure was not caused by a crooked hole or bent drillpipe. The recovered casing had relatively smooth, consistent wear. The casing failed in tension when the cross-sectional area was reduced by wear until there was insufficient metal to support the lower pipe section. The cause of failure was the smooth, hard banding on the drillpipe. This was the second, similar problem in the field with this type of hard banding, so it was not used again. Some consideration was given to uncemented casing causing the failures, but this type of problem did not occur in a number of wells with similar programs.

In a Cemented Section

Casing parted in a cemented section can be a severe problem and difficult to repair. The best repair is to cover the failure with another casing.

In a few cases the upper section may be free a short distance above the parted section. Subject to well conditions, part and recover the free casing. Then mill out the remaining parted casing and repair it as described for casing parted while running. This procedure is seldom used because of the difficulty of milling long sections of casing.

If the upper part of the intermediate casing is not cemented, it may be possible to part or back it off and sidetrack. This has the advantage of not reducing the hole size but causes increased cost and the loss of casing and hole.

The failed section can be squeezed, in many cases to a relatively high pore pressure, by a combination of block- and multistage squeezes. Another alternative may be to squeeze as noted and later cover the failure with production casing. If the intermediate casing is used as production casing, then the parted section may be packed off, depending upon when the failure occurs and the relative size of the two casings.

Otherwise plug, abandon, and redrill the well if justified.

Alignment Tool

Parted casing may be repaired in place with an alignment tool. The problems of running and cementing large-diameter, deep intermediate casing and later repairing a deep parted section with an alignment tool are described in the following example.

Repair Parted Casing with an Alignment Tool

A deep well had 30-in. conductor casing at 50 ft, 13 3/8-in. surface casing at 6,000 ft in a 17 1/2-in. hole, and a 12 1/4-in. hole drilled to 16,000 ft into a pressure transition zone. Intermediate, combination casing included 10 3/4-in. 55.5 lb/ft, ID 9.760 in., drift 9,604 in. and 60.7 lb/ft, ID 9.660 in., drift 9.504 in. with special flush joint-type couplings, fully inspected at the yard and again at the wellsite. It was run in 10.4 lb/gal, 47 sec, 4.4 cc mud. The casing was partially floated in and tagged bottom at 15,900 ft, indicating 100 ft of fill.

The casing did not circulate freely. It was worked slowly up and down a distance of about 10 ft while circulating at a reduced rate with about 1,500 psi. It began dragging heavily with somewhat reduced weight due to the pressure and plunger effect. This indicated cuttings settling around the pipe. It was set at the neutral point, and the pump pressure was reduced to about 600 psi. As the pump pressure and pipe weight began to stabilize, pressure was increased gradually to about 1,500 psi. This was the maximum pressure attainable without beginning to pressure-up and lift the casing with the plunger effect, restricting the mud flow.

The original cementing design provided for a circulating pressure of 2,000 psi in order to displace the cement and to allow about 1 hour before the cement began to set up. The circulating pressure of 1,500 psi did not allow for enough extra pumping time. There was no time to blend additional additives and some question of how long the pipe would continue to circulate at the 1,500 psi rate.

The cement program included high-temperature-resistant, highly retarded cement followed by retarded class G to tail in around the casing shoe. The lead slurry was mixed at a slightly lighter weight to increase pumping time. The slurry was mixed and displaced, and the plug was bumped with a low pressure. While preparations were made to land the casing, it parted at about 10,000 ft, failing at a safety factor of 4.3. The cause of the failure was unknown. After parting, the upper section of casing circulated and moved freely. A small amount of cement was circulated out, indicating that the lower section was fully cemented.

Casing and Liner Repair

There were several apparent alternative actions. One was to pull the free section and run a joint with a new pin to screw back in. Another was to connect with a casing bowl if the threads were parted. Either gave a strong connection, which was important since the well was planned as a deep well with over 15.5 lb/gal mud. This would require considerable time, including laying down and picking up the casing and probably substantial casing connector damage. Also, problem formations, in the worst case, could prevent running the casing to bottom, possibly causing a lost hole.

Another alternative included setting the casing down carefully, rotating slowly, and reconnecting it in case the casing had jumped a collar. This was eliminated because coupling failures almost invariably occur in the first thread at the critical thread area. It would still be necessary to perforate and squeeze since there was little chance of unscrewing and displacing cement and then screwing the casing back together. Also, regardless of how carefully the casing was set down, the lips of the casing could be damaged and might prevent running a close-tolerance alignment tool.

The upper section of casing could be set down on the lower section and cemented. If the upper casing were set down directly, it might not align with the lower section. After cementing, the failed section should be opened to full gauge with tapered reamers, followed by testing and squeezing as necessary.

The upper section of casing could also be set down using an alignment tool to align the two sections, followed by cementing and squeezing as necessary. This would allow a good chance of almost metal-to-metal contact. Reaming would be minimal. Both of these advantages would help improve the integrity of the repaired casing and the chances of squeezing and testing the point of failure to a higher pressure. The repaired casing might not be as strong, but the risk of losing the hole was almost negligible. The repair cost would be less.

The alignment tool used was a cement retainer with the check valve removed and an aluminum alignment section on-bottom. A 20-ft section of heavy-wall 2-in. aluminum pipe was screwed into the bottom of the retainer. A piece of 8-in. OD heavy-wall aluminum pipe was slipped over the 2-in. and centralized and held in place with vertical ribs. The lower end of the 8-in. pipe was orange peeled, tapered to a point, and heliarc-welded to the 2-in. pipe. The 2-in. by 8-in. annular space was filled with a mixture of cement and plaster of paris to increase tool weight and rigidity. Sections of aluminum angle were welded to the outside of the 8-in. pipe, vertically in rows on 90° phasing, with each lower row overlapped 1 ft and spaced 45° laterally from the upper row. These angles served for alignment and as annular flow channels (Fig. 7–22).

Figure 7-22 Alignment tool

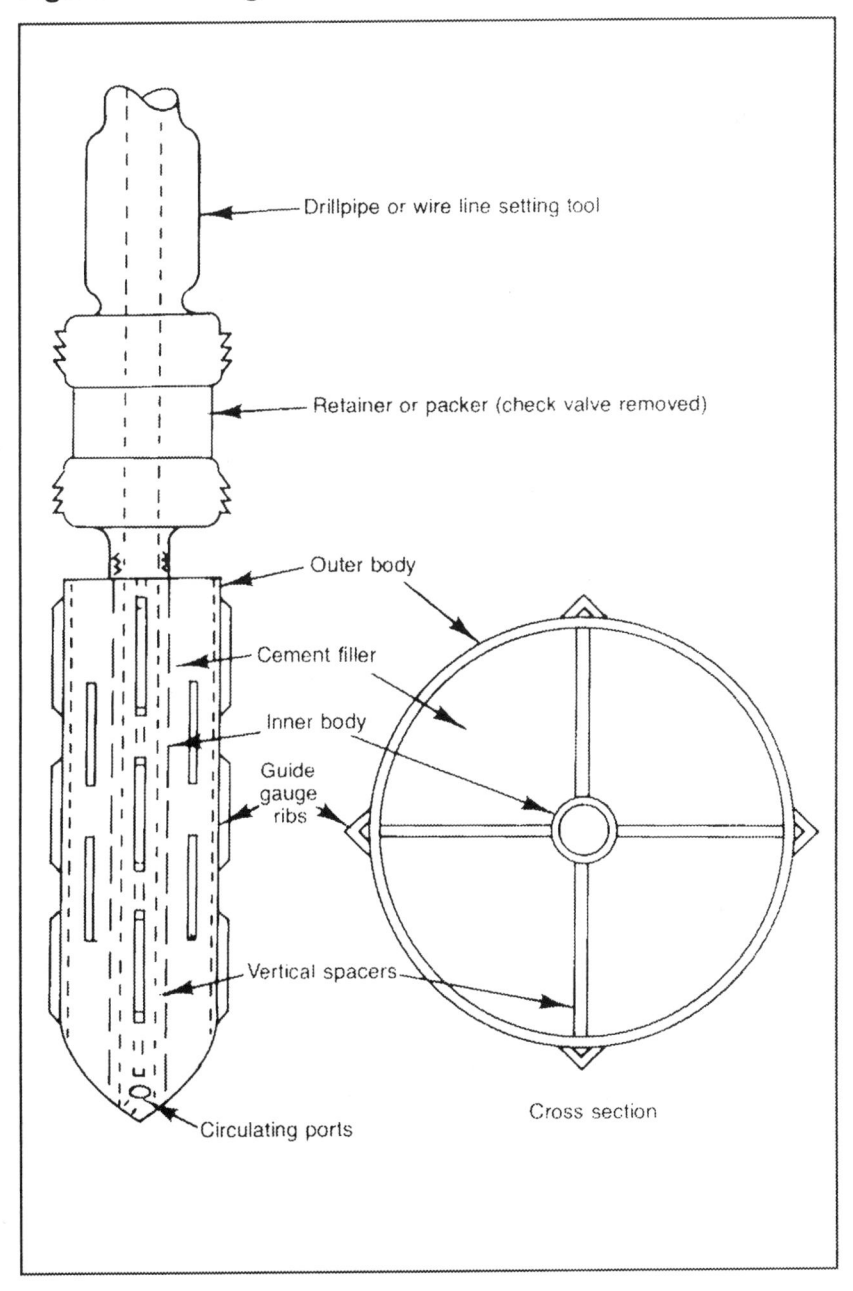

Casing and Liner Repair 515

The alignment tool can be positioned by two methods. One is to set the upper casing down on the lower casing with 5,000 to 10,000 lb of weight. Then run the alignment tool on drillpipe and set it with the midpoint of the alignment stringer located opposite the point of failure in the casing. This allows pushing the tool into the lower section of casing, especially if the parted sections are out of line. This procedure creates a problem of holding the casing. Use a slip-and-spider or side-door elevators set on top of the presenters, if there is space. Also there is the problem of moving the casing during cementing if needed and if well control is questionable.

The best procedure is to pull the free casing up about 20 ft above the lower casing. Run the alignment tool on a wireline and set the retainer with about one-half of the alignment stringer extending out the bottom of the free casing. Then set the casing down with 5,000 to 10,000 lb of weight. If the alignment tool sticks while running, there is little chance of pulling it free with the wireline. In this case, pull the line out of the rope socket. Lower the upper section of casing until the parted sections touch, and cement the casing. Displace the slurry by volumetric calculations and stop it with the slurry top well above the parted section. Hold pressure until it reaches an initial set to prevent flowback. Then drill out the alignment tool and cement, test, and squeeze as needed.

The hole was circulated slowly, and the casing was moved periodically while the alignment tool was being constructed. A 1 $^{11}/_{16}$-in. OD gamma-ray collar log was run inside the 10¾-in. casing to 15,600 ft, and the hole and casing collars were logged back to 6,000 ft. A 9.550-in. gauge ring was run to about 100 ft below the parted section.

The alignment tool was run on a retainer-setting tool, with a 3⅛-in. collar locator, to the bottom of the free casing without difficulty. But, the collar locator would not pick up or record the collars. This was because of the minimum amount of excess metal in the semi-flush joint collars compared to the casing body and because the collar locator was centered in the hole, not riding the inside edge of the casing like the smaller collar locator. The original plan was to use the collar locator and compare this with the collar log run earlier to position the alignment tool precisely in the upper casing. The alignment tool could have been run out of the end of the free casing so that the bottom of the casing could have been detected by the collar log, but this was not done because of the risk of not being able to pull the alignment tool back in the casing. The alignment tool was pulled.

A gamma-ray tool was run to reference the collars to the formation. Then the alignment tool was run with a gamma-ray tool and positioned without moving the casing. The casing was lowered and set on the lower casing with about 5,000 lb of weight. A collar locator verified the position of the alignment tool, and the stinger entered the lower casing correctly. The casing was then circulated (Fig. 7–23).

Figure 7-23 Repairing parted casing with an alignment tool
Note: The alignment tool in figure d was run on a wireline. It can run on pipe, but it is more difficult to position precisely, and there is the problem of holding the casing while running the pipe. If the tool is run on pipe subject to well conditions, the best procedure is to lower the casing until the ends are touching lightly and then run the tool on pipe and push it into place ensuring casing alignment.

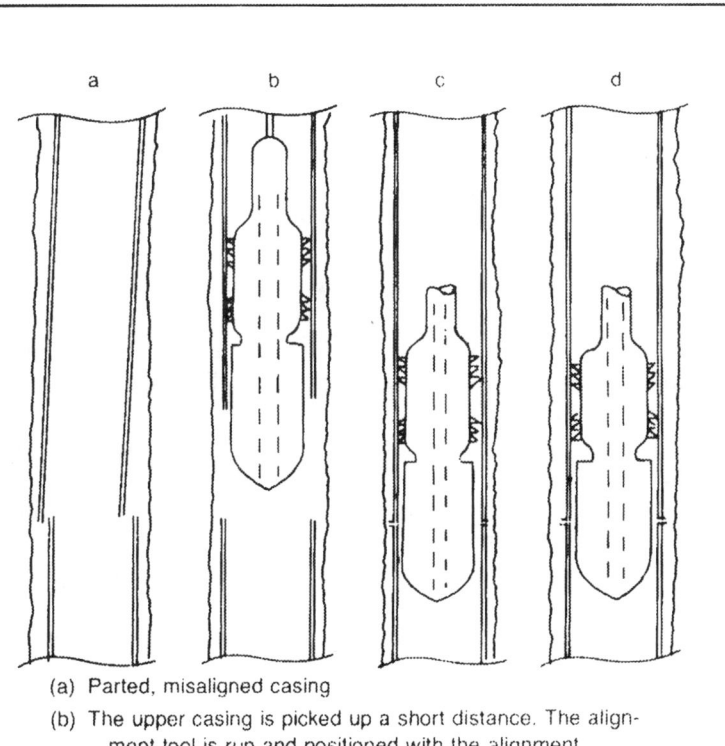

(a) Parted, misaligned casing

(b) The upper casing is picked up a short distance. The alignment tool is run and positioned with the alignment section, half in and half out of the casing. After checking the position of the tool, set the retainer or packer. (Check tool position with a collar locator, gamma log, etc.)

(c) Lower the upper casing, inserting the bottom half of the alignment tool into the lower casing. Circulate and condition hole and mud. Cement by volumetric displacement. If a wiper plug is used to separate mud and cement, allow for mud film accumulation in front of the plug.

(d) Repaired casing. Drill out cement, retainer, and alignment tool. Use a casing scrapper if conditions permit and make a run with a full-gauge tapered mill if necessary.

Casing and Liner Repair

The casing was cemented with lightweight cement, followed by class G, and was designed to fill the inside of the surface casing to 6,000 ft. The slurry was displaced based on volumetric calculations, with allowance for pump efficiency, and stopped with the top of the slurry about 500 ft above the failed section. A cement-wiper plug separated the cement and displacement mud to prevent displaced mud from bypassing and channeling through the slurry. After WOC, the casing was landed, and the preventers were nippled up.

The casing was tested and would not hold pressure. A retrievable cementing packer was set above the alignment tool. The annulus was tested and held 2,500 psi. Mud was pumped into the parted section at 6 BPM at 1,100 psi. The failure was squeezed (job 1) with 200 sacks, staged to 750 psi, and over-displaced 20 bbl, followed by squeezing (job 2) with 200 sacks to 800 psi and staged to 1,200 psi. The cement did not set properly, and after 8 hours, it was over-displaced with 800 psi. The casing was squeezed (job 3) with 200 sacks and staged to 2,500 psi. After cleaning out cement and testing the casing to 2,400 psi, the pressure broke back to 1,500 psi. The packer was rerun and the parted section was squeezed (job 4) with 75 sacks to 2,900 psi. A bit was run, and it cleaned out cement to 11,000 ft. The casing tested to 2,500 psi. The bit did not encounter any obstruction at the parted section. It was run in the hole and drilled the float collar. The entire string of casing was then pressure-tested to 2,600 psi.

The first stiff bottom-hole assembly took weight at the point where the casing had been repaired. It was pulled, and the section was reamed with a tapered, full-gauge mill. No further trouble was encountered in running tools through the section. A casing inspection log and a cement-bond log were run later. The casing was normal, and there was no indication of casing separation at the repaired section. The casing generally showed good bonding and was well cemented in the repaired section and in the lower hole. The parted section was exposed to over 16.0 lb/gal mud during later operations and did not cause any problems.

Loose Float and Shoe Joints

Loose float and shoe joints, collectively called shoe joints, are a potential problem common to all casing strings drilled through. One or more joints may back off and lie loose in the hole. Frequently they drop a few feet with separation between them and the upper casing.

Backing off one or more joints is very unlikely with the correct preventive measures. Backing off shoe joints often is blamed on drilling out the float collar, cement and float, or guide shoe, especially if drilling takes extended time. Sometimes the cause is attributed to the type of drilling assembly, including stabilizers, and roller or drag bits or mills. In most cases these are not correct if the equipment is connected properly.

Epoxy cement has replaced strapping or tack welding the float and shoe joints. Mix and apply it correctly. The joint must be clean and dry. Cement all threads below the top of the float collar or one joint higher, including pulling collars. Then tighten to specifications.

Cement the casing correctly with good slurry and use both top and bottom wiper-plugs. The importance of using a bottom plug often is overlooked. When the bottom plug is omitted (Fig. 7–24a) a film of mud may be left on the inner casing wall while circulating and cementing. The top plug, run behind the cement, wipes the mud film off the casing wall, and it accumulates ahead of the plug (Fig. 7–24b). The mud normally is deposited between the float collar and the casing shoe when the plug lands on the float collar. In Fig. 7–24c, the displacing cement is separated from the mud with a plug, and the bottom plug wipes away the film to ensure a good cementing job (Fig. 7–24d).

Figure 7–24 Cementing with one vs. two plugs

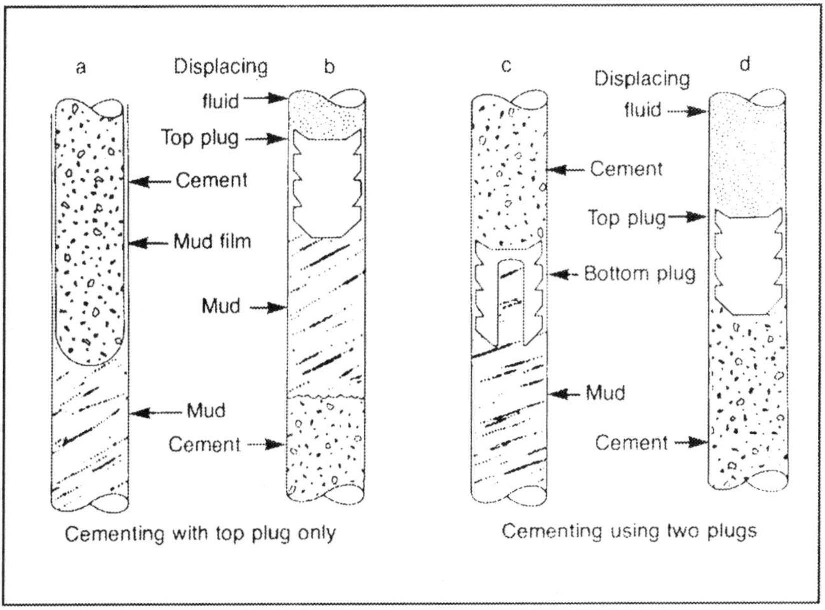

Cementing with top plug only

Cementing using two plugs

Casing and Liner Repair

Table 7-1 Volume of mud and feet of fillup for 10,000 ft of various sizes of casing

Casing OD, in.	4½	5½	7	9⅝
Casing wt, lb/ft	13.5	15.5	29.0	43.5
Casing ID, in.	3.920	4.950	6.184	8.755
Casing capacity, bbl/ft	0.0149	0.0238	0.0371	0.0744
¹⁄₆₄-in. layer, bbl	2.34	2.99	3.73	5.30
ft of fill	159	126	101	71
¹⁄₃₂-in. layer, bbl	4.72	5.96	7.46	10.58
ft of fill	316	250	201	142
¹⁄₁₆-in. layer, bbl	9.35	11.85	14.84	21.08
ft of fill	628	498	400	283
⅛-in. layer, bbl	18.40	23.40	29.39	41.86
ft of fill	1235	983	792	563

However, the volume of mud film can be more than the casing volume between the collar and the shoe, especially on long, large-diameter casing strings or where the condition of the mud creates a thick film. Large volumes first fill the space between the float and shoe, and the remainder displaces cement from around the bottom and outside of the casing, leaving it uncemented. The situation is worse if the mud channels through the cement, which is likely at this time because of the reduced pumping rate, before bumping the plug. This is a very good reason for running at least two joints below the float, especially with large-diameter casing. Except for special cases, always use the bottom plug. In questionable cases, run two or more joints between the float collar and float shoe to leave more volume for the mud film and ensure that the bottom joints are cemented properly.

Another cause of this failure is not mixing the tail or last slurry properly, most commonly when mixing on the fly. Consider batch-mixing if this is a potential problem. As a good precaution, always drop the top plug in good cement. Dropping it late may allow displacing thin, poor-quality cement around the shoe joints.

Drill the float collar, cement, and float or guide shoe smoothly. However, as noted, this seldom is the problem. Normally drill the float, cement, and shoe with the same bit and drilling assembly used for the formation below the casing. One exception is when using diamond bits.

Loose shoe joints may restrict tool movement. There may be very little indication, or the tools may hang up from lightly to severely. Marks and striations on top of the top drill collar and on the top and outside of bit shanks are an indication. Sometimes there will be few if any indications until the hole is drilled considerably below the casing shoe.

One action when loose shoe joints occur is to leave them alone and continue drilling. This may be possible if the loose joints do not cause a problem. However, there is always a risk that the shoe joints may shift after deeper drilling. In this case, either the bit cannot be run through the shoe joints, or the bit or other drill collar assembly equipment may hang up on a trip out. This could mean losing the hole and is not recommended.

If this option is elected, drill with a slick assembly. If problems arise in the future, then mill through with a tapered mill and open the joints. If the hole is deep enough, pull the pipe until it stops, connect a string reamer, run back in the hole, ream through the shoe joints, and pull out of the hole. If the pipe is in the hole and cannot be pulled, set it on bottom and back off. Make the repair, fish, and recover the pipe.

In most cases, cement loose shoe joints, fixing them in place, either when first observed or after they become a problem. One procedure is a very low-pressure squeeze. Displace the slurry into the open hole from about 100 ft below the loose joints to about 200 ft or more above them. Pick the open-ended pipe up so that the end is about 500 ft above top of the slurry. Pressure up to about ½ lb/gal below the leak-off pressure. Release and repressure periodically at about 15 min intervals until the pumping time expires. Gradually increase the pressure to or slightly below the leak-off pressure. Monitor volumes to ensure that slurry is left in the casing. Hold or maintain the final pressure for the equivalent of two pumping times. This procedure should place cement around and behind the shoe joint and help ensure that it is cemented solidly in place. Then clean out through the shoe joint with a bit and mills or string mills, if necessary, to ensure that it is full-gauge.

A special case may occur if a longer section of pipe backs off. In this case, run and set a permanent packer, WITHOUT THE LATCH IN, in the top of the loose joints. Sting into the packer with a slick assembly, establish circulation around the loose joints, and cement them with a primary-type job. Then, immediately pick up free of the cement. After the cement sets, drill out the packer and ream as necessary to ensure a full-gauge hole.

Casing and Liner Repair　　　　　　　　　　　　　　　　　　　　521

Normally squeeze the shoe joints with open-ended pipe. A retrievable packer may be used in deeper holes, but open-ended pipe is preferred in most cases for ease of use and safety.

One cementing technique is illustrated in the following example.

Cementing Loose Shoe Joints

An 8¾-in. hole was drilled to 9,500 ft, and 7-in. casing was run and cemented. A float collar was run on top of the first joint, and the first two joints were reportedly tack-welded. After waiting on cement, a 6⅛-in. bit was run. The float collar was drilled, and reportedly there was either soft or no cement above the casing shoe. The shoe was drilled on for about 2 hours, with the bit apparently spinning. The drilling-rate recorder indicated that at one point the shoe dropped, or the bit drilled about 3 ft and began spinning for another 45 minutes. About 20 ft of open hole was drilled.

When the pipe was pulled up to test the open hole, the bit had to be worked up into the casing. The pipe action indicated that the shoe joints were loose. When the drilling assembly was pulled, there were marks on the upper shanks of the bit and on the lip of the junk sub, also indicating loose shoe joints.

The drillpipe was run open-ended to the bottom of the hole and circulated for about 2 hours. Cement was mixed and spotted in the lower 400 ft of the hole. The drillpipe was picked up above the top of the slurry, the pipe rams were closed, and pressure — about 1 lb/gal less than the fracture pressure — was applied to the cement for a bradenhead squeeze. This was held until the cement reached an initial set. The top of the cement moved downhole about 75 ft during this period. The hole was cleaned out with a full-gauge bit, and the shoe joint did not cause any further problems.

Sometimes, the bottom joint(s) will back off while drilling several thousand feet below the casing shoe or back off early and cause problems later.

Opening Loose Shoe Joints with a String Reamer

An 8¾-in. hole was being drilled at 8,500 ft below 9⅝-in. surface casing set at 3,500 ft. On a trip to change bits, the bit could not be pulled up past about 3,700 ft but was free to move down. A string reamer was connected, the drillpipe was lowered, and the section at 3,500 ft was reamed, recovering metal cuttings. Tool action indicated that two or three joints of the 9⅝-in. casing had backed off and dropped downhole about 50–100 ft. The string reamer and stiff drilling assembly were pulled without difficulty after reaming.

Two alternatives were available. One was to stabilize the loose casing with cement, clean out and resume drilling. Since 7-in. intermediate casing was to be set at 9,000 ft, drilling operations were resumed with a slick assembly. The hole was drilled to 9,000 ft, and 7-in. casing was run and cemented without difficulty. This was slightly risky, but the results justified it.

Releasing Casing from a Bradenhead or Hanger Spool

Releasing casing from the bradenhead or casing hanger flange or spool — collectively called bradenhead — is one of the most consistent, perplexing, and aggravating problems in repairing casing failures. The classic example is the case of an operator trying to release a string of 5½-in. casing from a casing hanger spool on a rework. The well had some oil inside the casing and on the ground around the wellhead. After all methods failed — and somewhat in desperation — the operator took a strain on the casing with a casing spear, wrapped primer cord around the casing hanger spool, and detonated it. The predictable result was a rig partially destroyed by fire. After the fire was extinguished, the damaged workover rig removed, and another rig set up, it was found that the casing was still firmly seated in the casing hanger spool.

There are various methods of releasing casing from the bradenhead or casing hanger spool, including normal tensional pull, using casing jacks and cutting. Take precautions to shut off all high-pressure zones exposed in the wellbore before attempting to release the casing.

Normal Tensional Pull

A normal tensional pull with a casing spear is the most common method of releasing casing from a bradenhead or casing hanger assembly. This is successful about 60% of the time. Set a casing spear about 15 ft below the bradenhead. Start by pulling about 20,000 lb over the weight of the uncemented casing or the casing weight that was set on the casing slips.

If the casing is not released, hammer around the casing bowl and on the outer side of the flange. Do not hammer on the ring groove or directly on the face of the flange. Another procedure is bumping with a joint of drillpipe on a lifting line, if there is sufficient clearance. Use a metal impact-plate over the flange and a thread protector on a drillpipe pin.

Casing and Liner Repair

The next alternative is to increase the pulling force in a step-wise fashion to a maximum safe pull. To ensure that the workstring can safety pull the maximum, sometimes connect the spear directly to small drill collars. Normally the spear mandrel will be the limiting factor, but also consider the casing strength. Jars are seldom, if ever, used. Jarring is a questionable safety procedure at the surface, but a surface jar may be considered.

Casing Jacks

Casing jacks are used to release a casing stuck in a bradenhead. They can exert a very high tensional pull and are successful about 80% of the time without damaging the casing.

Catch the casing with a casing spear on drillpipe or drill collars. Place a heavy-duty slip-and-spider assembly or side-door elevator on the drillpipe, set at a convenient height above the jacks. Another method, normally used on heavier casings is to attach a piece of casing to the casing stub protruding above the head or hanger. Weld it securely, including strapping. Catch with a slip-and-spider or side-door elevator.

Set the jacks on a solid base such as the rotary. Apply tension in stages to the safe maximum pull. Vibrate or bump the casing head periodically if space permits.

Release by Cutting

The last method of releasing casing from the casing hanger spool is by cutting with a torch. This is successful 100% of the time. Cut the casing by one of several methods, depending on the type of casing, size, and type of head. Take precautions for fire and personnel safety if the casing drops a short distance or the possibility of a surface assembly failure exists.

One method is to lower the bradenhead and release the casing SET IN IT. Cut and remove a section of the supporting casing so that the bradenhead or casing spool drops, releasing the inner casing. This procedure is most commonly used when the outer casing is cemented to or near the surface. It may be necessary to dig a bell hole in order to cut the casing far enough below the bradenhead.

Catch the casing with a casing spear and pull tension about equal to the casing weight. Cut the external casing, usually surface, and remove a section of casing about 6 in. to 12 in. long. This should allow the bradenhead to drop, releasing the inner casing. Vibrate or bump the casing-head down if it does not drop. Cut the inner casing to remove the entire head in extreme cases. Usually replace the bradenhead with a slip-on welded type. It may be necessary to weld a spacer nipple on the surface casing stub and place the bradenhead on the nipple for the correct spacing. Then resume planned operations.

Another procedure often used for hanger flanges is to release the casing by cutting it internally at the slips. Catch the casing with a casing spear and pull tension of about 110% of the weight of uncemented casing. If the casing slips are set deep in the head, it may be necessary to cut off any casing extending above the slips so there is space for the head of the cutting torch. Other provisions for additional space may include removing the catch assembly or having the drillpipe body at the bradenhead or hanger-flange level. Temporarily plug the hole with rags to prevent slag or cut metal from falling down around the spear and sticking it.

Cut vertical slots in the casing on 90° phasing. Slot length is from three-fourths to the full length of the slips in the slip-and-seal assembly. Cut or wash the slots to a depth of about ½ to ⅔ of the casing-wall thickness. The casing releases by collapsing in the slip-and-seal area. Cut additional slots as necessary until the casing releases. After the casing collapses, remove the hanger flange, cut a clean top on the casing, and resume planned operations.

COMPLETION-PRODUCTION CASING

Completion or production casing preserves the hole, isolates productive formations, and serves other completion and production functions. It may be replaced by a production liner, sometimes run below a completion-type intermediate casing. Tie-back casing may cover worn casing and serve as completion casing.

Production casing normally is designed with higher safety factors, which helps reduce the risk of failure. It often holds maximum pressures or serves as a pressure back-up when high pressures are confined inside tubing. Production casing is less exposed to wear and corresponding associated failures since most drilling operations, a major cause of wear, are completed. Most types of casing failures occur in production casing, but the most common are collapse, splitting, and bursting.

Liners cause problems during running, including hanging off, cementing and liner top leaks. Drilling liners often serve the same purpose as deep intermediate casing and have similar failures. Otherwise most liners serve as production liners and have the same problems as the casing they replace.

Slim-hole completions have smaller-sized casing in the range of 4-in. to 5½-in. casing and are equipped with tubing. Casing failures generally are repaired by the various methods described in this or other sections in this chapter. Fishing and casing repairs in these smaller-sized, cased holes generally are more difficult because of the smaller hole. Tubingless completions use tubing in the range of 2⅜-in. to 3½-in. as casing. Repairs are described in the section "Tubingless Completions" in Chapter 6.

Casing and Liner Repair

Figure 7-25 Repairing split/burst 7-in. casing

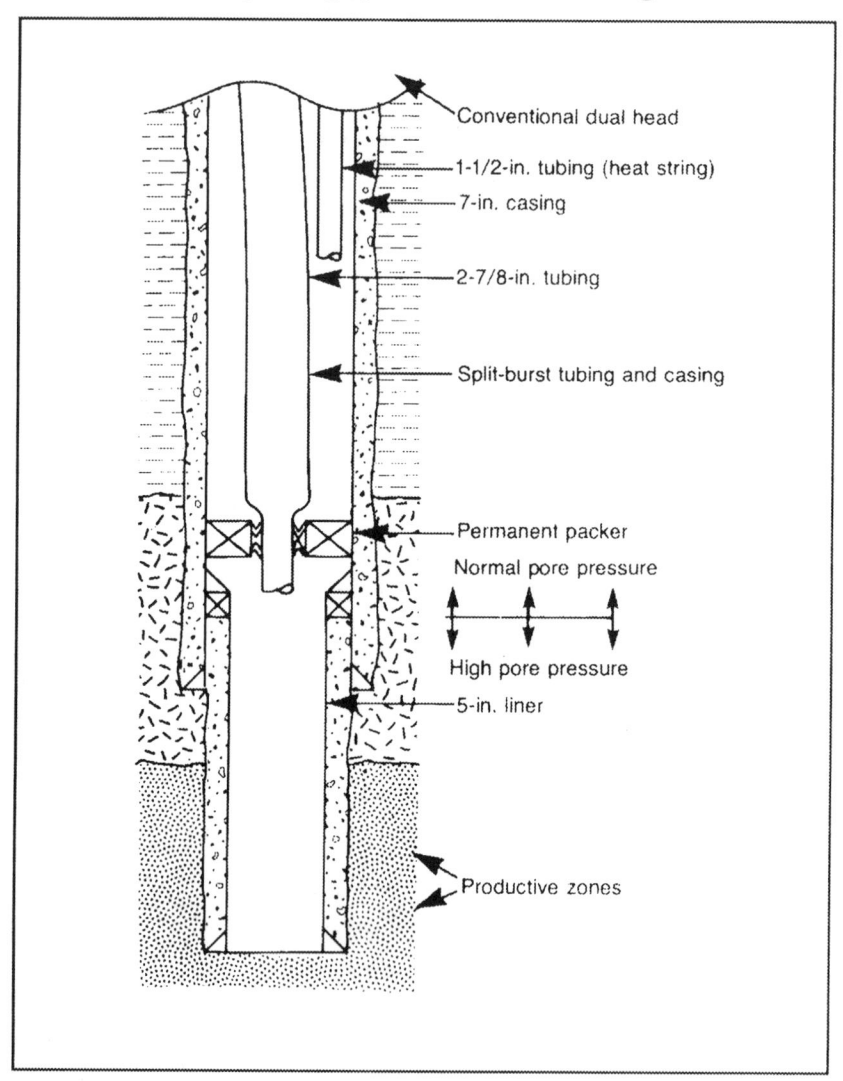

Split or Burst Casing

Split or burst casing is a common production casing failure, especially during the testing and completion phase. The most common cause is excessive internal pressure. Split casing usually is still connected and intact. Burst casing may be separated at the point of failure. In either case the casing may part by failure in tension. The inside diameter generally is not restricted. These failures tend to be more common in uncemented or poorly cemented sections and often deeper in the hole.

The easiest method of repairing split and burst casing is to repair it in place by squeezing. This option is seldom available except in a few special cases, such as covering later with another casing or liner.

Split or burst casing can be repaired by isolating it with a packoff. One method is to straddle the failure between two packers connected by tubing or smaller casing. Otherwise, split and burst production casing either in cemented or uncemented sections is repaired similarly to parted intermediate casing. The procedures, advantages and disadvantages are described in the section "Intermediate Casing."

Both tubing and casing may split and/or burst. A difficult fishing and repair operation is illustrated as follows (Fig. 7–25).

Repairing Split Tubing and Burst Casing

A development well had combination 7-in., N-80, 26 and 29 lb/ft casing at 11,000 ft with a 5-in., 18 lb/ft, N-80 flush joint liner at 14,000 ft. The well was completed by latching 2⅞-in. EUE, N-80 tubing with special couplings into a permanent packer set in the casing about 20 ft above the top of the liner. A parallel 1½-in. SFJ heat-string with tapered collars was landed open ended at 4,000 ft. The well was completed by perforating several zones in the 5-in. liner and commingling the low GOR, high pourpoint crude from various zones.

After perforating and flow testing, a pressure buildup test indicated skin damage and general low-formation permeability. The carbonate formations were to be treated with a deep stimulation with 30,000 gal of acid. Pumping equipment was hooked up and tested. The annulus was pressured to 1,500 psi, and the acid pumped with a surface injection pressure of about 9,000 psi. After about 5,000 gal of acid were pumped into the formation, the tubing lost pressure, the pumps were shut down, and the annulus pressurized almost instantly to 6,000 psi. The tubing and casing pressure equalized immediately, and then reduced slowly to 2,500 psi, indicating a tubing and casing failure.

The well was flowed back on the tubing side, recovering acid and acid gas, followed by clean oil. Flowing tubing pressure dropped to 2,000 psi, then increased and stabilized at 3,000 psi at a flow rate of about 250 bopd, compared to about 100 bopd before treatment. The well was shut in, and the surface pressure increased to 3,500 psi and stabilized.

Assuming a full column of 40° API oil, the 3,500 psi shut-in surface pressure represented a bottomhole pressure of about 8,500 psi at 13,000 ft, the average producing formation depth. The hole was reverse-circulated with 10.5 lb/gal saltwater. This would not kill the well, but it was immediately available and injected as a precaution to load the hole with a heavier fluid and reduce the casing pressure. Normally, the well would have been circulated down the tubing, but this would have caused additional delay in hooking up the pumping equipment. Also, the casing side was not equipped with a choke so that it could be flowed safely while circulating.

After pumping about 600 bbl of saltwater, the well was shut in, and the tubing and casing pressures equalized at about 2,600 psi. This pressure indicated parted tubing at about 7,000 ft, based on bottom-hole pressure and liquid densities. The 600 bbl of water pumped was greater than the casing/tubing volume above the packer, so some fluid was lost to the formation. Near the end of the pumping period, the last returns from the casing were still slightly oil-and-gas cut, so the actual failure was probably deeper than the 7,000 ft calculated.

A workover rig was moved in, and the hole was circulated under pressure with 16 lb/gal mud. A lower-weight mud would have killed the well if it could have been circulated to bottom, but the failed tubing only allowed a partial mud column. The well was not killed initially by pumping into the formation because of the high pressure required and fluid loss into the formation at the point of failure. About 50 bbl of mud were lost during circulating and before obtaining clean 16 lb/gal mud returns.

The well was shut in at 200 psi. About 50 bbl of mud were bullheaded down the casing and the well was killed. The fluid level was monitored for about 6 hr by pumping small amounts of mud down the casing, and it took fluid at about 2 bbl/hr. During this time a collar-locator log was run through the tubing. The tool stopped at 7,800 ft in the middle of a tubing joint, indicating intact tubing to that point. When the tool was picked up, there was no sticking, indicating collapsed tubing.

The well was in a relatively safe condition so the Christmas tree was removed, and the BOPs were installed without difficulty. Top plugs were set in both tubing strings. The 1½-in. split hanger and heat-string were pulled. The 2⅞-in. tubing was stuck, but there was enough stretch to remove the split hanger and install the BOPs. It was worked for about 4 hours, pulling up to about 90% of tensile strength. Then the tubing was backed off one full joint above the restriction at 7,800 ft and pulled. The upper tubing joints appeared to have some pulled threads, and some of the couplings were not as tight as would be expected, but these were not definitive. A 3½-in. workstring was picked up.

A standard fishing string was run and screwed into the tubing fish. It was worked by jarring, over-pulling and periodically bumping down. Working force levels were started at about 50% of the tubing-fish minimum yield and gradually increased to about 100 percent. After about 4 hours, the tubing came free with a sudden release, indicating parting, and there was a slight increase of about 1,000 lb of weight. The fishing string was pulled, and seven joints of tubing fish were recovered. The bottom joint was split for about 15 ft in the lower half of the joint and indicated a tension failure where the tubing parted near the upset due to working.

An impression block was run but did not give a definite impression. Overshots were run with different-sized grapples, but would not grasp the fish. One joint of washpipe was run with a mill shoe and milled about 6 in. The milling action indicated that the shoe was not on solid metal. It would tend to hang up, torque, and spin or run free. After pulling, the shoe was found worn on the outer side of the bottom cutting-face as if it had been milling on metal lying at an angle inside the casing. Another mill shoe was run with the same results. An impression block did not give a distinct impression but was marked on the outer peripheral face, indicating that the obstruction was at the edge of the hole.

The upper part of the tubing probably was lying over in a split in the casing. Several runs were made with washover shoes on a joint of washpipe. The shoes were dressed heavily with tungsten carbide. Each shoe milled about 6 in. before it wore out. The last mill shoe broke through and dropped about 15 ft. Then it stopped milling and torqued excessively. As the mill was pulled, it hung up about 15 ft from bottom, where the first heavy milling occurred. It was worked and jarred for about 30 minutes before re-leasing. When the mill was pulled, it had heavy markings on the outside, indicating the casing ID was restricted. An impression block showed a half-circle impression about the size of a piece of $2^7/_8$-in. tubing. The fish could not be caught with several overshot runs.

Milling was continued using heavily dressed mills on washpipe with tungsten carbide cutting material in the throat area and flush-joint mill shoes that had a slightly larger diameter than the washpipe. The first mill cut to 8,000 ft. After pulling, the badly worn throat area indicated milling on metal. Three mills were used to mill from 8,000 to 8,025 ft. The third mill was pulled without any sticking action and a 10-ft piece of badly twisted, bent, and milled tubing was recovered in the washpipe. The mills were not as efficient as expected. The tubing fish was either turning or had a spring action so that the mills were not cutting on a fixed surface.

Another mill was run on three joints of washpipe, and after milling for a short time at 8,025 ft, broke free and moved down to 8,055 ft, indicating that it was probably on a tubing collar. The mill was pulled out and an overshot run on the washpipe, dressed to catch a $2^7/_8$-in. collar. The overshot stopped at 8,025 ft and was worked down to 8,040 ft. It was pulled without any recovery. The grapples were broken, indicating a rough top on the fish.

The overshot was redressed with new grapples and fitted with a guide-type mill shoe. The collar at 8,055 ft was milled over lightly and caught. The fish was worked for about 3 hours, including jarring and rotating to release the latch-in from the packer. The tubing fish was released and pulled, recovering all of the tubing, including the latch-in device. About 30 ft of tubing inside the washpipe was badly cut, showing that it had been partially milled over. A full-gauge bit ran freely to the top of the packer.

The hole took fluid at an average rate of ½ bbl/hr during the entire time. The loss was probably at the casing failure, not in the producing zones, because of the pressure. At one point the hole was filled, and the well kicked. It was closed in and circulated dead through a choke with 16 lb/gal mud.

A retrievable bridge plug was run and set at 9,500 ft and 10 ft of sand was dumped on top of it to shut off the lower high-pressure zones. The rams were closed, and the well took fluid at 5 bbl/min at 500 psi. The casing was displaced with 10.5 lb/gal of saltwater.

A casing-inspection log showed a full joint of split casing. There were some traces on the log that could have indicated metal in the middle of the split or where the edge of the split had curled back toward the wellbore. A retrievable packer was run and set above the casing split. The upper casing was tested satisfactorily. The packer was reseated below the casing split and the bridge plug and lower casing were pressure tested.

The split casing was squeezed and tested to a final pressure equivalent of 13 lb/gal mud. The method was to squeeze, leave cement over the split, clean out, and test. About 7 squeeze jobs were performed in a 3-week period. The split should have been squeezed using the multiple-squeeze technique and to a higher squeeze pressure.

After the casing repair, the completion equipment was installed, the well was stimulated with heated fluid, cleaned up, placed on production, and it produced satisfactorily. During this time, care was taken to ensure that the production casing was not over-pressured.

In analyzing this failure, all tubulars were fully inspected in the yard and at the wellsite. The cold fluid used for stimulation probably caused tubing contraction, so excess tension contributed to bursting. However, the tubing failed at 8,000 ft, and the maximum tensile load was at or near the surface. Contributing factors may have been a defective casing coupling or insufficient slack-off when landing the tubing. The tubing failure exposed the annulus to a high-pressure surge, causing the casing failure. This occurred where the casing was well cemented. In summary, treating the well with a large volume of cold fluid probably was the basic cause of the problem.

Collapsed Casing

Collapsed casing is more common in production casing and is frequently associated with the completion operation. It is caused by external forces that exceed the collapse-pressure rating of the casing. The most common force is hydraulic due to excess external pressures. Mechanical force due to plastic formations, such as massive salt section and formation movement in fault zones, is less common.

Completion casing most commonly fails due to incorrect squeezing and stimulation procedures in combination with poor primary cementing. It is directly due to failure to pressurize the annulus. The treating pressure communicates through channels outside of the casing and uphole, applying an excess external pressure and collapsing the casing.

The first and preferable method of repairing a collapse failure is to roll and swage the casing back out to or near the original inside diameter. Normally this applies to pipe where the ID is restricted by about 25% of the diameter or less. Roll and/or swage the casing out to or near the original ID. Run the swage on a fishing string. Bump it into the collapsed section and jar it free repeatedly. Run a casing roller on a fishing string with extra drill collars for weight. Open the casing by rolling it out with heavy weight. Jar the roller free if it sticks. While using either tool, use multiple passes with progressively larger-sized tools to reduce the risk of splitting the casing. Pressure test the repaired section, squeezing if necessary.

Severely collapsed casing normally cannot be rolled or shredded out and usually must be milled to full gauge. First run a small diameter tapered mill only slightly larger in diameter than the work string. In a few cases it may be possible to mill through the casing with an under-gauge mill and then roll or swage the casing to full diameter. Otherwise, make subsequent runs with successively larger tapered mills. Run a full-gauge guide higher above the mill to help prevent sidetracking. Then squeeze to complete the repair if applicable. Cover the milled casing section with another string of casing if this option is available.

Casing may collapse around pipe such as tubing and stick it, but in most cases the tubing is free to move. However, it may have a larger diameter tool such as a packer on bottom, which prevents pulling the tubing. If the pipe is free to move, then run the tool to bottom and back off or cut pipe above the larger-diameter tool. Pull the pipe out of the hole, repair the collapsed section, and then recover the fish.

Figure 7-26 Repairing collapsed production casing

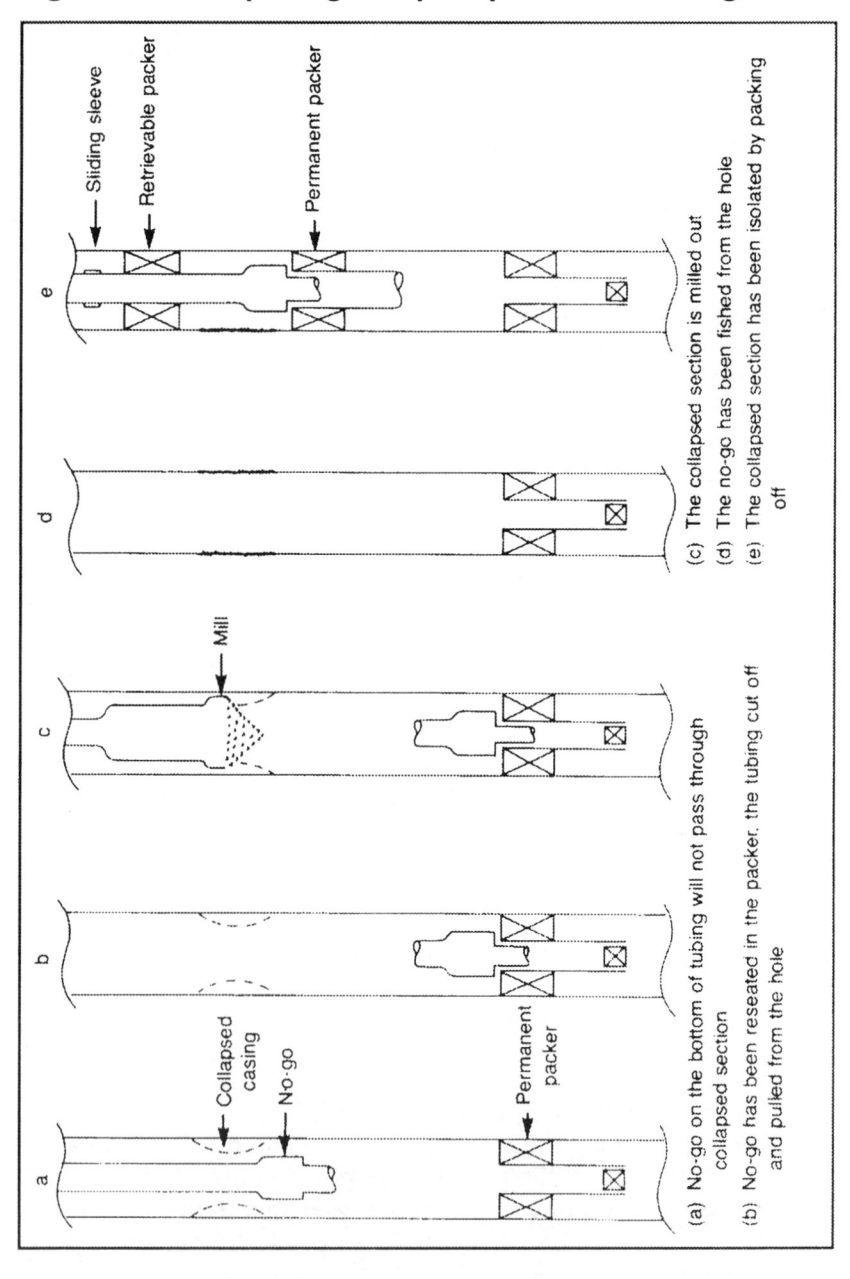

(a) No-go on the bottom of tubing will not pass through collapsed section
(b) No-go has been reseated in the packer, the tubing cut off and pulled from the hole
(c) The collapsed section is milled out
(d) The no-go has been fished from the hole
(e) The collapsed section has been isolated by packing off

If the casing cannot be repaired by the above methods, then use one of the procedures that may be applicable as described in the sections "Split or Burst Casing" and "Intermediate Casing."

The following example illustrates one procedure for repairing collapsed casing (Fig. 7–26).

Packing Off Milled Collapsed Casing

Seven-inch casing was set and cemented at 12,000 ft. Several zones were perforated below 9,000 ft, and a permanent packer with a mill-out extension was set at 9,000 ft. These were tested through 2⅞-in. EUE tubing stung into the permanent packer. They were questionably productive, so a plug was set in a nipple in the mill-out extension, and the tubing was started out of the hole, preparatory to testing a good prospective zone at 8,500 ft.

The tubing hung up at 8,000 ft. It was free to move down but would not pull up past 8,000 ft. When it was pulled into the tight place with considerable tension, it stuck and had to be bumped down to free it. This wedging action indicated collapsed casing. Apparently the larger diameter no go on the bottom of the tubing would not pass.

The tubing was run back downhole, reset in the packer at 9,000 ft and cut with a chemical cutter, leaving a 6-ft stub above the packer. The tubing was pulled without difficulty. An impression block was run, and it verified collapsed casing at 8,000 ft. The collapsed casing could not be rolled out with a casing roller. It was milled out to or near the full ID with a guided, tapered mill. An overshot recovered the tubing fish left in the packer at 9,000 ft.

The collapsed section was repaired by packing off. A permanent packer was set at 8,100 ft. The completion assembly included a stinger, 200 ft of 2⅞-in. tubing, a retrievable packer, and 2⅞-in. tubing to the surface. It was run, the stinger seals were seated in the permanent packer, and the retrievable packer was set at 7,900 ft. The tubing was landed and a Christmas tree nippled up. A plug was set in the mill-out extension of the packer at 9,000 ft, and the lower casing section was tested to ensure that it was isolated. The zone at 8,500 ft was perforated with a decentralized tubing gun, then stimulated, and the well was completed.

The cause of the collapse is unknown. The casing had sufficient strength based on standard design factors. In earlier completion operations, tubing was cemented in the hole at about 8,000 ft and required extensive milling before it was recovered. The milling most likely damaged the casing. The casing was not well-cemented as indicated by bond logs. The zones below 9,000 ft were stimulated at high pressures without holding pressure on the 2⅞-in. annulus. A combination of the weak casing and pressure communication to 8,000 ft behind the 7-in. casing probably caused the collapse.

The failure could have been repaired by other methods such as squeezing, setting a flush-joint liner or completing below a packer, but the method chosen was the most applicable under the circumstances.

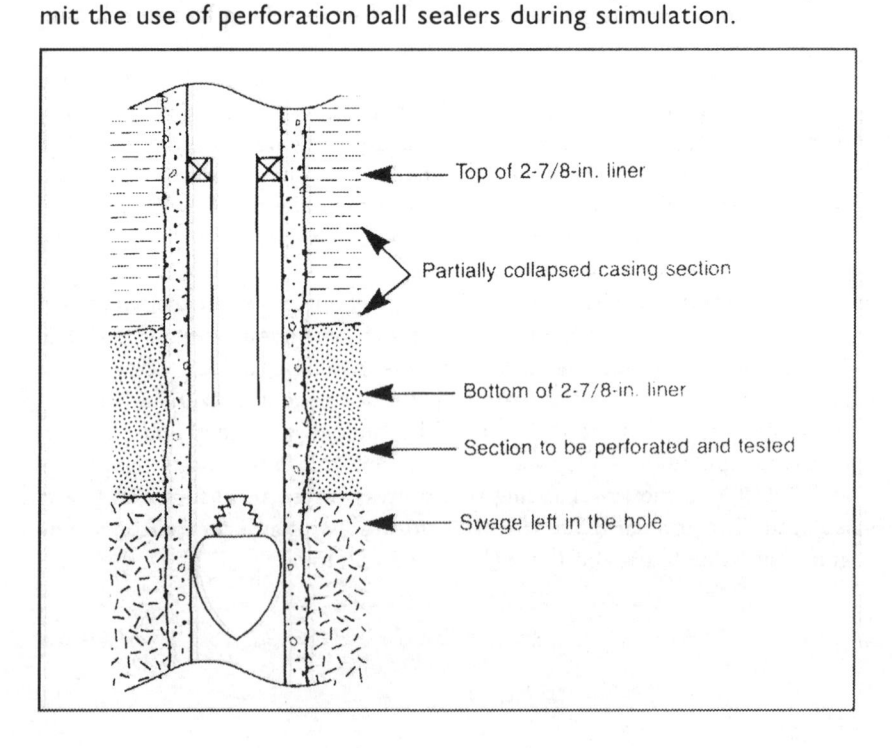

Running a liner through a collapsed section is illustrated in the following example (Fig. 7–27).

Figure 7–27 Repairing collapsed 5½-in. casing
Note: The liner bottom was landed above the perforations to permit the use of perforation ball sealers during stimulation.

Running a Liner Through Collapsed Production Casing

A wildcat well had 5½-in. casing set at 12,500 ft. Various zones were to be tested by isolating them with a retrievable packer and bridge plug. After testing several lower zones, the bridge plug was seated at 11,500 ft, and the packer at 11,250 ft. The zone was perforated and tested, and the packer and plug were pulled without difficulty. Breakdown pressures were in the range of 7,500 psi, and the last zone was balled out without pressurizing the annulus.

A new packer was run to continue testing and hung up at 11,100 ft. It was worked 6 hours before releasing. After pulling, the mandrel was found to be bent. A casing inspection tool stopped at 11,100 ft. A gauge run was made with a bit, and it also stopped at 11,100 ft. A casing scraper with spring-loaded blades was run, worked down to 11,130 ft, and pulled because of excessive torque. A 4½-in. roller swage was run under 6 3⅛-in. OD drill collars with 3⅛-in. jars and bumper sub. The swage was worked through the tight hole down to 11,400 ft, where it stopped. When the tools were pulled, the swage had backed off and was left in the hole.

The tools were rerun with a new box on bottom and could not screw into the swage. Another swage was run and worked through the tight spots to about 12,200 ft where it stuck, apparently pushing the first swage downhole. It was worked free and pulled after working and jarring for several hours. A casing-inspection log showed the section from 11,000 to 11,100 ft was partially collapsed and under-gauge from ⅛ in. to 1 in.

There was only one prospective section remaining to be tested below the collapsed zone, and it was considered marginal. Repairing the collapsed zone could be difficult, especially considering the limitations due to the depth and the small 5½-in. casing. The decision was made to perforate the prospective interval and run a liner through the collapsed section to ensure that it would remain open and permit testing.

Several zones were perforated in the gross interval from 11,200 to 11,300 ft. A 250-ft length of 2⅞-in. flush joint tubing was run on a liner hanger, which was set at 10,900 ft with the tailpipe at 11,200 ft. The zones were stimulated and production tested but were noncommercial. A bridge plug was set on top of the liner hanger, and the well was completed in several upper zones.

The 5½-in. casing was run in 13 lb/gal mud and should not have failed under normal conditions. There were no salt sections or flowing formations that could have collapsed the casing. The logical cause of collapse was pressure communication behind the casing when the zones at 11,250 ft were being broken down with over 7,000 psi and with no pressure in the annulus. However, the packer and plug were pulled without reported difficulty. There may have been problems encountered in the well that were not covered in the reports. Otherwise, the reports were not as clear and detailed as would be desired. This illustrates the importance of good reports.

Liners

Liners are subject to all casing failures. The most common are similar to those in the casings replaced by the liner. Problems may occur while running liners. Take normal precautions such as measuring out-of-hole on the last trip before running a liner. Rabbit the drillpipe before running the liner, usually on the last trip. String up extra lines for heavy liners. Insure that the liner weight does not exceed the drillpipe strength rating. Allow for 75,000–125,000 lb of overpull. Use all precautions in landing and cementing the liner.

Liner hangers may malfunction when running or setting and usually the malfunction is detected at that time. Hydraulic liner hangers may preset before the liner is on bottom. Bow springs or drag blocks may break, become smooth, or wear out during running, thus preventing hanging the liner. Also the hanger may preset due to slips hitting an obstruction while running. Mechanical liner hangers require torque and slight movement to actuate. It may be difficult to obtain the necessary torque to hang the liner in crooked or highly deviated holes. Hydraulic liner hangers are preferred in these situations.

Failed Casing above a Liner

The casing at the liner seat is subject to failure if it does not have sufficient strength to support the liner. The liner slips can push through the wall of the casing, especially with long, heavy liners. Provide for this in the casing design. However, the casing may wear during drilling, especially when drilling a long section below the casing shoe. Drillpipe rubbers may be ineffective since they are run higher to prevent them from passing into the open hole. Sometimes it is a good practice to inspect the casing where a hanger will be seated. Running a longer liner overlap section is a good preventive action.

Seating the liner may cause a failure in the casing due to either a combination of a heavy liner or worn casing. The most common method of repair is by running tie-back casing or a stub liner. In a few cases these are repaired by squeezing.

The liner may drop while hanging off, as described in the following example (Fig. 7–28).

Figure 7–28 Cementing a 7-in. liner after a hanger failure

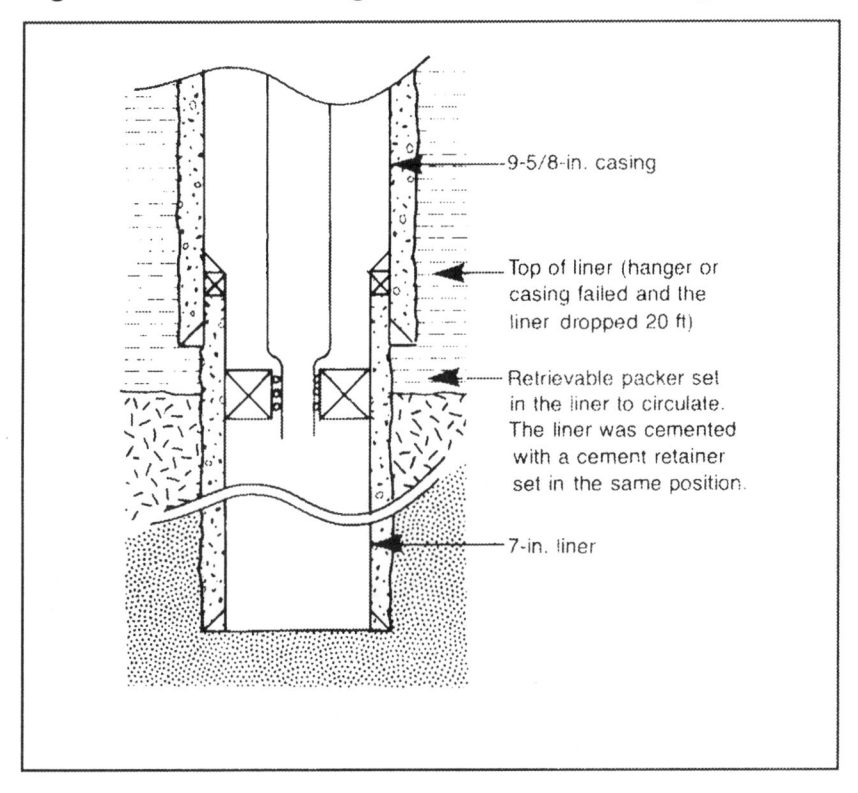

Labels in figure:
- 9-5/8-in. casing
- Top of liner (hanger or casing failed and the liner dropped 20 ft)
- Retrievable packer set in the liner to circulate. The liner was cemented with a cement retainer set in the same position.
- 7-in. liner

Cementing a Dropped Liner

A wildcat had $9\frac{5}{8}$-in. intermediate casing at 15,000 ft, an $8\frac{1}{2}$-in. hole was drilled to 21,000 ft, and a 6,500 ft long, 7-in. FJ liner was run on a combination string of $4\frac{1}{2}$- and $5\frac{1}{2}$-in. drillpipe. The bottom was tagged about 25 ft high, indicating some fill. The liner was picked up about 20 ft to ensure that the shoe joint was not buried in fill material and that the liner could be circulated satisfactorily.

While hanging the liner, it dropped, and the drillpipe lost the equivalent of the liner weight. The pipe was lowered slowly, and it tagged the top of the liner about 20 ft below the point where it was first landed. The best estimate was that the liner slips probably had penetrated the casing, sheared off, and allowed the liner to drop.

Casing and Liner Repair

The drillpipe was circulated, but there was no convenient way of ensuring that the stinger was sealed in the liner and the mud was circulating completely through and around it. A carbide lag indicated that mud was circulating from the top of the liner, and possibly some was circulating through the liner. Since there was no way of ensuring that the cement would be circulating through and around the liner, the drillpipe was pulled from the hole without cementing the liner.

The top of the liner was cleaned off with an 8½-in. bit, and a 6⅛-in. mill was run. It cleaned through the liner top and down to the liner float collar. The liner was run in good oil mud, so a packer trip was justified to see if the liner could be circulated and cemented with a primary cement job. A 6⅛-in. retrievable packer was run and set 70 ft below the liner top in the middle of the second joint of the liner. About 2,500 psi was used to break circulation. The liner was circulated slowly, and the rate increased gradually until the liner was circulating freely. Pipe weights did not indicate any liner movement during this time. About 72 hours had elapsed from the time the liner was first set until circulation was established. A carbide lag indicated that mud was circulating through the liner and around the shoe.

The liner was circulated until the hole was clean. The retrievable packer was pulled, and a cement retainer was run and set 70 ft below the liner top. Circulation was established and the liner was cemented conventionally, except that the slurry was underdisplaced to leave about 400 ft of cement in the liner above the float collar. After the cement was displaced, the drillpipe was pulled out of the retainer, and the excess cement above the top of the liner was reversed out.

The liner top was cleaned out and tested satisfactorily both for inflow and pressure. The bottom was cleaned out, 20 ft of new hole was drilled and tested, and drilling operations resumed.

The procedure to cement the liner was unconventional. However, under the circumstances it resulted in a good cement job and was more efficient than perforating and cementing.

Squeezing the Liner Top

A leaking liner top is the most common liner failure, almost always caused by incorrect cementing procedures, most often due to channeling in the overlap section (Fig. 7–29). Obtain a good liner-top cement job by flushing all the mud out of the overlap section and replacing it with a good quality cement slurry. Then ensure that the slurry does not move until it reaches an adequate compressive strength. This applies to both primary cementing and squeezing. Squeezing should not be required if the liner is cemented correctly. A good preventive measure is to use a longer liner overlap section of at least 300 ft so that there is more space for cementing and also to obtain the additional strength of cement backed by metal.

The method of testing is also important. Run the inflow test first. Otherwise, pressure-testing first may temporarily block the channels and, at the same time, plug them by compressing mud and cement particles so that the liner tests satisfactorily. This also can prevent an inflow leak, temporarily giving a false favorable test. Then, over time, the leak develops again with fluid movement through the overlap section. During the inflow (if there is one) and assuming safe operating conditions, let the leaking liner-top flow for a reasonable period of time to help flush out the channels and remove the less-mobile mud in the overlap. During pressure testing, and unless there are special reasons preventing it, always use a pressure that is in excess of the fracture pressure. The fluid has no reasonable place to go except into the formation, which in most cases means fracturing.

Figure 7–29 Liner overlap channeling

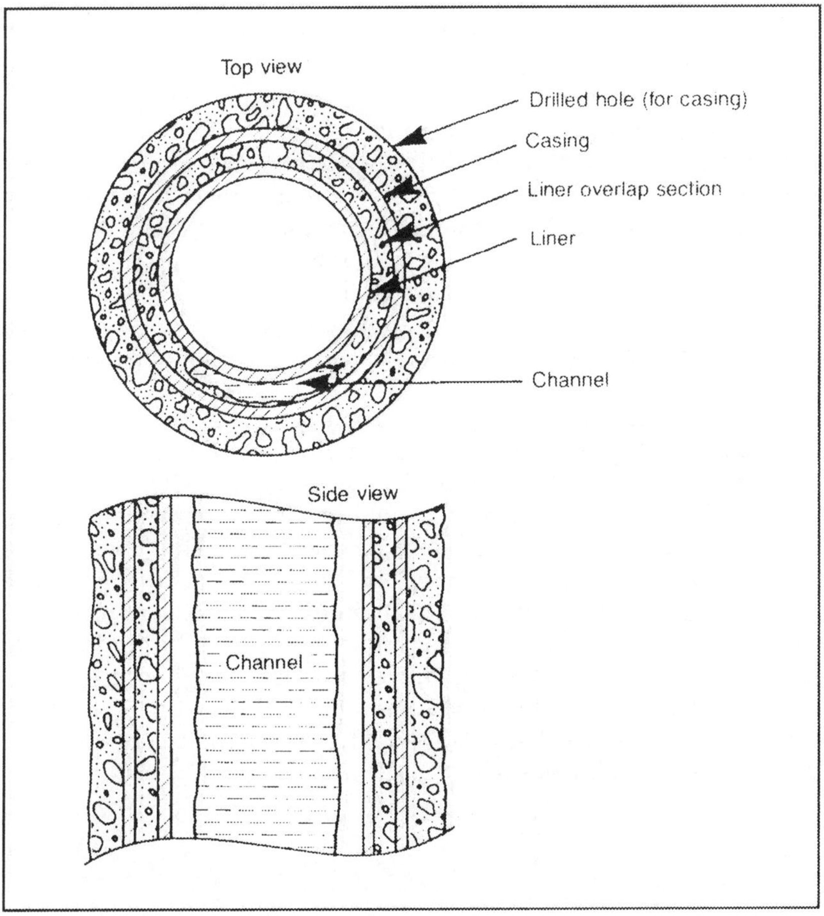

Casing and Liner Repair

Therefore, if the liner top leaks and assuming there are no tubular failures, cement must have channeled through the mud in the liner overlap section, or moved after cementing. Since minimum cement volumes are commonly used when cementing liners, this is understandable but not acceptable. Design the squeeze to both remove the mud channels in the overlap and replace them with good cement slurry. Pressure is secondary except that it must be sufficient to serve these purposes.

One of the first steps to obtaining a good squeeze job is to determine the breakdown pressure and the pump-in rate and pressure. After the breakdown, a main concern is a high injection rate to help clean the overlap section. The test normally is done with water, so flush the liner top with a reasonably large volume of water during testing, usually at least five or more liner overlap volumes. However, low viscosity water may tend to flow through the mud channel without replacing all the mud. If conditions are favorable, the overlap may be cleaned better by breaking down and pumping in water, followed with higher viscosity mud, and then displacing the mud with water. Follow this with the higher viscosity slurry to obtain the optimum fill in the overlap section.

Squeeze through either a fixed or retrievable cement retainer. Each has advantages and disadvantages, as described in the section "Squeeze Cementing." Squeeze through a retainer if moderately high pressures are expected or the top of the liner is relatively deep. The retainer will hold back pressure on the cement, preventing flowback — even accidentally — while reversing-out excess slurry. There is a risk of collapsing the casing above the squeeze tool. Prevent or reduce the risk by maintaining pressure in the annulus. Set the packer about 250 ft above the liner top or the equivalent of 1 to 2 liner overlap annular volumes (Fig. 7–30).

The general squeeze procedure is similar to conventional or stage squeezing with some modifications to allow for the special conditions of the liner. Several general precautions may help to ensure sealing the liner top. Use a good-quality cement with the required minimum of additives or no retarders. Have extra slurry volume to help flush the overlap free of mud and fill it with good slurry. As a guide, have a slurry volume of 10–20 times the liner overlap annular volume, with lower multiples for higher pressure, or a minimum of 100 sacks or about 20 bbls, whichever is larger. Design for a thickening time of 2 to 3 times the projected pumping time. Higher viscosity slurries help remove channels but may require higher pressures, so evaluate the benefits of each.

Unweighted spacers are most common. Spearhead with a minimum of 20 bbls — more is better — and at least 10 bbls in the tail. Use a twin-pump truck on low-pressure squeezes of 3,000 psi or less and two twin-pump trucks for higher pressures.

Displace the first half of the slurry into the overlap section at a maximum rate and stage the last part of the squeeze job to reach a higher squeeze pressure and obtain one or more pressure breaks. The slurry moves through the channels, causing the leak. Nevertheless, the cement may be channeling through a larger channel filled with mud. As long as the pumping pressure and rate do not change, the cement probably is moving through the same channel. This could leave a mud channel that would leak later. Staging or changing pumping rates may help remove the residual mud and replace it with slurry, improving displacement.

Figure 7–30 Squeezing a liner top

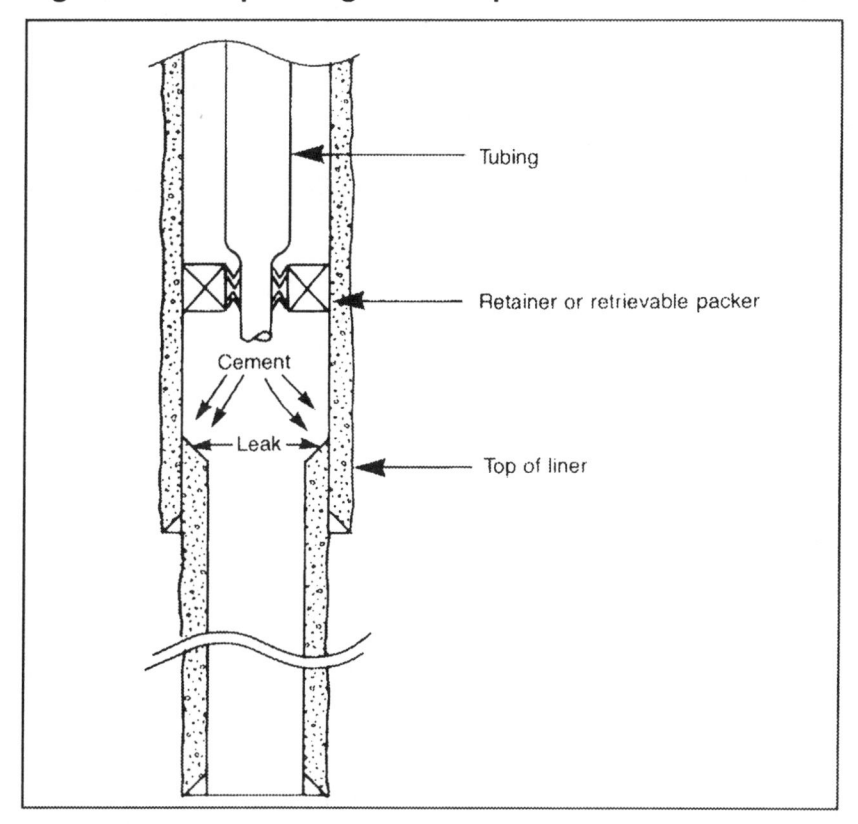

OLD WELLS

Common abandonment procedures for old wells included pulling the tubing and cutting and recovering production casing at the deepest point. The casing frequently is parted with dynamite or other explosives, usually at or below the bottom of the surface casing. Other casing strings may be removed similarly. The surface casing normally is cemented to the surface and may be intact. Set cement plugs at various points in the hole both before and after recovering part of the casing. Final abandonment commonly includes cutting off the surface casing below-ground and welding a plate over it, sometimes with a well monument. These wells are relatively easy to reenter. One exception is junk in the hole, so records should be researched carefully.

Start the reentry by digging a bell hole around the casing. Cut off the plate and weld on a nipple and bradenhead. Then clean out the original hole to the top of the next casing string. Normally there are few problems inside surface casing. In some cases the hole may contain metal junk, which may not justify milling, fishing, and cleanout. Excluding junk, cleanout seldom presents a problem except for hazardous formations which usually are well stabilized. The hole may be filled with heavy mud, highly jelled due to dehydration. Often break circulation several times and dilute as required during cleanout.

Production casing in old wells is subject to the same type of casing failures as those occurring in the production casing of new wells. After the well is completed, the risk of damage during the completion process is eliminated. However, the older well casing may be corroded or worn due to fluid erosion and tubing movement in pumping wells. Therefore, reduce the strength rating of the casing based on judgment and experience in the area.

Note that the casing may be in good condition in some older wells. There are many cases of pulling casing that is in almost-new condition, some with the factory stencil intact, after the casing has been in the ground for 20 to 40 years. A good example of this is in the Appalachian Basin. Some older wells have oxygen corrosion in the top part of the casing if the surface is not well sealed.

Run casing-inspection logs to provide additional information. One general rule is to apply the minimum stress required to perform the job when testing old casing, and use minimum pressures and rates when stimulating. Similarly, apply the minimum weight necessary to seat and hold a packer in place. Packer slips will convert the vertical force to a radial force that could exceed the burst-rating of older, possibly worn casing. When older wells are recompleted, take all precautions to prevent damaging the casing. Generally inspect the casing in older wells with a casing-inspection log before drilling deeper.

There is another, intangible, reason for being cautious when working on old wells. Generally, they do not have the productive potential of newer wells. There may be a problem with unreliable records and unknown well status. If a casing failure occurs, evaluate the cost of repairing the casing compared to the remaining income-generating potential of the well. The cost of repairing the casing failure may not be justified and frequently represents a high risk. If repair is justified, make the most economical repair to accomplish the objective.

Leaks are one of the more common failures in the casing of older wells. They are less common in the lower part of the casing which usually is cemented, Leaks often occur at shallower depths due to oxygen corrosion, exposure to corrosive groundwaters, and electrolytic corrosion.

Casing leaks in older wells generally are repaired by squeezing, similar to any other casing leak. Do not subject the casing to excess pressure. If the leak occurs in an uncemented section of the casing, consider a primary cement job. Set a plug in the casing below the leak and a packer on tubing above the leak. Pressurize the annulus to help determine whether there is communication around the packer and to prevent casing collapse above the packer.

It is a good practice to run a casing scraper for drilling out cement (Fig. 7–31). This removes any cement particles on the casing walls that could tear up packer rubbers or stick tools that are run on a wireline. The casing scraper also removes corrosion products like scale or rust that accumulate and stick to the wall of old casing.

Except for special cases, freshwater or saltwater should not be used as the leading, or spearhead, fluid when circulating the casing. It tends to pick up any cuttings in the annular space. These settle in the water and may bridge off, increasing pressure, possibly breaking down the formation, and may prevent circulation behind the casing. The recommended procedure is to spearhead about 25 bbl of gel water containing about 20 lb of bentonite per barrel. It is not necessary to wait for the gel to hydrate, since the concentration of bentonite will give the fluid a satisfactory gel. This will lift the cuttings, hold them in suspension, and help prevent bridging when the casing circulates.

Although the casing is relatively stable after a well is placed on production, older casing strings may collapse. This can be caused by casing wear or corrosion and may be associated with a treatment or stimulation. Depending on the nature of the failure, continue producing the well if it is possible and safe. If the casing must be repaired, and the repair is justified, then repair it by one of the procedures described in the section on collapsed casing.

Figure 7–31 Casing scraper (courtesy Homco)

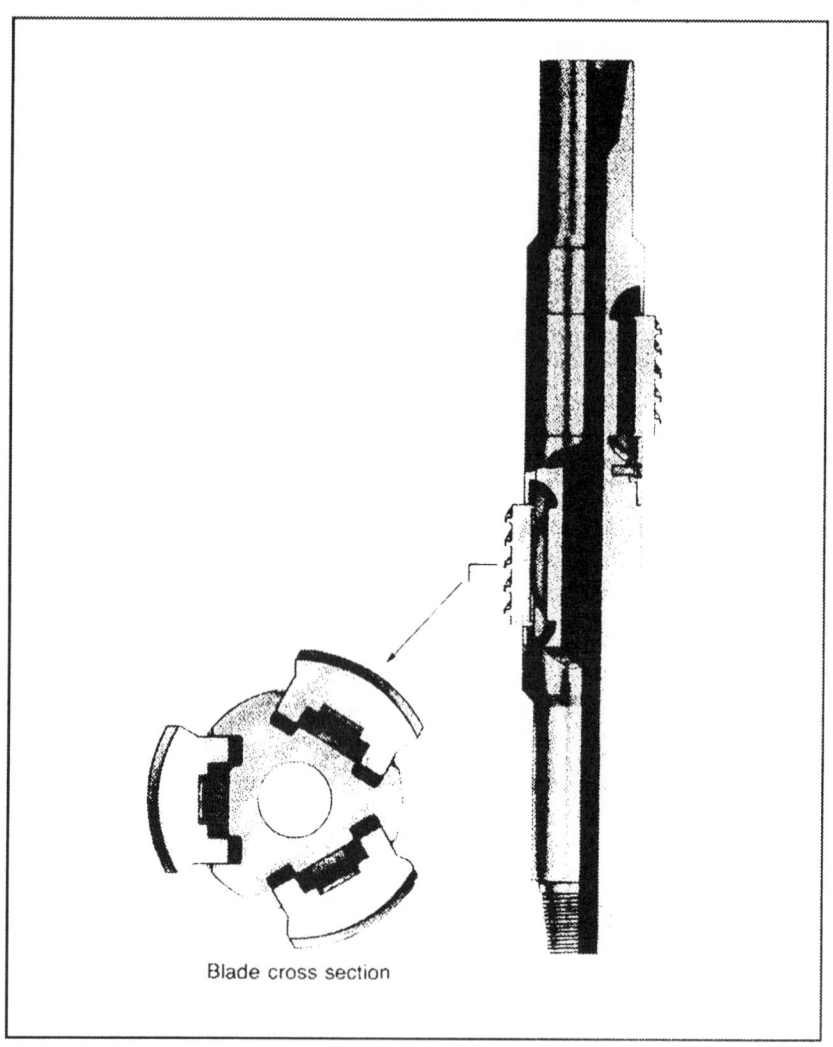

Blade cross section

    Reconnect cut-off casing with a casing-bowl run on new casing if necessary — to make the connection if the casing was cut off with a clean cut — using procedures described for repairing parted casing. It may be necessary to run a short section of washpipe with a mill shoe and clean out over the casing stub left in the hole for clearance for the casing-bowl. If the casing was cut off at or near a collar, it may be necessary to make another cut or to mill off a short section of the casing. Note that when milling casing collars, the collar may crack and push down over the casing and require fishing. Then connect the casing with a casing-bowl and cement with a primary job, or squeeze it.

When the casing has been blown apart with an explosive, cleaning out and reconnecting the strings is more complex. The top of the casing usually will be shredded and must be cleaned to give a smooth top to receive the casing-bowl. This process often requires extensive milling and considerable fishing to recover all of the pieces of the casing. Make a casing-bowl connection after cleaning the casing top by milling or cutting.

A relatively simple, strong repair can be made if a low-pressure connection is satisfactory. Clean out as necessary and mill a full-gauge hole through the section of casing originally left in the hole. Run casing, align the casings with an alignment tool, and cement the upper string with a primary cement job, leaving some cement in the casing. Then drill out the alignment tool and test the reconnected section, squeezing if necessary.

Another procedure may be applicable, depending upon well conditions. It can be used on casing with a split top where it is difficult to mill a clean top for a casing-bowl connection. Clean off the top of the fish as much as is reasonably possible. Then run several joints of washpipe on casing. Cut a 45° taper on the bottom of the washpipe or use an open-type mill shoe. Work the washpipe down over the top of the casing in the hole, preferably for 50–100 ft, and then cement it. The metal-to-metal backing helps obtain a better cement seal. Mill and smooth the inside of the casing and resume operations.

Depending on the condition of the lower casing, another alternative is to run a casing roller with an outside diameter one inch greater than the inside diameter of the casing in the hole. Dress or smooth off the top of the casing stub and bell or taper it slightly. Then run a casing pin, seat it in the bell, and cement. This procedure is seldom used.

The procedure for reconnecting casing during an old hole reentry is described in the following example.

Reentering an Old Well

A plugged and abandoned well was to be reentered and recompleted (Fig. 7–32). The well originally had 9⅝-in surface casing set at 1,200 ft and 7-in. production casing set at 8,500 ft. When the well was abandoned, tubing was pulled from the hole and the production casing was shot off at 5,500 ft and pulled. Various cement plugs were spotted in the hole. The surface casing was cut off 4 ft below ground level and a plate was welded on the top of the casing.

Casing and Liner Repair

Figure 7–32 Reentering a plugged and abandoned hole

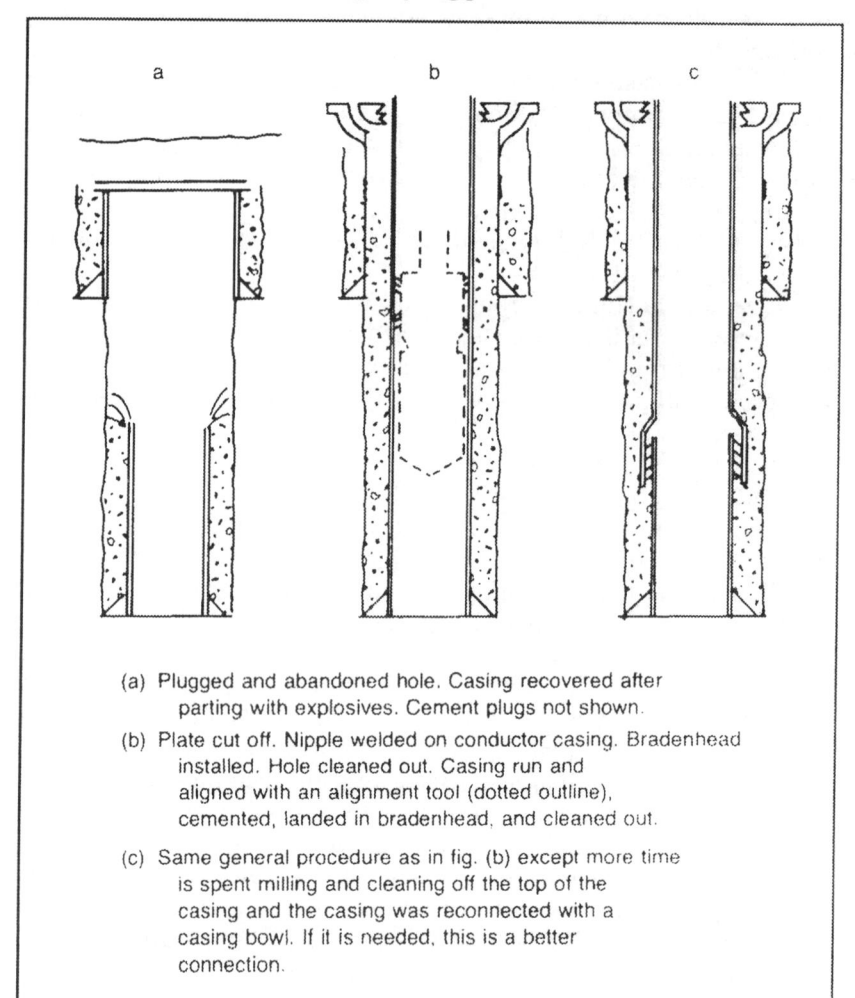

(a) Plugged and abandoned hole. Casing recovered after parting with explosives. Cement plugs not shown.

(b) Plate cut off. Nipple welded on conductor casing. Bradenhead installed. Hole cleaned out. Casing run and aligned with an alignment tool (dotted outline), cemented, landed in bradenhead, and cleaned out.

(c) Same general procedure as in fig. (b) except more time is spent milling and cleaning off the top of the casing and the casing was reconnected with a casing bowl. If it is needed, this is a better connection.

On the reentry, a backhoe located the casing and dug out around it. After the plate was cut off, a 9⅝-in. nipple was welded onto the original surface casing, and a slip-on bradenhead was welded onto the nipple. A rig was moved in and set up, and BOPs nippled up on the bradenhead. The hole was cleaned out without difficulty to the top of the parted casing at 5,500 ft with an 8 ¾-in. bit. A 6½-in. bit was run, and it cleaned the inside of the 7-in. casing to about 7,000 ft. Bit action indicated that there was some junk in the hole. A tapered mill was run, and the top of the casing at 5,500 ft was milled out to full gauge. This was not difficult since the explosion blew all of the metal radially outward away from the center.

An alignment tool was constructed inside the bottom of a joint of 7-in. casing. This was run on 7-in. casing and seated inside the lower section of 7-in. casing with about 20,000 lb of extra weight to ensure that both sections of casing were touching. The casing was cemented with a primary job. A wiper plug separated the slurry from the displacing mud. The slurry was displaced, based on volumetric calculations, to about 500 ft above the top of the alignment tool. After the cement hardened, the cement and alignment tool were drilled out. The parted section was tested, and it leaked at the equivalent of 12 lb/gal mud. The section was squeezed with four consecutive, multiple-type squeezes, cleaned out, and tested to the equivalent of 18.5 lb/gal mud. The hole was cleaned out to the original TD, the producing zone was reperforated and stimulated, and the well was placed on production through tubing set on a packer.

This method of reconnecting the casing was faster than a casing-bowl connection, but the casing-bowl connection is stronger. The repaired casing was satisfactory for this well.

If original casing is relatively large and a smaller casing is satisfactory, then clean the hole to the depth needed and run a complete string of casing that is cemented with a primary cement job. Use this procedure and make a tubingless completion if applicable. In a few cases, it may be possible to clean out the old hole and run tubing and a packer. Seat the packer in the original cleaned-out casing. This type of repair is used only in limited cases, such as reopening shallow gas wells.

Consider cleaning out to the top of the casing stub, plugging back and sidetracking above the casing stub. This has several options, such as running a larger casing for a deeper well and diverting the hole to a more prospective geological location. Use the standard "plug back and sidetrack" procedure. Otherwise, make the most economical casing connection when reentering an old hole.

Casing and Liner Repair

REFERENCES AND SUGGESTED READING

Archuletta, Jim, Bruce Bitler, Mark Binford, and John Modi, "Cooperative Corrosion Control and Treatment Program Proves Effective," *Oil & Gas Journal*, Aug. 6, 1990: 60–61.

Brouse, Mike, "How to Run Casing and Drill Out," *World Oil*, Feb. 1, 1983: 35–40.

Bruno, M.S., "Subsidence-Induced Well Failure," *Society of Petroleum Engineers*, June 1992: 148–152.

Dewar, John, and Douglas Halkett, "Procedures Control Total Mud Losses While Drilling in Deep Water," *Oil & Gas Journal*, Nov. 1, 1993: 104–108.

French, F.R., and M.R. McLean, "Development-Drilling Problems in High-Pressure Reservoirs," *Journal of Petroleum Technology*, Aug. 1993: 772-777.

"Halliburton Cementing Tables." Halliburton Services, a Halliburton Company. Little's, Duncan, OK.

Howard and Fast, "Hydraulic Fracturing," *Society of Petroleum Engineers*, 1970.

Keller, Rod, "Downhole Repair Permits Kickoff Probe from Old Hole," *Oil & Gas Journal*, April 20, 1992: 90–92.

Leturno, Richard, Clay Bretches, and Johnny Shields, "Casing Leak Is Repaired Using Reverse Cementing Procedure," *World Oil*, Nov. 1992: 44–45.

Love, Barney D., Michael N. Gilstrap, and Hugh Hay-Roe, "Pumping Sized LCM Sequentially Can Restore Circulation," *Oil & Gas Journal*, May 2, 1994: 104–108.

MacEachern, Doug, and Stephan C. Young, "Ultrafine Cement Seals Slow Leak in Casing Collar," *Oil & Gas Journal*, Sept. 7, 1992: 49–51.

O'Brien, John, and Ian Lerche, "Understanding Subsalt Overpressuring May Reduce Drilling Risks," *Oil & Gas Journal*, Jan. 24, 1994: 28–34.

INDEX

A

abandonment, 1, 79, 402, 542
accept the risk, 487
acetic acid, 371
acid gas, 527
A-frame leg, 492
air cushion, 173
alignment tool, 469, 509
aluminum pipe, 485
American Petroleum Institute, 42
angle building, 320
anhydride, 64
API Specifications, 213
API top-lock hold down, 401
assemblies, wall stuck, 351
    back off, 272
    high angle, 217
    horizontal, 217
    packed-hole pendulum, 355
    slick, 521
    stiff drill collar, 331

B

babbitt, 205, 208
back off, 100, 144, 211
    joint, 177
back-up pads, 147
back-up tongs, 8
backed by metal, 481
backhoe, 546
bailer, chipping, 387
barb, 189, 190, 193
barite, 302, 44
barite sag, 334

batch treatments, 120
bell hole, 501, 524
bell, 23
    nipple, 410
belly board, 391
bent box sub, 181
bird nest, 195, 308, 430, 434
bit, record, 80
    balled, 329
    blow jets, 365
    bull nose, 113
    diamond, 19, 217, 361
    drag 13, 217
    fixed blade, 22
    insert, 13
    life, 230
    pilot, 113, 225
    pin, 160
    plugged, 120, 258
    polycrystalline 14, 217, 305
    roller, 13, 23
    shear cones, 348
    wall off, 82
    unplugging, 334
blades, 170, 283, 284
blind box, 421
blind rams, 404
blooie line, 32, 59
blowout, 1, 98, 102, 103, 109, 111, 208, 373, 403, 488, 499
    conditions, 284
    drills, 209
    pipeline, 2
    underground, 367
borehole, erosion, 26
bottom clutch, 215
bottom-hole pressure, 70
boulders, 492

buoyancy, 87
bradenhead, 380, 499
   weld on, 500, 502
breakdown pressure, 328
bridge, 42, 55, 62, 100
   cuttings, 287, 328
   sand, 363
bridging agent, 365, 399
bullet nose, 20
bullheaded, 393
bumping or jamming, 280
burrs, 143
burst testing, 451
burst tubing, 391
bypassing the fish, 431
cable guided method, 435

C

calcite, 61
calcium chloride, 42
caliper, logs, 130
   three-arm, 130
   two-arm, 130
carbide lag, 538
carbide, 242
carbon dioxide, 57
carbon steel, 419
carbonate formation, 527
casing, bowl, 469
   jacks, 414
   patch, 469
   scraper, 388
   spool, 499
   salvaging, 472
   tie back, 536
caterpillar tractor, 16
catline, 3
caving, 362
cement, 43
   accelerators, 304
   batch mix, 484, 304, 520
   bullheading, 90, 496, 478

channeling, 16
contaminated, 16
dump, 60, 457
epoxy, 17, 519
excess retarder, 303
green, 16, 54, 257, 480
initial set, 485
lost circulation, 53
plug, 16, 56, 138
puddle, 5, 387
pumping time, 484
scavenger, 38
slurry, 304, 498
thickening time, 304, 473, 484
thixiotropic, 53
chain out, 84, 98, 240, 241, 296
change in scope, 446
channel, 462, 520, 540
chert, 13
chloride, 109
choke manifold, 32
Christmas tree, 427, 528
circulating ports, 192
   reverse, 18
   valve, 479
close-contact tool, 192
close-tolerance tools, 190, 378
coal beds, 64
coil tubing, 391
cold fluid, 530
collet-type latch, 427
color locator, 427
commingled zones, 443
completion, tubingless, 392
compression, 266
concentric, 393
   casing strings, 507
concrete pad, 494

connection, casing bowl,
    509
  coarse-threaded, 179
  dry, 11
  fine thread, 33, 144, 291
connectors, left hand,
    139
  right hand, 139
contact area, 351, 354
core of formation, 149
core strand, 419
corrosion, electrolytic, 543
  oxygen, 542
corrosive groundwaters, 543
cost effective, 445
course-measuring instrument,
    441
current, alternating, 24
  direct, 24
cut-lip guide, 181, 183
cutting pattern, 13
  torch, 409, 417, 524, 525
cuttings, 36
  metal, 19
  mill, 307

D

dart valve, 198
de-gassers, 36
decentralized gun, 533
dehydration, 542
depth capacity, 3
development well, 437
diesel, 242, 351, 356, 372
diesel, weathered, 105
displacing mechanism, 477
diverter, 31, 59
dogleg, 62, 72, 87, 94, 113,
    117, 118, 121, 126, 131, 306
  absolute, 27, 505
  angle, 75

double bowl overshot, 416
double elevators, 403
double rubbered, 502
downtime, 25
dragging slips, 408
dressing off, 161, 237
drill collars, bottle neck, 22,
    213, 222
  bit, 214
  donuts, 213
  double box, 214
  fluted, 213
  nonmagnetic, 213, 219
  pony, 213, 219
  spiral, 213
  square, 213
  weights, 213
drill tubing, 392
drillable material, 381
drilling line, cut, 100, 403
  slip, 100
drilling, directional, 212, 220
  horizontal, 212, 220
  time, 306
  vertical, 220
drillpipe, bent, 15
  breaks, 14
  combination, 27
  hardbanding, 85
  plastic coated, 210
  rubbers, 19, 121, 233
  tapered, 119
  wear groove, 85
dry drill, 48, 55, 58, 63
Dutchman, 223, 296, 331, 385,
    388

E

earthquake, 64, 464
electric wirelines, 424
electrical conduits, 269

Index 551

elevator latch spring, 454
equalizer pin, 386, 403, 443
   valve, 383
erosion, 239
evaporites, 61
explosives, primer chord, 272
   dynamite, 284
extra string of casing, 446

F

failures, bow springs, 378
   drag blocks, 378
   hydraulic holddowns, 378
   sheared j-slots, 378
   surface, 403
   tool joint, 404
   tubular, 403
fatiguing, 406
fault 448, 531
field experience, 355
field proven, 119
fill up, automatic, 23
fire, 409
fish, clean top, 237
   second, 6
   third, 6
fishing costs, 416
fishing neck, 166, 237, 378
   neck, 440
fishing specialist, 80
fishing tool operator, 80
flapper valve, 387
flexure failures, 421
flow surges, 38
fluid erosion, 144
   passages, 166
   spring, 173
   surface, 421
flush joint, 5
force level, 529
   higher, 273
   working, 264

force, lateral, 74
   maximum, 326, 330
   tensile, 184
formation breakdown, 299
fracture, 45
   gradients, 474
   initiation gradient, 476
   initiation pressure, 68, 474
   pressure, 69
   propagating gradient, 476
   propagating pressure, 68, 474, 477
free mandrel travel, 267
free point, 82, 102, 266
   electric log, 324, 326
   free travel, 261
   movement, 243
   stretch type, 324, 341
fulcrum, 355
full opening catch, 287

G

gas, background, 105
   connection, 105
   detector, 242
   dissolved, 97
   drill cuttings, 105
   drilled, 61, 105
   gas flows, 55
   pressure, 386
   shows, 105
   swabbed, 105
gauge ring, 420, 427
grab, 431, 434
   two-prong 428
granulated material, glass
   beads, 119, 120
   plastic beads, 119, 120
   walnut hulls, 119, 120
graphite, 126
guide bushing, 307
guide shoe, thin lip, 396
gypsum, 64

H

half-moon, 114
hard banding, 211, 499, 502
heat string, 528
hematite, 43, 44
hexagon, 417
high-pressure flow, 373
high risk, 156, 298, 392, 399, 410
hole
    cleaning, 39
    crooked, 87, 112, 117, 129, 284
    deviated, 1
    erosion, 41
    full-gauge, 7, 216
    high-angle 1, 43, 93
    higher-risk, 325
    horizontal, 1, 43, 93
    in-gauge, 270
    inclined, 158
    opener, 225
    out-of-gauge, 60, 61 249, 252, 270, 284, 292
    pilot, 58, 306, 387
    straight, 1, 73, 118
    tapered, 216, 264
    tight, 19, 64 113
    under-gauge, 7, 19, 319
hook load, 121, 221
hook weight, 118
hostile environment, 508
hydraulic, hold down, 267
    pulling tool, 385
    actuated, 408
hydrochloric acid, 371
hydrogen sulfide, 1, 57, 210
    embrittlement, 210
hydrostatic head, 386, 462
hydrostatic lift, 144

I

illmenite, 43, 44

inflow test, 539
inside casing cutter, 306
inside preventer, 403
inspections, 450
    visual, 450
internal spear, 327
inverted position, 299
inverted slips, 384
isopropyl alcohol, 353, 358

J

jar bumper, 176, 327
jar, accelerator, 380
    double-drilling bumper, 219
    oil, 172
    surface, 177
    trip, 172, 261
    tripping level, 261
jerk line, 5
jet charge, 388
joint alignment, 456
joint, float, 17
    loose, 17
    shoe, 17
junk sub, 150
junk, rotating, 231

K

kelly, 11, 403
    heavy-duty, 11
    long, 11
keyseat, growing, 129
    wiper, 132
kickoff point, 303, 305
kick, 16, 27, 32, 33, 36, 47, 52, 59, 63, 92, 99, 107, 109, 111, 364, 454
    fluid, 110, 111
    severity, 111
kickoff, 20
kill well, 499
knives, 170, 283, 284
knock-out, 24

L

lag time, 242
large junk, slip segments, 323
   small subs, 323
   tong links, 323
last resort, 52, 114, 156, 231, 306
lead, 205, 208
leak-off test, 105, 474
leak, surface, 37
left-hand helix, 142
leaf spring, 112
lift nipple, 325
lifting bar, 415
lifting cables, 412
line, cat, 325
   failure, 429
   lifting, 23, 325, 523
   stretch, 195
liner hanger, mechanical, 536
liner, overlap, 539, 540
   production, 525;
   stub, 536
   tie back, 471, 525
loading, impact, 252
   shock, 252
log, 1, 8
   base inspection, 451
   bond, 534
   cased hole, 504
   casing inspection, 488, 501, 528, 542
   induction, 295
   open-hole, 504
   stuck pipe, 335, 353, 509
loose ball, 430
lost circulation material (LCM), 44
   bottom fell out, 45
   fibrous 49
   gilsonite, 55
   induced, 44
   LCM pills, 47, 364
   mica, 49
   natural, 44, 45
   paper, 49
   particle sizes, 51
   pill, 50, 52
lost hole, 507
lower fish, 361
lubricator, 427

M

macaroni tubing, 391
magnesium chloride, 42
magnet, 192, 245, 323, 428
   bar, 307
   flow line, 501
   insert, 146
   skirted, 235
magnetic, interference, 338
magnetite, 44
mandrel travel, 215
margin of safety, 138, 343
marked magnetically, 422
mast, 227
master bushing, 3
matrix material, 161
matting boards, 495
Measurement While Drilling, 105, 120
mechanical malfunction, 377
mechanical slip joint, 176
metal cuttings, 161, 170, 499
metal fatigue, 114
   fibers, 412
   fragments, 153
metal impact plate, 523
metal stock, 331
metal-to-metal backing, 545
mill
   concave, 320
   drag toothed, 166, 168
   section, 168, 306

mill marks, 450
mill-out extension, 5, 157, 282, 386, 388, 437
mill over, 16
minimum yield, 101
miscellaneous causes, 313
mousehole, 408
mud, 47
   abrasive, 239
   aerated, 36
   API funnel seconds, 307
   balanced system, 240
   bucket, 16
   collection pan, 16
   courses, 161
   dispersed polymer, 27
   drain pan, 38
   equivalent circulating density, 41
   film, 520;
   gas cut, 18, 105, 454
   handling equipment, 209
   losses, 38
   lubricity, 126
   mixing plant, 89
   motor, 212
   pH range, 211
   plant, 46
   rheology, 239
   salvage loss, 39
   screening, 378
   solids, 39
   tripping losses, 39
   underbalanced, 27
   viscous, 257
   wall cake, 38, 125
muleshoe, 183
multiple squeezing, 486

N

near bit, 216
neutral point, 278, 513

neutron source, 439
nitrogen, 18, 360
no-go, 24, 101, 154, 401
nylon fibers, 422

O

oil spills, 2
open-hole logging, 133
open perforations, 387
order of stands, 14
out of alignment, 181, 182
out of round, 153
over-cautious, 89
over-torquing, 161
overlap section, 483
overpull, 102, 118, 170, 210, 219, 251, 356
   excess, 330
overstressed, 404, 421

P

pack off, 142, 144, 154, 241
   elements, 142
   overshot, 403
packer, cement retainer, 469
   compression, 5, 467
   elements, 18, 267, 383
   hookwall, 467
   mandrel, 368
   permanent, 467
   plucker, 156, 382
   retriever, 165
   seat, 18
   tailpipe, 443
   wireline, 5
paint, 241
parallel, 393
particles, abrasive, 300
pass tool joints, 342
pendulum, 215

penetrating fluid, 474
perforation ball sealer, 381, 335
permafrost section, 464
piano wire, 419
pick-up line, 456
pick-up tool, 3
pigtail, 272
pill, 37;
   barite, 89, 363
   volume, 358
pilot error, 222
pilot light, 32
pin failure, 96
pin sub, 158
pinching, 19
pinned collar, 422
pipe cheater, 369
pipe misalignment, 492
pipe rams, 3
pipe stub, 414
piston effect, 258
pit volume totalizer, 99
plan, alternate, 88
   fishing action, 96
   long range, 88, 91
   of action, 87
plastic formation, 456
plowing effect, 121
plug, 101
   barite, 89, 139, 372
   bridge, 372
   cast iron, 381
   dressed off, 303
   immobile, 303
   isolating, 303
   magnesium, 381
   pump out, 92
   retrievable 378, 380
   seating nipple, 437
   slurry, 304
point of failure, 516

pore pressure, 93
   plot, 67, 107
positive displacement motor, 212, 220
potassium chloride, 42
premature setting, 377
premature squeeze pressure, 479
pressure lubricator, 443
pressure relief pin, 386, 388
pressure reversal, 103
pressure shock, 18
pressure, leak-off, 260
preventer, inside blowout, 31
prong, 189, 190, 193, 431
pulsation dampener, 240
pump
   crippled, 335
   main, 27
   on line, 28
   second, 28
   standby, 27
pumping fish out, 370
pyrites, 13

Q

quick connector, 436

R

rams
   blind, 31, 63
   pipe, 31
rasp, 443
rathole digger, 490
reamer, 226
reaming, 9
   off-bottom, 116
record, cutoff, 29
   ton mile, 29
recorder, depth, 14

red ink, 2
redrill, 79
reef, pinnacle, 63
refinery fires, 2
reheaded, 422
reports, drilling, 80
retainer setting tool, 516
rework, 523
rig, diesel electric, 25
    hoisting system, 30
    lines, 30
    silicon-controlled rectifier, 24
    top drive, 240
    workover, 254, 401, 528
    unmatched, 25
right and left lay, 419
right and left twist, 419
right-hand release, 142
ring out, 13
root thread, 483
rope socket, 195
    crippled, 201, 269, 418
rule of thumb, 195, 224
running out of hole, 448

S

S-shaped, 126
S-type, 114
safety factor, 249
safety, 522
salt, 64
    formation, 371, 446, 535
    water flow, 64, 373, 109
sand, line, 241
    settling pit, 120
    stingers, 19, 71, 105, 364
scalloped, 333
scraper, 420
second fish, 79, 99
shale shaker, 38, 119

shale, chemistry, 57
    montmorillonite, 59
    popping, 60
    geopressured, 62
shear joint, 270
shear pin release, 193, 208, 227, 444;
    shear down, 227
    shear pin release, shear up, 227
sheared off, 416
sheave, 421
    unhooded, 427
shifting sleeves, 418
shipwrecks, 2
shock loading, 507
shoe, joints, 518
    split, 139
site specific, 410
skin damage, 527
skirt, thin, 333
slag inclusion, 449, 461
slick lines, 419
slick tool, 195
sliding sleeve, 5
slim hole, 392
    completion, 525
slip and spider, 407
slip, area, 9
    slips, inverted, 297
    segment, 22, 368, 403, 459
slip-type elevator, 407
sloughing, 362
slug, trip, 37
slugging, 16
small junk
    bolts 2, 234
    hand tools, 21, 234
    nuts, 21, 234
    tong dies, 21, 234
small tools, 4
snap and bounce, 253, 269

snapping procedure, 440
snapping the line, 426
snubbing unit, 386, 393
soaking, agents, 351, 357
   period, 355
sonic, 222
spacer, 43, 541
   tail, 479
spalling, 61
spear, 434
spiral-type grapples, 414
split body, 139
   casing, 368
   washpipe shoe, 366
sprung grab, 190
spud, 1, 101
squeeze, block, 510
   bradenhead, 522
   gunk, 56
   multistage, 512
stabilizer, 226
   integral blade (IBS), 295
stage squeezing, 480
stages, 480
standoff, 188
sticking point, 324
stinger, 387, 402
storm chokes, 2
strap, 3, 519
strength, minimum yield, 248, 249
   ultimate yield, 248
stress reversals, 243
string shot, 243, 380
   back-off, 142
stripping over, 426, 440
structural integrity, 462
stub liner, 464
stuck point, 256, 257, 258, 266
sub, 219
submersible electric pump, 369
substructure shims, 495
suction stroke, 199

surging, 481
survey, deviation, 298
   spinner, 47
swab line, 430

T

taper, 23
tapered, collars, 394
   strings, 209
tectonic forces, 103
test, leak-off, 334
Texas Pattern shoe, 454
thimble, 208
thin stringer, 57
tieback casing, 472, 473
tong area, 9
tool failure, 377
tool joint
   box-type, 161
   butt-shouldered, 211
   pin-type, 161
top drive, 209, 220
torque, 319
   back off, 118, 119, 121, 179
   excess, 328
   fluctuating, 11, 13
   left hand, 401
   low level, 11
   rounds of, 118
   uneven, 11
transition zones, 108, 109, 447
trapdoor socket, 395
   full opening, 396
traveling block, 403
trip, cleanout, 335
   hole wiper, 15
   slug, 82
tubing contraction, 530
tubingless completion, 525
tungsten carbide, 19, 143, 161, 170, 211
turbine, 212, 220
twin pump truck, 541

twin well, 5
tying slip handles, 341

U

ultimate yield, 101
ultraviolet light, 222
uncontrolled flow, 403
under-displaced, 483
uneven wall thickness, 461
upside-down slips, 154

V

V-door, 456
valves, double, 32
visual inspection, 209, 224, 237

W

wash over, 16
water flow, 412
water-oil interface, 198
wear bushing, 233, 491

wear ring, 307
wear, casing, 300
wedging action, 533
weight, free-pipe, 263
   set-down, 251, 279
welding procedure, 500
well, fire, 208
   development, 5, 408, 412
   monument, 542
   on-structure development, 282
   on-structure wildcat, 96, 116, 349, 384, 448, 535
   plan, 1, 5, 68
whipstock, carrying ring, 389
   modified, 389
wireline, fatigue, 426
   fish, 377
   flags, 421
   balled, 258
work string, 139, 172, 248, 486

Z

zone isolation, 94